AF598080

METHODS IN MOLECULAR BIOLOGY™

Series Editor
John M. Walker
School of Life Sciences
University of Hertfordshire
Hatfield, Hertfordshire, AL10 9AB, UK

For other titles published in this series, go to
www.springer.com/series/7651

Antimicrobial Peptides

Methods and Protocols

Edited by

Andrea Giuliani

Spider Biotech, Colleretto, Giacosa, Italy

Andrea C. Rinaldi

University of Cagliari, Monserrato, Italy

Editors
Andrea Giuliani
Spider Biotech S.r.l.
Via Ribes, 5
10010 Colleretto Giacosa TO
Bioindustrypark del Canavese
Italy
agiuliani@spiderbiotech.com

Andrea C. Rinaldi
Università di Cagliari
Dipto. di Scienze e Tecnologie
Biomediche
09042 Monserrato
Cittadella Universitaria
Italy
rinaldi@unica.it

ISSN 1064-3745 e-ISSN 1940-6029
ISBN 978-1-60761-593-4 e-ISBN 978-1-60761-594-1
DOI 10.1007/978-1-60761-594-1
Springer New York Dordrecht Heidelberg London

Library of Congress Control Number: 2009943247

© Springer Science+Business Media, LLC 2010
All rights reserved. This work may not be translated or copied in whole or in part without the written permission of the publisher (Humana Press, c/o Springer Science+Business Media, LLC, 233 Spring Street, New York, NY 10013, USA), except for brief excerpts in connection with reviews or scholarly analysis. Use in connection with any form of information storage and retrieval, electronic adaptation, computer software, or by similar or dissimilar methodology now known or hereafter developed is forbidden.
The use in this publication of trade names, trademarks, service marks, and similar terms, even if they are not identified as such, is not to be taken as an expression of opinion as to whether or not they are subject to proprietary rights.
While the advice and information in this book are believed to be true and accurate at the date of going to press, neither the authors nor the editors nor the publisher can accept any legal responsibility for any errors or omissions that may be made. The publisher makes no warranty, express or implied, with respect to the material contained herein.

Cover illustration: Inset image adapted from Figure 2 of Chapter 13. Background image adapted from Figure 3E of Chapter 15.

Printed on acid-free paper

Humana Press is part of Springer Science+Business Media (www.springer.com)

Dedicated to the memory of Silvano Fumero, poet and scientist.
(Andrea Giuliani)

On April 6, 2009, an earthquake devastated the town of L'Aquila, in Italy, taking the lives of more than 300 people and marking indelibly those who survived. This book is dedicated to the memory of the victims, and, as a warm wish, to all those, including some of the authors of the present volume, that will have to build their future from scratch.
(Andrea C. Rinaldi)

Preface

As our knowledge of the innate immune systems in multicellular organisms has grown steadily in the last two decades, so has our comprehension of the basic role played by gene-encoded, ribosomally synthesized antimicrobial peptides (AMPs) in this ancient line of defense against infections. Indeed, although the ability of many prokaryotes to produce and release peptidic or proteinaceous substances with antibacterial and antifungal activity has been known for quite a long time, the presence of AMPs in humans and animals (and later plants) has been recognized only more recently, in the 1980s, with the identification of cecropins in the pupae of *Hyalophora cecropia* moth, defensins in human neutrophils, and magainins in skin secretions of the frog *Xenopus laevis* (1–3); these findings, though, were grounded on previous work conducted in the 1960s and 1970s, for example, by Hussein Zeya and John Spitznagel. Since then, the quest for new AMPs has gifted us with many hundreds of biologically active peptides, extremely varied in sequence and structure, isolated from virtually every multicellular organism where they have been looked for (4). Visiting the several available web-based repositories dedicated to AMPs – such as AMSDb (http://www.bbcm.univ.trieste.it/~tossi/pag1.htm), ANTIMIC (http://research.i2r.a-star.edu.sg/Templar/DB/ANTIMIC/), and APD2 (http://aps.unmc.edu/AP/main.php) – can help in getting a more precise idea of what "diversity" means in this case. By the end of the 1990s, AMPs research was already blooming. The first Gordon Research Conference on Antimicrobial Peptides was held in 1997 in Ventura, California, and these meetings have since been a privileged point of discussion and the cradle of many ideas around these intriguing molecules. The same year the GRC series on AMPs started, the first *Methods in Molecular Biology* volume dedicated to AMPs, edited by William Shafer, was published (5).

From those early days, we have gone a long distance ahead, although the journey seems by no means to come to an end soon. Much has been learned about the mechanism of target selection and cell killing, always with the aim of transforming an evolutionary, successful, antimicrobial shield into the next century's antibiotic drugs. Thanks to years of intense study in many public and private research institutions worldwide, many AMPs (in particular those of mammals) began to be recognized as endowed with more than simply direct antimicrobial properties, by acting at the interface between innate and adaptive immunity, for example, as immunomodulants, immunostimulants, and/or inducers of proinflammatory cytokines or chemokines, to the point that today many prefer to refer to these substances with the more general term of "host defense peptides." Notwithstanding these major efforts, AMPs have not yet fulfilled their main promise, i.e., to be able to truly represent the next-generation antibiotics with new modes of actions. This is for a complex variety of causes that have collectively hampered clinical progress to date. Some of these are related with the peptidic nature of AMPs which translates, for instance, to poor bioavailability, poor proteolytic stability, and cost of goods issues, or with their mechanism of action, which might bear some potential for unwanted toxicity. A helping hand, in this sense, would probably come in the near future by the development of synthetic analogues of AMPs, with improved characteristics (and lower cost) with respect to their natural counterparts (6).

For one that approaches this research area from outside, the truly multidisciplinary scenario it offers might at first sight be surprising and even a bit confusing. Immunologists, infectivologists, dermatologists, microbiologists, cell biologists, molecular geneticists, biochemists, peptide chemists, biophysicists, and many others with different scientific backgrounds have entered the arena, each one approaching the challenges on the table from a distinct point of view, while elaborating at the same time original perspectives for the field's development. The medical potential of which AMPs are inherently endowed with certainly lies at the core of this interest, but it does not explain everything. In our opinion, a good fraction of the AMPs' ability in catalyzing such a diversified attention relies on their privileged position at the center of a perfect storm, where scientifically mature fields such as antibiotics and immunology research on one side and membrane/peptide biochemistry and biophysics on the other one positively clash. On this stage, each field is in search of a new object of study for unleashing the potential of perfectionated equipment and theories, a reachable goal with important practical consequences, and, ultimately, an occasion for rejuvenation.

In the present volume of *Methods in Molecular Biology*, we have attempted to provide an updated overview of this burgeoning field, by offering snapshots of the many different approaches leading scientists in the field follow to crack AMPs' most hidden attributes and properties, with the ultimate end of harnessing the potential these compounds do display for applicative purposes. As in the series' style (and name), most of the contributing chapters deal with experimental protocols, presented "in readily reproducible, step-by-step fashion." In addition, several review chapters discuss in a more classic manner selected issues pertaining to different aspects of AMPs, namely the forms of interaction of AMP's with the lipopolysaccharide (LPS) component of Gram-negative bacteria (Chapter 10), the diverse dynamic transitions peptides (not only AMPs, but also, for example, fusogenic peptides) may undergo in membranes (Chapter 13), and the therapeutic potential of host defense peptides (Chapters 19 and 20).

As a reading guide, chapters are organized around three main themes: the isolation, purification, and production in recombinant form of AMPs from natural sources and the design and synthesis of unnatural compounds; the methodologies – of either biophysical, biochemical, or computational nature – usually applied to define AMPs' molecular features and to explore their mechanism(s) of action; the studies focused on the biological activities (antimicrobial, anti-inflammatory, immunomodulatory, and so on) of AMPs, and on their use as therapeutic agents. Contributions gathered in the first part will lead the reader through the discovery of novel AMPs from sources as different as frog skin secretions (Chapter 1) and human skin (Chapter 2), and of lantibiotics from Gram-positive bacteria (Chapter 3). The production of recombinant antimicrobial peptides in bacteria is one of the possible routes to achieve an in-depth structure–activity analysis of AMPs, through selective and systematic removal and/or substitution of given residue positions, and a suitable strategy for the development of a less expensive peptide-production platform. Chapters 4 and 5 will thus discuss the production of AMPs by recombinant approaches in *Escherichia coli*. Approaches for the rational design of AMPs and for the production of peptide variants by solid-phase synthesis will be described in Chapters 6 and 7, respectively. Finally, the SPOT synthesis technique – a powerful tool offering the opportunity of synthesizing and screening a large number of peptides arrayed on a planar cellulose support – comes under focus in two different contributions (Chapters 8 and 9). With membranes apparently being an obligate passage in the action of all AMPs – whether exerting their activity through destabilization of the membrane itself or directed toward an intracellular target – it is not surprising that all contributions of the second part of the volume deal, in a way or the other, with membrane interactions of peptides. These can be characterized by taking advantage of sophisticated membrane model systems or live cells, tryptophan and lipid fluorescence, and a few selected fluorescent dyes, a combination that permits to carefully measure several aspects of membrane-binding and membrane-perturbing activities of selected AMPs and peptidomimetics, as described in Chapters 11 and 12. Structural studies of AMPs, either alone or when embedded in a membrane environment, are unquestionably of pivotal importance, and in this context solid-state NMR is gaining momentum as a very informative technique, as underlined in Chapter 14 (see also Chapter 13 for more details on NMR studies). Besides NMR, microscopic techniques can also offer important clues in order to disclose more on AMPs function and cell specificity, providing a "whole-cell" perspective. Chapter 15 depicts how atomic force microscopy can be used to dissect the behavior of some AMPs (in this specific case, the so-called Sushi peptides from horseshoe crab), whereas Chapter 16 describes the use of a mix of fluorescence/electron microscopy techniques to get insight into the damage caused by AMPs on the morphology and membrane structure of intact bacterial cells. This section terminates with the important contributions coming from computational resources, in the form of molecular dynamics simulations (Chapter 17) and quantitative structure–activity relationships (QSAR) analysis of antimicrobial peptides (Chapter 18 for this approach see also Chapter 6). The third and last section of the book sketches just a few, although very significant, approaches put in force so far to disclose the medical and therapeutic potential of AMPs as anti-infective and immunomodulant agents. Single contributions deal with infection models and activity assay systems (Chapters 21, 22, 23, and 24), and with the characterization of the antimicrobial activity of peptides against the protozoan pathogen *Leishmania* (Chapter 25).

Within the limitations inherent to the selection process of topics to be included in the present volume, which necessarily left over many important issues that would have deserved the same covering, we hope that the readers – either expert of the field or newcomers – will find in these pages both an authoritative guide for their own lab work on AMPs or related substances and a good load of thought stimuli to inspire their scientific endeavors.

References

1. Steiner, H., Hultmark, D., Engstrom, A., Bennich, H., and Boman, H. G. (1981) Sequence and specificity of two antibacterial proteins involved in insect immunity. *Nature* **292**, 246–268.
2. Ganz, T., Selsted, M. E., Szklarek, D., Harwig, S. S. L., Daher, K., Bainton, D. F., and Lehrer, R. I. (1985) Defensins. Natural peptide antibiotics of human neutrophils. *J. Clin. Invest.* **76**, 1427–1435.
3. Zasloff, M. (1987) Magainins, a class of antimicrobial peptides from *Xenopus* skin: isolation, characterization of two active forms, and partial cDNA sequence of a precursor. *Proc. Natl. Acad. Sci.* USA **84**, 5449–5453.
4. Zasloff, M. (2002) Antimicrobial peptides of multicellular organisms. *Nature* **415**, 389–395.
5. Shafer, W. M., ed. (1997) Antibacterial Peptide Protocols. *Methods in Molecular Biology*, vol. 78. Humana Press, Totowa, NJ, USA.
6. Giuliani, A., Pirri, G., Bozzi, A., Di Giulio, A., Aschi, M., and Rinaldi, A. C. (2008) Antimicrobial peptides: natural templates for synthetic membrane-active compounds. *Cell. Mol. Life Sci.* **65**, 2450–2460.

Andrea Giuliani
Andrea C. Rinaldi

Contents

Contributors

SERGII AFONIN • *Karlsruhe Institute of Technology, Institute for Biological Interfaces (IBG-2); Institute of Organic Chemistry, Karlsruhe, Germany*

PIERPAOLO AIMOLA • *Dipartimento di Biologia di Base ed Applicata, Università di L'Aquila, L'Aquila, Italy*

CHRISTOPHER AISENBREY • *Institut de Chimie, CNRS, Université de Strasbourg, Strasbourg, France*

PAULO F. ALMEIDA • *Department of Chemistry and Biochemistry, University of North Carolina Wilmington, Wilmington, NC*

MOJTABA BAGHERI • *Leibniz Institute of Molecular Pharmacology (FMP), Berlin, Germany*

ANNELISE E. BARRON • *Department of Bioengineering, Stanford University, Stanford, CA*

BURKHARD BECHINGER • *Institut de Chimie, CNRS, Université de Strasbourg, Strasbourg, France*

PHILIPPE BERTANI • *Institut de Chimie, CNRS, Université de Strasbourg, Strasbourg, France*

NATHANIEL P. CHONGSIRIWATANA • *Department of Chemical and Biological Engineering, Northwestern University, Evanston, IL*

J. MICHAEL CONLON • *Department of Biochemistry, Faculty of Medicine and Health Sciences, United Arab Emirates University, Al-Ain, UAE*

GILL DIAMOND • *Department of Oral Biology, UMDNJ-New Jersey Dental School, Newark, NJ*

JEAK LING DING • *Department of Biological Sciences, National University of Singapore, Kent Ridge, Singapore*

STEFANO DONADIO • *NAICONS Scrl., Milano, Italy*

TIMOTHY J. FALLA • *Helix Biomedix Inc., Bothell, WA*

SHAREL FIGUEREDO • *Department of Pathology and Laboratory Medicine, School of Medicine, College of Health Sciences, University of California, Irvine, CA*

MARIA GIAMMATTEO • *Centro di Microscopie, Università di L'Aquila, L'Aquila, Italy*

ANDREA GIULIANI • *Research and Development Unit, SpiderBiotech S.r.l, Colleretto Giacosa, Italy*

STEPHAN L. GRAGE • *Karlsruhe Institute of Technology, Institute for Biological Interfaces (IBG-2); Institute of Organic Chemistry, Karlsruhe, Germany*

ROBERT E.W. HANCOCK • *Centre for Microbial Diseases and Immunity Research, University of British Columbia, Vancouver, BC, Canada*

KAI HILPERT • *Institute of Biological Interfaces – IBG 2, KIT (Karlsruhe Institute of Technology), Forschungszentrum Karlsruhe, Eggenstein-Leopoldshafen, Germany*

BOW HO • *Department of Microbiology, Yong Loo Lin School of Medicine, National University of Singapore, Kent Ridge, Singapore*

DANIELA JABES • *NAICONS Scrl., Milano, Italy*

HÅVARD JENSSEN • *Department of Science, Systems, and Models, Roskilde University, Roskilde, Denmark*

ROMAN JERALA • *Department of Biotechnology, National Institute of Chemistry, Ljubljana, Slovenia*

YIANNIS N. KAZNESSIS • *Department of Chemical Engineering and Materials Science, University of Minnesota, Minneapolis, MN*

ALLISON LANGHAM • *Department of Chemical Engineering and Materials Science, University of Minnesota, Minneapolis, MN*

ANG LI • *Singapore-MIT Alliance (SMA), National University of Singapore, Kent Ridge, Singapore*

CHWEE TECK LIM • *Division of Bioengineering and Department of Mechanical Engineering, National University of Singapore, Kent Ridge, Singapore*

KAREN LOVETRI • *Kane Biotech Inc., Winnipeg, MB, Canada*

JUAN ROMÁN LUQUE-ORTEGA • *Centro de Investigaciones Biológicas (CSIC), Madrid, Spain*

SRINIVASA MADHYASTHA • *Kane Biotech Inc., Winnipeg, MB, Canada*

MARIA LUISA MANGONI • *Istituto Pasteur-Fondazione Cenci Bolognetti, Dipartimento di Scienze Biochimiche, Azienda Ospedaliera S. Andrea, Università "La Sapienza", Roma, Italy*

LUDOVICA MARCELLINI • *Istituto Pasteur-Fondazione Cenci Bolognetti, Dipartimento di Scienze Biochimiche, Università "La Sapienza", Roma, Italy*

JENNIFER R. MASTROIANNI • *Department of Pathology and Laboratory Medicine, School of Medicine, College of Health Sciences, University of California, Irvine, CA*

LAURA MCMAHON • *Department of Oral Biology, UMDNJ-New Jersey Dental School, Newark, NJ*

RALF MIKUT • *Institute for Applied Computer Science (IAI), Forschungszentrum Karlsruhe GmbH, Karlsruhe, Germany*

ANDRÉ J. OULLETTE • *Department of Pathology and Laboratory Medicine; Department of Microbiology and Molecular Genetics, School of Medicine, College of Health Sciences, University of California, Irvine, CA*

CHRISTOPHER PASETKA • *MIGENIX Inc., Vancouver, BC, Canada*

GIOVANNA PIRRI • *Research and Development Unit, SpiderBiotech S.r.l, Colleretto Giacosa, Italy*

ANTJE POKORNY • *Department of Chemistry and Biochemistry, University of North Carolina Wilmington, Wilmington, NC*

ISAURA RIGO • *Department of Oral Biology, UMDNJ-New Jersey Dental School, Newark, NJ*

ANDREA C. RINALDI • *Department of Biomedical Sciences and Technologies, Cittadella Universitaria, University of Cagliari, Monserrato, CA, Italy*

LUIS RIVAS • *Centro de Investigaciones Biológicas (CSIC), Madrid, Spain*

EVELINA RUBINCHIK • *MIGENIX Inc., Vancouver, BC, Canada*

JENS-M. SCHRÖDER • *Clinical Research Unit "Cutaneous Inflammation", Department of Dermatology, University Hospital Schleswig-Holstein, Kiel, Germany*

AGNES SONNEVEND • *Department of Medical Microbiology, Faculty of Medicine and Health Sciences, United Arab Emirates University, Al-Ain, UAE*

OLIVIER TABOUREAU • *Department of Systems Biology, Center for Biological Sequence Analysis, Technical University of Denmark, Lyngby, Denmark*

KENNETH P. TAI • *Department of Pathology and Laboratory Medicine, School of Medicine, College of Health Sciences, University of California, Irvine, CA*
ANNE S. ULRICH • *Karlsruhe Institute of Technology, Institute for Biological Interfaces (IBG-2); Institute of Organic Chemistry, Karlsruhe, Germany*
DIRK F.H. WINKLER • *Brain Research Centre, University of British Columbia, Vancouver, BC, Canada V6T 2B5*
SUNGHAN YIM • *Department of Oral Biology, UMDNJ-New Jersey Dental School, Newark, NJ*
LIJUAN ZHANG • *Helix Biomedix Inc., Bothell, WA*
MATEJA ZORKO • *Department of Biotechnology, National Institute of Chemistry, Ljubljana, Slovenia*

Section I

Isolation, Purification, Design, and Synthesis of Antimicrobial Peptides

Chapter 1

Antimicrobial Peptides in Frog Skin Secretions

J. Michael Conlon and Agnes Sonnevend

Abstract

Skin secretions from many species of anurans (frogs and toads) are a rich source of peptides with broad-spectrum antimicrobial activities that may be developed into agents with therapeutic potential, particularly for topical applications. This chapter describes the use of norepinephrine (injection or immersion) to stimulate peptide release from granular glands in the skin in procedures that do not appear to cause distress to the animals. The peptide components in the secretions are separated using reversed-phase HPLC on octadecylsilyl-silica (C_{18}) columns after partial purification on Sep-Pak C_{18} cartridges. Peptides with antimicrobial activity are then identified by demonstration of their abilities to inhibit growth of Gram-negative (*Escherichia coli*) and Gram-positive (*Staphylococcus aureus*) bacteria in liquid phase microtiter plate assays. Individual peptides with activity are purified to near homogeneity by further chromatography on butylsilyl-(C_4) and diphenylmethylsilyl-silica columns and characterized structurally by automated Edman degradation and mass spectrometry.

Key words: Frog skin secretions, antimicrobial peptide, reversed-phase HPLC, *Escherichia coli*, *Staphylococcus aureus*.

1. Introduction

The emergence in all regions of the world of strains of pathogenic bacteria and fungi with resistance to commonly used antibiotics constitutes a potentially serious threat to public health and has necessitated a search for novel types of antimicrobial agent to which the microorganisms have not been exposed (1). Analysis of skin secretions from different species of Anura (frogs and toads) has led to the characterization of a wide range of peptides with biological activity that may have potential for drug development (2). Peptides with broad-spectrum antibacterial and antifungal activities and with the ability to lyse mammalian cells

A. Giuliani, A.C. Rinaldi (eds.), *Antimicrobial Peptides*, Methods in Molecular Biology 618,
DOI 10.1007/978-1-60761-594-1_1, © Springer Science+Business Media, LLC 2010

are synthesized in the skins of species from certain, but by no means all, anuran families. These peptides represent a component of the system of innate immunity that defends the animal against invasion by pathogenic microorganisms (3) and have excited interest as candidates for development into therapeutically valuable anti-infective agents (4). Structural characterization of the frog peptides has shown that they comprise between 10 and 48 amino acid residues and a comparison of their amino acid sequences reveals the lack of any conserved domains that are associated with biological activity. However, the frog skin peptides, with few exceptions, are cationic because of the presence of several lysine residues and also contain a preponderance of hydrophobic amino acids particularly leucine and isoleucine. Circular dichroism and NMR studies have shown that they generally lack stable secondary structure in aqueous solutions but have the propensity to form an amphipathic α-helix in the environment of a phospholipid vesicle or in a membrane-mimetic solvent such as 50% trifluoroethanol–water (5).

It is a common misconception that antimicrobial peptides are synthesized in the skins of all frog species. At this time, cationic α-helical antimicrobial peptides have been identified in the skins of frogs from species belonging to the Alytidae, Bombinatoridae, Hylidae, Hyperoliidae, Leiopelmatidae, Leptodactylidae, Myobatrachidae, Pipidae, and Ranidae families but several well-studied species from the Bufonidae, Ceratophryidae, Dicroglossidae, Microhylidae, Pelobatidae, Pyxicephalidae, Rhacophoridae, and Scaphiopodidae families do not appear to synthesize these peptides (reviewed in (6)). Skin secretions from South American and Australian frogs belonging to the family Hylidae (7) and North American and Eurasian frogs belonging to the family Ranidae (8) have proved to be particularly rich sources of antimicrobial peptides.

The dermal peptides are stored in granular glands present in the skin and are released into skin secretions, often in very high concentrations, in a holocrine manner upon stress or injury as a result of contraction of myocytes surrounding the glands. In the laboratory or in the field, intradermal injection of norepinephrine (9) or immersion of the animal in a solution of norepinephrine (10) are effective methods for stimulating release of peptides into skin secretions that are well tolerated by animals and are the procedures that will be described in this chapter. Following initial fractionation by reversed-phase HPLC, peptides with antimicrobial activity are identified by measuring their abilities to inhibit the growth of reference strains of Gram-negative (*Escherichia coli*) and Gram-positive (*Staphylococcus aureus)* bacteria in liquid phase in vitro assays. These peptides are purified to near homogeneity (>98% purity) by reversed-phase HPLC using columns containing a range of different packing materials. The primary structures of the purified peptides are determined by automated Edman degradation and the proposed amino acid sequences confirmed by MALDI-TOF mass spectrometry. A general strategy for purification of peptides to near homogeneity by reversed-phase HPLC (11) and characterization of peptides isolated from frog skin by MALDI-TOF mass spectrometry (12) has been presented in recent articles.

The major obstacles to the use of frog skin antimicrobial peptides as useful anti-infective drugs are their toxicities, particularly if they are to be administered systemically, and their short half-lives in the circulation. However, peptides applied to infected skin or skin lesions in the form of sprays or ointments can penetrate into the *stratum corneum* to kill microorganisms so that future therapeutic applications are more likely to involve topical rather than systemic administration. Similarly, strategies have been developed for the design of analogs of naturally occurring peptides that show reduced cytolytic activity against mammalian cells while maintaining potency against microorganisms (13, 14).

2. Materials

2.1. Collection and Partial Purification of Skin Secretions

1. Norepinephrine bitartrate salt (Sigma-Aldrich, St Louis, MO, USA).
2. Distilled water.
3. Concentrated hydrochloric acid (36% w/v) (BDH, Poole, UK). This reagent is corrosive.
4. Sep-Pak C_{18} cartridges (Waters Associates, Milford, MA, USA).
5. Minipuls 3 peristaltic pump (Gilson, Middleton, WI, USA).
6. Speed-Vac concentrator with multi-tube rotor (Thermo Fisher Scientific, Waltham, MA, USA).

2.2. Reagents and Equipment for Preparative HPLC

1. Acetonitrile (HPLC Spectro grade, Thermo Fisher Scientific). This reagent is toxic.
2. Water (Milli-Q purified; 18.2 MΩcm^{-1}) or HPLC Spectro grade (Thermo Fisher Scientific).
3. Trifluoroacetic acid (99.8% purity; Sequenal grade, Thermo Fisher Scientific). This reagent is extremely corrosive and hand and eye protection should be worn.
4. An HPLC system capable of generating a binary gradient using pumps operating in the flow rate range 0.1–10 mL/min, a Rheodyne 7125 injection system equipped with a 2 mL loop, a detector capable of simultaneously monitoring at two wavelengths (typically 214 and 280 nm) (*see* **Note 1**).

5. HPLC columns: (a) (2.2 cm × 25 cm) Vydac 218TP1022 preparative C_{18} column (Grace, Deerfield, IL, USA), (b) (1.0 cm × 25 cm) Vydac 218TP510 semi-preparative C_{18} column, (c) (1.0 cm × 25 cm) Vydac 214TP510 semi-preparative (C_4) column, (d) (1.0 cm × 25 cm) Vydac 219TP510 semi-preparative diphenylmethyl column. Preparative columns are used at a flow rate of 6 mL/min and semi-preparative columns at a flow rate of 2 mL/min (*see* **Note 2**).
6. Helium cylinder for degassing of solvents (*see* **Note 3**).
7. Gastight model 1002 2.5 mL injection syringe (Hamilton, Reno, NV, USA).
8. Polypropylene tubes 12 mm × 75 mm and 15 mm × 100 mm (Nunc brand by Thermo Fisher Scientific) (*see* **Note 4**).
9. 1.5 mL clear polypropylene Eppendorf tubes (e.g., Nunc).
10. Model 206 fraction collector (Gilson).
11. Freezone 4.5 lyophilizer (Labconco, Kansas City, MO, USA).

2.3. Reagents and Equipment for Antimicrobial Assays

1. Spectrophotometer able to measure at $\lambda = 600$ nm.
2. Microtiter plate reader able to measure at $\lambda = 630$ nm.
3. Incubator or 37°C constant temperature room for culturing microorganisms in air.
4. 15 mL sterile tubes with cap (Corning, MA, USA).
5. 96 well cell culture plates with lid (Nunclon Surface, Nunc).
6. Disposable semi-micro polystyrene cuvettes (1.5 mL) (BrandTech Scientific, Essex CT).
7. Mueller Hinton broth (Oxoid, Basingstoke, UK).
8. Tryptic Soy Agar solid media plates (Oxoid).
9. Bacterial strains: *E. coli* ATCC 25922 and *S. aureus* ATCC 25923 (American Type Culture Collection, Manassas, VA, USA).

3. Methods

3.1. Collection of Skin Secretions

3.1.1. Injection Procedure

1. Inject animals individually at two sites within the dorsal sac with a freshly prepared solution of norepinephrine in water (2 nmol/g body weight in a volume of 200 μL) (*see* **Note 5**).
2. Immediately after injection, place the animal for a period of 15 min in water contained in a covered glass beaker (typically 100 mL with a greater volume for particularly large specimens) (*see* **Note 6**).

3. Immediately acidify the solution containing the secretions after collection with concentrated hydrochloric acid (final concentration 1% v/v) and store at –20°C until time of analysis (*see* **Note 7**).

3.1.2. Immersion Procedure

1. Immerse the animals for 15 min in a covered glass beaker to which has been added a sufficient volume of distilled water containing 200 μM norepinephrine to completely cover the frog.
2. Process the solution containing the secretions as in **Section 3.1.1**, Step 3.

3.2. Sample Preparation Using Sep-Pak Cartridges

1. Prepare HPLC solvent A by adding 1.2 mL trifluoroacetic acid to 1,000 mL water (*see* **Note 8**).
2. Prepare HPLC solvent B by adding 1.0 mL trifluoroacetic acid to 700 mL acetonitrile/300 mL water.
3. Activate Sep-Pak cartridges by pumping acetonitrile (2 mL per cartridge) at a flow rate of 2 mL/min using a peristaltic pump or manually using a 20 mL plastic syringe, followed by HPLC solvent A (2 mL/cartridge; *see* **Section 3.2**, Step 1). Up to 10 Sep-Pak cartridges may be connected in series depending on the volume and amount of peptide material in the extract to be processed.
4. Centrifuge the combined skin secretions and washings (5,000*g* for 30 min) (*see* **Note 9**).
5. Pump the supernatant through the Sep-Pak cartridges at a flow rate of 2 mL/min. If a peristaltic pump is not available, a polypropylene syringe (50 mL) can be used to apply the solution manually.
6. Pump HPLC solvent A (4 mL/cartridge) is pumped at a flow rate of 4 mL/min and discard the eluate.
7. Pump HPLC solvent B (2 mL/cartridge; *see* **Section 3.2**, Step 2) at a flow rate of 1 mL/min and collect the eluate into polypropylene tubes.
8. Reduce the volume of eluate to approximately 1.5 mL under reduced pressure in a Speed-Vac concentrator (*see* **Note 10**). With an efficient vacuum pump, this step can be accomplished in approximately 60 min at room temperature or in 30 min with external heating. Sample may be stored at 4°C overnight or frozen at –20°C for several days.

3.3. Peptide Separation by Reversed-Phase HPLC

1. Degass HPLC solvent A and solvent B (*see* **Section 3.2**, Steps 1 and 2) with helium for 1 min. This time should not be exceeded.

2. Column preparation. Before injecting the sample, it is necessary to "condition" whatever column is to be used in order to improve resolution by following these steps: irrigate the column at an appropriate flow rate (*see* **Section 2.2**, Step 5) with HPLC solvent B for 20 min; decrease the concentration of solvent B to 0% over 10 min using a linear gradient; increase the concentration of solvent B to 100% over 10 min; decrease the concentration of solvent B to 0% over 10 min; equilibrate the column with HPLC solvent A for 20 min.
3. Centrifuge the sample for 5 min at 13,000*g* in a 1.5 mL polypropylene Eppendorf tube to ensure clarity of solution (*see* **Note 11**).
4. Program the HPLC system to perform chromatography under the following conditions using linear gradients for elution: (a) increase concentration of solvent B from 0 to 30% over 10 min; (b) increase concentration of solvent B from 30 to 90% over 60 min; (c) increase concentration of solvent B from 90 to 100% over 1 min and hold at 100% until UV-absorbance returns to baseline value (*see* **Note 12**).
5. Inject the sample onto the column. Up to 1.5 mL may be injected into a 2 mL loop.
6. Increase the concentration of solvent B according to the elution program listed in **Section 3.3**, Step 4. Collect fractions (1 min) into (15 mm × 100 mm) polypropylene tubes using a fraction collector (*see* **Note 13**).
7. Take aliquots (20–200 μL depending on the size of the UV-absorbing peak) of each fraction for determination of antimicrobial activity and dry under reduced pressure in 1.5 mL Eppendorf tubes using a Speed-Vac concentrator.

3.4. Purification of Antimicrobial Peptides to Near Homogeneity by Reversed-Phase HPLC

1. Replace the preparative C_{18} column used in **Section 3.3** with a (1.0 cm × 25 cm) Vydac 214TP510 semi-preparative C_4 column equilibrated with 30% HPLC solvent B (*see* **Section 3.2**, Step 2) at a flow rate of 2 mL/min. (*see* **Note 14**).
2. Reduce the volume of the fractions containing antimicrobial peptides (**Section 3.3**, Step 6) to approximately 1 mL in a Speed-Vac concentrator. Inject each sample onto the C_4 column.
3. Raise the concentration of solvent B to 80% using a linear gradient (*see* **Note 14**). Collect major individual UV-absorbing peaks by hand into polypropylene tubes.
4. At this point, the peaks containing the peptide with antimicrobial activity are identified by the assay method described in **Section 3.5** and purified further as described in the following steps.

5. Replace the C_4 column by a (1.0 cm × 25 cm) Vydac 219TP510 diphenylmethyl column equilibrated with 30% solvent B at a flow rate of 2 mL/min.
6. Reduce the volume of the fraction from **Section 3.4**, Step 3 to approximately 1.0 mL in a Speed-Vac concentrator. Inject the sample onto the column. Follow the same elution conditions as in **Section 3.4**, Step 3.
7. Collect the major UV-absorbing peak by hand into a polypropylene tube (*see* **Note 15**).
8. Freeze dry and weigh the purified peptide (*see* **Note 16**).

3.5. Detection of Peptides with Antimicrobial Activity

1. Inoculate reference bacterial strains (*S. aureus* ATCC25923 and *E. coli* ATCC25922) into Mueller Hinton broth (5.0 mL) in separate 15 mL sterile tubes and culture at 37°C for 4 h (logarithmic phase of growth). Sterile pipette tips must be used in all procedures. Determine the cell count (colony forming unit, CFU/mL) of the bacterial cultures by measurement of absorbance at 600 nm (*see* **Note 17**).
2. Dilute the cultures with Mueller Hinton broth to a 10^6 CFU/mL cell count.
3. Redissolve the Speed-Vac dried fractions from **Section 3.3**, Step 7 in 10 μL of 0.1% trifluoroacetic acid–water to which 100 μL of Mueller-Hinton broth is added. Distribute aliquots (50 μL) in duplicate into the wells of a 96 well microtiter plate (*see* **Note 18**). Fill at least four wells with Mueller-Hinton broth only (100 μL) for sterility control and fill at least another four wells with Mueller-Hinton broth only (50 μL) to serve as growth control.
4. Add the *S. aureus* ATCC 25923 suspension (50 μL 10^6 CFU/mL) to each individual well of one plate containing the reconstituted fractions and to the four wells containing 50 μL Mueller-Hinton broth only (growth control).
5. Add the *E. coli* ATCC 25922 suspension (50 μL; 10^6 CFU/mL) to each individual well of a second plate containing the reconstituted fractions and to the four wells containing 50 μL Mueller-Hinton broth only (growth control).
6. Incubate the microtiter plates at 37°C overnight (approximately 16 h) in a humidified atmosphere of air in an incubator or in a room at constant temperature (*see* **Note 19**).
7. Determine the bacterial growth by measuring absorbance at either 600 or 630 nm (depending on the filter installed) using a microtiter plate reader. The absorbance of the sterility control wells containing Mueller-Hinton broth only is taken as zero value. The absorbance of the growth control wells containing bacteria only is taken as the 100%

value. Peptides in chromatographic fractions that inhibit the growth of either bacterium by ≥50% are purified further, as described in **Section 3.4** (*see* **Note 20**).

In the example shown in **Fig. 1.1**, norepinephrine-stimulated skin secretions obtained from the Asian frog *Hylarana picturata*, after partial purification on Sep-Pak cartridges, were chromatographed on a (2.2 cm × 25 cm) Vydac 218TP510 preparative C_{18} column equilibrated with HPLC solvent A at a flow rate of 6 mL/min. The concentration of acetonitrile in the eluting solvent was increased as described in **Section 3.3**, Step 4. Aliquots (100 μL) of the major peaks in the chromatogram were subjected to antimicrobial analysis (**Section 3.5**) and the peaks designated 1–5 were shown to inhibit growth of both *E. coli* and *S. aureus* (15).

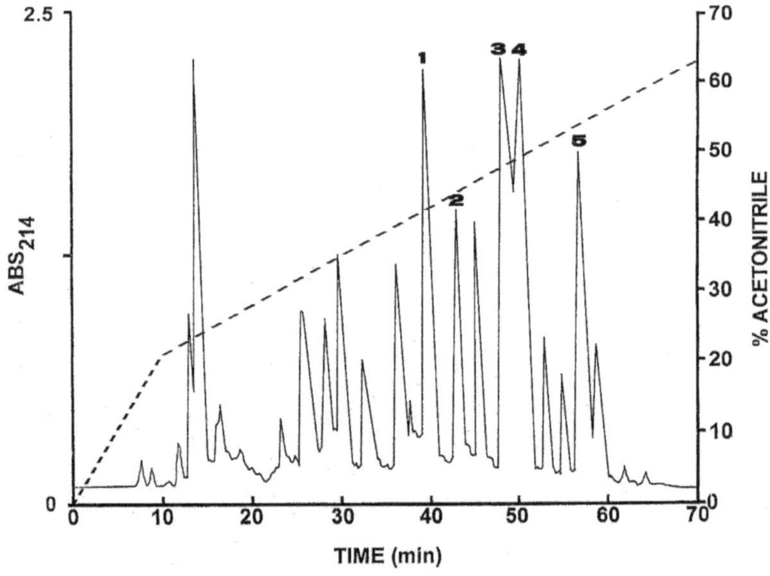

Fig. 1.1. Elution profile on a preparative Vydac C_{18} column of skin secretions from *Hylarana picturata* after partial purification on Sep-Pak cartridges. The peaks designated 1–5 were shown to contain peptides that inhibited the growth of both *E. coli* and *S. aureus*. The *dashed line* shows the concentration of acetonitrile in the eluting solvent. Subsequent structural analysis demonstrated that peak 1 contained brevinin-2PTa, peak 2: brevinin-2PTb; peak 3: brevinin-PTa and brevinin-2PTc; peak 4: brevinin-1PTb, brevinin-2PTd, and temporin-PTa, and peak 5: brevinin-2PTe. (Reproduced from (14) with permission from Elsevier Science).

4. Notes

1. A dual pen flatbed chart recorder (Kipp and Zonen, Delft, The Netherlands) to supplement any on-screen computer

recording system is useful especially when there is a significant delay in the response of the absorbance detector and the appearance of the peak on the computer screen.

2. The decision whether to use a preparative or a semi-preparative C_{18} column for initial fractionation depends upon the peptide content of the secretions. For samples containing > 10 mg peptide material, a (2.2 cm × 25 cm) preparative column should be used. The approximate amount of peptide material may be estimated using a BCA (bicinchoninic acid) protein assay reagent kit following the recommended procedure of the manufacturer (Thermo Fisher Scientific).
3. Degassing of solvents may be achieved although less efficiently by using an ultrasonic bath but degassing by reducing the pressure is not recommended as it may lead to relative loss of volatile components.
4. Polypropylene tubes should be used throughout, not glass or polystyrene, in order to minimize irreversible binding of peptides to the tubes.
5. The amount of norepinephrine injected is not critical but the agent is unstable in solution and so must be dissolved in water immediately before use. Animals that appear to be agitated or highly mobile can be partially anesthetized by immersion for 5 min in crushed ice.
6. Norepinephrine injection is generally well tolerated by the animals without signs of discomfort and fatalities are extremely rare. However, in the case of very small frogs, tadpoles, or protected or endangered species for which it may be impossible to obtain a permit to carry out an invasive procedure, the immersion procedure may be used. The agent is absorbed through the frog's skin and stimulates peptide release, although with reduced effectiveness compared with injection. Alternative methods used to stimulate secretions include electrical stimulation (typically 10 V DC, pulse duration 3 ms) at multiple sites on the dorsal region (16) and confining the animal in a closed container containing ether (17) but these methods appear to cause distress to the frog.
7. If secretions are collected in the field, it may not be possible to freeze the sample immediately. In this case, secretions can be kept at 0°C in an ice bath for several hours until access to a freezer is obtained.
8. A slightly greater concentration of trifluoroacetic acid in HPLC solvent A than in solvent B produces a flat baseline under HPLC gradient elution conditions.

9. Frog skin secretions contain lipid components but it is generally unnecessary to remove them by extraction with an organic solvent prior to partial purification on Sep-Pak cartridges.
10. The sample should not be dried completely except when this is unavoidable (e.g., for shipment to a different laboratory) as this can lead to formation of insoluble material and irreversible binding to plastic surfaces.
11. Filtering the sample is not recommended unless absolutely necessary as it can lead to appreciable loss of peptide by irreversible binding to the filter material.
12. At the end of the experiment, wash the column with acetonitrile (50 mL) and store in this solvent.
13. Fractions may be stored in stoppered tubes at –20°C for up to several months. However, peptides containing methionine and tryptophan residues may oxidize on prolonged storage, even at low temperature.
14. The majority of frog skin antimicrobial peptides are eluted from a Vydac C_{18} column in the concentration range of 55–85% HPLC solvent B. To purify the more hydrophobic peptides on a Vydac C_4 column, raise the initial solvent concentration to 40% B and the final concentration to 90% B.
15. Peptide purity is assessed by a symmetrical peak shape measured at two wavelengths (214 and 280 nm) and by mass spectrometry.
16. It is necessary to use a freeze drier to obtain the purified peptide in a flocculent powder form that is suitable for weighing. This generally is not achieved when solutions are taken to dryness in a Speed-Vac concentrator. When amino acid sequence analysis is to be carried out in the chromatographer's own laboratory or institute, it is preferable not to dry the fractions completely but reduce the volume to approximately 30 μL in order to minimize losses of peptide due to irreversible binding to the plastic tubes. When samples are to be analyzed by an external core facility, lyophilization to dryness cannot be avoided.
17. The first time a new bacterial strain is used, a standard curve of OD_{600} vs. CFU/mL is constructed using the following procedure: Prepare a series of eight double dilutions of a dense suspension (2.5 mL; $OD_{600} > 1.5$) of the bacterium in sterile 0.9% saline (1:2, 1:4, up to 1:256 dilution). Measure the absorbance at 600 nm of each dilution using disposable plastic cuvettes. Prepare a series of eight 10-fold dilutions of the same initial suspension (1:10, 1:10^2 up to

1:10^8 dilution). Spread aliquots (200 μL) of the 10^5, 10^6, 10^7, and 10^8 dilutions in duplicate onto Tryptic Soy Agar plates. Incubate the plates overnight at 37°C and count the numbers of colonies on each plate. Count only plates with >50 colonies. Calculate the cell count in the original bacterial suspension by multiplying the average number of colonies by the dilution factor. For example, if 78 and 86 colonies are counted on the plates prepared from the 10^6 dilution, the original cell count in the suspension is $82 \times 5 \times 10^6 = 4.1 \times 10^8$ CFU/mL (the factor 5 arises from the fact that only a 200 μL aliquot was used). Construct a standard curve where the logarithmic abscissa (X) represents the cell count (CFU/mL) and the linear ordinate (Y) represents the corresponding OD_{600} values.

18. It is better to use separate plates for different microorganism to avoid contamination.
19. The microtiter plates may be placed in a sterile polyethylene box with lid containing a paper towel moistened with water.
20. If required, the minimum inhibitory concentrations (MIC values) of the purified antimicrobial peptides can be determined by the microdilution method described in detail in (18).

References

1. Norrby, S. R., Nord, C. E., and Finch, R. (2005) Lack of development of new antimicrobial drugs: a potential serious threat to public health. *Lancet Infect. Dis.* **5**, 115–119.
2. Rinaldi, A. C. (2002) Antimicrobial peptides from amphibian skin: an expanding scenario. *Curr. Opin. Chem. Biol.* **6**, 799–804.
3. Hancock, R. E. (2001) Cationic peptides: effectors in innate immunity and novel antimicrobials. *Lancet Infect. Dis.* **1**, 156–164.
4. Conlon, J. M. (2004) The therapeutic potential of antimicrobial peptides from frog skin. *Rev. Med. Microbiol.* **15**, 17–25.
5. Powers, J. P. and Hancock, R. E. (2003) The relationship between peptide structure and antibacterial activity. *Peptides* **24**, 1681–1691.
6. Conlon, J. M., Demandt, A., Nielsen, P. F., Leprince, J., Vaudry, H., and Woodhams, D. C. (2009) The alyteserins: two families of antimicrobial peptides from the skin secretions of the midwife toad *Alytes obstetricans* (Alytidae). *Peptides* **30**, 1069–1073.
7. Amiche, M., Ladram, A., and Nicolas, P. (2008) A consistent nomenclature of antimicrobial peptides isolated from frogs of the subfamily Phyllomedusinae. *Peptides* **29**, 2074–2082.
8. Conlon, J. M., Kolodziejek, J., and Nowotny, N. (2009) Antimicrobial peptides from the skins of North American frogs. *Biochim. Biophys. Acta* **1788**, 1556–1563.
9. Nutkins, J. C. and Williams, D. H. (1989) Identification of highly acidic peptides from processing of the skin prepropeptides of *Xenopus laevis*. *Eur. J. Biochem.* **181**, 97–102.
10. Davidson, C., Benard, M. F., Shaffer, H. B., Parker, J. M., O'Leary, C., Conlon, J. M., and Rollins-Smith, L. A. (2007) Effects of chytrid and carbaryl exposure on survival, growth and skin peptide defenses in foothill yellow-legged frogs. *Environ. Sci. Technol.* **41**, 1771–1776.
11. Conlon, J. M. (2007) Purification of naturally occurring peptides by reversed-phase HPLC. *Nat. Protoc.* **2**, 191–197.
12. Conlon, J. M. and Leprince, J. (2010) Identification and analysis of bioactive peptides

in amphibian skin secretions. In *Peptidomics Protocols and Methods Handbook*. Soloviev, M. (Ed.), Totowa, NJ: Humana, in press.

13. Conlon, J. M., Al-Ghaferi, N., Abraham, B., and Leprince, J. (2007) Strategies for transformation of naturally-occurring amphibian antimicrobial peptides into therapeutically valuable anti-infective agents. *Methods* **42**, 349–357.
14. Conlon, J. M., Galadari, S., Raza, H., and Condamine, E. (2008) Design of potent, non-toxic antimicrobial agents based upon the naturally occurring frog skin peptides, ascaphin-8 and peptide XT-7. *Chem. Biol. Drug. Des.* **72**, 58–64.
15. Conlon, J. M., Kolodziejek, J., Nowotny, N., Leprince, J., Vaudry, H., Coquet, L., Jouenne, T., and King, J. D. (2008) Characterization of antimicrobial peptides from the skin secretions of the Malaysian frogs, *Odorrana hosii* and *Hylarana picturata* (Anura:Ranidae). *Toxicon* **52**, 465–473.
16. Tyler, M. J., Stone, D. J., and Bowie, J. H. (1992) A novel method for the release and collection of dermal, glandular secretions from the skin of frogs. *J. Pharmacol. Toxicol. Methods* **28**, 199–200.
17. Lai, R., Zheng, Y. T., Shen, J. H., Liu, G. J., Liu, H., Lee, W. H., Tang, S. Z., and Zhang, Y. (2002) Antimicrobial peptides from skin secretions of Chinese red belly toad *Bombina maxima*. *Peptides* **23**, 427–435.
18. Clinical and Laboratory Standards Institute (2006) *Methods for Dilution Antimicrobial Susceptibility Tests for Bacteria that Grow Aerobically* (7th ed.). Wayne, PA: CLSI.

Chapter 2

Purification of Antimicrobial Peptides from Human Skin

Jens-M. Schröder

Abstract

Human skin is a rich source of human antimicrobial peptides. Its cellular source is the keratinocyte, which terminally differentiates in the uppermost parts of the skin, eventually forming the stratum corneum, the horny layer. The easy availability of human stratum corneum makes it possible to identify and characterize human antimicrobial peptides with a biochemical approach. Moreover, the availability of lesional scales of patients with psoriasis, an inflammatory skin disease, allows the identification of human-inducible peptide antibiotics, which are absent in healthy skin. With this strategy, the beta-defensins hBD-2 and hBD-3, RNase-7 as well as psoriasin/S100A7 have been discovered as human antimicrobial peptides and proteins. A detailed description of the strategies and methods is presented, which allowed a successful identification and characterization of human antimicrobial peptides. We used various HPLC techniques, combined with antimicrobial testing as read-out system. In parallel, SDS-PAGE analyses as well as electrospray ionization mass spectrometry were used for further biochemical characterization as well as purity assessment.

Key words: Antimicrobial peptides, defensin, RNase-7, S100A7, skin, stratum corneum, purification, electrospray ionization mass spectrometry, high-performance liquid chromatography.

1. Introduction

Human skin is always in contact with the environment, but despite the presence of many favourable conditions for microbe growth, it is rarely infected. This unexpected observation led to the hypothesis that the uppermost skin layer contains a "chemical defence shield" consisting in antimicrobial peptides. Recent evidence indicates that antimicrobial peptides are synthesized by fully differentiated keratinocytes in the surface-exposed epidermis, a self-renewing stratified squamous epithelium composed of several layers of keratinocytes. Keratinocytes in the outmost

A. Giuliani, A.C. Rinaldi (eds.), *Antimicrobial Peptides*, Methods in Molecular Biology 618,
DOI 10.1007/978-1-60761-594-1_2, © Springer Science+Business Media, LLC 2010

living cells containing layer (stratum granulosum) terminally differentiate and eventually form a physical barrier consisting of the horny layer (stratum corneum, SC). SC cells are sloughed off and replaced by newly differentiated cells originating from epidermal stem cells located in the basal layer. Thus, stratum corneum should be a rich source of human antimicrobial peptides (AMPs) and should allow its detection and characterization in an unbiased approach.

Purification of AMPs from skin requires special strategies. First of all, the detection system should be useful for screening to detect the antimicrobial peptide at low concentration. Several assay systems can be used for screening (1), e.g. radial diffusion assay (RDA) or microbroth dilution assay. The RDA assay is a highly sensitive assay which consumes only minimal amounts of peptides for testing (1). Unfortunately the microbroth dilution assay requires approx. 10 times more peptide than the RDA assay (1). Thus, due to the limited availability of human material, the RDA assay might be more useful for successful AMP purification from skin samples.

1.1. Extraction of AMPs from Stratum Corneum

SC is a rich source of various polar and non-polar lipids which make it less efficient to extract peptides and proteins. It might be thus useful to wash the material with a non-polar organic solvent such as ethyl acetate prior to peptide extraction. To achieve an optimal peptide extraction from SC, acidic buffers need to be used. The buffer system should be carefully chosen according to the strategy used for AMP purification: When reversed phase (RP)-HPLC is used as first and sole separation step (when amounts of only a few milligrams of SC are available), SC extraction with acetonitrile containing acidic buffers is recommended. When amounts > 500 mg SC are available, ethanolic citric acid buffers have been proven to be useful for extraction.

1.2. Chromatographic Separation

The SC amounts available will determine the purification strategy. Extracts from a few milligrams of SC will only allow the (partial) purification of the principle AMPs by RP-HPLC. This should be done with a micro-HPLC system. When amounts > 500 mg SC are available, a different, more complex purification protocol, which also includes cation exchange-HPLC, will allow the purification of the principal AMPs as well as most of the less abundant AMPs. Identification of less abundant AMPs requires an enrichment step by pooling the material from different HPLC runs.

For enrichment of cationic and amphipatic antimicrobial peptides (which represent by far the majority of all AMPs in SC), heparin-affinity chromatography is useful. The heparin-bound material is then separated by preparative RP-HPLC followed by micro-cation exchange-HPLC and finally micro-RP-HPLC.

1.3. Bioassays

Antimicrobial activity in HPLC fractions can be tested with various designer assays (1). HPLC solvents are usually not compatible with the AMP bioassays. In the case of volatile solvents (RP-HPLC steps), lyophilization of the samples will be sufficient to get rid of the problem. In the case of cation exchange-HPLC, salts are required for protein elution. Because several AMPs are salt sensitive, there will be a principal risk to get false-negative results.

In some cases it is possible to overcome this problem by using unconventional, pragmatic conditions for cation exchange-HPLC using volatile buffers such as ammonia salts at an acidic pH.

It is further important to choose the right bacterial target. Many of the AMPs have a restricted activity profile. It is therefore important to use the bacterial target of interest.

1.4. Biochemical Analyses

HPLC fractions should be analysed also by analytical biochemical techniques. Due to the highly cationic behaviour, conventional sodium dodecyl sulphate polyacryl amide electrophoresis (SDS-PAGE) analyses often did not work with AMPs. Instead, acidic acryl amide electrophoresis was initially used for checking purity of the AMPs (2). The Tricine–SDS-PAGE system (3) has been now successfully used for determination of the AMP's size (4–6). It is important to choose electrophoresis conditions which allow sufficient loading of SDS to the highly cationic AMPs and minimize formation of dimers and oligomers. Tricine buffers containing urea were found to be optimal.

Due to its high sensitivity, silver staining is preferred for detection of AMPs in the gel, because only nanogram amounts of AMPs are required.

RP-HPLC fractions can also be monitored by mass spectrometry. Electrospray ionization mass spectrometry (ESI-MS) is useful to directly determine the exact mass of AMPs in RP-HPLC fractions. Characteristic charge state patterns of AMPs upon ESI-MS analyses reveal further structural information about the AMPs isolated from SC. Because ESI-MS is highly sensitive towards the presence of salt, analyses of ion exchange-HPLC fractions requires elimination of anorganic salts, either by the use of volatile buffers or RP cartridges for each separate fraction.

HPLC solvents are usually not compatible with the AMP bioassays. In the case of volatile solvents (RP-HPLC steps), lyophilization of the samples will be sufficient to get rid of the problem. In the case of cation exchange-HPLC, salts are required for protein elution. Because several AMPs are salt sensitive, there will be a principal risk to get false-negative results.

In some cases it is possible to overcome this problem by using unconventional, pragmatic conditions for cation exchange-HPLC using volatile buffers such as ammonia salts at acidic pH.

2. Materials

2.1. Extraction of AMPs from SC

1. Suspend stratum corneum or lesional psoriatic scales (e.g. 2.5 g in 100 mL of 0.1 M aqueous citric acid containing 50% (v/v) of 96% ethanol [denaturated with heptane]) (*see* **Note 1**).
2. Homogenizer (Ultraturrax®): use 2,000 rpm for 60 min and chilling in an ice–water mixture.
3. Ultrafiltration: YM3 filters, cutoff of 3 kDa (Millipore, Danvers, MA), are useful (*see* **Note 2**).

2.2. Chromatographic Separation of Antimicrobial Peptides

Use HPLC-quality solvents. All chromatographic steps are performed at room temperature (*see* **Note 3**).

1. Any HPLC or FPLC machine containing a pump, gradient mixer, UV detector, HPLC columns, and fraction collector can be used. UV detection should be done at either 215, 280 nm, or both. Use solvents that do not show absorbance at 215 nm (e.g. acetonitrile, water, and trifluoroacetic acid [TFA]) (*see* **Note 4**). For micro-HPLC we use a Smart®HPLC system (GE Healthcare Biosciences AB, Uppsala, Sweden) or an ETTAN LC®Purifier system (GE Healthcare).
2. For HPLC separation the following columns have been used:
   a. Heparin Sepharose cartridge (Hi Trap, 10 × 5 mm, 1 mL volume, Amersham) (*see* **Note 5**).
   b. Preparative wide-pore (300 Å) reverse phase (RP-8)-HPLC column (C8 Nucleosil with endcapping, 250 × 12.6 mm, 7 mm particle size, Macherey-Nagel, Düren, Germany) (*see* **Note 6**).
   c. Micro-MonoS®-HPLC column for Smart®HPLC system (GE Healthcare) (*see* **Note 7**).
   d. Micro-RP-18 (C2/C18) column for Smart®HPLC system (GE Healthcare) (*see* **Note 8**).

2.3. SDS-PAGE Analysis

1. The method of Schägger and von Jagow (3) is used. As sample buffer 50 mM Tris–HCl, 4% (w/v) SDS, 12% (w/v) glycerol, pH 6.8, containing 8 M urea is used (*see* **Note 9**).
2. Gels with the dimension 130 × 100 × 1 mm are used and electrophoresis is done in the presence of 8 M urea for 2 h at 10 mA current, 30 V power (power limit 10 W) at room temperature (*see* **Note 10**).

3. Fixation of antimicrobial peptides is done for 30 min with aqueous 2-propanol (30% [v/v]) containing 10% (v/v) acetic acid and 0.3% (v/v) glutaraldehyde (*see* **Note 11**).
4. Proteins are stained with 0.03% (w/v) silver nitrate in deionized water followed by developing with a solution of 10% saturated aqueous Na_2CO_3 solution containing 0.1% (v/v) of saturated aqueous formaldehyde (40% [v/v]). Development is terminated by acetic acid (3% [w/v] in water) (*see* **Note 12**).

2.4. Electrospray Ionization Mass Spectrometry (ESI-MS)

1. We used a Q-Tof™ II mass spectrometer (Waters Micromass, Milford, MA).
2. As ESI gas nitrogen was used.
3. Nanoelectrospray ionization was used when sample amounts were limited or for MS/MS peptide sequencing experiments and were carried out after fitting the Q-Tof™ II mass spectrometer with a nano-Z spray source (*see* **Note 13**).
4. For nanoelectrospray MS analysis nanoflow-electrospray needles (Borosilicate metal-coated glass capillary, Waters Micromass, Milford, MA) were used. Salt containing samples are desalted using self-prepared μ-columns using constricted GelLoader tips (Eppendorf, Wesseling-Berzdorf, Germany) packed with a 2:1 mixture of Poros® 50R2 and Oligo R3 reversed phase medium (Applied Biosystems, Forster City, CA).
5. Acquisition and data analysis are performed using the MassLynx 4 software package supplied by Waters Micromass, Milford, MA.

2.5. Antimicrobial Activity Assay

The radial diffusion assay was used according to Steinberg and Lehrer (1) (*see* **Note 14**).

1. Trypticase soy broth (TSB, Difco, Detroit, MI): Full strength broth (30 g per litre deionized water), autoclaved, and stored at room temperature.
2. Sterile phosphate buffer, pH 7.4 or 6.5 (or as desired).
3. Agarose (Sigma): The use of a low electroendosmosis (EEO)-type agarose in place of standard agar is *critical* to limit electrostatic interactions between cationic AMPs and sulphated moieties of the agaropectin component of standard agar (1).
4. Underlay gels (1): Mix 50 mL of 100 mM sodium phosphate buffer with 5 mL full-strength trypticase soy broth, add 5 g agarose, and bring the volume to 500 mL with deionized distilled water. Adjust the final pH with NaOH or HCl, place the suspension on a hot plate, and stir it under heat until the

agarose dissolves. Dispense 50-mL aliquots, autoclave them at 121°C for 20 min, and store the sterilized media at room temperature. Before use, the solidified medium should be fluidized (we use a microwave oven) and placed it into a water bath maintained at 42°C.

5. Overlay agar (1): This agar contains 60 g (twice the customary amount) of trypticase soy broth (TSB, Difco) and 10 g agarose (Sigma) per litre of deionized water and is similarly treated as the underlay gels.
6. Microorganisms: *Escherichia coli*, *Staphylococcus aureus*, and *Candida albicans* were used for routine AMP testing after determination of optical density (OD)/colony forming unit (CFU) ratio (*see* **Note 15**).
7. Peptides are introduced into 3 mm sample wells within the underlay gel.

3. Methods

3.1. Extraction of AMPs from Stratum Corneum

1. Wash 1 g heel stratum corneum (SC) with 50 mL ethylacetate and dry it under vacuum (*see* **Note 16**).
2. Suspend SC in acidic buffer, homogenize, and then add ethanol (*see* **Note 17**).
3. Centrifuge and use supernatant (*see* **Note 18**).
4. Concentrate the supernatant to ~ 5–10 mL, adjust to pH 8.0, then again to pH 4–4.5, centrifuge it again, and freeze it (at –20°C) until further use (*see* **Note 19**).
5. For further purification, thaw the sample, adjust the pH to pH 8.0, centrifuge it, and apply it to a heparin affinity column (*see* **Note 20**).
6. Wash the column with 3 volumes of equilibration buffer: 10 mM Tris–citrate, pH 8.0.
7. Strip the bound material from the column by the use of 3 mL 2 M NaCl in equilibration buffer (*see* **Note 21**).
8. Add trifluoroacetic acid (TFA) and adjust to pH 3.

3.2. Preparative RP-8-HPLC

1. Apply the heparin-bound material (from **Section 3.1**, Step 8.) to a preparative RP-8-HPLC column by the use of a 2–5 mL loop (*see* **Note 22**).
2. Separate proteins by elution with a gradient of increasing concentrations of acetonitrile. Turn the detector on 215 nm and – when possible also 280 nm – and choose the integrator

attenuation appropriate to protein amounts you expect (*see* **Note 23**).

3. Separate peaks – when possible – manually according to the appearance of UV-absorbing peaks and shoulders (*see* **Note 24**).
4. Place fractions immediately in the refrigerator.
5. Take off an aliquot of each fraction for the bioassay, SDS-PAGE analysis, or ESI-MS analyses using a microtitre plate (*see* **Note 25**).

3.3. Micro-MonoS® Cation Exchange HPLC

1. Combine fractions of RP-8-HPLC, which contain the same AMP and lyophilize them.
2. Dissolve the residue in 50–200 μL 20 mM ammonium formate buffer, pH 4.0, which contains 25% (v/v) acetonitrile.
3. Apply the sample to a micro-MonoS®-HPLC column, which has been equilibrated with the same acetonitrile-containing ammonium formate buffer, pH 4.0 (*see* **Note 26**).
4. Elute with a gradient (0–20 min) of increasing concentrations of NaCl (0–1 M) (*see* **Note 27**).
5. Take off an aliquot of each fraction for the bioassay, SDS-PAGE analysis, or ESI-MS analyses using a microtitre plate (*see* **Note 25**).

3.4. Micropore C2/C18-RP-HPLC

1. Combine fractions containing the same AMP and apply them directly onto a micropore C2/C18-RP-HPLC column (*see* **Note 28**).
2. Elute proteins with increasing concentrations of acetonitrile. Turn the UV detector on 215 nm (and 280 nm, if possible) and separate peaks manually.
3. Take off an aliquot of each fraction for the bioassay, SDS-PAGE analysis, or ESI-MS analyses using a microtitre plate (*see* **Note 25**).

3.5. Radial Diffusion Antimicrobial Activity Assay

1. Prepare organisms for the assay as well as the gel as outlined in **Section 2.5**.
2. Add 30 μL aliquots of preparative RP-8-HPLC fractions or 2 μL aliquots of micro-HPLC fractions to the wells of a round-bottom well microtitre plate. Lyophilize samples and add to each well 8 μL 0.01% (v/v) aqueous acetic acid and transfer 5 μL aliquots of the various samples into each well in the underlay gel.
3. Incubate the gel for 3 h in a 37°C incubator, then add the overlay, and incubate at 37°C overnight.

4. Calculate the results by measuring the diameter of the clearing zone (*see* **Note 29**).

3.6. SDS-PAGE Analyses

1. Mix fractions (30 μL of RP8-HPLC fractions, 2 μL of MonoS®-HPLC, and C2/C18-micro-HPLC fractions) with 10 μL sample buffer and boil for 10 min (*see* **Note 30**). Then load sample on the stacking gel and separate electrophoretically in the presence of 8 M urea (30 min at 100 V, 400 mA; then 90 min at 120 V, 400 mA) (*see* **Note 31**).
2. Fix peptides in the gel with fixation solution.
3. Stain proteins with silver nitrate solution (*see* **Note 32**).

3.7. Electrospray Ionization Mass Spectrometry (ESI-MS) Analyses

1. For routine MS analysis, dilute aliquots (2–10 μL) of HPLC fractions containing the sample with 100 μL carrier solvent (50:50 acetonitrile:water containing 0.2% (v/v) formic acid) and infuse it into the electrospray source at a rate of 10–20 μL/min (*see* **Note 33**). Sodium iodide can be used for mass calibration for a calibration range of *m/z* 100–2,000. The capillary potential is set to 3.5 or 4 kV and cone voltages between 25 and 75 V are chosen; cone temperature is set to 80°C; desolvation temperature is 150°C. The charge-to-mass ratio of ions is scanned within the range of 280–2,000.
2. For nanoelectrospray MS analysis a 1–3 μL aliquot of salt-free sample (*see* **Note 34**) is loaded into the nanoflow-electrospray needle. Salt containing samples are desalted using self-prepared μ-columns. Formic acid is added to the samples just before desalting to give a final concentration of 5% (v/v). After loading the column with the sample the column should be washed two times with 10 μL 5% formic acid. Peptides are eluted with 1–3 μL 60% methanol/5% formic acid directly into the nanoelectrospray needle and measured by nanoelectrospray ionization MS using a capillary voltage between 1 and 1.4 kV.
3. For mass mapping (mass fingerprint) when necessary, cysteine residues are reduced (by incubation of HPLC fractions containing 10–20 μg protein in 85 μL 100 mM NH_4HCO_3, 10 mM dithiothreitol (DTT) for 20 min at 60°C) and subsequently alkylated (by incubation in 200 mM iodoacetamide for 20 min at RT) in the dark. Prior to digestion samples are desalted using 10 μL C_{18} Poros®pipette tips or HPLC with C_2/C_{18} columns.
   HPLC sample fractions are either directly or after reduction and alkylation of cysteine residues subjected to trypsin digestion: lyophilized samples containing 2–4 μg protein are dissolved in 20 μL 100 mM NH_4HCO_3, pH 8; tryptic digestion is started by adding 50 ng (with 1% acetic acid)

activated trypsin (modified trypsin, sequencing grade, Roche; 1:50 final molar enzyme:substrate ratio) and is allowed to proceed for at least 4 h at 37°C. Digests are either analysed directly or stored frozen until subsequent mass mapping: tryptic fragment masses are determined after desalting using self-prepared μ-columns (**Section 2.5**, Step 4) using ESI-MS.

4. Peptide sequencing by MS/MS is done by analyses of tryptic peptide fragments using a Q-TOF 2 mass spectrometer (Waters Micromass, Milford, MA) with nanoelectrospray ionization (*see* **Note 35**). Cone voltage and collision energy used to perform MS–MS are optimized for each peptide analysed. Collision energy varied between 17 and 40 eV with argon (14 psi) as the collision gas. Data for MS/MS are acquired over the mass range *m/z* 80–2,000.
5. Mass spectra are averaged typically over 2–10 scans (20–2 s/scan). The multiply charged raw data of intact proteins are background-subtracted and deconvoluted using MaxEnt1 to obtain singly charged ion mass spectra to determine average molecular masses of intact proteins.

The raw combined spectral data from small peptides or tryptic protein fragments or obtained after MS/MS fragmentation of selected precursor ions are background subtracted and subjected to Maximum Entropy 3 ("MaxEnt3") deconvolution to determine monoisotopic molecular masses. Sample identity is determined by database search analysis of peptide fragment mass patterns and/or after de novo sequencing of tryptic peptide fragments. All mass fingerprint and MS/MS data are searched against the human protein database using the Mascot program (Matrix Science, Boston, MA). Peptide sequences are directly ("de novo") determined from MS/MS data using the software program PepSeq from the MassLynx4 software package (Waters Micromass, Milford, MA). PepSeq-derived peptide sequences are analysed with the NCBI-BLAST protein database search program.

4. Notes

1. Heptane-denaturated ethanol is cheaper than non-denaturated ethanol. Highly volatile heptane does not interfere with any purification step. Instead of ethanol, acetonitrile can also be used. In the absence of organic solvents we were confronted with severe problems with the extracts, which gave turbid solutions after centrifugation

with a high content of fines (lipid drops mixed with solid particles). Ignoring this phenomenon has usually resulted in HPLC problems, such as high-pressure error and giving "memory compounds" on the HPLC column (7).

Protease inhibitor cocktails have not been added to the stratum corneum prior to extraction. We never observed proteolytic digests of the AMPs which we had isolated so far from stratum corneum. This includes hBD-2, hBD-3, RNase-7 (4–6), and psoriasin (8). In particular RNase-7 and psoriasin are extremely sensitive towards proteolysis. We therefore speculate that in skin endogenous protease inhibitors prevent proteolysis of AMPs.

2. The recovery of AMPs using Amicon (Danvess, MA) filters was found to be highest when acidic solutions containing a small percentage (20–30%) of water-soluble organic solvents were used. Use the right diameter of the filter to avoid losses of AMPs due to an un-appropriate ratio of filter diameter and (final) sample volume. Small extract volumes (100 μL–1 mL) should be better either lyophilized or concentrated by vacuum evaporation or directly applied to the RP-HPLC column for separation. Losses of material seen in some cases usually come from sticking to the surfaces rather than degradation. Once cationic AMPs sticked to surfaces (glass, siliconized glass, plastic), we have been unable to reverse the process by washing with organic solvents. Therefore, our strategy to avoid sticking is to add organic solvents whenever possible at acidic pH to AMP-containing solutions.

3. It is our experience that chromatography at low temperature is not necessary for AMP purification. Skin-derived AMPs are remarkably stable at ambient temperature.

4. UV detection at 215 nm allows quantification of protein content and thus estimation of the AMP amounts in purified peaks. Although we used ubiquitin for calibration, the absolute amounts of known AMPs can be estimated from its calculated extinction coefficient.

5. For extracts from >10 g stratum corneum a 5 mL heparin–sepharose cartridge is recommended.

6. A wide variety of reverse-phase columns obtained from different manufacturers can be used. Take care to use wide pore (300 Å) columns with endcapping. Most of the AMPs are very cationic. When there is no endcapping (which blocks the free silanol groups of the silica bead material), the highly cationic AMPs will show broadened peaks and often dramatic losses of recovery are observed. Choose RP-HPLC columns which have been optimized for the separation of basic compounds.

7. Do not use oversized cation exchange columns! For extracts obtained from >1 g stratum corneum we use the micro-MonoS®-HPLC column. For smaller amounts we are using a MiniS®-HPLC column.
8. C2/C18-μHPLC columns give excellent resolution for several AMPs. When only little stratum corneum material is available and a single HPLC analysis is possible, a (micro)column with this material is recommended. In our hands classical wide pore C8-RP columns do not always allow complete separation of lysozyme from hBD-2. The use of a μC2/C18 column leads to a retention time difference of about 10 min!
9. Addition of urea is essential for several AMPs to get bands at the expected size in the gel. For example, hBD-2 shows in "Phast®-System-high density gels", which lack urea, a 14 kDa band, just below that of lysozyme, whereas in Tricine gels with urea a band at 4 kDa can be observed.
10. With this method we obtained highest resolution of bands. We were able to separate the different 77, 72, and 69 residues containing forms of interleukin-8 (9).
11. Fixation can be a problem for low MW AMPs. The use of glutaraldehyde is compelling for detection of low amounts (< 10 ng per lane) of AMPs.
12. When highly sensitive detection is necessary, the gels can be destained with $K_3[Fe(CN)_6]$ until background is cleared. Bands should never be completely destained! After rinsing the destaining solution, staining can be repeated as described. The band intensity does not reflect the relative amounts of the applied proteins.
13. Nanospray modus is recommended when very low peptide amounts are available. RP-HPLC fractions should be lyophilized and the (invisible) residue dissolved in 5 μL ESI-MS solvent (*see* **Section 3.7**, Item 1). It is essential to completely evaporate the HPLC solvent, because otherwise remaining TFA will affect sensitivity. Take care to avoid the presence of detergents, which make analyses impossible.
14. This assay is highly sensitive and needs only minimal amounts of the preparation being tested. With appropriate solvent controls it is possible to test micro-HPLC fractions without any lyophilization. In some cases it could be necessary to protect HPLC fractions from losses by irreversible adherence to the plate. When this is a problem, add 5 μL 0.1% (w/v) bovine serum albumin in 0.01% aqueous acetic acid to each well of the microtitre plate prior to adding HPLC fractions.

15. Vigorous mixing of bacterial suspensions (especially for *Pseudomonas aeruginosa* and after centrifugation) is important to disperse any clumps and provide an even inoculum suspension (1).
16. It is our experience that ethylacetate treatment of stratum corneum enhances recovery of AMPs, possibly by dissolving non-polar lipids and therefore increasing access of aqueous solutions to stratum corneum particles, which contain high amounts of lipids. Treatment with acetone is not recommended because most of the AMPs are soluble in acetone and thus would be lost. Alternatively diethylether could be used, but this requires extreme care!
17. Instead of ethanol, acetonitrile can also be used. In the absence of organic solvents, we often have had problems with the extracts giving turbid solutions after centrifugation with a high content of fines (lipid drops mixed with solid particles). Ignoring this phenomenon has usually resulted in HPLC problems, such as high-pressure error and giving "memory compounds" on the HPLC column.
18. The recovery of AMPs using Amicon (Danvess, MA) filters was found to be highest when acidic (pH < 3) solutions containing a small percentage (20–30%) of water-soluble organic solvents were used. Use the right diameter of the filter to avoid losses due to inappropriate ratio of filter diameter and (final) sample volume.
19. pH adjustment to pH 8 has always been done after thawing of frozen acidic samples. It is our experience that AMP recovery from samples which have been stored at pH 8 were always lower than those stored at acidic pH. After adjustment of extracts to an acidic pH often very fine precipitates occur. It is recommended to centrifuge these concentrated extracts again to avoid problems with HPLC analyses.
20. The size of the heparin affinity cartridge should be appropriate to the amount of extracted material. For extracts obtained from stratum corneum amounts < 1 g we always used a 1 mL cartridge. A 5 mL cartridge was used for higher amounts of extracts. It could be important to perform a second heparin affinity chromatography to see whether the whole heparin-binding material was indeed bound in the first chromatography step. Initially we had diafiltered the extracts to omit low MW compounds. Today we directly apply the extracts onto the heparin column. If extracts are directly (without diafiltration) applied, be aware that also low MW cationic compounds (which could represent antibiotic substances) will bind to the heparin

column! If your antimicrobially active HPLC fractions do not show strong silver stained bands and no masses > 2 kDa upon ESI-MS, look for the raw data (single charged species) of the ESI-MS analyses to identify possibly a low MW antibiotic compound!

21. It may occur that not all AMPs are efficiently stripped from the column with 2 M NaCl. We have observed that most of the hBD-3 and parts of HNP-1 sticked to the heparin column. hBD-3 could be eluted with acidic buffers like 0.1 M glycine/HCl buffer, pH 2 and HNP-1 with 0.1 M NaOH. Be aware that a repeated use of these eluents may destroy AMP-binding capacities of the heparin columns when often re-used!
22. Instead of concentrating the samples to a very low volume we used injection loops with the capacity for several millilitre sample volume. Broadening of peaks was never found to be a problem, except when the samples contained organic solvents >10%. When very polar AMPs are expected, it is recommended to lyophilize or "SpeedVac" the (acidic) sample and dissolve the residue in a small volume of water.
23. Detection at 215 nm is recommended, because some of the AMPs show no or a very low absorbance at 280 nm. When possible, absorbance at three wavelengths (215, 254, 280 nm) should be measured because the absorbance ratio is very characteristic for each peptide. This UV absorbance profile together with the retention time will often give sufficient initial information about the AMP that eluted in given HPLC fractions.
24. We use manual separation of HPLC fractions. Often AMPs do not represent the major UV-absorbing peaks, but instead shoulders within the peak. These can be easily separated manually and then analysed for its specific AMP activity. If this is higher in the shoulder than in the major UV-absorbing peak, it indicates that most of the contaminating protein has already been separated. It is important for a successful purification of AMPs to separate as early as possible contaminating non-active impurities.
25. We take, just after the RP-HPLC step, aliquots of each HPLC fraction and place it into round-bottomed microtitre plates (for each assay system one plate to avoid repeated freezing and thawing), which then will be stored frozen (–20°C). Immediately before biological testing will be performed, samples are lyophilized and then dissolved in appropriate solvents.

26. We use unusual conditions for cation exchange-HPLC for AMP separation! These conditions were found to be optimal for separation of various skin proteins (4–8) and allowed us good recoveries of highly cationic AMPs. It is our experience that textbook conditions (neutral pH, anionic buffers) gave lower AMP resolution (broad peaks!) and lower recoveries. Highly cationic AMPs, which are eluting only with high NaCl content, often cannot be detected in antibacterial assay systems due to AMP's salt sensitivity. Therefore, it has been sometimes our strategy to separate all, at high NaCl eluting proteins by subsequent (C2/C18)-RP-HPLC and to test RP-HPLC fractions for AMP activity.
27. Instead of acetonitrile also 30% (v/v) ethanol (denaturated with heptane) can be used. Addition of organic solvents to the elution buffer is essential for high AMP recovery. It is believed that amphipathic AMPs otherwise stick to the cation-exchanger matrix.
28. Fractions need not to be concentrated to decrease the volume, when an injection loop of appropriate size is available. It is recommended to add 1 μL TFA/1 mL fraction.
29. Depending on the peptide, sometimes a completely clear zone, a clear zone with bacterial growth within the clearing zone or a complete clear zone, which is surrounded by a concentric zone of partial clearing owing to a reduction in microcolony density, was observed. We found the different patterns as AMP-specific parameters, which also depended on the concentration of the AMP as well as the assay condition, including the test organism.
30. We always performed SDS-PAGE analyses in the absence of reducing agents, because we never saw better band resolution in its presence. In contrast, some AMPs showed broader bands in the presence of dithiothreitol or mercaptoethanol.
31. Urea is necessary to reduce or prevent formation of dimers and oligomeric peptides. HBD-2 (MW ~ 4 kDa) migrates in the absence of urea like a 16 kDa protein. HBD-3 (MW ~ 5 kDa) still migrates in the presence of urea like a 9 kDa peptide.
32. Note a different band shape (even, convex, or concave) and band staining colours of different AMPs, e.g. defensins always showed a grey colour, whereas psoriasin and lysozyme were stained red-brown.
33. Take care that samples do not contain too much TFA, which would cause low signal intensity. It may be helpful to lyophilize (or "SpeedVac") the sample prior to

ESI-MS analyses and dissolve it in ESI-MS solvent. Samples originating from MonoS®-HPLC cannot be measured directly upon ESI-MS (due to the salt content). Intensive signals of anorganic salt clusters do not allow detection of far less intensive protein signals. NaCl-containing cation exchange-HPLC fractions therefore require a desalting process, otherwise no protein signals are visible. Alternatively, cation exchange-HPLC can be performed with volatile buffers (e.g. 1 M ammonium formate) as eluents. When adducts of ammonium formate adducts are still observed, we recommend to repeatedly add water to the samples and repeatedly lyophilize them until no crystalline residue (from salt) is visible.

34. Nano-ESI-MS is recommended when only low AMP amounts are expected. Because for nanospray-ESI-MS analyses only a few microlitres are required, due to the high TFA content it is essential to lyophilize the HPLC fractions prior to analyses! Dissolve the residue in sample solvent.
35. Peptide digests should be passed through a self-prepared μ-column and then eluted with acetonitrile. This procedure gives far less background and therefore much better results than the direct application of digests due to enrichment as well as elimination of contaminants.

Acknowledgments

This work was supported by Deutsche Forschungsgemeinschaft, SFB 617.

References

1. Steinberg, D. A. and Lehrer, R. I. (1997) Designer assays for antimicrobial peptides. Disputing the “one-size-fits-all” theory. *Methods Mol. Biol.* **78**(1), 69–186.
2. Ganz, T., Metcalf, J. A., Gallin, J. I., Boxer, L. A., and Lehrer, R. I. (1988) Microbicidal/cytotoxic proteins of neutrophils are deficient in two disorders: Chediak-Higashi syndrome and “specific” granule deficiency. *J Clin. Invest.* **82**, 552–556.
3. Schägger, H. and von Jagow, G. (1987) Tricine-sodium dodecyl sulfate-polyacrylamide gel electrophoresis for the separation of proteins in the range from 1 to 100 kDa. *Anal. Biochem.* **166**, 368–379.
4. Harder, J., Bartels, J., Christophers, E., and Schröder, J. M. (1997) A peptide antibiotic from human skin. *Nature* **387**, 861.
5. Harder, J., Bartels, J., Christophers, E., and Schröder, J. M. (2001) Isolation and characterization of human beta -defensin-3, a novel human inducible peptide antibiotic. *J. Biol. Chem.* **276**, 5707–5713.
6. Harder, J. and Schröder, J. M. (2002) RNase 7, a novel innate immune defense antimicrobial protein of healthy human skin. *J. Biol. Chem.* **277**, 46779–46784.

7. Schröder, J. M. (2000) Purification of chemokines from natural sources. *Methods Mol. Biol.* **138**, 1–10.
8. Gläser, R., Harder, J., Lange, H., Bartels, J., Christophers, E., and Schröder, J. M. (2005) Antimicrobial psoriasin (S100A7) protects human skin from *Escherichia coli* infection. *Nat. Immunol.* **6**, 57–64.
9. Schröder, J. M., Mrowietz, U., Morita, E., and Christophers, E. (1987) Purification and partial biochemical characterization of a human monocyte-derived, neutrophil-activating peptide that lacks interleukin 1 activity. *J. Immunol.* **139**, 3474–3483.

Chapter 3

Strategies for the Isolation and Characterization of Antibacterial Lantibiotics

Daniela Jabes and Stefano Donadio

Abstract

Lantibiotics are biologically active peptides produced by several strains from the phyla *Firmicutes* and *Actinobacteria*. They are ribosomally synthesized and undergo posttranslational modifications that endow them with the characteristic (methyl)-lanthionine residues. As a result, lantibiotics contain a variable number of rings, each carrying one thioether link. Many lantibiotics inhibit growth of Gram-positive bacterial strains by interfering with peptidoglycan formation. Because they bind to the key intermediate lipid II at a site not affected by clinically used antibiotics, they are effective against multidrug-resistant strains. We describe a bioassay-based method suitable for finding antibacterial lantibiotics from actinomycete strains and provide selected procedures for characterizing newly discovered lantibiotics for their antibacterial properties.

Key words: Actinomycetes, antibiotics, bioactive peptides, lantibiotics, MRSA, screening.

1. Introduction

1.1. Definition of Lantibiotics

The term lantibiotics refers to a family of ribosomally synthesized, posttranslationally modified peptides, which are characterized by the presence of (methyl)-lanthionine residues (1, 2). These amino acids (**Fig. 3.1**) result from the formation of thioether bridges between cysteine thiols and dehydroalanine or dehydrobutyrine residues, which in turn derive from dehydration of serine and threonine, respectively. These thioether bridges impart conformational constraints to these molecules generating a series of rings (usually numbered in alphabetical order), which are essential for biological activity. In addition to these features,

A. Giuliani, A.C. Rinaldi (eds.), *Antimicrobial Peptides*, Methods in Molecular Biology 618,
DOI 10.1007/978-1-60761-594-1_3, © Springer Science+Business Media, LLC 2010

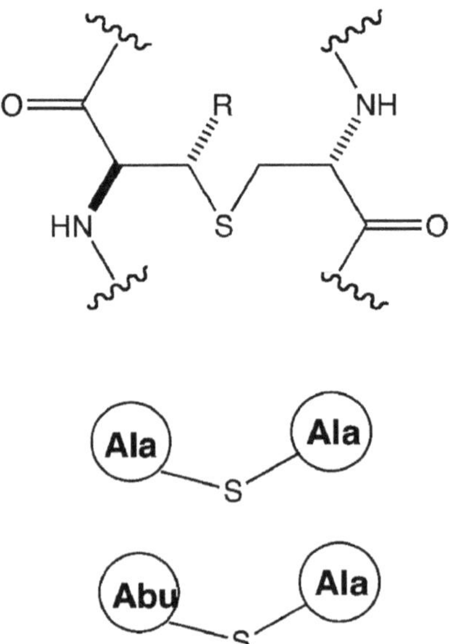

Fig. 3.1. Structure of lanthionine (X=H) and of methyl-lanthionine (X=CH_3) and their schematic representations (*bottom*) in lantibiotics.

lantibiotics may contain other posttranslational modifications (1, 2). Within the lantibiotics discovered so far, modifications may include 2,3-dehydroalanine (Dha) and (*Z*)-2,3-dehydrobutyrine (Dhb), derived from serine and threonine residues, respectively, not recruited in thioether bridges; a C-terminal *S*-aminovinyl-D-cysteine (or its 3-methyl analog), derived from decarboxylation and cross-linkage of the C-terminal cysteine residue; N-terminal oxopropionyl and 2-oxobutyryl, resulting from spontaneous dehydration of N-terminal Dha and Dhb residues; a lysinoalanine bridge; hydroxylated aspartic acid or proline residues; or a chlorinated tryptophan residue.

Lantibiotics are produced by Gram-positive bacteria and have been reported so far only from the phyla *Firmicutes* and *Actinobacteria*. It should be noted that most lantibiotics were identified using bioactivity-based screens for antibacterial activity. The recent discovery that lantibiotics participate in morphological development in *Streptomyces* spp. (2) suggests that additional biological activities may be uncovered within this class. Similarly, inspection of the increasing number of microbial genomes for genes typical of lantibiotic biosynthesis may lead to the identification of additional lantibiotics in a bioassay-independent manner.

1.2. Biological Activity of Lantibiotics

Although the great majority of known lantibiotics have antimicrobial activity, this might be a consequence of the assays used to discover them. Indeed, some compounds display other bioactivities, including potentially interesting pharmacological properties. For example, cinnamycin (**Fig. 3.2**) and the related duramycins are

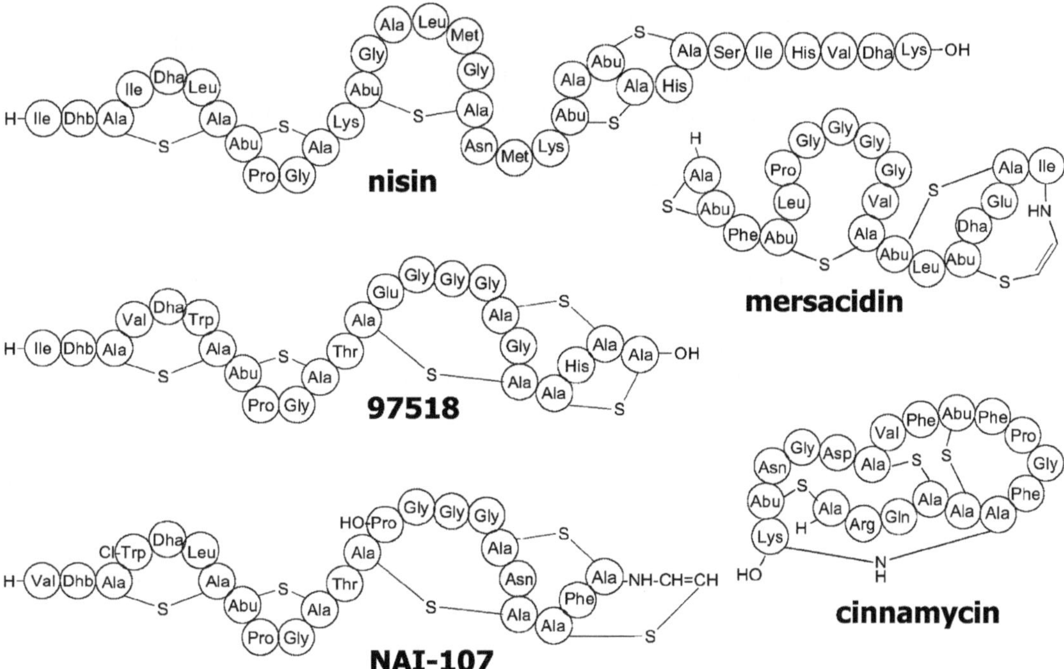

Fig. 3.2. Some lantibiotics mentioned in the text.

potent inhibitors of phospholipase A2 (3). Moreover, duramycin has been shown to increase chloride secretion in lung epithelium (4) and has been in clinical trials to evaluate its efficacy in clearing mucus secretions associated with cystic fibrosis and other airway diseases (5). Another example is cytolysin that, in addition to having antimicrobial activity, functions as a virulence factor, lysing erythrocytes and polymorphonuclear leukocytes (6). Finally, the peptides SapB and SapT, produced by the filamentous soil microbes *Streptomyces coelicolor* and *S. tendae*, respectively (7, 8), are hydrophobic, surface-active compounds believed to function as biosurfactants, releasing the surface tension at the colony–air interface to facilitate the emergence of nascent aerial hyphae (9).

1.3. Mode of Action of Antibacterial Lantibiotics

Traditionally, lantibiotics possessing antibacterial activity have been classified into two subgroups: type-A lantibiotics, exemplified by nisin (**Fig. 3.2**), are elongated, amphiphilic peptides; while type-B lantibiotics, exemplified by mersacidin (**Fig. 3.2**), are compact and globular (10). More recently, a biogenetic classification has been proposed for antibacterial lantibiotics, which places them in either class I or class II on the basis of the pathway by which maturation of the peptide occurs (2, 11). [A class III has also been proposed for the morphogenic lantibiotics, which probably require a different pathway for peptide maturation (2).]

This designation is less dependent on the structure of the mature lantibiotic and can accommodate, for example, globular-type lantibiotics such as NAI-107 (also known as 107891) and 97518 (**Fig. 3.2**), which share nonetheless the same pharmacophore as nisin (12, 13).

Despite differences in shape and primary structure, both nisin- and mersacidin-type lantibiotics interact with the membrane-bound peptidoglycan precursor lipid II. However, binding to this essential intermediate in peptidoglycan biosynthesis involves different portions in the two classes of lantibiotics, and further events may occur after it. For example, it has been demonstrated that the two N-terminal rings of nisin dock on lipid II, followed by oligomerization of the resulting complex after membrane insertion of the C-terminal portion of the lantibiotic. In contrast, globular lantibiotics such as mersacidin are incapable of forming pores.

1.4. Biosynthesis of Lantibiotics

Lantibiotics are synthesized ribosomally from a pre-propeptide, often designated with the generic name LanA, which consists of a leader peptide preceding the structural segment (the propeptide). The propeptide undergoes the modifications required for lantibiotic maturation, while the leader peptide acts as a recognition sequence for the lantibiotic processing enzymes and is eventually removed from the mature lantibiotic (1, 2). Serine and threonine residues in the propeptide are dehydrated to Dha and Dhb, respectively. Subsequent intramolecular Michael-type addition by cysteine thiols to the unsaturated side chains yields the corresponding thioethers (**Fig. 3.3**). In class I lantibiotics, the dehydration and cyclization steps are catalyzed by two distinct enzymes, designated LanB and LanC, respectively, whereas a bifunctional LanM enzyme is responsible for both reactions in class II lantibiotics. Further posttranslational modifications may also occur during or after thioether formation. Eventually, the leader peptide is removed to produce the mature lantibiotic. For most lantibiotics, the ABC-type transporter LanT transports the modified peptides out of the cytoplasm.

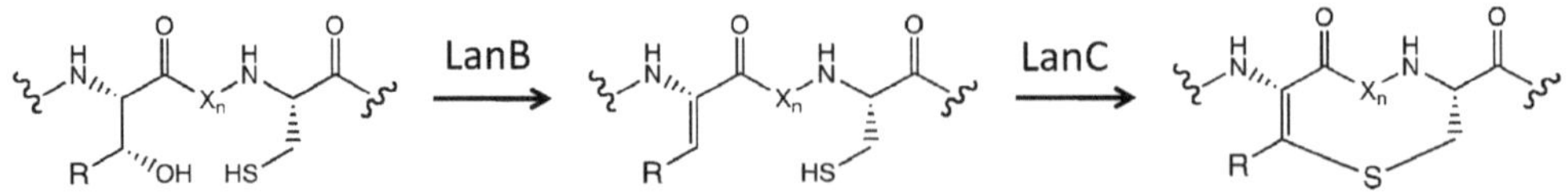

LanB LanC

Fig. 3.3. Formation of lanthionines.

1.5. Scope

This chapter describes detailed procedures for the discovery of antibacterial lantibiotics and for the evaluation of their potential as antibiotics. Recent advances in the in vitro and in vivo manipulation of lantibiotic biosynthesis provide effective tools for generating variants of existing lantibiotic classes. The reader is referred to recent publications on how to employ lantibiotic biosynthesis enzymes for generating new variants (14) or on the genetic manipulation of lantibiotic producer strains (15).

2. Materials

2.1. Bacterial Strains

1. The microbial strains used in this work derive from the NAICONS strain collection. For the screening test to detect lantibiotics, the strains used were *S. aureus* ATCC 6538P and its L-form, prepared as described (16). The other strains described here and their relevant phenotypes are reported in **Tables 3.1** and **3.2**. However, strains equivalent to those described here can be obtained from public collections.

Table 3.1
An initial panel of bacterial strains for evaluating lantibiotics

Medium	Strain
MHB	*Staphylococcus aureus* MSSA 6538P
MHB	*Staphylococcus aureus* Met^R 1400
THB	*Streptococcus pyogenes*
THB	*Streptococcus pneumoniae* P203
MHB	*Streptococcus haemolyticus*
MHB	*Listeria monocytogenes* ATCC 4428
MHB	*Enterococcus faecium* Van^S
MHB	*Enterococcus faecium* VanA
MHB	*Enterococcus faecalis* Van^S
MHB	*Enterococcus faecalis* VanA
MHB	*Escherichia coli* K12
7H9	*Mycobacterium smegmatis* mc^G 155

Frozen stocks are prepared by collecting mid-exponential cultures in the appropriate medium, adding glycerol to 20% (v/v), and dispensing the cultures as small aliquots in different vials. After freezing, one vial is thawed and used for the determination of colony-forming units (CFU). Strains can be maintained at –80°C for prolonged periods of time.

2.2. Media

The media described here were all from Difco, unless otherwise specified, and were prepared according to the manufacturer's instructions:

Table 3.2
Suggested strains, media, and culture conditions for further characterization of lantibiotics

Strain	Culture medium	Incubation
Neisseria meningitidis ATCC 13090 ATCC	Supplemented GCB	35°C, 24 h, 5% CO_2
Haemophilus influenzae 19418	HTM	35°C, 24 h
Moraxella catarrhalis ATCC 8176	MHB	35°C, 24 h
Corynebacterium diphtheriae	MHB	35°C, 24 h
Propionibacterium acnes	BB	35°C, 24 h, anaerobic atmosphere
Clostridium difficile	Oxyrase-supplemented WCB	35°C, 24 h, anaerobic atmosphere
Lactobacillus spp.	LB	35°C, 24 h
Bacillus subtilis	AM 3	35°C, 24 h

1. Brain Heart Infusion (BHI).
2. Cation-adjusted Mueller Hinton Broth (MHB), with $CaCl_2$ and $MgCl_2$ to final concentrations of 20 and 10 mg/L, respectively.
3. Mueller Hinton Agar (MHA).
4. Todd Hewitt Broth (THB).
5. MHSA is MHB containing 0.7% agar.
6. Middlebrook 7H9 Broth (7H9), supplemented with 10% oleic acid–albumin–dextrose catalase (Beckton Dickinson) and 0.2% glycerol.
7. Gonococcus-Supplemented Broth (GCB), supplemented with 1% isovitalex (Beckton Dickinson) and 5 mg/L hemin solution (Sigma).
8. Haemophilus Test Medium (HTM); Brucella Broth (BB) supplemented with 5 mg/L hemin, 1 mg/L vitamin K_1, and 5% lysed horse blood (Oxoid).
9. Wilkins–Chalgren Broth (WCB) supplemented with 1:25 v/v oxyrase (Oxyrase Inc.).
10. Lactobacillus Broth (LB).
11. Antibiotic Medium 3 (AM3).

2.3. Other Chemicals

1. All antibiotics used for determination of minimal inhibitory concentration (MIC) are dissolved in DMSO to obtain

stock solutions of 10 mg/mL. Then, they are appropriately diluted with the specific growth medium to obtain working solutions, from which serial 1:2 dilutions are performed.

2. Bovine serum albumin (BSA) for well coating is prepared as a 0.02% solution in distilled water, filter sterilized, and kept at 4°C for up to 1 week.
3. Human serum (Sigma Chemical) is filter sterilized and kept at 4°C for up to 1 week.
4. The β-lactamase mixture consists of 0.001 and 0.002 U/mL type-1 and type-2 penicillinases, respectively, from *Bacillus cereus*; and 0.0025 and 0.5 U/mL type-3 and type-4 penicillinases from *Enterobacter cloacae*. All enzymes were purchased from Sigma.

3. Methods

3.1. Identification of Antibacterial Lantibiotics

Genomic-based approaches for discovering lantibiotic gene clusters have been described and will not be mentioned here (14). We will focus on the use of a bioassay that, according to our experience, resulted in a high frequency of lantibiotics among fermentation broth extracts derived from different actinomycete genera. Indeed, when this screening program was applied to a library of ca. 120,000 extracts, lantibiotics represented over 60% of observed hits. It remains to be determined whether this high percentage resulted from biases in the type of strains (e.g., a relatively low percentage of *Streptomyces* spp. and a high incidence of *Actinoplanes* spp.) or in the over-representation of samples derived from just a few lantibiotic-producing strains. In any case, others have observed the relative abundance of lantibiotics among actinomycetes (17). The following procedure describes a series of four tests in a bioassay-based screening for discovering cell-wall-acting lantibiotics from microbial fermentation extracts. Extracts can be prepared by different methods, provided that care is taken to ensure sufficient diversity of the strains from which fermentation extracts are derived (18).

1. Dispense 10 μL aliquots of samples to be tested in each well of a 96-well microtiter plate for a total of 80 samples, leaving the remaining wells for positive and negative controls (*see* **Note 1**).
2. Prepare the inoculum by dispensing an appropriate amount of the frozen *S. aureus* ATCC 6538P stock into BHI medium to obtain a final titer of approximately 10^5 CFU/mL.
3. Add 90 μL of the bacterial inoculum to each well of the microtiter plate from Step 1 and incubate for 24 h at 35°C.

4. Measure the turbidity of each well at 590 nm in a suitable multiplate reader (e.g., Tecan SpectraFluor) and express results as

$$I_i = 100 \times [1 - (T_i - T_c)/(T_p - T_c)],$$

where I_i is the percent growth inhibition of sample *i*, with T_i representing the absorbance values of sample *i*, and T_c and T_p the averages of negative and positive controls, respectively.
5. Identify those samples showing an I_i value ≥ 90 in Step 4 (*see* **Note 2**).
6. Prepare two identical plates, each containing 10 μL aliquots of 10 different positives in the upper row, and serial 1:2 dilutions of each sample in the remaining rows.
7. Use one plate to repeat Steps 2–4, determining for each sample and its dilutions the corresponding I_i.
8. Prepare the L-form inoculum, by diluting 1:2 an overnight culture in BHI with fresh medium, and add 90 μL per well to the second plate from Step 6. Use positive and negative controls in the remaining wells (*see* **Note 3**). Incubate at 35°C for 24 h.
9. For each sample and its dilutions from Step 8, measure absorbance as in Step 4, expressing results as

$$G_i = 100 \times [1 - (A_i - A_c)/(A_p - A_c)],$$

where G_i is the percent growth inhibition of sample *i*, with A_i representing the absorbance values of sample *i*, and A_c and A_p the averages of negative and positive controls, respectively.
10. Determine for each sample D_I and D_G, i.e., the highest dilutions giving I and G values ≥ 80 and ≤ 40, respectively (*see* **Note 4**).
11. Identify those samples showing an I value ≥ 80 and a D_I/D_G ratio ≥ 8 in Step 10 (*see* **Note 5**).
12. Prepare two new sets of identical plates, each containing 10 μL aliquots of different positives.
13. To one plate from Step 12, add 10 μL of the mixture of penicillinases and incubate for 1 h at 37°C, before adding the *S. aureus* inoculum and incubating as in Step 3. Record I_i values as in Step 4.

14. To the other plate from Step 12, add 10 μL of ε-amino-caproyl-D-alanyl-D-alanine, before adding the *S. aureus* inoculum and incubating as in Step 3. Record I_i values as in Step 4.

15. Identify as hits those samples retaining I_i values ≥80 in both Steps 13 and 14 (*see* **Note 4**).

The above procedure actually consists of a primary assay (inhibition of *S. aureus* growth), followed by three secondary assays (lack of inhibition of L-form growth; persistence of *S. aureus* inhibition after treating the sample with penicillinases; and persistence of *S. aureus* inhibition in the presence of an excess of ε-amino-caproyl-D-alanyl-D-alanine). This screening algorithm selects for cell-wall-acting antibiotics, discarding two of the most frequent inhibitor classes found in microbial fermentation extracts (β-lactams and glycopeptides), empirically providing a substantial enrichment in cell-wall-acting lantibiotics. However, as any screening algorithm, it is not going to be infallible and a number of false positives may go through the screening. These can be best discarded by chemical means, identifying characteristic properties during further processing of the hits.

It would be impossible to provide detailed protocols for the identification and characterization of newly discovered lantibiotics, since procedures would have to be adjusted to the producer strains and to the properties of the compound(s) under investigation. Instead, we provide some useful tips on the recognition of lantibiotics in fermentation broth extracts and in the subsequent structural elucidation of interesting compounds.

Since lantibiotics are relatively high molecular weight compounds, they can be identified by their UV absorption and mass spectrometry properties. Usually, an HPLC fractionation (18) allows the identification of fractions containing the bioactive compounds, to which UV and MS data can be assigned. Most lantibiotics have a molecular mass above 2,000 and display a typical peptide UV spectrum. The presence of lanthionine residues and dehydroamino acids can be revealed by selective modification (19). Structural elucidation is best performed by employing a combination of MS and NMR spectroscopy, possibly assisted by selective chemical derivatization and genomic-derived sequence of the *lanA* gene. As the size of the lantibiotic and the number of thioether rings increase, a combination of techniques should be employed to avoid misinterpretation of the data derived from a single analysis. A recent example demonstrates a substantial revision of a lantibiotic structure reported just 2 years before (13).

3.2. Evaluation of Antibacterial Lantibiotics

The scope of this section is to provide methods for evaluating antibacterial lantibiotics useful for treating bacterial infections in humans. Once a new lantibiotic has been selected for further evaluation, its activity against a panel of representative microorganisms should be determined in standard and ad hoc experimental conditions. The initial selection of test microorganisms should be based on literature data documenting the antibacterial spectrum of different lantibiotics. Numerous studies have shown that lantibiotics have low MICs against many clinically relevant Gram-positive bacterial species, belonging to the genera *Enterococcus*, *Listeria*, *Staphylococcus*, *Streptococcus*, *Bacillus*, and others (20, 21). Although lantibiotics share a somewhat similar profile, not all compounds possess the same potency and spectrum. For example, *Listeria monocytogenes* is relatively insensitive to nisin (MICs between 200 and 1,000 mg/L), while highly susceptible to mersacidin and NAI-107 (12). Furthermore, because Gram-positive bacteria do produce lantibiotics, some strains may produce a lantibiotic related to the one under study, and thus be naturally resistant to the compound under investigation.

Lantibiotics such as mersacidin and actagardine display considerable activity against *S. aureus*, including methicillin-resistant strains (MRSA). While the glycopeptide antibiotic vancomycin can be used to treat MRSA and enterococcal infections, vancomycin-insensitive and vancomycin-resistant strains have emerged. Independent of the mechanism of vancomycin resistance (e.g., redirection of cell-wall biosynthesis as in enterococci or a thickened cell wall as in staphylococci), vancomycin-resistant strains remain sensitive to mersacidin (22–24) and to other lantibiotics, such as NAI-107 (unpublished data), indicating that cross-resistance does not represent an immediate problem.

Several lantibiotics, including nisin and NAI-107, have excellent activity against antibiotic-resistant streptococci. In particular, both compounds are very active against clinical isolates of *Streptococcus pneumoniae*, including penicillin-resistant strains (21). Some lantibiotics have shown activity against *Mycobacterium tuberculosis* (25), an aspect that deserves further evaluation.

Since the actual antibacterial spectrum of a newly discovered lantibiotic is unknown, strains representative of the most clinically relevant pathogens should be included in the first panel of microorganisms to be used for susceptibility assays. An example of an initial microbial panel is provided in **Table 3.1**.

MIC determinations are performed by broth microdilution assays in sterile, polystyrene round-bottom 96-well microtiter plates, according to CLSI procedures (26, 27), as detailed below:

1. Use a different 96-well microtiter plate (*see* **Note 6**) for each strain. Dispense 10 μL of serial 1:2 dilutions of compounds to be tested in each row of the microtiter plate, using the top row for the growth control (no compound added) and the bottom row for a reference antibiotic.

2. Prepare the inoculum by dispensing an appropriate amount of the frozen stock into the appropriate medium of **Table 3.1** (*see* **Note** 7) to obtain a final titer of 1–5 × 10^6 CFU/mL for *Mycobacterium smegmatis* and 1–5 × 10^5 CFU/mL for the other strains (*see* **Note 8**).
3. Add 90 μL of the bacterial inoculum to each well of the microtiter plate from Step 1 and incubate plates at 37°C for 72 h (*M. smegmatis*) or for 20–24 h (all other strains).
4. For each compound and each strain, the MIC is defined as the lowest concentration causing complete suppression of visible bacterial growth.

Further characterization of new lantibiotics should include some Gram-negative pathogens and anaerobic Gram-positive bacteria. A list of suggested microorganisms and relative culture conditions is reported in **Table 3.2**. To improve the outcome of the in vitro profiling of new lantibiotics, we recommend albumin precoating of plastic wells, measuring activity at different pHs (*see* **Note** 7) and using cation-adjusted media (*see* **Note 8**). The addition of human or rat serum may cause a reduction in the activity of some lantibiotics, presumably due to significant binding to serum protein(s). However, this is not a general phenomenon, and odd behaviors can also be observed; for example, serum proteins reduce the activity of mutacin against *S. pneumoniae*, but increase its activity against *S. aureus*. Sera from different sources should be added at a 10–30% concentration.

In general, lantibiotics are bactericidal against susceptible organisms, but their bactericidal activity might be limited. It is therefore recommended to evaluate the bactericidal activity of new lantibiotics by time-kill experiments using different strains, such as *S. aureus* (including MRSA), *S. pneumoniae*, *Enterococcus faecalis*, or *Enterococcus faecium*. The following method for time-kill experiments is recommended by CLSI (28):

1. Pick several colonies of the desired strain from an MHA plate incubated overnight at 37°C and disperse them into MHB medium.
2. Incubate cultures under shaking conditions at 37°C until an OD_{600} of 0.2 is reached.
3. Dilute cultures to 1–5 × 10^6 CFU/mL (approximately 1:10 of the culture obtained in Step 2) in fresh MHB medium, distribute them into flasks containing the compound at different concentrations (*see* **Note 9**), and incubate flasks under shaking conditions at 37°C (*see* **Note 10**).
4. Withdraw duplicate 0.1 mL aliquots at various time intervals for up to 24 h, including also a time 0 control.
5. Prepare serial 1:10 dilutions of each aliquot in MHB medium, mix with 4 mL MHSA medium, and spread onto MHA plates.

6. Score colonies after 24 h at 35°C.
7. Calculate the CFU in the original culture from the different dilutions, average out the results from the duplicate aliquots, and plot for each antibiotic concentration the observed CFU/mL as a function of time.

Once the in vitro antimicrobial profile and the bactericidal activity of the selected lantibiotic have been defined, it is important to confirm that the compound is equally active on a larger number of strains, including fresh clinical isolates. At least 10 different isolates should be used for each species to be investigated.

4. Conclusions

The examples described above provide detailed methods for discovering lantibiotics from actinomycetes and other bacterial strains and for characterizing the antibacterial properties of newly discovered lantibiotics. We have also tried to provide general considerations to guide the reader in understanding the procedures and the context in which they are executed. Additional characterizations will depend on the properties of the lantibiotic under investigation. Furthermore, as our understanding of this poorly explored class expands, additional bioactivities may be discovered and new tools may become available.

5. Notes

1. Positive and negative controls are represented by 10% DMSO and 0.1 μg/mL penicillin G, respectively. Any other cell-wall-active antibiotic can be used in place of penicillin G, as long as it does not interfere with growth of L-forms (*see* also **Note 3**).
2. The >90% value as a positive threshold is a suggested value based on the authors' experience. Because assay variability is strongly dependent on the equipment used, it is recommended to run the assay system through a pilot study to establish statistically relevant thresholds for accepting or discarding a sample after each test. Refer to specialized literature on the topic (29).
3. Positive and negative controls are represented by 0.1 μg/mL penicillin G and 5 μg/mL chloramphenicol, respectively. Any other antibiotic active on L-forms can be used in place of chloramphenicol.

4. The ≥80% value for reconfirmation of a positive and the ≤40% value for accepting a sample because it does not affect growth of the L-form are again suggestions based on our experience. *See* also **Note 2**.
5. With microbial fermentation extracts, the probability that a sample retains a signal above threshold decreases with its dilution, so in many cases the measured D_I/D_G ratio may only be ≥2. These samples may be retained for further analysis, but they may not confirm their lack of activity against L-forms when more concentrated preparations become available.
6. Since lantibiotics may bind to plastic surfaces, it is recommended to pre-coat the wells with 0.02% BSA. To this end, 100 μL are distributed in each well and, after 30 min at room temperature, excess liquid is removed and the microtiter plates are briefly dried in a sterile hood.
7. The stability of different lantibiotics to pH may vary. MICs against selected organisms (preferably *S. aureus*, *S. pneumoniae*, *L. monocytogenes*) should be determined at different pHs, by adjusting the pH of the corresponding medium (*see* **Table 3.1**) with 1 N HCl or NaOH, as appropriate. It is recommended to check and, if necessary, adjust the pH also after autoclaving the medium.
8. Because of the charged nature of lantibiotics, the use of cation-adjusted medium is advisable, since the presence of $CaCl_2$ enhances the activity of some lantibiotics (30, 31).
9. Use three or four different concentrations of the lantibiotic under investigation, starting at the MIC value for that strain and increasing the concentration up to 10× MIC.
10. Use one flask as growth control (no antibiotic added) and a second flask as positive control, adding a reference antibiotic at a single concentration.

Acknowledgments

We are grateful to Wilfred van der Donk and Mike Dawson for sharing information ahead of publication.

References

1. Chatterjee, C., Paul, M., Xie, L., and van der Donk, W. A. (2005) Biosynthesis and mode of action of lantibiotics. *Chem. Rev.* **105**, 633–684.
2. Willey, J. M. and van der Donk, W. A. (2007) Lantibiotics: peptides of diverse structure and function. *Annu. Rev. Microbiol.* **61**, 477–501.

3. Märki, F., Hänni, E., Fredenhagen, A., and van Oostrum, J. (1991) Mode of action of the lanthionine-containing peptide antibiotics duramycin, duramycin B and C, and cinnamycin as indirect inhibitors of phospholipase A2. *Biochem. Pharmacol.* **42**, 2027–2035.
4. Cloutier, M. M., Guernsey, L., Mattes, P., and Koeppen, B. (1990) Duramycin enhances chloride secretion in airway epithelium. *Am. J. Physiol.* **259**, C450–C454.
5. Molina y Vedia, L. M., Stutts, M. J., Boucher, J. R. C., and Henke, D. C. (1995) U.S. Patent No. 5,716,931.
6. Cox, C. R., Coburn, P. S., and Gilmore, M. S. (2005) Enterococcal cytolysin: a novel two component peptide system that serves as a bacterial defense against eukaryotic and prokaryotic cells. *Curr. Protein Pept. Sci.* **6**, 77–84.
7. Kodani, S., Hudson, M. E., Durrant, M. C., Buttner, M. J., Nodwell, J. R., and Willey, J. M. (2004) The SapB morphogen is a lantibiotic-like peptide derived from the product of the developmental gene *ramS* in *Streptomyces coelicolor*. *Proc. Natl. Acad. Sci. USA* **101**, 11448–11453.
8. Kodani, S., Lodato, M. A., Durrant, M. C., Picart, F., and Willey, J. M. (2005) SapT, a lanthionine-containing peptide involved in aerial hyphae formation in the streptomycetes. *Mol. Microbiol.* **58**, 1368–1380.
9. Tillotson, R. D., Wösten, H. A., Richter, M., and Willey, J. M. (1998) A surface active protein involved in aerial hyphae formation in the filamentous fungus *Schizophyllum commune* restores the capacity of a bald mutant of the filamentous bacterium *Streptomyces coelicolor* to erect aerial structures. *Mol. Microbiol.* **30**, 595–602.
10. Kellner, R., Jung, G., Josten, M., Kaletta, C., Entian, K. D., and Sahl, H. G. (1989) Pep5: structure elucidation of a large lantibiotic. *Angew. Chem.* **101**, 618–621.
11. Pag, U. and Sahl, H. G. (2002) Multiple activities in lantibiotics – models for the design of novel antibiotics? *Curr. Pharm. Des.* **8**, 815–833.
12. Castiglione, F., Lazzarini, A., Carrano, L., Corti, E., Ciciliato, I., Gastaldo, L., Candiani, P., Losi, D., Marinelli, F., Selva, E., and Parenti, F. (2008) Determining the structure and mode of action of microbisporicin, a potent lantibiotic active against multiresistant pathogens. *Chem. Biol.* **15**, 22–31.
13. Maffioli, S. I., Potenza, D., Vasile, F., De Matteo, M., Sosio, M., Marsiglia, B., Rizzo, V., Scolastico, C., and Donadio, S. (2009) Structure revision of the lantibiotic 97518. *J. Nat. Prod.* DOI: 10.1021/np800794y.
14. Li, B., Cooper, L. E., and van der Donk, W. A. (2009) In vitro studies of lantibiotic biosynthesis. *Meth. Enzymol.* **458**, 533–558.
15. Cortés, J., Appleyard, A. N., and Dawson, M. J. (2009) Whole-cell generation of lantibiotic variants. *Meth. Enzymol.* **458**, 559–574.
16. Somma, S., Merati, W., and Parenti, F. (1977) Gardimycin, a new antibiotic inhibiting peptidoglycan synthesis. *Antimicrob. Agents Chemother.* **11**, 396–401.
17. Dodd, H., Gasson, M., Mayer, M., and Narbad, A. (2006) Identifying lantibiotic gene clusters and novel lantibiotic genes. WO 2006/111743.
18. Donadio, S., Monciardini, P., and Sosio, M. (2009) Approaches to discovering novel antibacterial and antifungal agents. *Meth. Enzymol.* **458**, 3–28.
19. Meyer, H. E., Heber, M., Eisermann, B., Korte, H., Metzger, J. W., and Jung, G. (1994) Sequence analysis of lantibiotics: chemical derivatization procedures allow a fast access to complete Edman degradation. *Anal. Biochem.* **223**, 185–190.
20. Brumfitt, W., Salton, M. R., and Hamilton-Miller, J. M. (2002) Nisin, alone and combined with peptidoglycan-modulating antibiotics: activity against methicillin-resistant *Staphylococcus aureus* and vancomycin-resistant enterococci. *J. Antimicrob. Chemother.* **50**, 731–734.
21. Goldstein, B. P., Wei, J., Greenberg, K., and Novick, R. (1998) Activity of nisin against *Streptococcus pneumoniae*, in vitro, and in a mouse infection model. *J. Antimicrob. Chemother.* **42**, 277–278.
22. Brötz, H., Bierbaum, G., Markus, A., Molitor, E., and Sahl, H. G. (1995) Mode of action of the lantibiotic mersacidin: inhibition of peptidoglycan biosynthesis via a novel mechanism? *Antimicrob. Agents Chemother.* **39**, 714–719.
23. Brötz, H., Bierbaum, G., Reynolds, P. E., and Sahl, H. G. (1997) The lantibiotic mersacidin inhibits peptidoglycan biosynthesis at the level of transglycosylation. *Eur. J. Biochem.* **246**, 193–199.
24. Brötz, H., Bierbaum, G., Leopold, K., Reynolds, P. E., and Sahl, H. G. (1998) The lantibiotic mersacidin inhibits peptidoglycan synthesis by targeting lipid II. *Antimicrob. Agents Chemother.* **42**, 154–160.

25. Bavin, E. M., Beach, A. S., Falconer, R., and Friedman, R. (1952) Nisin in experimental tuberculosis. *Lancet* **259**, 127–129.
26. NCCLS document M100-S16. (2006) Performance standards for antimicrobial susceptibility testing.
27. NCCLS document M7-A7. (2008) Methods for dilution antimicrobial susceptibility tests for bacteria that grow aerobically.
28. NCCLS document M26-A. (2005) Methods for determining bactericidal activity of antimicrobial agents.
29. Zhang, J. H., Chung, T. D., and Oldenburg, K. R. (1999) A simple statistical parameter for use in evaluation and validation of high throughput screening assays. *J. Biomol. Screen* **4**, 67–73.
30. Barrett, M. S., Wenzel, R. P., and Jones, R. N. (1992) In vitro activity of mersacidin (M87-1551), an investigational peptide antibiotic tested against Gram-positive bloodstream isolates. *Diagn. Microbiol. Infect Dis.* **15**, 641–644.
31. Schneider, T. R., Kärcher, J., Pohl, E., Lubini, P., and Sheldrick, G. M. (2000) Ab initio structure determination of the lantibiotic mersacidin. *Acta Crystallogr. D Biol. Crystallogr.* **56**, 705–713.

Chapter 4

Expression and Purification of Recombinant α-Defensins and α-Defensin Precursors in *Escherichia coli*

Sharel Figueredo, Jennifer R. Mastroianni, Kenneth P. Tai, and André J. Ouellette

Abstract

Recombinant expression of α-defensins can be obtained at efficient levels in *Escherichia coli*. Amplified α-defensin or pro-α-defensin coding cDNA sequences are cloned directionally between *EcoRI* and *SalI* sites of the pET-28a expression vector and expressed in *E. coli* BL21 RIS cells. Cells growing exponentially in nutrient-rich liquid medium are induced to express the recombinant protein by addition of 50 mM isopropyl β-D-1-thiogalactopyranoside for 3–6 h. After bacterial cells collected by centrifugation are lysed in 6 M guanidine-HCl under non-reducing conditions, the expressed defensin fused to its 6xHis-34 amino acid N-terminal fusion partner is purified by affinity chromatography on nickel-NTA columns. A Met codon introduced at the N terminus of expressed Met-free peptides provides a unique CNBr cleavage site, enabling release of the α-defensin free of ancillary residues by sequential C18 RP-HPLC. Molecular masses of C18 RP-HPLC purified peptides are confirmed by MALDI-TOF mass spectrometry, and peptide homogeneity is assessed using analytical RP-HPLC and acid-urea polyacrylamide gel electrophoresis. α-Defensins prepared in this manner are biochemically equivalent to the natural molecules.

Key words: pET-28a, directional cloning, affinity chromatography, reverse-phase high-performance liquid chromatography, matrix-assisted laser desorption ionization time-of-flight mass spectrometry, acid-urea polyacrylamide gel electrophoresis.

1. Introduction

Defensins are one of two antimicrobial peptide (AMP) superfamilies, the other being the cathelicidins, with roles in the mammalian innate immune response (1). Members of the α-defensin subfamily occur at two primary sites: as abundant constituents of neutrophil azurophil granules (2, 3) the dense core secretory

A. Giuliani, A.C. Rinaldi (eds.), *Antimicrobial Peptides*, Methods in Molecular Biology 618,
DOI 10.1007/978-1-60761-594-1_4, © Springer Science+Business Media, LLC 2010

granules of Paneth cells in small intestinal crypts from which they are secreted into the small intestinal lumen (4, 5). Typically, these ~ 4 kDa bactericidal peptides are cationic, amphipathic, and constrained by a highly conserved tridisulfide array with Cys^{I}-Cys^{VI}, Cys^{II}-Cys^{IV}, and Cys^{III}-Cys^{V} disulfide pairings (1, 6). Their mode of action involves disruption of bacterial cell membranes mediated in part by selective interactions with electronegative target cell envelopes (7–9). Myeloid and Paneth cell α-defensins derive from ~10 kDa pre-propeptides that contain canonical signal sequences, acidic proregions, and a ~4 kDa mature α-defensin peptide in the C-terminal portion of the precursor (10–13).

Despite having highly diverse primary structures (14, 15), α-defensins uniquely retain conserved biochemical features that include the invariant disulfide array (16), a canonical Arg-Glu salt bridge, a conserved Gly residue at position Cys^{III+8}, and high Arg content relative to Lys (17, 18). Although functional determinants of α-defensin microbicidal activity have been suggested by inspection of sequences or interpreted from comparisons of bactericidal peptide assays reported independently by numerous laboratories, few α-defensins have been subjected to a systematic structure–activity analysis in which residue positions critical for peptide function have been tested by induction of loss of function following site-directed mutagenesis. This latter approach has been pursued by production of peptide variants by solid-phase synthesis (18–21) or by using recombinant approaches (7, 20, 22–24). Here, we will consider the production of α-defensins by recombinant approaches in *Escherichia coli*, because the method is inexpensive and flexible and readily can be adapted to the expression of precursor forms which exceed the chain length for optimal synthesis in solid phase.

2. Materials

2.1. Expression Cloning of α-Defensin Coding Sequences

1. pET-28a plasmid DNA (EMD/Novagen, EMD Chemicals, Gibbstown, NJ, USA).
2. GeneAmp PCR Core Reagents (Applied Biosystems, Foster City, CA, USA).
3. QIAEX II Gel Extraction Kit (Qiagen, Inc., USA, Valencia, CA, USA).
4. TOPO TA Cloning Kit, with pCR2.1-TOPO vector (Invitrogen, Carlsbad, CA, USA).
5. DNA Ligation Kit Ver. 2, Buffer I (Takara Bio obtained via Fisher Scientific, Pittsburgh, PA, USA).

6. Kanamycin (Fisher Scientific) stock solution: dissolve 100 mg/mL in deionized water, filter sterilize, and store at –20°C.
7. Trypticase Soy Agar plates: combine 12 g Bacto™ agar, 18 g Bacto™ Tryptic Soy Broth with 600 mL H_2O. Stir the mixture for 30 min, sterilize by autoclaving, stir aseptically, cool to ~ 50°C, add kanamycin from stock solution to a final concentration of 50 μg/mL.
8. Qiaprep Spin Miniprep Kit (Qiagen).
9. *E. coli* XL-2 Blue Ultracompetent cells (Stratagene, La Jolla, CA, USA).
10. *E. coli* BL21(DE3)pLysS competent cells (Stratagene).

2.2. Expression of Recombinant α-Defensins

1. Six 2,800 mL, wide-mouth PYREX® Fernbach-style culture flasks with triple baffles located at the center of the flask bottom to achieve maximal oxygen transfer to culture medium.
2. Stock solutions for Terrific Broth (TB) culture medium: (A) 12 g Bacto tryptone (BD Biosciences, San José, CA, USA), 24 g Bacto yeast extract (BD), 4 mL glycerol, 900 mL deionized water. Dissolve all solids and autoclave for 45 min. (B) Potassium phosphate buffer: dissolve 23.1 g potassium phosphate (KH_2PO_4, Fisher Scientific) and 125.4 g potassium phosphate dibasic (K_2HPO_4 anhydrous, Fisher Scientific) in deionized water to a final volume of 1 L and autoclave for 15 min. (C) 30% Dextrose: dissolve 30 g anhydrous dextrose (Fisher Scientific) in deionized water to a final volume of 100 mL and autoclave for 15 min. (D) Kanamycin (Fisher Scientific) stock solution (*see* **Section 2.1**, Step 4).
3. Beckman model GS-6R benchtop centrifuge.
4. Incubator shaker, Model C25KC (New Brunswick Scientific, Edison, NJ, USA), or comparable instrument with capacity for six 2,800 mL Fernbach-style flasks.
5. Isopropyl-β-D-1-thiogalactopyranoside (IPTG, Fisher Scientific) solution: dissolve 25 g IPTG in 105 mL deionized water. Store as 1.2 mL aliquots at 4°C.
6. Nalgene 1,000 mL wide-mouth polycarbonate (PP GF20 resin) centrifuge bottles with sealing cap for FIBERLite F8-4X1000y rotor (FIBERLite Centrifuge, Santa Clara, CA, USA).
7. Sorvall RC-26 Plus superspeed centrifuge.
8. Lysis buffer: to prepare 6 M guanidine-HCl and 100 mM Tris-HCl, dissolve 573.2 g guanidine hydrochloride (Calbiochem, EMD Chemicals, Gibbstown, NJ, USA) and

12.1 g Tris (Bio-Rad, Hercules, CA, USA) in deionized water, adjust to pH 8.1, and add water to 1 L.

9. Branson Sonifier 450 (Branson Ultrasonics Corp, Danbury, CT, USA), constant cycle, output control set at 5.
10. FIBERLite rotor F21-8X50, rotor code SA-600.

2.3. Cleavage of Fusion Proteins with CNBr

1. Nickel-nitrilotriacetic acid (Ni-NTA Superflow, Qiagen).
2. Column elution buffer: 6 M guanidine-HCl (Calbiochem), 20 mM Tris-HCl (pH 8.1).
3. A standard 100 mL chromatography column, in our case a 2.5 × 23 cm Kontes column (Kimble-Chase, Vineland, NJ, USA).
4. 1 M imidazole, 6 M guanidine-HCl, 20 mM Tris-HCl (pH 6.4): prepare by dissolving 69.08 g imidazole (Sigma), 573.2 g guanidine hydrochloride (Calbiochem, EMD Chemicals, Gibbstown, NJ, USA), and 2.42 g Tris (Bio-Rad, Hercules, CA, USA) in deionized water, adjust to pH 8.1, and add water to 1 L final volume.
5. Spectra/Por 3 membranes (Spectrum Laboratories, Inc., Rancho Dominguez, CA) immersed in deionized water for at least 1 h before use.
6. N_2 gas cylinders (Airgas West, Santa Ana, CA, USA).
7. Cyanogen bromide (CNBr) (Fisher Scientific) solution: dissolve fresh at 20 mg/mL in 80% formic acid (Sigma-Aldrich, St. Louis, MO, USA). All procedures should be performed in chemical fume hood.
8. 5% acetic acid: dissolve 50 mL glacial acetic acid (Fisher Scientific, ACS certified) in 950 mL dH_2O.

2.4. Analytical Acid-Urea Polyacrylamide Gel Electrophoresis (AU-PAGE)

1. Acrylamide solutions. Solution A (60% acrylamide and 1.6% bisacrylamide): dissolve 1,200 g acrylamide (Fisher Scientific) and 32.0 g bisacrylamide (Fisher Scientific) with H_2O in a 2 L beaker to a volume of ~ 1,800 mL. Place beaker in 50°C water bath, dissolve solids with vigorous stirring, adjust volume to 2 L, filter solution through Whatman #4 paper, aliquot, and store at ambient temperature in foil-wrapped bottles. Solution B (43.2% acetic acid, 4% N,N,N′,N′-tetramethylethylenediamine, TEMED): for 1 L combine 432 mL glacial acetic acid, 40 mL TEMED (Fisher Scientific), and 528 mL water.
2. Ammonium persulfate solution: 0.50 g in 5 mL water, made fresh.
3. AU-PAGE electrode buffer: 5% acetic acid in deionized water.

4. Sample loading buffer (3X): dissolve 54 g urea (ultra pure, MP Biomedicals, LLC, Solon, OH, USA) in H_2O to a volume of 95 mL. When dissolved, add ~1 teaspoon AG 501-X8, 20–50 mesh mixed bed resin (Bio-Rad), stir 5 min, filter decant supernatant, add 5 mL glacial acetic acid and methyl green (Sigma-Aldrich) until the solution is a dark blue-green color. Store in 1 mL aliquots at –20°C.
5. 0.9% agarose: 1.35 g agarose (Cambrex Bio Science Rockland, Rockland, ME, USA) up to 150 mL with H_2O.

2.5. Reversed-Phase High-Pressure Liquid Chromatography (RP-HPLC)

1. Preparative RP-HPLC: Waters PrepLC 25 mm Module column model 25 M configured with a Waters Delta Prep 4000 Preparative PrepLC System Controller and a Waters 486 Tunable Absorbance Detector (Waters Corporation, Milford, MA; USA).
2. Semi-Preparative and analytical RP-HPLC: Vydac Semi-Prep C18 column 218TP510 and Vydac Analytical C18 column 218TP54 (Grace, Deerfield, IL, USA) configured with Waters Automated Gradient Controllers, Waters Model 510 pumps, and Waters 484 Tunable Absorbance Detectors.
3. Water (HPLC) from Fisher Scientific and acetonitrile (Fisher Scientific, Optima grade), both containing 0.1% trifluoroacetic acid (TFA, HPLC/Spectro grade; Pierce, Thermo Fischer Scientific, Rockford, IL, USA) as the ion-pairing agent.
4. Coomassie stain solution (54% H_2O, 30% methanol, 15% formalin, 1% Brilliant Blue): add 1.0 g Brilliant Blue R-250 (Fisher Scientific, electrophoresis grade) to 300 mL methanol (Fisher Scientific) and 550 mL H_2O. When dissolved, add 150 mL formalin (Fisher Scientific, 10% solution). Stir until mixed completely and store in a glass bottle at room temperature.
5. AU-Gel destain solution (74% H_2O, 25% methanol, 1% formaldehyde): to prepare 3 L, combine 1,850 mL H_2O, 650 mL methanol, and 25 mL formaldehyde (Fisher Scientific, ACS certified, 37% w/w). Store in a glass bottle at room temperature.
6. AU-Gel Soak Solution (42% H_2O, 40% methanol, 10% acetic acid, 5% formalin, 3% glycerol): for 2 L, combine 840 mL H_2O, 800 mL methanol, 200 mL glacial acetic acid, 100 mL formalin, and 60 mL glycerol (Fisher Scientific). Store in a glass bottle at room temperature.
7. Bio-Rad GelAir dryer.

2.6. Matrix-Assisted Laser Desorption Ionization Time-of-flight (MALDI-TOF) Mass Spectrometry

1. α-Cyano Matrix: 10 mg/mL α-cyano-4-hydroxycinnamic acid (Sigma-Aldrich) dissolved in 60% acetonitrile with 0.1% TFA and stored in 0.5 mL aliquots at –20°C.
2. Voyager-DE MALDI-TOF mass spectrometer (P-E-Biosystems, Foster City, CA, USA).

3. Methods

3.1. Expression Cloning of α-Defensin Coding Sequences

1. Prepare recombinant mouse α-defensins and pro-α-defensins using the pET-28a expression system to produce N-terminal 6xHis-tagged fusion proteins (*see* **Section 2.1**, Step 1). For α-defensins that lack Met, a unique Met-coding trideoxynucleotide is incorporated 5′- of the first codon of the desired peptide by PCR amplification and cloned in-frame with the amino terminal 6XHis in the *Eco*RI/*Sal*I sites of pET-28a (25, 26).
2. Perform amplification reactions using the GeneAmp PCR Core Reagents (*see* **Section 2.1**, Step 2) by incubating reaction mixtures at 95°C for 5 min, followed by successive cycles at 60°C for 1 min, 72°C for 1 min, and 94°C for 1 min for 40 cycles. For example, for cloning proCrp4, forward primer pC4-F [5′-GCGCG AATTC ATGGA TCCTA TCCAA AACAC A] was paired with reverse primer pC4-R [5′-ATATA TTGT TCAGC GGCGG GGGCA GCAGT ACAA], corresponding to nucleotides 104 to 119 and 301 to 327 in pre-proCrp4 cDNA (27). The underlined codon in the pC4-F primer denotes the Met codon (ATG) that introduces the unique CNBr cleavage site immediately downstream of the GAATTC *EcoR*I site. Underlined residues in the pC4-R cloning primer denote the *Sal*I recognition site.
3. Following PCR amplification, electrophorese samples (25 μL) of individual reactions in 2% agarose gels, stain amplification products with ethidium bromide, excise them, and extract them using QIAEX II according to the manufacturer's protocol.
4. Clone amplification products in pCR2.1-TOPO (**Section 2.1**, Step 4). Screen plasmid minipreps for inserts by digestion with *EcoR*I and subject positive clones to DNA sequence analysis to confirm that the expression sequence is in-frame.
5. Prepare the expression vector by sequential incubation of pET-28a plasmid DNA with *Sal*I overnight followed by 4 h incubation with *EcoR*I and deproteinize it. Determine vector

background by transforming competent *E. coli* XL-2 Blue cells with mock-ligation reactions containing 20 ng samples of *EcoRI* and *SalI*-digested pET-28a but without amplified DNA and plating on TSB plates containing 50 μg/mL kanamycin (*see* **Note 1**).

6. Digest plasmids containing defensin-coding inserts in pCR2.1-TOPO clones with *EcoRI* and *SalI*, and gel-purify *EcoRI/SalI*-digested fragments.
7. Ligate fragments into the *EcoRI/SalI*-digested pET-28a plasmid DNA and transform into *E. coli* XL-2 Blue cells for archiving and into *E. coli* BL21(DE3)pLysS cells for expression.

3.2. Expression of Recombinant α-Defensin Fusion Proteins

1. For large-scale preparation of recombinant α-defensins, streak BL21(DE3)pLysS cells transformed with pET-28a α-defensin constructs onto TB plates containing 70 μg/mL kanamycin to obtain isolated colonies. Inoculate 100 mL of kanamycin-TB broth with a single colony and grow overnight at 37°C in a model C25KC incubator shaker.
2. After overnight growth, transfer the culture to two 50 mL Falcon tubes, centrifuge at 3,000 rpm for 10 min in a Beckman GS-6R centrifuge, and resuspend the bacterial cell pellet in 6 mL TB as concentrated inoculum.
3. To each autoclaved 2,800 mL Fernbach-style flask, add 900 mL of sterile Bacto Tryptone, yeast extract, glycerol solution, 100 mL of potassium phosphate buffer, 5 mL 30% dextrose solution, and 10 mg/mL kanamycin from stock solution to a final concentration of 70 μg/mL kanamycin (*see* **Section 2.2**, Step 2). Prewarm medium to 37°C, inoculate each flask with 1 mL of the concentrated inoculum, and grow cells at 37°C in the model C25KC incubator shaker at a setting of 250 rpm. Monitor the optical density of the culture at 600 nm until the exponentially growing cells reach an A_{600} value of 0.70–0.80, at which point production of recombinant proteins is induced by addition of 100 μM IPTG under kanamycin selection (*see* **Note 2**).
4. After 6 h at 37°C, transfer cultures to 1,000 mL wide-mouth polycarbonate (PP GF20 resin) centrifuge bottles with sealing caps and centrifuge for 5 min at 5,000 rpm in a FIBER-Lite F8-4X1000y rotor in a Sorvall RC-26 Plus superspeed centrifuge. Decant supernatants from deposited bacterial cells, decontaminate with bleach, and store deposited bacterial cell pellets at –20°C overnight (*see* **Note 3**).
5. Lyse cells by resuspending the bacterial cell pellets in lysis buffer (*see* **Section 2.2**, Step 8), followed by sonication at

70% power and 50% duty cycle for 2 min using a Branson Sonifier 450. Clarify lysates by centrifugation at 30,000×*g* for 30 min at 4°C in a FIBERLite F21-8X50 rotor using rotor code SA-600 in a Sorvall RC-26 Plus superspeed centrifuge. Fusion proteins are purified immediately after lysate clarification (*see* **Section 3.3**, below).

3.3. Affinity Purification and Cleavage of Fusion Protein

1. Recombinant His-tagged α-defensin fusion peptides are purified by nickel-nitrilotriacetic acid (Ni-NTA) resin affinity chromatography (22). Subsequently, α-defensins are cleaved from fusion proteins with CNBr cleavage, and α-defensins are recovered and purified by RP-HPLC. Incubate cell lysates with Ni-NTA resin at a ratio of 25:1 (v/v) in 6 M guanidine-HCl, 20 mM Tris-HCl (pH 8.1) (*see* **Section 2.3**, Step 2) for 4 h at 4°C. Pour the mixture into a 2.5 × 23 cm Kontes column fitted with a polyethylene drain tube and a small thumb-clamp closure to maintain a flow rate of ~ 1 mL/min.
2. Wash the resin lysate mixture with 10–20 column bed volumes of 6 M guanidine-HCl, 20 mM Tris-HCl (pH 8.1) to remove proteins and peptides non-specifically associated with the resin. Fusion proteins, containing a 6xHis tag near the fusion protein N terminus, a 26 amino acid spacer, and the α-defensin (**Fig. 4.1**) are then eluted from the resin with 2 bed volumes of 1 M imidazole in 6 M guanidine-HCl, 20 mM Tris-HCl (pH 6.4; *see* **Section 2.3**, Step 4), dialyzed in Spectra/Por 3 membranes against 3 × 10 L changes of 5% acetic acid and lyophilized.

CNBr cleavage site

Linker sequence for pET-28a:
34 amino acids upstream of the EcoRIsite:
MGSS**HHHHHH**SSGLVPRGSH**M**AS**M**TGGQQ**M**GRGS(EF)**MET**- **peptide** TGATGA*SalI*
6x-His *EcoRI* Stops

Fig. 4.1. General features of pET-28a/α-defensin fusion proteins. The endogenous initiating methionine codon of the pET-28a codes for a 34 amino acid fusion protein immediately N-terminal of the expressed defensin peptide. Residue positions 5–10 form the 6xHis affinity tag, and the *arrow* indicates the position of the introduced Met residue that provides the unique CNBr cleavage site. Methionine residues in the pET-28a fusion protein are shown in *bold.*

3. The Met residue added immediately before the N terminus of each α-defensin or pro-α-defensin peptide, as exists in human pro-α-defensin-1 (13), provides a cyanogen bromide (CNBr) cleavage site for separating the recombinant α-defensin molecule from the remainder of the His-tagged fusion peptide (**Fig. 4.1**). To this end, dissolve lyophilized fusion proteins in 80% formic acid, transfer to foil-wrapped polypropylene tubes to which solid CNBr is added to a final concentration of 10 mg/mL CNBr, purge with N_2, then

seal tubes. Incubate samples under N_2 overnight at 25°C in a chemical fume hood. Cleavage reactions are terminated by addition of 10 vol. H_2O followed by freezing and lyophilization. Then dissolve the CNBr-cleaved peptide samples in 5% acetic acid and store at 4°C prior to purification by RP-HPLC (*see* **Section 3.5**).

3.4. Preparation of AU-PAGE gels

1. AU-PAGE is a superior system for separation of small cationic peptides on the basis of their electropositive charge-to-size ratios (28). Running gels are poured in a standard vertical slab gel system. Because longer gels give better resolution, a custom system using 18 × 25 cm plates and 18 × 1 cm spacers of 1.5 mm thickness is used routinely. Plates are clamped with binder clips and sealed with molten agarose (*see* **Section 2.4**, Step 5).
2. To prepare five 20% polyacrylamide running gels, combine 48 g urea, 47.45 mL H_2O, 53.45 mL solution A (**Section 2.4**, Step 1), 20 mL solution B (**Section 2.4**, Step 1), and 3 mL 10% ammonium persulfate solution (**Section 2.4**, Step 2). Dissolve completely and pour into the plate assembly, leaving approximately 3 cm above the gel solution free for addition of the stacking gel. The freshly poured gel is overlain with water and polymerizes in ~ 60 min.
3. To prepare 7.5% polyacrylamide stacking gel solution for five gels, combine 12 g urea, 20.25 mL H_2O, 5 mL solution A (**Section 2.4**, Step 1), 5 mL solution B (**Section 2.4**, Step 1), 0.75 mL AP solution (**Section 2.4**, Step 2), and dissolve all solids.
4. Remove water layer from the running gel, pour stacker, insert combs, overlay with water. The stacking gel polymerizes in ~ 60 min after which combs are removed, and gels are "pre-run" overnight using 5% acetic acid electrode buffer at 150 V. Because proteins migrate through AU gels as fully protonated forms, the polarity is reversed compared to SDS-PAGE, that is, the anode is the negative electrode (*see* **Note 4**). After the pre-run, the electrode buffer is discarded. Gels may be used immediately or assembled gels may be wrapped in Saran Wrap and stored at 4°C for several days.

3.5. Peptide Purification Using RP-HPLC

1. α-Defensins are purified to homogeneity using a reiterative C18 RP-HPLC approach (29). After CNBr cleavage, Crp4 peptides are separated from the 36 amino acid 6xHis-tag fusion partner by chromatography on a Waters PrepLC 25 mm Module column model 25 M. After sample injection, proteins and peptides are eluted using a 10–60% linear acetonitrile gradient developed for 25 min at a flow rate of 20 mL/min. Samples of individual fractions are lyophilized

and analyzed in 20% AU-PAGE gels (*see* **Section 3.4**) and stained to identify column fractions that contain putative α-defensins based on co-migration with α-defensin or pro-α-defensin peptide standards. Protein samples are lyophilized prior to electrophoresis and dissolved in ≤ 15 μL 1X sample buffer and electrophoresed with fresh 5% acetic acid running buffer at 100 V for sample loading, then increase to 200 V for 3–6 h. After electrophoresis, gels are fixed, stained with Coomassie stain solution and destained (*see* **Section 2.5**, Step 4 and 5) submerged in AU-Gel Soak Solution (*see* **Section 2.5**, Step 6) for 1 h, and dried using a Bio-Rad GelAir dryer (*see* **Section 2.5**, Step 7 and **Note 5**). The putative defensin-containing column fractions from preparative RP-HPLC (**Fig. 4.2a**, boxed region) are then concentrated in a high-capacity Speed-Vac to evaporate the acetonitrile, pooled, lyophilized, dissolved in 5% acetic acid, and subjected to further RP-HPLC purification.

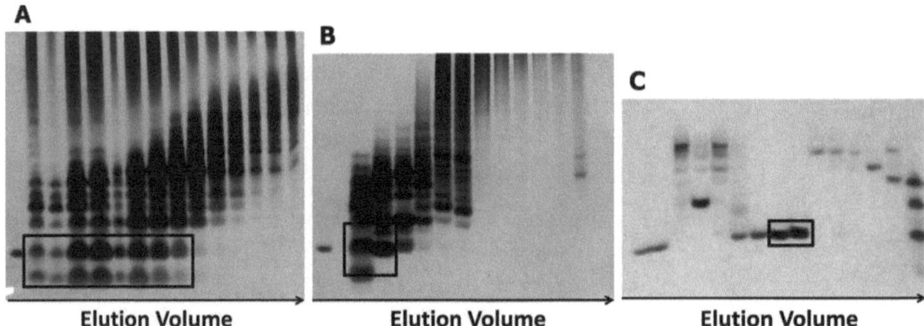

Fig. 4.2. AU-PAGE analysis of recombinant Crp4 purification. Samples of individual fractions from preparative (**a**), semi-preparative (**b**), and analytical C18 RP-HPLC (**c**) were lyophilized for analysis in 20% AU-PAGE gels. After lyophilization, protein samples were dissolved in ≤ 15 μL 1X sample buffer, electrophoresed, fixed, and stained with Coomassie stain solution. Boxed regions denote the position of the recombinant Crp4 peptide at each purification stage. Recombinant control Crp4 standard peptides are shown in the left lane of each gel.

2. Recombinant peptides are purified to homogeneity by sequential semi-preparative and analytical RP-HPLC. After injection of the pooled putative α-defensin sample from preparative RP-HPLC, the Vydac 218TP510 (C18) column is developed using a 20–40% linear acetonitrile gradient developed for 60 min at a flow rate of 2 mL/min. As before, aliquot samples of individual column fractions are lyophilized and assayed for putative α-defensins by co-migration with appropriate α-defensin peptide standards in AU-PAGE analyses (**Fig. 4.2b**, boxed region). MALDI-TOF MS determinations (*see* **Section 3.6** below) are made to confirm whether column fractions with apparent recombinant proteins by AU-PAGE contain peptides with appropriate molecular masses. Column fractions containing recombinant peptide (**Fig. 4.2b**, boxed region) are combined and

concentrated by Speed-Vac evaporation of acetonitrile and subsequent lyophilization as before. Samples are subjected to the final purification separation on analytical RP-HPLC. A Vydac analytical 218TP54 column is developed with a linear 15–25% acetonitrile gradient for 60 min at a 1 mL/min flow rate. Again, aliquot samples of individual column fractions are lyophilized and the homogeneity of the purified recombinant α-defensin is assessed by co-migration with appropriate α-defensin peptide standards in AU-PAGE analyses (**Fig. 4.2c**, boxed region). The concentrations of purified recombinant α-defensin and pro-α-defensin solutions are quantified by amino acid analysis services provided, in our case, by the UCLA (http://proteomics.ctrl.ucla.edu/) or UC Davis (http://proteomics.ucdavis.edu/) Proteomics Core Facilities, or spectrophotometrically at 280 nm using extinction coefficients obtained at ExPASy Proteomics Server (http://ca.expasy.org/). For example, $\varepsilon = 3355\ M^{-1}$ cm^{-1} for mouse Crp4.

3.6. Laser Desorption Ionization Time-of-flight (MALDI-TOF) Mass Spectrometry

1. As noted above, matrix-assisted laser desorption ionization time-of-flight mass spectrometry (MALDI-TOF MS) is used extensively to identify and monitor the purification of recombinant α-defensins. For this purpose, apply dilute peptide samples (1 μL) to stainless steel MALDI sample plates with numbered circles. After application, 1 μL aliquots of α-cyano matrix is applied to each sample, mixed, and allowed to dry.
2. At this time, perform measurements on a Voyager-DE MALDI-TOF (in our case, available in the Mass Spectroscopy Facility at the Department of Chemistry, University of California Irvine, CA, USA). In the example shown (**Fig. 4.2c**, boxed region), homogeneously pure recombinant mouse Crp4 had an experimental mass of 3756.3 a.m.u. by MALDI-TOF MS, consistent with the theoretical Crp4 mass of 3755.55 a.m.u.

The methods described have proven to be generally successful for the recombinant production of α-defensins and pro-α-defensins from mouse and human Paneth cells as well as those from rhesus macaque neutrophils and Paneth cells (24, 30–35) (*See* **Note 6**).

4. Notes

1. If more than 50 colonies/10 ng of vector DNA are recovered, *EcoR*I and *Sal*I digestion of vector DNA sample should be repeated until the background is suitably low.

2. Although the standard temperature for induction of recombinant protein expression is 37°C, recovery of certain defensin peptides may be low when inductions are performed at this temperature. In certain instances, slowing the rate of overall protein production by reducing the temperature during the period of induction frequently improves recovery of fusion protein and properly folded defensins. The optimal induction temperatures are determined empirically by incubating test cultures at temperatures ranging from 30 to 4°C for 18 h following addition of IPTG. In this respect, an incubator shaker with refrigeration capability (*see* **Section 2.2**, Step 4) may be critical to success in expressing certain defensin peptides or their variants.
3. This freezing step greatly improves subsequent cell lysis as compared to cells that are not frozen.
4. Failure to pre-run gels for a sufficient time under these conditions results with all proteins stacking about half-way down the gel.
5. Stain or destain solutions that lack formaldehyde result in defensin loss from the gel. Also, low sample volume (10–20 μL) in 1X sample buffer is critical to good AU-PAGE resolution. Up to 1 mg of α-defensin or total cellular protein can easily be dissolved in 20 μL under these conditions. Lyophilization is a strongly recommended method for concentration of protein samples, because vacuum centrifugation to dryness, i.e., Speed-Vac, may lead to formation of insoluble gels.
6. The use of CNBr to separate the expressed α-defensin limits the method described to molecules that are methionine-free, a problem that may be circumvented by replacement of the Met codon with codons that code for enzymatic cleavage sites. For methionine-containing α-defensins and pro-α-defensins, the Met residue introduced by the forward cloning primer (**Fig. 4.1**) may be replaced with enzymatic cleavage sites such as Asp-Asp-Asp-Asp-Lys, recognized by enterokinase such as the EnterokinaseMax system (Invitrogen).

References

1. Selsted, M. E., and Ouellette, A. J. (2005) Mammalian defensins in the antimicrobial immune response. *Nat. Immunol.* **6**, 551–557.
2. Faurschou, M., Sorensen, O. E., Johnsen, A. H., Askaa, J., and Borregaard, N. (2002) Defensin-rich granules of human neutrophils: characterization of secretory properties. *Biochim. Biophys. Acta* **1591**, 29–35.
3. Borregaard, N., Sorensen, O. E., and Theilgaard-Monch, K. (2007) Neutrophil granules: a library of innate immunity proteins. *Trends Immunol.* **28**, 340–345.
4. Selsted, M. E., and Ouellette, A. J. (1995) Defensins in granules of phagocytic and

non-phagocytic cells. *Trends Cell. Biol.* **5**, 114–119.
5. Selsted, M. E., Miller, S. I., Henschen, A. H., and Ouellette, A. J. (1992) Enteric defensins: antibiotic peptide components of intestinal host defense. *J. Cell. Biol.* **118**, 929–936.
6. Ouellette, A. J. (2005) Paneth cell alpha-defensins: peptide mediators of innate immunity in the small intestine. *Springer Semin. Immunopathol.* **27**, 133–146.
7. Hadjicharalambous, C., Sheynis, T., Jelinek, R., Shanahan, M. T., Ouellette, A. J., and Gizeli, E. (2008) Mechanisms of alpha-defensin bactericidal action: comparative membrane disruption by Cryptdin-4 and its disulfide-null analogue. *Biochemistry* **47**, 12626–12634.
8. White, S. H., Wimley, W. C., and Selsted, M. E. (1995) Structure, function, and membrane integration of defensins. *Curr. Opin. Struct. Biol.* **5**, 521–527.
9. Wimley, W. C., Selsted, M. E., and White, S. H. (1994) Interactions between human defensins and lipid bilayers: evidence for formation of multimeric pores. *Protein Sci.* **3**, 1362–1373.
10. Wilson, C. L., Ouellette, A. J., Satchell, D. P., Ayabe, T., Lopez-Boado, Y. S., Stratman, J. L., Hultgren, S. J., Matrisian, L. M., and Parks, W. C. (1999) Regulation of intestinal alpha-defensin activation by the metalloproteinase matrilysin in innate host defense. *Science* **286**, 113–117.
11. Ayabe, T., Satchell, D. P., Pesendorfer, P., Tanabe, H., Wilson, C. L., Hagen, S. J., and Ouellette, A. J. (2002) Activation of Paneth cell alpha-defensins in mouse small intestine. *J. Biol. Chem.* **277**, 5219–5228.
12. Ghosh, D., Porter, E., Shen, B., Lee, S. K., Wilk, D., Drazba, J., Yadav, S. P., Crabb, J. W., Ganz, T., and Bevins, C. L. (2002) Paneth cell trypsin is the processing enzyme for human defensin-5. *Nat. Immunol.* **3**, 583–590.
13. Valore, E. V., and Ganz, T. (1992) Posttranslational processing of defensins in immature human myeloid cells. *Blood* **79**, 1538–1544.
14. Ouellette, A. J. (2006) Paneth cell alpha-defensin synthesis and function. *Curr. Top. Microbiol. Immunol.* **306**, 1–25.
15. Ouellette, A. J., and Bevins, C. L. (2001) Paneth cell defensins and innate immunity of the small bowel. *Inflamm. Bowel Dis.* **7**, 43–50.
16. Selsted, M. E., and Harwig, S. S. (1989) Determination of the disulfide array in the human defensin HNP-2. A covalently cyclized peptide. *J. Biol. Chem.* **264**, 4003–4007.
17. Lehrer, R. I. (2007) Multispecific myeloid defensins. *Curr. Opin. Hematol.* **14**, 16–21.
18. Zou, G., de Leeuw, E., Li, C., Pazgier, M., Zeng, P., Lu, W. Y., Lubkowski, J., and Lu, W. (2007) Toward understanding the cationicity of defensins. Arg and Lys versus their noncoded analogs. *J. Biol. Chem.* **282**, 19653–19665.
19. de Leeuw, E., Burks, S. R., Li, X., Kao, J. P., and Lu, W. (2007) Structure-dependent functional properties of human defensin 5. *FEBS Lett.* **581**, 515–520.
20. Pazgier, M., Li, X., Lu, W., and Lubkowski, J. (2007) Human defensins: synthesis and structural properties. *Curr. Pharm. Des.* **13**, 3096–3118.
21. Wu, Z., Li, X., Ericksen, B., de Leeuw, E., Zou, G., Zeng, P., Xie, C., Li, C., Lubkowski, J., Lu, W. Y., and Lu, W. (2007) Impact of pro segments on the folding and function of human neutrophil alpha-defensins. *J. Mol. Biol.* **368**, 537–549.
22. Shirafuji, Y., Tanabe, H., Satchell, D. P., Henschen-Edman, A., Wilson, C. L., and Ouellette, A. J. (2003) Structural determinants of procryptdin recognition and cleavage by matrix metalloproteinase-7. *J. Biol. Chem.* **278**, 7910–7919.
23. Maemoto, A., Qu, X., Rosengren, K. J., Tanabe, H., Henschen-Edman, A., Craik, D. J., and Ouellette, A. J. (2004) Functional analysis of the α-defensin disulfide array in mouse cryptdin-4. *J. Biol. Chem.* **279**, 44188–44196.
24. Figueredo, S. M., Weeks, C. S., Young, S. K., and Ouellette, A. J. (2009) Anionic amino acids near the pro-alpha-defensin N terminus mediate inhibition of bactericidal activity in mouse pro-cryptdin-4. *J. Biol. Chem.* **284**, 6826–6831.
25. Satchell, D. P., Sheynis, T., Shirafuji, Y., Kolusheva, S., Ouellette, A. J., and Jelinek, R. (2003) Interactions of mouse Paneth cell alpha-defensins and alpha-defensin precursors with membranes: prosegment inhibition of peptide association with biomimetic membranes. *J. Biol. Chem.* **278**, 13838–13846.
26. Satchell, D. P., Sheynis, T., Kolusheva, S., Cummings, J., Vanderlick, T. K., Jelinek, R., Selsted, M. E., and Ouellette, A. J. (2003) Quantitative interactions between cryptdin-4 amino terminal variants and membranes. *Peptides* **24**, 1795–1805.
27. Ouellette, A. J., Darmoul, D., Tran, D., Huttner, K. M., Yuan, J., and Selsted, M. E. (1999) Peptide localization and gene structure of cryptdin 4, a differentially expressed mouse Paneth cell alpha-defensin. *Infect. Immun.* **67**, 6643–6651.

28. Selsted, M. E. (1993) Investigational approaches for studying the structures and biological functions of myeloid antimicrobial peptides. *Genet. Eng. (NY)* **15**, 131–147.
29. Selsted, M. E. (1997) HPLC methods for purification of antimicrobial peptides. *Methods Mol. Biol.* **78**, 17–33.
30. Tanabe, H., Yuan, J., Zaragoza, M. M., Dandekar, S., Henschen-Edman, A., Selsted, M. E., and Ouellette, A. J. (2004) Paneth cell alpha-defensins from rhesus macaque small intestine. *Infect. Immun.* **72**, 1470–1478.
31. Tanabe, H., Qu, X., Weeks, C. S., Cummings, J. E., Kolusheva, S., Walsh, K. B., Jelinek, R., Vanderlick, T. K., Selsted, M. E., and Ouellette, A. J. (2004) Structure-activity determinants in Paneth cell alpha-defensins: loss-of-function in mouse cryptdin-4 by charge-reversal at arginine residue positions. *J. Biol. Chem.* **279**, 11976–11983.
32. Tanabe, H., Ouellette, A. J., Cocco, M. J., and Robinson, W. E., Jr. (2004) Differential effects on human immunodeficiency virus type 1 replication by α-defensins with comparable bactericidal activities. *J. Virol.* **78**, 11622–11631.
33. Weeks, C. S., Tanabe, H., Cummings, J. E., Crampton, S. P., Sheynis, T., Jelinek, R., Vanderlick, T. K., Cocco, M. J., and Ouellette, A. J. (2006) Matrix metalloproteinase-7 activation of mouse Paneth cell pro-alpha-defensins: SER43↓ILE44 proteolysis enables membrane-disruptive activity. *J. Biol. Chem.* **281**, 28932–28942.
34. Rosengren, K. J., Daly, N. L., Fornander, L. M., Jonsson, L. M., Shirafuji, Y., Qu, X., Vogel, H. J., Ouellette, A. J., and Craik, D. J. (2006) Structural and functional characterization of the conserved salt bridge in mammalian Paneth cell alpha-defensins: solution structures of mouse Cryptdin-4 and (E15D)-Cryptdin-4. *J. Biol. Chem.* **281**, 28068–28078.
35. Kamdar, K., Maemoto, A., Qu, X., Young, S. K., and Ouellette, A. J. (2008) In vitro activation of the rhesus macaque myeloid alpha-defensin precursor proRMAD-4 by neutrophil serine proteinases. *J. Biol. Chem.* **283**, 32361–32368.

Chapter 5

Production of Recombinant Antimicrobial Peptides in Bacteria

Mateja Zorko and Roman Jerala

Abstract

Large quantities of antimicrobial peptides are required for investigations and clinical trials, therefore suitable production method alternative to traditional chemical synthesis is necessary. Production of recombinant antimicrobial peptides in prokaryotic systems has successfully demonstrated the viability of this approach. Production of antimicrobial peptides in *Escherichia coli* is potentially limited due to their toxicity to host cells and susceptibility to proteolytic degradation, which can be avoided using fusion protein approach. We describe antimicrobial peptide production in *E. coli* based on forcing antimicrobial peptides into inclusion bodies, which is affective for the production of large quantities of antimicrobial peptides. Chemical reagents for cleaving peptide bond between antimicrobial peptides and fusion proteins such as cyanogen bromide and diluted acid are selective and provide antimicrobial peptides for biological studies in short time.

Key words: Antimicrobial peptide, recombinant fusion protein, pET vector, ketosteroid isomerase fusion protein, inclusion bodies, chemical cleavage.

1. Introduction

During the past decade, many small cationic antimicrobial peptides have been successfully prepared by recombinant DNA methods (1–7). In contrast to preparation and expression of heterologous proteins in bacteria, as by far the simplest and most inexpensive means to produce large amounts of recombinant protein, expression of antimicrobial peptides faces large potential difficulties because of their cytotoxicity to host cells and sensitivity to proteolytic degradation of nonstructured peptides. Expression systems with an antimicrobial peptide fused to a partner protein are able to avoid those problems because of the reduced

A. Giuliani, A.C. Rinaldi (eds.), *Antimicrobial Peptides*, Methods in Molecular Biology 618,
DOI 10.1007/978-1-60761-594-1_5, © Springer Science+Business Media, LLC 2010

toxicity against host cells and enhanced product stability since the fusion protein facilitates product recovery and purification (8). Several biological expression systems have been developed by fusing antimicrobial peptides to a partner protein to mask the intrinsic antimicrobial activity of the peptide inside bacterial cells (shown in **Table 5.1**), and variety of host cells such as bacteria (3, 9, 10), yeast (11), or insect cells (12) have been used. The most commonly used host cell for recombinant protein production is *Escherichia coli* because of its high growth rate and well-established expression system (6). On the other hand, a biological expression system using genetic techniques based on fusion technology is also economically viable for large-scale production. Such fusion proteins generally lack antimicrobial activity if they form insoluble products or tightly interact with antimicrobial peptides. The peptide is released from the fusion by chemical or enzymatic cleavage. Chemical reagents for cleavage of dipeptide between fusion protein and antimicrobial peptide are shown in **Table 5.2**.

Carrier molecules generally take into account the following criteria (13):

- The size of the selected carrier molecule should be three to four times larger than that of the expressed peptides of interest.
- The isoelectric point of the expressed molecules should be higher or lower than that of the expressed molecule by at least 2 pH units.
- There should be a unique chemical or protease cutting site at the interface of fusion partners.
- It must not be toxic to the expression host.

Table 5.1
Fusion system used for recombinant antimicrobial protein production

Fusion protein partner (5)	Molecular weight (kDa)	pI
Prochymosin (21)	38.0	4.6
Green fluorescence protein (22)	26.8	5.7
Maltose-binding protein (MBP) (23)	44.4	9.2
Ketosteroid isomerase (KSI) (15, 17)	13.7	4.8
Glutatione S-transferase (GST) (14)	15.8	5.3
Thioredoxin (Trx) (24)	11.8	4.6
Intein (25)	35.3	5.3
Baculoviral polyhedrin (26)	28.7	5.8
Ubiquitin (27)	8.5	6.5
Anionic peptide (4)	2.5	3.1

Table 5.2
Chemical cleavage of the dipeptide between fusion protein and antimicrobial peptide

Cleavage site	Reagent	Cleavage conditions
Met–X (7)	CNBr	70% TFA, CNBr of 100-fold molar excess over methionyl residues. RT, 24 h
Asp–Pro (15)	HCl or formic acid	90 mM HCl; 85°C, 2 h
Asn–Gly (26)	Hydroxylamine	Buffer pH 9.0, 1.7 M hydroxy-lamine-HCl, 55°C, 24 h
Cys–X (28)	NTCB (2-nitro-5-tiocyanatobenzoic acid)	20 mM Tris–HCl buffer pH 9.0, NTCB of 20-fold molar excess of the total cysteine content, 37°C, 16 h

Although the antimicrobial peptides can be produced via solid-phase chemical synthesis, this method is less suitable to obtain uniformly or selectively isotopically (^{13}C, ^{15}N, and ^{2}H) labeled peptides, which are required to elucidate biophysical properties of the ligand–receptor interaction by high-resolution liquid or solid-state NMR spectroscopy. Recombinant expression of peptides is therefore an alternative to chemical synthesis with the opportunity to incorporate isotopic labels (7, 14).

In our laboratory, we have prepared several antimicrobial peptides by recombinant technology. Here, optimized protocol for production of recombinant antimicrobial peptides and selective chemical cleavage from ketosteroid isomerase (KSI) that forms inclusion bodies are described. We describe a protocol for chemical cleavage between fusion protein and antimicrobial peptide using CNBr or acidic cleavage. For this purpose, a Met or Asp-Pro has to be engineered between the fusion partners. The selected strategy of the Asp–Pro cleavage results in an N-terminal proline residue of peptide, which is in fact even favorable for some peptides. We have demonstrated the essential structural role of the N-terminal proline residue for the conformation of antimicrobial peptide adopted in the membrane mimetic environment (15).

2. Materials

1. pET31b(+) (Novagen) (**Fig. 5.1**), dissolved in sterile MQ.
2. *E. coli* BL21 (BL3) pLysS (Invitrogen) stored in 15% glycerol at 80°C.

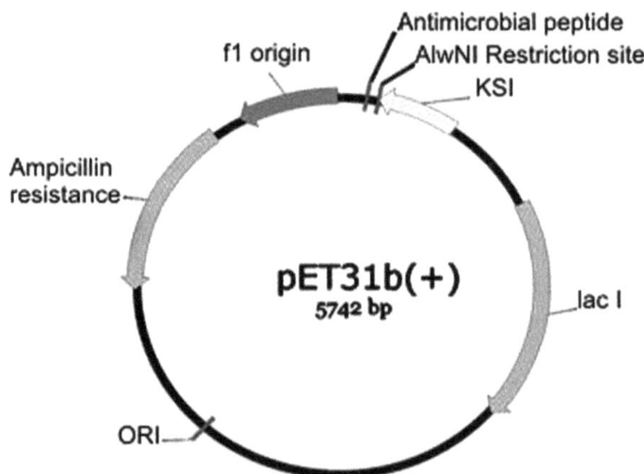

Fig. 5.1. Schematic representation of the expression vector pET31(+) which indicates antimicrobial peptides and KSI fusion partner.

3. TB (Terrific Broth) (Sigma) and LB (Luria-Bertani) medium (Sigma): for 1 L LB medium, dissolve 25 g LB mixture in deionized water and autoclave; for 1 L TB medium, suspend 47.6 g TB mixture in 900 mL of deionized water and add 10 g of glycerol (Sigma). Add deionized water to bring volume to 1 L and autoclave for 20 min at 121°C.
4. Ampicillin (Sigma): prepare stock solution of 50 mg/mL in sterile MQ water and store in aliquots at –20°C.
5. Isopropyl β-D-thiogalactoside (IPTG) (Inalco Pharmaceuticals): prepare 1 M stock solution in sterile MQ water and store in aliquots at –20°C.
6. Guanidine-HCl is dissolved at 6 M in MQ water.
7. Hydrochloric acid (Aldrich): prepare 90 mM solution in MQ water.
8. Cyanogen bromide (Aldrich) is stable for months in dry, dark, refrigerated story. Warm to room temperature before opening. Use only white and not yellow crystals.
9. Lysis buffer: 10 mM Tris–HCl pH 8.0, 1 mM EDTA (ethylenediamine tetraacetic acid), and 0.1% DOC (deoxycholate).
10. Wash buffers: wash with 10 mM Tris–HCl pH 8.0, 1 mM EDTA, 0.1% DOC, twice; with 10 mM Tris–HCl pH 8.0, 1 mM EDTA, 0.1% DOC, 2 M urea, twice, and with 20 mM Tris–HCl pH 8.0 three times.
11. Ligation buffer: 20 mM Tris–HCl pH 7.5, 10 mM $MgCl_2$, 1 mM dithiotreitol.

12. Digestion buffer: 50 mM potassium acetate, 20 mM Tris–acetate, 10 mM magnesium acetate, 1 mM dithiotreitol pH 7.9.

3. Methods

3.1. Gene Construct Preparation

1. Design the sense and antisense synthetic oligonucleotide encoding the antimicrobial peptides that fuse to the C-terminus of the bacterial ketosteroid isomerase gene. Insertion of coding sequence for peptide is achieved through the recognition site for the restriction endonuclease *AlwNI* which produces an overhang of three arbitrary nucleotides (in this case methionine coding ATG) bounded on each side by trinucleotide recognition site shown in **Fig. 5.2**.
2. Denature primers at 95°C for 10 min and anneal by slow cooling to the room temperature in the ligation buffer.

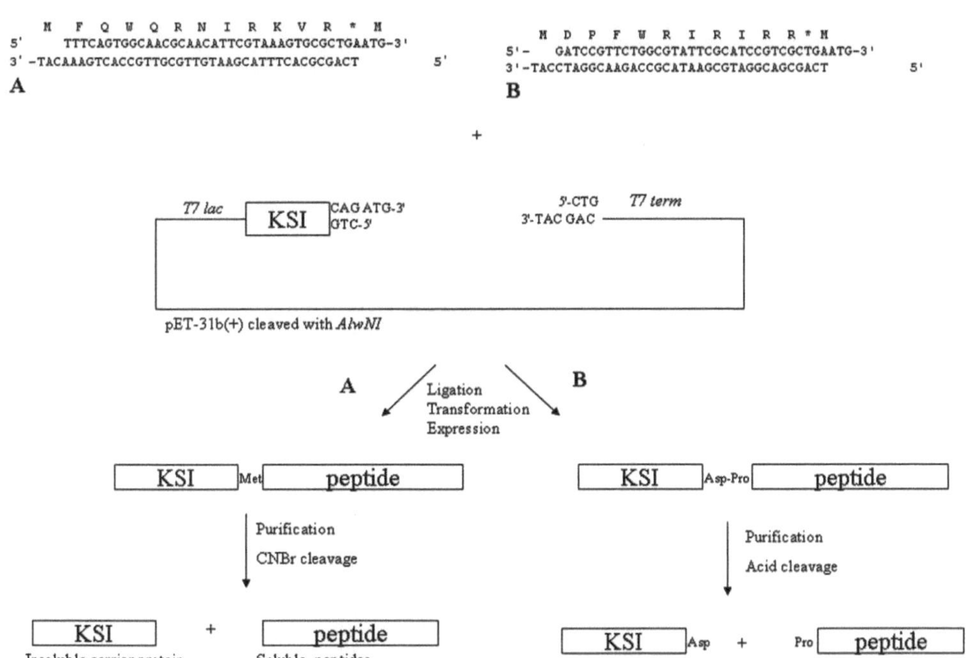

Fig. 5.2. (**a**) Strategy for the construction, production, purification, and CNBr cleavage of ketosteroid isomerase–peptide fusion protein. Oligonucleotides encoding both strands of the antimicrobial peptides were annealed and ligated into the *Alw*NI site of pET31(+) to create ketosteroid–peptide fusion gene. The gene was placed under the control of T7 lac promotor, which is recognized as a transcription start site by the T7 RNA polymerase additionally controlled by lac repressor. (**b**) Strategy for the construction, production, purification, and diluted acid cleavage of ketosteroid isomerase–peptide fusion protein.

3. Pipette into a microfuge tube 1 μg pET31b(+), 2 μL 10 × digestion buffer, 1 μL AlwNI enzyme, and add water to final volume 20 μL. Incubate at 37°C for 1 h. Take 2–5 μL of the digested sample and run on agarose gel to check the result.
4. Mix 100 ng of *AlwNI* digested pET31b(+) expression vector and synthetic oligonucleotides for antimicrobial peptides with sticky ends in 3:1 molar ratio of insert to vector, 2 μL ligation buffer, 1 μL T4 DNA ligase; and add water to final volume 20 μL and incubate at 16°C for 18 h (*see* **Note 1**).
5. Transform the ligated plasmid into the DH5 α strain of *E. coli* as described:
   - Place competent cells on ice after removing them from the –80°C storage. As the cells thaw, add 5 ng of plasmid to 60 μL of competent cells. Flick the tubes to mix the contents and store immediately on ice for further 30 min.
   - Remove tubes from ice and incubate for 4 min in a 42°C water bath for a heat shock. Heat shock facilitates the plasmid uptake by bacteria.
   - Put tubes for 2 min on ice, afterward add 900 μL of sterile LB medium to tube and incubate at 37°C for 60 min.
   - Plate 10 μL of culture on LB + Amp plate using a glass spreader. Allow the plate to absorb the inoculums for 5 min, then invert and incubate at 37°C for 16 h or until colonies grow to the desired size.
   - Examine the plates and determine the efficiency of ligation and transformation in comparison to the plate of ligation without added insert and identify the colonies harboring the plasmid by PCR using primers KSI (GGCAAGGTGGTGAGCATC) and T7 term (TGCTAGTTATTGCTCAGC) (7).
   - Isolate the plasmid and determine the nucleotide sequence of the insert.

3.2. Media and Culture Conditions

1. Transform the protein production strain of *E. coli* BL21 (BL3) pLysS with pET31b(+) encoding KSI-antimicrobial peptide as described in Paragraph 5 in **Section 3.1**.
2. From a freshly streaked plate, transfer a single bacterial colony to 10 mL LB broth containing ampicillin (100 μg/mL). Incubate at 37°C overnight in a shaking incubator at 160 rpm.
3. Dilute the overnight inoculums 1:1,000 into the TB broth containing ampicillin and incubate in a shaker at 37°C and 220 rpm.
4. Add 0.4 mM IPTG when culture reaches an OD_{600} of 0.8.

5. Four hours after induction, centrifuge the bacterial culture at 5,000*g* force for 10 min and carefully decant supernatant and hold tube, inverted, for several seconds to drain the residual medium. Blot mouth of inverted tube on paper towel to remove any residual medium.

3.3. Purification of the Expressed Proteins and Cleavage from Fusion Protein

1. Suspend the cell pellet in the 10 mL lysis buffer and lyse by ultrasound sonicator in the water bath cooled with ice for 20 min with 3 or 4 s pulses at 50% duty cycle.
2. Centrifuge the bacterial lysate at 15,000*g* force for 15 min at 4°C.
3. Wash the insoluble inclusion bodies with a succession of wash buffers. Centrifuge the mixture at 15,000*g* force for 10 min at 4°C after every wash.
4. Dissolve the insoluble inclusion bodies in 6 M guanidine-HCl and centrifuge.
5. Dialyze the soluble supernatant against the deionized water. Centrifuge the precipitate of KSI–peptide fusion protein at 15,000*g* force for 10 min at 4°C (*see* **Note 2**).
6. Analyze the produced protein by SDS-PAGE (**Fig. 5.3**).

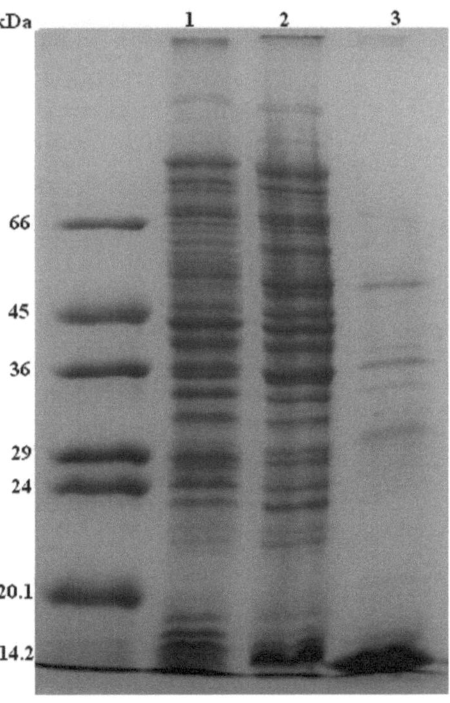

Fig. 5.3. SDS-PAGE gel in which lanes 1 and 2 represent total bacterial lysate of non-induced bacteria (1) and bacteria 4 h after IPTG induction (2). Lane 3 represents inclusion bodies after purification.

3.3.1. Cleavage of the Fusion Protein Containing Met Residue Between the KSI and Peptide

1. Dissolve the dried sample in 70% TFA to a concentration of the fusion protein KSI-antimicrobial peptide of 10 mg/mL.
2. Add cyanogen bromide to the sample solution at 100-fold molar excess over the concentration of methionyl residues in the fusion protein (*see* **Notes 5** and **6**).
3. Bubble nitrogen through the solution for 1 min, wrap the flask in aluminum foil to protect from light, and incubate at room temperature for 24 h.
4. Dry the reaction product under vacuum in a rotary evaporator at 30°C and wash three times with water and evaporate.
5. Centrifuge the suspension at 15,000*g* force for 15 min at 4°C.
6. Concentrate the soluble antimicrobial peptide with rotary evaporation and analyze with RP-HPLC.

3.3.2. Acid Cleavage of the Asp–Pro Bond Between the KSI and Peptide

1. Asp–Pro bond is acid labile and can be cleaved by incubation at pH < 2. Resuspend the fusion protein in 90 mM HCl 10% (v/v), mix the suspension for 2 h at 85°C (*see* **Notes 3** and **7**).
2. Neutralize the reaction suspension to natural with NaOH solution.
3. Centrifuge and concentrate the soluble peptide with rotary evaporation.
4. Purify the peptide with HPLC on a C5 RP-HPLC column using a gradient of 5% acetonitrile to 95% acetonitrile and 5 mM HCl (*see* **Note 4**).
5. Check the molecular weight of the peptide fractions by mass spectrometer.
6. Typical yield is 15 mg/L of bactericidal culture.

3.4. Production of ^{15}N-Enriched Recombinant Antimicrobial Peptides

1. Production of ^{15}N-enriched peptides is performed in M9 minimal medium, where $^{15}NH_4Cl$ is used as the sole source of nitrogen. **Table 5.3** shows the component of 1 L of the medium for the production of isotope-labeled peptides (*see* **Note 8**). The rest of the procedure is the same.
2. Isotope labeling of the recombinant peptide was exacted by MS or NMR.

4. Notes

1. An *E. coli* expression vector should contain, apart from the gene of interest, an origin of replication, a gene that confers antibiotic resistance, a promoter, and a transcription ter-

Table 5.3
Components of 1 L of the medium for the production of ^{15}N-antimicrobial peptide

1. 985 mL of M9-15 N medium, pH 7.4 (autoclaved)	per L
$Na_2HPO_4 \cdot 2H_2O$	7.5 g
KH_2PO_4	3.0 g
NaCl	0.5 g
2. 0.5 mL of trace salt (sterile filtered)	per L
$CuSO_4 \cdot 7H_2O$	6.0 g
NaI	0.08 g
$MnSO_4 \cdot 7H_2O$	3.0 g
$Na_2MoO_4 \cdot 2H_2O$	0.2 g
H_3BO_3	0.02 g
$CoCl_2$	0.5 g
$ZnCl_2$	20.0 g
$FeSO_4 \cdot 7H_2O$	65.0 g
Biotin	0.2 g
H_2SO_4	5.0 mL
3. 1 g $_{15}NH_4Cl$	
4. 10 mL glucose (300 mg/L) (sterile filtered)	
5. 3 mL 1 M $MgSO_4$ (autoclaved)	
6. 2 mL ampicillin (100 mg/mL)	

minator. The pET vector system, which is one of the most common expression system used in laboratory-scale cultivations, has an ampicillin resistance which is mostly used only on laboratory scale, because β-lactamase, the enzyme that confers the resistance, degrades ampicillin and thus the selective pressure is lost after a few generations of cell growth. The pET vector system has a T7 promoter, which is transcribed only by T7 RNA polymerase and must be used in a strain carrying a chromosomal T7 polymerase gene which is under the control of a *lac* promoter (16). The T7 RNA polymerase is under the control of isopropyl β-D-thiogalactopyranoside (IPTG)-inducible promoter. We, used the pET31b(+) vector (**Fig. 5.1**), which is designed for cloning and high-level expression of peptide sequences fused with the ketosteroid isomerase protein. Sequence coding for an antimicrobial peptides is subcloned into *Alw*NI site downstream of the ketosteroid isomerase (KSI) fusion partner. This KSI has been modified to remove all methionine residue in protein (17). Therefore, between the fusion protein and antimicrobial peptide we inserted either the residue Met or a dipeptide Asp-Pro, which are unique and allow chemical cleavage between the KSI and peptide. Optimal growth and expression condition for the protein production was established by using TB medium, where the production of the protein was five times higher than in LB medium.

2. Intracellular degradation of mostly unstructured short antimicrobial peptides or its toxicity to host cell is prevented by fusion of peptides to larger proteins. We used a system with ketosteroid isomerase, a protein that is highly insoluble in water as well as in the cytoplasm of bacteria (7). Like all fusion systems that use larger proteins, the absolute productivity of the system is relatively low because of the intrinsic limitation that only small part of the produced recombinant protein is actually the antimicrobial peptides of interest, so that even when fusion protein expression is very high, antimicrobial peptide expression can at best be only moderate (5). In our case approximately 9% of the whole fusion proteins represent antimicrobial peptides. This can be remedied by insertion of peptide concatamers to increase the fraction of the peptide product within the fusion protein (18). Additionally, forcing proteins into inclusion bodies can also be a useful strategy for purification. After cleaning and removal of residues of bacterial wall, we dissolved inclusion bodies in guanidine-HCl, and after dialysis against water more than 90% pure fusion protein is precipitated. Participation occurs rapidly, but we dialyzed the fusion protein overnight to precipitate the maximal quantity of the fusion protein.

3. Peptide bonds of aspartyl residues may be cleaved in diluted acid at rates at least 100 times greater than other peptide bonds, and moreover the aspartyl–prolyl bonds are the most labile of the aspartyl–peptide bonds. An advantage of this type of peptide cleavage is that it is not necessary to solubilize the inclusion bodies in order for this reaction to be effective. Fusion partner (KSI) in this reaction is in insoluble form while the liberated peptide is soluble. Conditions of temperature, incubation time, concentration of hydrochloric acid, and weight/volume concentration of inclusion bodies affect the efficiency of acid hydrolysis. Our results show that optimal conditions for maximal acid cleavage of peptides from fusion protein were achieved at 90 mM HCl in 2 h at 85°C and 10% weight/volume concentration of inclusion bodies. In case that the weight/volume concentration of inclusion bodies is increased the reaction rate decreased significantly. The optimal pH for acid hydrolysis is around 1.5–2. Increase

of the incubation time in 90 mM HCl at 85°C above 2.5 h decreased the proportion of the liberated antimicrobial peptides, probably due to the unspecific cleavage the other peptide bonds in the antimicrobial peptide, as shown in **Fig. 5.4**. The mechanism of action of HCl on Asp-containing peptides is shown in **Fig. 5.5**.

4. Traditionally, TFA is used in the mobile phase for RP-HPLC separations, because of its effectiveness in solubilizing hydrophobic peptides. However, although it is a volatile component, it remains in sample after drying, suggesting that it is strongly bound to the cationic peptides (19). After an analysis of the product, from the RP-HPLC (**Fig. 5.6**) column, on the mass spectrophotometer, we detect two products. The molecular mass of the first product corresponds to the peptide, but there is an additional product with the molecular weight increased for a unit of TFA. Since acidic conditions are needed in the hydrophobic steps of purification, we replaced TFA with HCl. Our results demonstrate that the use of HCl as a replacement for TFA in the purification procedure did not affect peptide recovery shown in **Fig. 5.7**. Moreover, HCl can be removed easily from the sample, while removal of TFA from

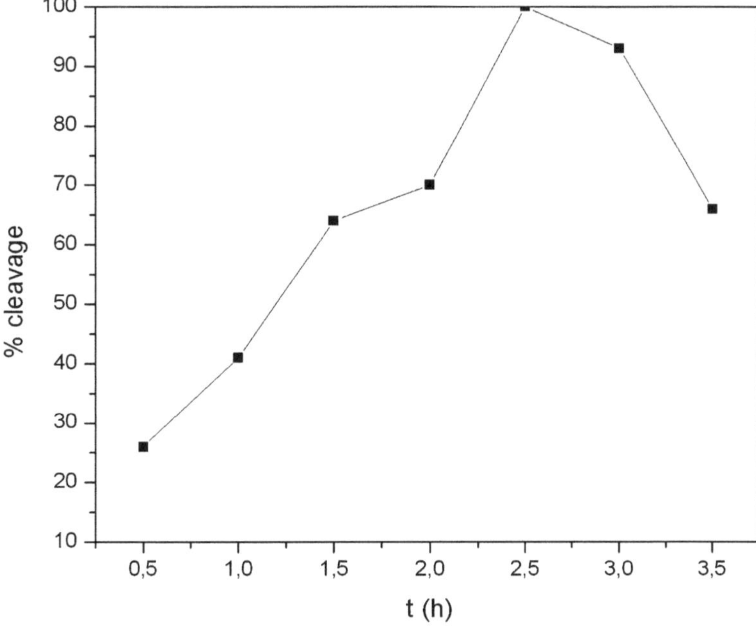

Fig. 5.4. Time dependence of peptide cleavage from fusion protein. The fusion protein KSI–antimicrobial peptides was suspended in diluted acid (90 mM HCl) and mixed at 85°C to cleave the aspartyl–proline bond between the fusion protein and peptide.

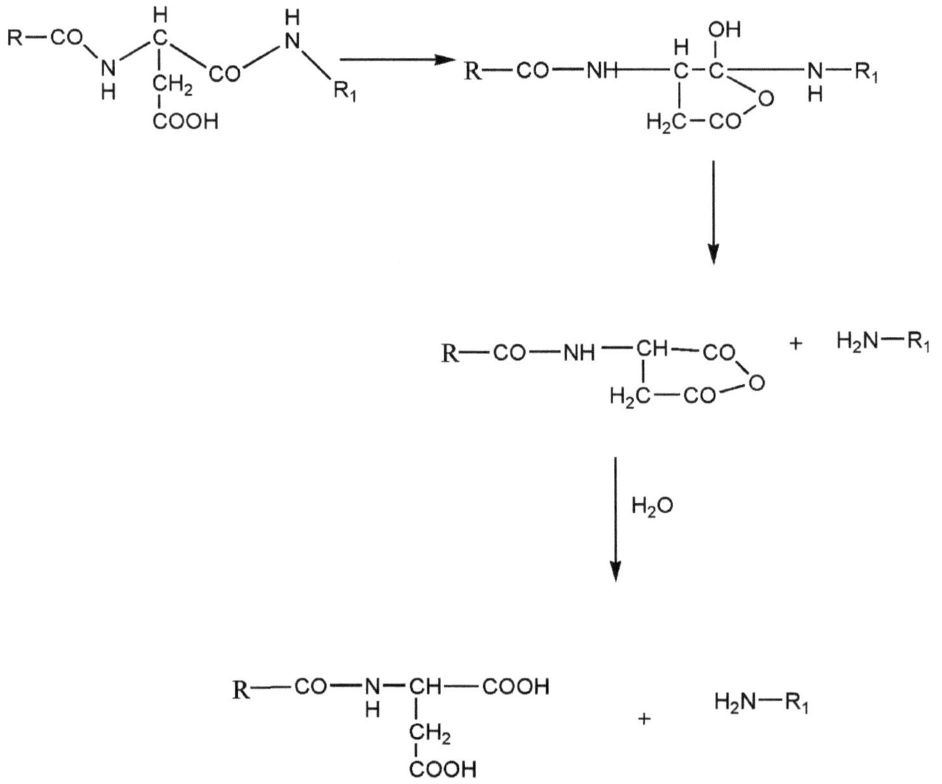

R—CO—NH—CH(CH_2COOH)—CO—NH—R_1 → R—CO—NH—CH—C(OH)(—O—CO—H_2C)—NH—R_1

↓

R—CO—NH—CH—CO—O—CO—CH_2 (anhydride) + H_2N—R_1

↓ H_2O

R—CO—NH—CH(—CH_2—COOH)—COOH + H_2N—R_1

Fig. 5.5. Mechanism of the cleavage of bonds to the COOH site of aspartyl residues in dilute acid, R1 amino acid residue represents proline.

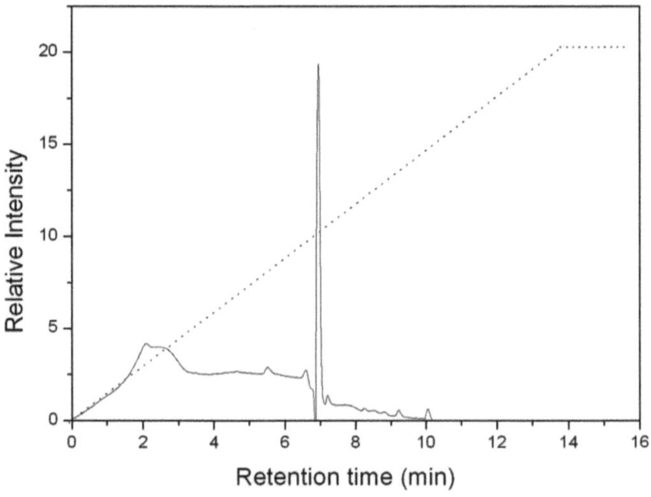

Fig. 5.6. Reversed-phase HPLC purification of recombinant antimicrobial peptides after HCl cleavage of the KSI–antimicrobial peptide fusion. The peak at 7 min represents purified recombinant peptide.

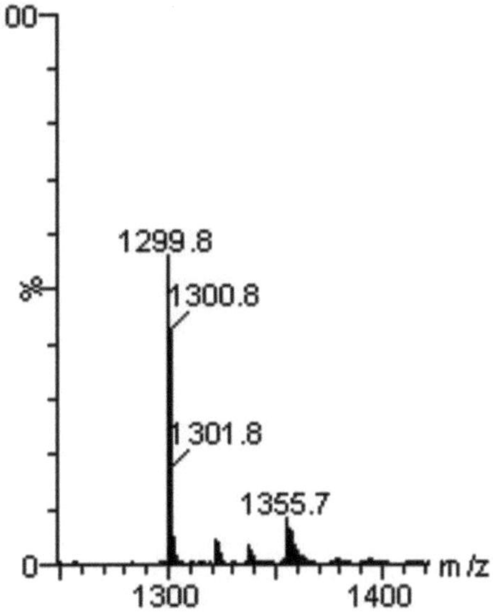

Fig. 5.7. Mass spectra of recombinant antimicrobial peptide (theor. 1,299.5 Da) after purification with HPLC.

samples required washing in dialysis membranes by several steps.

5. Note that cleavage at the methionine residues with CNBr depends on pH. In the alkaline pH range, cyanogen bromide reacts with basic groups in proteins, but under neutral and acidic conditions, only methionine and cysteine are attacked by cyanogen bromide. The conditions for cleavage should be mild enough to avoid nonspecific cleavage of other peptide bonds. After cleavage with CNBr, the KSI precipitated in aqueous solution, while the soluble peptide was separated from other contaminants by RP-HPLC. The mechanism of reaction of cyanogen bromide with methionine-containing peptides is shown in **Fig. 5.8**.
6. In comparison to the method of cleaving peptide bonds with HCl, which is fast and also not toxic, the method to liberate antimicrobial peptides from fusion protein with CNBr takes more time and caution due to its toxicity. Cyanogen bromide is volatile and readily absorbed through the skin or gastrointestinal tract. Therefore, toxic exposure may occur by inhalation, physical contact, or ingestion. It is acutely toxic, causing a variety of nonspecific symptoms.
7. Proteases can also be used to cleave the peptide bond between the fusion protein and antimicrobial peptides but are not always appropriate because some of them leave one or two amino acid residues at the N-terminus of the peptide and cleavage sometimes occurs at site distinct from the enzyme's nominal specificity (20). Additionally, proteases act

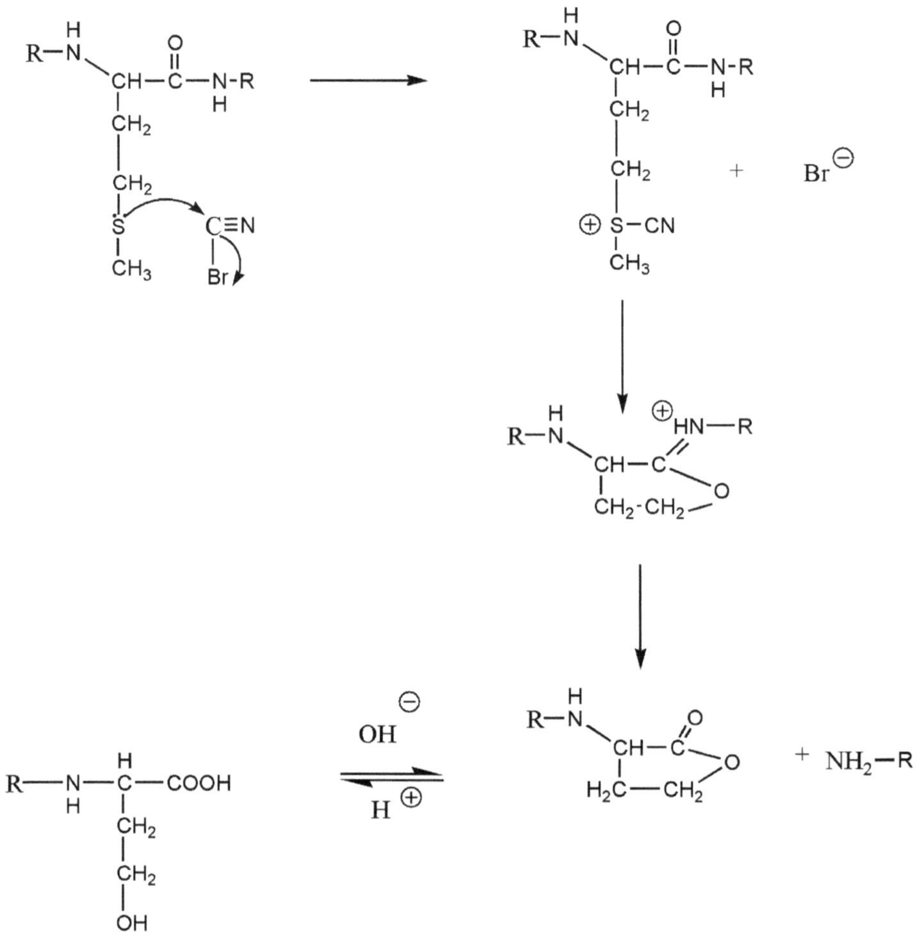

Fig. 5.8. Mechanism of cleavage of the Met–X bonds by cyanogen bromide.

poorly against insoluble substrates and are sensitive to strong protein solubilizers, such as SDS, TFE, or guanidine-HCl.

8. Note that growth and protein production of bacteria in M9 minimal medium is significantly slower than in complex medium such as LB or TB medium and bacteria take some time to adapt to the growth in the minimal medium. Other procedures in the preparation of isotopically labeled peptides remain identical.

References

1. Hara, S. and Yamakawa, M. (1996) Production in *Escherichia coli* of moricin, a novel type antibacterial peptide from the silkworm, *Bombyx mori*. *Biochem. Biophys. Res. Commun.* **220**, 664–669.
2. Shen, Y., Lao, X. G., Chen, Y., Zhang, H. Z., and Xu, X. X. (2007) High-level expression of cecropin X in *Escherichia coli*. *Int. J. Mol. Sci.* **8**, 478–491.
3. Cipakova, I., Gasperik, J., and Hostinova, E. (2006) Expression and purification of human antimicrobial peptide, dermcidin, in *Escherichia coli*. *Protein Expr. Purif.* **45**, 269–274.
4. Kim, H. K., Chun, D. S., Kim, J. S., Yun, C. H., Lee, J. H., Hong, S. K., and Kang, D. K. (2006) Expression of the cationic antimicrobial peptide lactoferricin fused with

the anionic peptide in *Escherichia coli. Appl. Microbiol. Biotechnol.* **72**, 330–338.
5. Ingham, A. B. and Moore, R. J. (2007) Recombinant production of antimicrobial peptides in heterologous microbial systems. *Biotechol. Appl. Biochem.* **47**, 1–9.
6. Xu, Z. N., Peng, L., Zhong, Z. X., Fang, X. M., and Cen, P. L. (2006) High-level expression of a soluble functional antimicrobial peptide, human beta-defensin 2, in *Escherichia coli. Biotechnol. Progress.* **22**, 382–386.
7. Majerle, A., Kidric, J., and Jerala, R. (2000) Production of stable isotope enriched antimicrobial peptides in *Escherichia coli*: an application to the production of a N-15-enriched fragment of lactoferrin. *J. Biomol. Nmr.* **18**, 145–151.
8. Wei, Q. D., Kim, Y. S., Seo, J. H., Jang, W. S., Lee, I. H., and Cha, H. J. (2005) Facilitation of expression and purification of an antimicrobial peptide by fusion with baculoviral polyhedrin in *Escherichia coli. Appl. Environ. Microb.* **71**, 5038–5043.
9. Hwang, S. W., Lee, J. H., Park, H. B., Pyo, S. H., So, J. E., Lee, H. S., Hong, S. S., and Kim, J. H. (2001) A simple method for the purification of an antimicrobial peptide in recombinant *Escherichia coli. Mol. Biotechnol.* **18**, 193–198.
10. Lee, J. H., Kim, J. H., Hwang, S. W., Lee, W. J., Yoon, H. K., Lee, H. S., and Hong, S. S. (2000) High-level expression of antimicrobial peptide mediated by a fusion partner reinforcing formation of inclusion bodies. *Biochem. Biophys. Res. Commun.* **277**, 575–580.
11. Reichhart, J. M., Petit, I., Legrain, M., Dimarcq, J. L., Keppi, E., Lecocq, J. P., Hoffmann, J. A., and Achstetter, T. (1992) Expression and secretion in yeast of active insect defensin, an inducible antibacterial peptide from the fleshfly phormia-terraenovae. *Invertebr. Reprod. Dev.* **21**, 15–24.
12. Andersons, D., Engstrom, A., Josephson, S., Hansson, L., and Steiner, H. (1991) Biologically-active and amidated cecropin produced in a baculovirus expression system from a fusion construct containing the antibody-binding part of protein-A. *A. Biochem. J.* **280**, 219–224.
13. Rao, X. C., Li, S., Hu, J. C., Jin, X. L., Hu, X. M., Huang, J. J., Chen, Z. J., Zhu, J. M., and Hu, F. Q. (2004) A novel carrier molecule for high-level expression of peptide antibiotics in *Escherichia coli. Protein Expr. Purif.* **36**, 11–18.
14. Moon, J. Y., Henzler-Wildman, K. A., and Ramamoorthy, A. (2006) Expression and purification of a recombinant LL-37 from *Escherichia coli. Biochim. Biophys. Acta.* **1758**, 1351–1358.
15. Zorko, M., Japelj, B., Hafner-Bratkovic, I., and Jerala, R. (2009) Expression, purification and structural studies of a short antimicrobial peptide. *Biochim. Biophys. Acta* **1788**, 314–323.
16. Jonasson, P., Liljeqvist, S., Nygren, P. A., and Stahl, S. (2002) Genetic design for facilitated production and recovery of recombinant proteins in *Escherichia coli. Biotechol. Appl. Biochem.* **35**, 91–105.
17. Kuliopulos, A. and Walsh, C. T. (1994) Production, purification, and cleavage of tandem repeats of recombinant peptides. *J. Am. Chem. Soc.* **116**, 4599–4607.
18. Lee, J. H., Skowron, P. M., Rutkowska, S. M., Hong, S. S., and Kim, S. C. (1996) Sequential amplification of cloned DNA as tandem multimers using class-IIS restriction enzymes. *Genetic Anal. Biomol. Eng.* **13**, 139–145.
19. Gaussier, H., Morency, H., Lavoie, M. C., and Subirade, M. (2002) Replacement of trifluoroacetic acid with HCl in the hydrophobic purification steps of pediocin PA-1: a structural effect. *Appl. Environ. Microb.* **68**, 4803–4808.
20. Kohno, T., Kusunoki, H., Sato, K., and Wakamatsu, K. (1998) A new general method for the biosynthesis of stable isotope-enriched peptides using a decahistidine-tagged ubiquitin fusion system: an application to the production of mastoparan-X uniformly enriched with N-15 and N-15/C-13. *J. Biomol. Nmr.* **12**, 109–121.
21. Haught, C., Davis, G. D., Subramanian, R., Jackson, K. W., and Harrison, R. G. (1998) Recombinant production and purification of novel antisense antimicrobial peptide in *Escherichia coli. Biotechnol. Bioeng.* **57**, 55–61.
22. Skosyrev, V. S., Rudenko, N. V., Yakhnin, A. V., Zagranichny, V. E., Popova, L. I., Zakharov, M. V., Gorokhovatsky, A. Y., and Vinokurov, L. M. (2003) EGFP as a fusion partner for the expression and organic extraction of small polypeptides. *Protein Expr. Purif.* **27**, 55–62.
23. Fassina, G., Merli, S., Germani, S., Ciliberto, G., and Cassani, G. (1994) High-yield expression and purification of human endothelin-1. *Protein Expr. Purif.* **5**, 559–568.

24. Xu, X. X., Jin, F. L., Yu, X. Q., Ji, S. X., Wang, J., Cheng, H. X., Wang, C., and Zhang, W. Q. (2007) Expression and purification of a recombinant antibacterial peptide, cecropin, from *Escherichia coli. Protein Expr. Purif.* **53**, 293–301.
25. Morassutti, C., De Amicis, F., Bandiera, A., and Marchetti, S. (2005) Expression of SMAP-29 cathelicidin-like peptide in bacterial cells by intein-mediated system. *Protein Expr. Purif.* **39**, 160–168.
26. Wei, Q. D., Kim, Y. S., Seo, J. H., and Cha, H. J. (2005) Facilitation of expression and purification of antimicrobial peptide by fusion with baculoviral polyhedrin in *Escherichia coli. Appl. Environ. Microbiol.* **71**, 5038–5043.
27. Moon, W. J., Hwang, D. K., Park, E. J., Kim, Y. M., and Chae, Y. K. (2007) Recombinant expression, isotope labeling, refolding, and purification of an antimicrobial peptide, piscidin. *Protein Expr. Purif.* **51**, 141–146.
28. Tang, H. Y. and Speicher, D. W. (2004) Identification of alternative products and optimization of 2-nitro-5-thiocyanatobenzoic acid cyanylation and cleavage at cysteine residues. *Anal. Biochem.* **334**, 48–61.

Chapter 6

Methods for Building Quantitative Structure–Activity Relationship (QSAR) Descriptors and Predictive Models for Computer-Aided Design of Antimicrobial Peptides

Olivier Taboureau

Abstract

Antimicrobial peptides are ubiquitous in nature where they play important roles in host defense and microbial control. More than 1,000 naturally occurring peptides have been described so far and those considered for pharmaceutical development have all been further optimized by rational approaches. In recent years, high-throughput screening assays have been developed to specifically address optimization of AMPs. In addition to these cell-based in vivo systems, a range of computational in silico systems can be applied in order to predict the biological activity of AMPs for specific bacteria. Among them, quantitative structure–activity relationships (QSARs) method, which attempts to correlate chemical structure to biological measurement, has shown promising results in the optimization and discovery of peptide candidates. Therefore, this chapter is devoted to describe the QSAR method and recent progress applied in AMP.

Key words: QSAR, molecular descriptors, antimicrobial peptide design, PLS and machine learning methods.

1. Introduction

Antimicrobial peptides (AMPs) constitute a diverse group of compounds that serve a common goal, that is, host organism defense from infection. They are widely distributed in animals, plants, and microbes and are among the most ancient host defense factors. In microbes, they eliminate competing microorganisms occupying the same ecological niches. In higher organisms, they constitute part of the immune system aiding the regulation of endogenous

A. Giuliani, A.C. Rinaldi (eds.), *Antimicrobial Peptides*, Methods in Molecular Biology 618,
DOI 10.1007/978-1-60761-594-1_6, © Springer Science+Business Media, LLC 2010

microbial flora and the response against invading pathogenic microorganisms (1, 2).

Due to their antimicrobial properties, these molecules attract practical interest as potential antibiotics for medical and veterinary use. In vitro and in vivo screening systems have been developed in order to optimize the therapeutic effect of specific AMPs (3). Another desirable goal is to develop a computational (or in silico) system capable of automatically evaluating very large compound libraries, thereby limiting the vast sequence space to be screened by these more traditional methods. In silico approaches based on computer-aided design and quantitative structure–activity relationship (QSAR) analysis can rationalize the design of potent therapeutic AMPs (4, 5). The main idea behind QSAR is that any chemical sequence or structure can be described with measured or calculated physico-chemical parameters, also called descriptors, and correlated to their respective biological activity using diverse mathematical models. Then, these models can be used for the prediction of AMP analogs with a potentially higher biological activity (**Fig. 6.1**). Of course, different aspects have to be taken into consideration in order to develop accurate predictive model. Information about the AMP studied, sequence or structure information, the choice of descriptors, the choice of mathematical model can have some major effect in the accuracy of the QSAR model prediction. Description of the QSAR methods and aspects to consider will be covered in this issue for the design of new AMPs.

2. Materials

2.1. AMP Structure

In general, AMPs range in size from 6 amino acids up to around 60. To date, more than 1,000 different AMPs have been isolated which can be divided into several classes based on bioactivity, structural features, and/or amino acid composition (6). The simplest structures are small, elongated α-helical peptides. Other AMPs fold into looped structures, β-sheets, or form compact and rigid, disulfide-bridged tertiary structures. Short linear or cyclic amphiphilic peptides that contain both L-amino acids and D-amino acids are also created in synthetic laboratory. To guide lead optimization, high-resolution X-ray crystallography or nuclear magnetic resonance (NMR) structure has been done for some AMPs (3). Knowledge of the spatial location of specific residues as well as the conformation and direction of the side chains can be used in connection with computer-aided design tools and QSAR, to further facilitate rational optimization of the lead backbones.

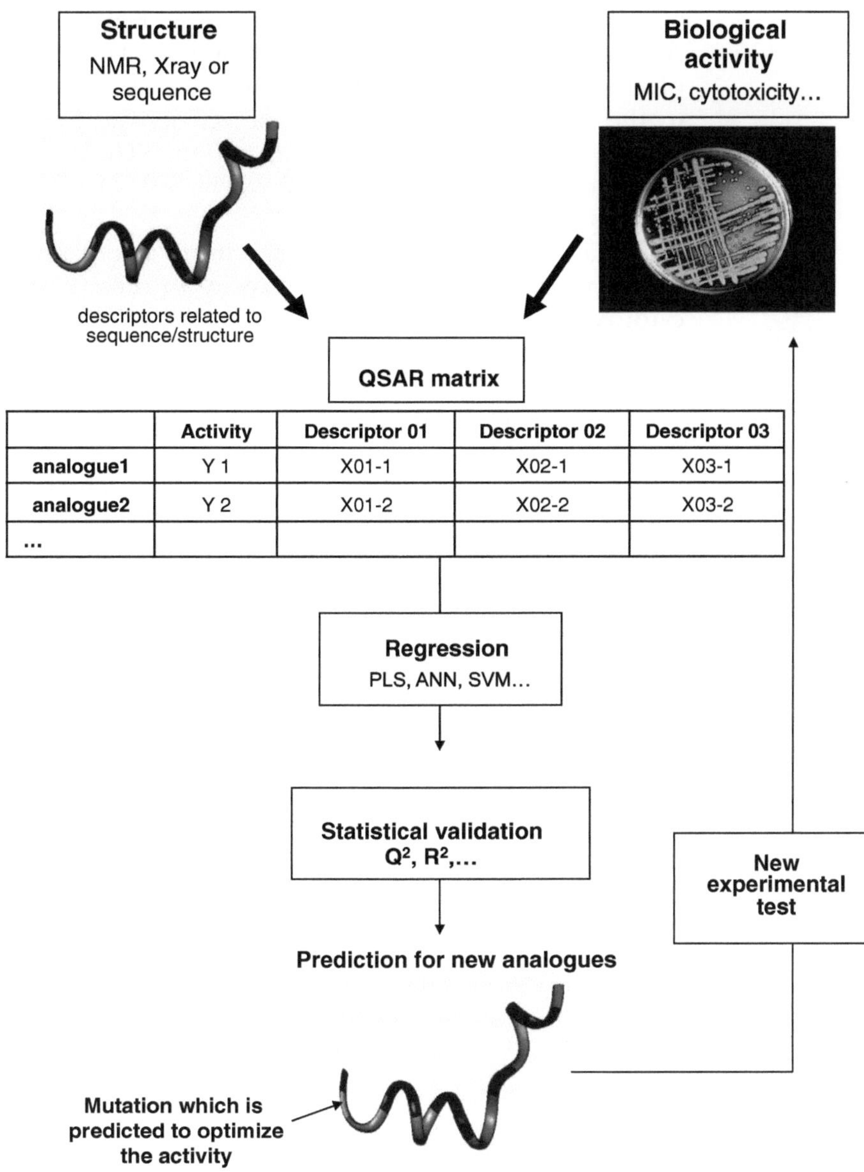

Fig. 6.1. General overview of the material and method in order to develop a QSAR model.

2.2. AMP Biological Activity

Biological activity of AMPs is required to develop QSAR models. A typical biological activity measurement of AMPs is to determine the efficacy of AMP in order to reduce the growth or to kill a microbe. The antimicrobial activity can be determined by measuring the minimal inhibitory concentration (MIC), which is the lowest concentration of AMP that reduces growth by more than 50% after overnight incubation.

Another activity often considered is the hemolytic activity as AMPs show some blood cell activities as well. Therefore, the lysis of erythrocyte is an activity measurement used also in parallel to the MIC activity. The hemolytic activity can be determined by a "half-maximal effective concentration (EC_{50})," which refers to the concentration of an AMP where 50% of the population exhibits a response.

Finally, AMPs are usually not specific to an organism but they have a tendency to show some toxicity to cells (7). Therefore, cytotoxicity assays, based on EC_{50}, are also used to evaluate the toxicity of AMPs to cells (*see* **Note 1**).

2.3. AMP Properties (Descriptors)

To generate a QSAR model, the sequence or preferably the molecular structure of the peptide, using the methods described in **Section 2.1**, must be described by diverse physico-chemical parameters. Although sequence and structure of AMPs can be very diverse, almost all AMPs are cationic and amphipathic, features that promote interaction with the negatively charged bacterial and fungal membrane. Therefore, these features and parameters related to amphipathicity can be a relevant starting point in the development of QSAR (**Fig. 6.2**) (8, 9).

Basically, the diversity of the descriptors is large. They can be related to not only experimental data such as molecular weight, partition coefficient, or HPLC retention time, but also theory based on the 2D or 3D structure of AMP studied.

In this area, the first theoretical descriptors associated with antibacterial and antimicrobial peptides were based on the work of Hellberg et al. (10). They have designed 29 experimental parameters, like the molecular weight, log P, pKa, for each of the 20 natural amino acids and have concatenated them into three parameters, called "z scores," using a principal component analysis (PCA) (11). The main advantage is that the descriptors are based on the amino acid sequence and the molecular structure is not needed.

Taking advantage of the 3D structure information, more complex descriptors can be designed in order to develop more robust QSAR model. Frecer et al. (7). used a combination of 14 descriptors including features such as charge, lipophilicity, and molecular surface area to analyze protegrin analogs. Mee et al. (12). combined more of 1,000 descriptors based on molecular properties to describe a 15-residue cecropin/melittin peptide hybrid (CAMEL 0). More recently, QSAR models based on molecular interaction fields (4) and inductive QSAR descriptors (5) showed promising results. Whereas, the molecular interaction field descriptors intend to mimic the binding energy between the antimicrobial peptides and its environment (represented by atomic probes), the inductive QSAR descriptors are based on the intramolecular steric effects, molecular capacitance,

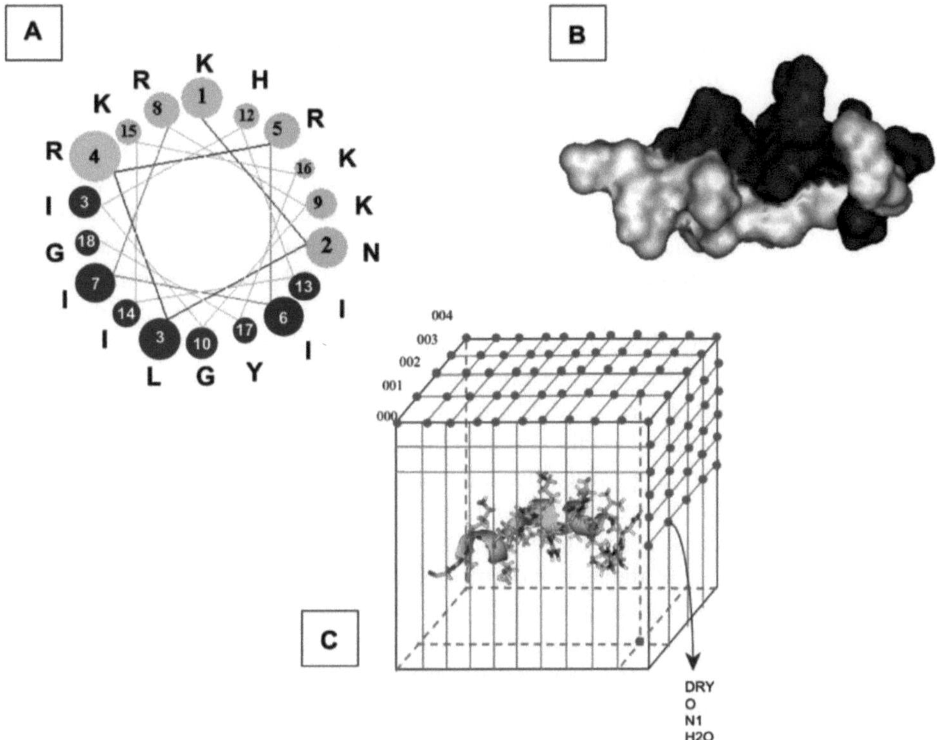

Fig. 6.2. Example of descriptors which can be used with Novispirin. (**a**) Descriptors based on the sequence and amphipathicity. The sequence is represented with a so-called wheel helix. In grey are polar residues and in black nonpolar residues. (**b**) Based on the AMP structure the surface and volume of hydrophobic area (*white*) and hydrophilic residue (*black*) can be computed. (**c**) Descriptors based on the molecular interaction field of the AMP and probes (defined by atoms). On each grey dot interaction energy between a probe and the AMP is computed.

and electronegativities. It should be noted, both methods are 3D-sensitive and depend on the structure of AMPs (*see* **Note 2**).

3. Methods

3.1. QSAR Methods (Linear/Nonlinear)

Historically, the QSAR concept was elaborated by Hammett in the 1940s and Taft in the 1950s. But it became really popular with the work on Hansch in the drug design area in the 1970s. QSAR are considered state-of-the-art methods for prediction of new substrate or new drug in drug design and are routinely used for understanding the relationships between chemical and their biological, pharmacological, and toxicological actions (13). With AMP, it is of relevance to understand how modifications in peptide sequences may lead to better performing drugs and such understanding is possible through mathematical models such as QSAR (*see* **Note 3**). To develop such correlation between structure and biological activity, many statistical learning methods and multivariate approach can be applied, categorized as linear or nonlinear.

3.1.1. Linear

Linear method was primarily used with AMP (14, 15). In linear method, principal component analysis (PCA), partial least squares projections to latent structures (PLS), and multivariate design are usually associated with each other. Typically, the development of a QSAR model start with the assembling of information, where for a certain set of AMP analogs, biological activity are characterized (defined as Y space) and relevant descriptors reflecting their chemical and structural properties (defined as X space) are associated with each AMP. Such a matrix can be difficult to handle as it is. Therefore, PCA is a mathematical procedure which is able to transform a number of possibly correlated variables into a smaller number of uncorrelated variables called principal components. It is a way to explore the relationships among AMP analogs and among variables. Then the relation between the biological activity and these principal components associated for each AMP can be handled using a PLS regression method. PLS model tries to find the multidimensional direction in the X space that explains the maximum multidimensional variance direction in the Y space. The main advantage of PLS models is the possibility to interpret the influence of the descriptors at the output prediction.

3.1.2. Nonlinear

Recently, QSAR approaches based on nonlinear methods have been used with AMPs. These methods are essentially artificial neural network (ANN) and support vector machines (SVM) (16, 17).

ANN is an artificial intelligence which tries to simulate some properties of neural networks. In AMP, ANN involves a network of descriptors which can be defined as input nodes or neurons. These nodes are connected together and form a network which after transformation in a hidden layer sum up into an output node (**Fig. 6.3a**).

SVM is a statistical learning method which minimizes the so-called structural risk in prediction. SVM is essentially used as a classification method, where two classes are separated by a hyperplane (H). The optimized plane is defined as the one that maximizes the margin between the two classes which is measured by the distances between plane H and the planes cutting the nearest sample points on both sides of H, namely, H1 and H2. In particular, the sample points located exactly on planes H1 and H2 are defined as support vectors (**Fig. 6.3b**).

Both methods are considered to give better results, but the interpretation of the models is rather hard and complex compared to linear models (*see* **Note 4**).

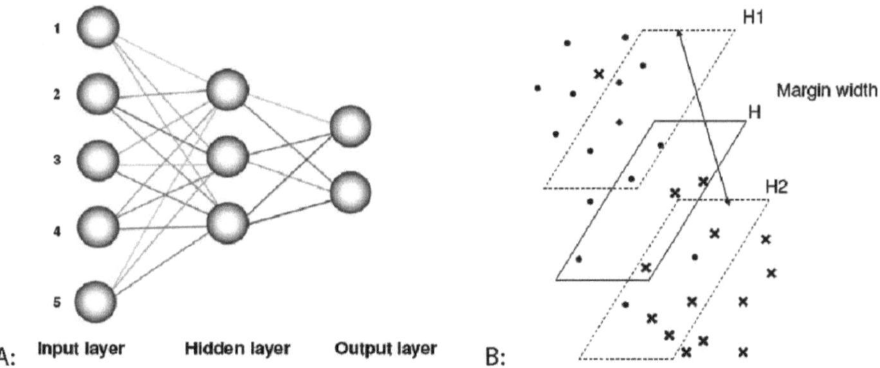

Fig. 6.3. (**a**) Artificial neural network: the input layer represents the descriptors. The hidden layer represents the transformations of the input layer to a reduced level. The output is the final value which can be associated with the AMP considered. (**b**) Support vector machine: H represents the classification of hyperplane. H1 and H2 hyperplanes are defined by the *dots* and the *crosses* which allow the best separation between two classes (also called support vector). The margin width is the distance between the hyperplanes H1 and H2.

3.2. Normalization and Descriptor Selection

In most of the cases, normalization of the descriptor set is recommended for a QSAR analysis. The diversity of descriptors and the scale of values associated with them can be varied from a factor of 1,000 or even more. Therefore, normalization of all the descriptors to a similar scale can help in the accuracy of the model.

Some descriptors are not suitable and do not give any information to the model. Some other descriptors are redundant or show a very low variance between each AMP analog. Instead of providing relevant information, they can reduce the quality of the QSAR model prediction. Therefore, in order to remove the noisy variables and improve the predictive abilities of the PLS model, descriptor selection procedure can be performed using methods such as genetic algorithm (GA) or D-optimal design (18). The genetic algorithm is based on the evolutionary theory of Darwin (19). Each AMP is defined by an initial vector (called also initial population) encoding all available descriptors in a bit-string. Then, a new population is generated randomly, covering the entire range of possible solutions. During each successive generation, a proportion of the existing population is selected to breed a new generation. Individual solutions are evaluated through a fitness or cost function before being selected for the next generation. The procedure stops when a solution that satisfies the minimum criteria (usually correlation coefficient or cross-validation) is found. D-optimal design, instead, is more suitable for calibrating linear model. It is an iterative search algorithm which seeks to minimize the covariance of the descriptors estimated for a specified model. This is equivalent to maximizing the determinant $\mathbf{D} = |\mathbf{X}^{\mathrm{T}}\mathbf{X}|$ where **X** is the design matrix of descriptors evaluated at specific treatments in the design space (the set of AMPs). By selecting variables with higher spread and less correlation,

D-optimal design will conserve more complementary information (20). D-optimal design was used, for example, for a selection of a representative set of cecropin/melittin peptide according to the information spanned in the descriptor space considered.

3.3. Quality Assessment

The quality and the predictive ability of the QSAR models are assessed, in general, with the leave-one-out cross-validation q^2 and the correlation coefficient r^2 (defined also as Pearson correlation coefficient (PCC)) (*see* **Note 5**).

Matthews correlation coefficient (MCC) is another measure used for the assessment of binary classification. It takes into account the correct class prediction (i.e., active/non-active) and is essentially associated with SVM and ANN methods.

$$\mathrm{PCC} = \frac{\sum xy - \frac{\sum x \sum y}{n-1}}{\sqrt{\left(\sum x^2 - \frac{(\sum x)^2}{n-1}\right)\left(\sum y^2 - \frac{(\sum y)^2}{n-1}\right)}}$$

where x is experimental bioactivity, y predicted activity, and n number of data points.

$$\mathrm{MCC} = \frac{(\mathrm{TP} \times \mathrm{TN}) - (\mathrm{FP} \times \mathrm{FN})}{\sqrt{(\mathrm{TP}+\mathrm{FP})(\mathrm{TP}+\mathrm{FN})(\mathrm{TN}+\mathrm{FP})(\mathrm{TN}+\mathrm{FN})}}$$

where TP represents the correct predicted active class (True Positive), TN correct predicted non-active (True Negative), FP is incorrect predicted active class (False Positive), and FN incorrect predicted non-active class (False Negative).

Both statistic measurements return a value between –1 and +1, with a perfect prediction at +1. QSAR models with PCC and MCC values above 0.4 are considered to be predictive.

4. Notes

1. Integration of experimental biological activity from different studies of the development of a QSAR model has to be performed with caution. The variation of the protocols and techniques to obtain a biological activity can influence drastically the accuracy of a QSAR model. Instead, a classification model based on SVM, ANN, or PLS-DA can be utilized as it uses a binary system (active/non-active). Nevertheless, consistent biological activity measurement is preferred in the development of a QSAR model.

2. If the information about the 3D structure of AMPs is available, descriptors which take into account this information, like molecular interaction field descriptors or inductive descriptors, will be favored prior to the sequence or amino acid descriptors. Previous studies have shown that one mutation can change drastically the 3D conformation of an AMP, and 3D descriptors can integrate this information into the QSAR model.
3. Number and diversity of AMP analogs with biological activity are quite important for the development of a QSAR model. The success of this method requires as much information as possible. The majority of the positions in the AMP sequence should be at least mutated to another residue to see their influence on the biological activity. Previous studies have developed QSAR models with 15–20 AMP analogs and it should be the minimum required.
4. Most of the QSAR modeling of AMPs have mainly been successful in designing AMP analogs of the same size. It is still quite challenging to develop such robust models with peptides from highly diverse sizes. However, with the recent advances in high-throughput peptide synthesis and the combination of advanced learning methods like ANN or SVM, QSAR methods should be able to in future studies overcome the size and diversity limitation.
5. Splitting the data set into a training set (for the development of the QSAR model) and a test set for the validation of the model performance is recommended. These models can easily be overtrained, resulting in a very unstable and inaccurate prediction.

References

1. Zasloff, M. (2002) Antimicrobial peptides of multicellular organism. *Nature* **415**, 389–395.
2. Mygind, P. H., Fischer, R. L., Schnorr, K. M., Hansen, M. T., Sonksen, C. P., Ludvigsen, S., Raventos, D., Buskov, S., Christensen, B., De Maria, L., Taboureau, O., Yaver, D., Elvig-Jørgensen, S. G., Sørensen, M. V., Christensen, B. E., Kjærulff, S., Frimodt-Moller, N., Lehrer, R. I., Zasloff, M., and Kristensen, H. H. (2005) Plectasin is a peptide antibiotic with therapeutic potential from a saprophytic fungus. *Nature* **437**, 975–980.
3. Raventos, D., Taboureau, O., Mygind, P. H., Nielsen, J. D., Sonksen, C. P., and Kristensen, H. H. (2005) Improving on nature's defenses: optimization and high throughput screening of antimicrobial peptides. *CCHTS* **8**, 219–233.
4. Taboureau, O., Olsen, O. H., Nielsen, J. D., Raventos, D., Mygind, P. H., and Kristensen, H. H. (2005) Design of novispirin antimicrobial peptides by quantitative structure-activity relationship. *Chem. Biol. Drug. Des.* **68**, 48–57.
5. Cherkasov, A. and Jankovic, B. (2004) Application of inductive QSAR descriptors for quantification of antibacterial activity of cationic polypeptides. *Molecules* **9**, 1034–1052.
6. Fjell, C. D., Hancock, R. E. W., and Cherkasov, A. (2007) AMPer: A database and an automated discovery tool for antimicrobial peptides. *Bioinformatics* **23**, 1148–1155.

7. Frecer, V., Ho, B., and Ding, J. L. (2004) De novo design of potent antimicrobial peptides. *Antimicrob. Agents Chemother.* **48**, 3349–3357.
8. Friedrich, C. L., Moyles, D., Beveridge, T. J., and Hancock, R. E. (2000) Antibacterial action of structurally diverse cationic peptides on gram-positive bacteria. *Antimicrob. Agents Chemother.* **44**, 208–209.
9. Tossi, A. (2005) Host defense peptides: roles and applications. *Current Protein Pept. Sci.* **6**, 1–3.
10. Hellberg, S., Sjostrom, M., Skagerberg, B., and And Wold, S. (1987) Peptide quantitative structure-activity relationships, a multivariate approach. *J. Med. Chem.* **30**, 1126–1135.
11. Wold, S., Esbesen, K., and Geladi, P. (1987) Principal component analysis. *Chemom. Intell. Lab. Syst.* **2**, 37–52.
12. Mee, R. P., Auton, T. R., and Morgan, P. J. (1997) Design of active analogues of a 15-residue peptide using D-optimal design, QSAR and a combinatorial search algorithm. *J. Pept. Res.* **49**, 89–102.
13. Dudek, A. Z., Arodz, T., and Galvez, J. (2006) Computational methods in developing quantitative structure-activity relationships (QSAR): a review. *CCHTS* **9**, 213–228.
14. Lejon, T., Strom, M. B., and Svendsen, J. S. (2001) Antibiotic activity of pentadecapeptides modeled from amino acid descriptors. *J. Pept. Sci.* 7, 74–81.
15. Lejon, T., Stiberg, T., Strom, M. B., and Svendsen, J. S. (2004) Prediction of antibiotic activity and synthesis of new pentadecapeptides based on lactoferricins. *J. Pept. Sci.* **10**, 329–335.
16. Cherkasov, A., Hilpert, K., Jenssen, H., Fjell, C. D., Waldbrook, M., Mullaly, S. C., Volkmer, R., and Hancock, R. E. (2009) Use of artificial intelligence in the design of small peptide antibiotics effective against a broad spectrum of highly antibiotic-resistant superbugs. *ACS Chem. Biol.* **4**, 65–74.
17. Lata, S., Sharma, B. K., and Raghava, G. P. (2007) Analysis and prediction of antibacterial peptides. *BMC Bioinformatics* **8**, 263–273.
18. Gonzalez, M. P., Teran, C., Saiz-Urra, L., and Teijeira, M. (2008) Variable selection methods in QSAR: an overview. *Curr. Top. Med. Chem.* **8**, 1606–1627.
19. Willett, P. (1995) Genetic algorithms in molecular recognition and design. *Trends Biotechnol.* **13**, 516–521.
20. Montgomery, D. C. (2000) *Design and Analysis of Experiments* (5th ed.). New York: Wiley.

Chapter 7

Synthesis and Thermodynamic Characterization of Small Cyclic Antimicrobial Arginine and Tryptophan-Rich Peptides with Selectivity for Gram-Negative Bacteria

Mojtaba Bagheri

Abstract

One promising strategy to combat the proliferation of bacteria resistance toward current antibiotics is the development of peptide-based drug. Among these compounds is a group of small cyclic peptides rich in arginine (Arg) and tryptophan (Trp) residues with selective toxicity toward Gram-negative bacteria. The small size of these peptides with improved toxicity toward Gram-negative bacteria makes them an interesting candidate to understand the forces responsible for their selectivity and paves the way to develop new therapeutics with potent activity toward multi-resistant Gram-negative bacteria. To reach this goal, isothermal titration calorimetry (ITC) is a useful technique which may provide the complete set of thermodynamic parameters of the interaction of peptides with lipid bilayers mimicking the properties of bacterial membranes within a few hours. The purpose of this chapter is to describe the synthesis of this group of small synthetic antimicrobial peptides together with the application of ITC to study their interaction with lipid membranes.

Key words: RW-rich antimicrobial peptides, cyclization, bacterial selectivity, solid-phase peptide synthesis, isothermal titration calorimetry, thermodynamics, unilamelar vesicles.

1. Introduction

The development of effective therapeutics for the treatment of infections caused by multidrug-resistant Gram-negative bacteria is a challenge worldwide. Among the available antibiotics those with killing capability against a broad variety of bacterial species and little toxicity toward human tissues are in an urgent need. Due to the non-specific membrane-disturbing mode of action and the absence of immune response, cationic antimicrobial peptides

A. Giuliani, A.C. Rinaldi (eds.), *Antimicrobial Peptides*, Methods in Molecular Biology 618,
DOI 10.1007/978-1-60761-594-1_7, © Springer Science+Business Media, LLC 2010

(CAPs) have become highly attractive as promising therapeutics to overcome bacterial resistance. However, unresolved problems of toxicity against eukaryotic cells, the limited stability against proteolytic digestion in vivo, and high production costs that prevent using them orally and systemically require the development of new classes of CAPs with improved activity profiles (1).

One class of CAPs includes those that are rich in particular amino acids like Trp and Arg residues. Tritrpticin and indolicidin as Trp-rich cathelicidin natural antimicrobial peptides (2) or the synthetic hexapeptide Ac-RRWWRF-NH_2 identified by screening synthetic combinatorial hexapeptide libraries (3) belong to this group. All these peptides adapt an amphipathic structure in membrane-mimetic environment with hydrophobic residues on one side and charged residues on the other side. Biophysical studies suggest that the charges favor accumulation of the peptides at the negatively charged bacterial membrane surface and the nonpolar face facilitates the penetration of peptides deeper into the cell membrane, leading to membrane destabilization and cell death. Recently, it was shown that introduction of conformational constraints by head-to-tail cyclization of the synthetic hexapeptide RRWWRF distinctly enhanced the selectivity of the sequence toward Gram-negative *Escherichia coli*. Biophysical studies confirmed that the cyclic peptide is able to permeabilize lipid bilayers (4). However, in contrast to the high antimicrobial activity, the cyclization-related increase in the membrane permeabilizing activity was low. Structural modifications of the cyclic parent peptide demonstrated that single amino acid substitutions or replacement of L-amino acids with D-enantiomers can enhance or abolish the antimicrobial activity toward Gram-negative and/or Gram-positive bacteria (5). The sequence diversity and conformational similarity of micelle-bound cyclic hexapeptides and mimetics (6), but different activities, raise the question whether the specific amino acid residues are responsible for the effects or whether the amphipathicity conferred upon the cyclic compounds by the indole and guanidino moiety is sufficient to explain the activity spectra. It seems that not only the charge-driven binding of the peptides at the membranes of *E. coli* bacteria but also the exact nature of the interaction of aromatic Trp residues at the interface zones of phospholipids in the inner target membrane and the outer lipopolysaccharide layer is important to be understood. To shed light onto this issue, ITC as a technique to study the interaction between the peptides and lipid bilayer mimicking the bacterial membrane is useful. ITC can provide the comprehensive thermodynamic characterization of the binding of antimicrobial peptides to different lipid vesicles (7). The thermodynamic parameters in terms of entropy, enthalpy, and binding constant are useful to understand the mechanism of action of CAPs and their selectivity toward different lipid membranes.

This chapter includes an efficient protocol for the synthesis of the small linear peptides and their head-to-tail cyclization in solution together with the thermodynamic characterization of the interaction of peptide with different lipid vesicles. To describe the protocol, the highly active antimicrobial peptide, cyclo-RRRWFW with three adjacent aromatic residues was taken as an example (**Fig. 7.1**). The effect of cyclization is most pronounced for this peptide and it binds strongly to model membranes compared to the less active linear sequence Ac-RRWFWR-NH_2. The synthesis protocol can be applied for the synthesis and cyclization of different variety of pharmaceutically interesting sequences (8, 9). For simplification of the thermodynamic characterization of the lipid–peptide interaction by means of ITC, we applied the Langmuir adsorption isotherm to extract the complete thermodynamics profile of the interaction. Based on this assumption that there is one single set of energetically equivalent sites called the "One Set of Sites" binding model, the evaluation and analysis of thermodynamic data was performed.

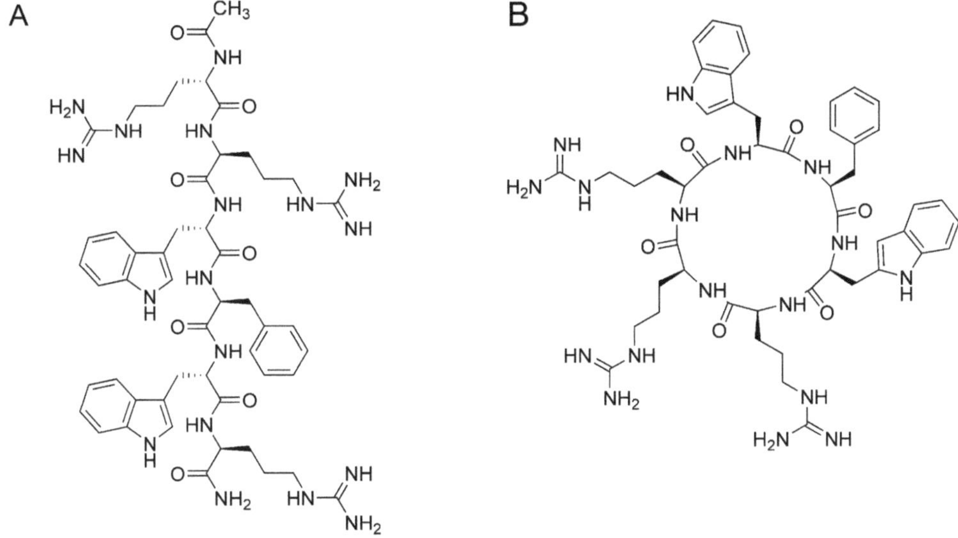

Fig. 7.1. Chemical structures of linear (**A**) Ac-RRWFWR-NH_2 and (**B**) cyclo-RRRWFW sequences.

The protocol as described here is turned out to be simple and can be applied to study the interaction of a variety of cell-permeabilizing antimicrobial peptides with vesicles of varying lipid composition to better understand the mode of action of CAPs.

2. Materials

2.1. Synthesis of Linear and Cyclic Peptides

1. Resins for peptide synthesis: TentaGel S RAM resin, particle size 90 μm, capacity 0.25 mmol/g (Rapp Polymere GmbH, Germany) for the synthesis of peptide amide; (2′-chloro)-chlorotrityl polystyrene resin, particle size 200–400 mesh, 1% divinylbenzene, capacity 1.03 mmol/g (Rapp Polymere GmbH, Germany) for the synthesis of peptide acid.
2. Equipments for the synthesis: Shaker (IKA-Labortechnik, Germany), magnetic stirrer (Heidolph, Germany), plastic syringe equipped with a porous frit, mini-membrane vacuum pump (KNF Laboport, Germany), magnetic stir bar, centrifuge (model 5416, Eppendorf, Germany), centrifuge tubes, 100 mL 1-neck round glass flask, Freeze Dryer ALPHA 2-4 LSC (Atr Biotech, Germany), analytical and preparative reversed-phase high-performance liquid chromatography (RP-HPLC) equipments (Jasco, Japan) with reverse phase C_{18} columns, measuring pipette, electrospray ionization-mass spectrometry (ESI-MS) in the positive ionization mode using an ACQUITY UPLC system with LCT *Premier* mass spectrometer (Waters Corporation, Milford, MA, USA), rotary evaporator (Heidolph LABOROTA Efficient evaporators; and VACUUBRAND PC3001 VARIO vacuum pump, Germany), UV cell (light path 10 mm), and spectrophotometer (Lambda 9, Perkin-Elmer, Germany).
3. 9-Fluorenylmethoxycarbonyl (Fmoc)-amino acid derivatives: Fmoc-Arg[2,2,4,6,7-pentamethyldihydrobe- nzofuran-5-sulfonyl (Pbf)]-OH, Fmoc-phenylalanine (Phe)-OH, and Fmoc-Trp[*tert*-butyloxycarbonyl (*t*-Boc)]-OH (GLS, China).
4. Peptide cleavage solvents: Trifluoroacetic acid (TFA), dichloromethane (DCM), and diethyl ether (Et_2O).
5. Fmoc cleavage solvents: Dimethylformamide (DMF) and 20% solution (v/v) of piperidine in DMF.
6. Coupling reagent: *O*-benzotriazol-1-yl-*N,N,N′,N′*-tetramethyluronium hexafluorophosphate (HBTU) (IRIS Biotech, Germany).
7. Cyclization reagent: 1-(1-pyrrolidinyl-1*H*-1,2,3-triazolo [4,5-*b*] pyridine-1-ylmethylene) pyrrolidinium hexafluorophosphate *N*-oxide (HAPyU) (**3** in **Fig. 7.2**) (*see* **Note 1**).
8. Coupling and cyclization solvents: DMF and *N,N*-diisopropylethylamine (DIEA).

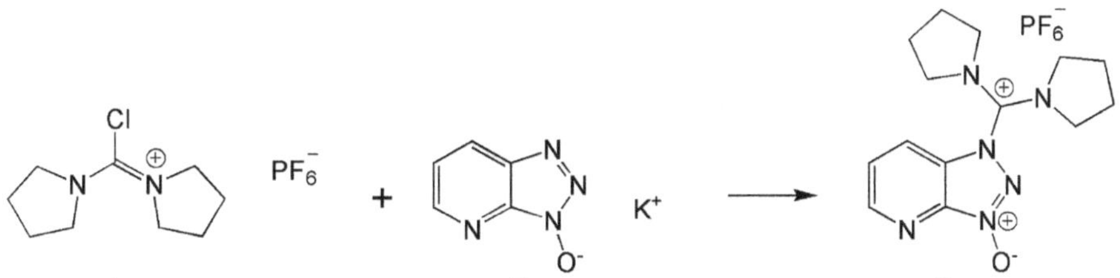

Fig. 7.2. Synthesis of HAPyU.

9. Scavengers: Thioanisole (Fluka, Germany), 1,2-ethandithiol (EDT) (Fluka, Germany), and phenol (Honeywell Riedel-de Haen, Germany).
10. Acetylation solvents: Acetic anhydride (Ac_2O) (Fluka, Germany), DIEA, and DMF.
11. Kaiser test solution: Ninhydrin (Sigma-Aldrich, Germany), pyridine (Fluka, Germany), and potassium cyanide (KCN) (Fluka, Germany). *Solution A*. 2 mL aqueous solution of 1 mM KCN in 98 mL pyridine; *Solution B*. 5 g ninhydrin in 100 mL EtOH (%5 w/v).
12. Lyophilization solvents: Water and acetonitrile (ACN).

2.2. Thermodynamic Characterization Using ITC

1. Equipments and devices for ITC experiment: High-sensitivity isothermal titration calorimeter (VP-ITC; MicroCal, Northampton, USA), thermovacuum (MicroCal Incorporated, USA, to facilitate thermostatting and degassing samples), 2.5-mL long-needle Hamilton syringe, disposable culture tubes borosilicate glass (KIMBLE GLASS INC. SIZE/CAP 6 × 50 mm, USA), ITC small tubes (for preincubation of the samples and degassing, 2-mL eppendorf vial, measuring pipette and pipette tips, magnetic stir bar (< 10 mm length), and 0.5% Decon in water.
2. Lipid: 1-Palmitoyl-2-oleoyl-*sn*-glycero-3-phosphocholine (POPC) (molecular mass: 760.1 g/mol) and 1-palmitoyl-2-oleoyl-*sn*-glycero-3[phospho-*rac*-(1-glycerol)] sodium salt (POPG) (molecular mass: 770.507 g/mol) lipids (Avanti Polar Lipids; Alabaster, Inc., Alabaster, AL, USA) (*see* **Notes 2** and **3**).
3. Small unilamelar vesicles (SUVs) preparation: Vacuum pump (~10^{-2} to 10^{-3} mbar) (Leybold TRIVAC, Germany), argon gas source, dynamic light scattering (DLS) on an N4 Plus particle sizer (Beckman Coulter, Fullerton, USA) equipped with a 10-mW helium/neon laser with a wavelength of 632.8 nm at a scattering angle of 90° to check the size and polydispersity of the vesicles, titanium-tip ultrasonicator (Labsonic L, B. Braun Biotech, Melsungen; and Sonopuls

HD 2070, Bandelin electronic, Germany) with clamp, rotary evaporator, cuvettes (polystyrene 10 × 10 × 45 mm) (SARSTEDT, Aktiengesellschaft & Co., Germany), round-bottomed glass tubes (~10 mL) with lids, volumetric glass pipette, disposal glass pasteur pipette (long form; 230 mm total length), and 2-mL eppendorf vial (for centrifugation).

4. Loading buffer: 10 mM sodium dihydrogen phosphate/disodium monohydrogen phosphate [(NaH_2PO_4)/(Na_2HPO_4)], 154 mM sodium fluoride (NaF), pH 7.4 (*see* **Notes 4** and **5**).
5. Analysis of raw data: VP-ITC ORIGIN 5.0 (MicroCal Software, Northampton, USA; for baseline correction, peak integration, and adjustment).

3. Methods

3.1. Synthesis of Linear and Cyclic Peptides

The methods outline the synthesis of Ac-RRWFWR-NH_2 and cyclo-RRRWFW peptides based on Fmoc/*t*-Boc strategy. The linear peptide is amidated and acetylated at C and N terminus, respectively. It contains three Arg and three aromatic residues. The cyclic peptide which is the result of ring closure of the unprotected peptide (*see* **Note 6**) at the site of Arg and Trp residues positioned at N and C terminus, respectively (*see* **Note 7**), has the same number of charges. This is particularly important when the activities of both peptides are compared in terms of the ability to inhibit the growth of bacteria and their affinity toward lipid bilayers.

3.1.1. Solid Phase Synthesis of Linear Peptide

3.1.1.1. Peptide Amide

The synthesis of a linear sequence Ac-RRWFWR-NH_2 was performed manually by solid-phase peptide synthesis (SPPS). Fmoc-based chemistry was used for the preparation of the peptide amide using TentaGel S RAM resin. The synthesis protocol was as follows:

1. Place resin in an appropriate plastic syringe equipped with a porous frit.
2. Let the resin swell for 30 min after addition of DMF (2 mL DMF for every 100 mg resin).
3. Remove the solvent by filtration under vacuum.
4. Add 20% piperidine in DMF for two times (1 mL of solvent for every 100 mg resin; 5+10 min with agitation) to the vessel to remove the first Fmoc group attached to the resin.

5. Wash the resin with DMF (1 mL DMF for every 100 mg resin; 5 times for 1 min).
6. Coupling of the first amino acid. Allow 0.5 mmol protected amino acid, 0.5 mmol HBTU, and 1 mmol DIEA in 1.25 mL DMF (for 500 mg resin) to react in reaction vessel while it shakes for 30 min.
7. Wash the resin with DMF (1 mL DMF for every 100 mg resin; 4 times for 1 min).
8. Perform Kaiser test (*see* **Note 8**).
9. Remove the Fmoc protecting group of the first-coupled amino acid with 20% piperidine in DMF for two times (1 mL of solvent for every 100 mg resin; 5 + 10 min with agitation).
10. Wash the resin with DMF (1 mL DMF for every 100 mg resin; 5 times for 1 min).
11. Repeat the Steps 4–8 until the completing of the synthesis.
12. Acetylation of the N terminus. Add 3.5 mL DMF, 1 mL Ac_2O, and 0.5 mL DIEA (for 500 mg resin) to the reaction vessel and shake it for about 15 min.
13. Wash the resin with DMF (1 mL DMF for every 100 mg resin; 3 times for 1 min).
14. Wash the resin with DCM (1 mL DCM for every 100 mg resin; 3 times for 1 min).
15. Dry the resin using the membrane vacuum pump for 10 min (before peptide cleavage).

3.1.1.2. Peptide Acid

Use (2′-chloro)-chlorotrityl polystyrene resin for the synthesis of RRRWFW-OH as the precursor of the cyclic peptide (*see* **Notes 9** and **10**).

1. Place 500 mg of the resin in a reaction vessel and let it swell with DCM for 30 min (*see* **Note 11**).
2. Remove the solvent by filtration under vacuum.
3. First coupling of amino acid. Dissolve 0.5 mmol protected amino acid and 0.5 mmol DIEA in 2.5 mL DCM and add it to the resin and let the mixture agitate on a shaker for 2 h.
4. Drain the vessel. Add 8 mL DCM, 2 mL MeOH, and 1 mL DIEA (v/v/v; 2 times for 10 min) to the vessel to block any unreacted chloro functional group remained on resin (*see* **Note 12**).
5. Wash the resin with DCM (1 mL DCM for every 100 mg resin; 5 times for 1 min).
6. Let the resin dry under vacuum.

7. Quantify the loading of the first amino acid on the resin by UV absorption of the released Fmoc (*see* **Note 13**).
8. Coupling of the second amino acid. Follow Steps 4–8 as discussed for the synthesis of peptide amide until the completing of the synthesis.
9. Wash the resin with DCM (1 mL DCM for every 100 mg resin; 3 times for 1 min) and let the resin dry.

3.1.1.3. Cleavage and Deprotection with TFA

Linear peptides are cleaved off from the resin by using TFA containing different compositions of scavengers as follows:

1. Transfer the dried resin (500 mg) to a 100 mL 1-neck round glass flask with a magnetic bar.
2. Add 0.5 g phenol together with addition of 4.25 mL TFA to the glass flask and stir it for 3 h (*see* **Notes 14, 15,** and **16**).
3. Filter the resin and wash it with TFA followed by DCM (5 mL of solvents for 1 time).
4. Combine the cleavage mixture and subsequent washes and concentrate it under vacuum using the rotary evaporator.
5. Add 50–60 mL Et_2O to the remaining mixture to precipitate the peptide.
6. Transfer the suspended peptide in Et_2O to a tube, centrifuge down the peptide (5,000 × *g* for 5 min), and let it dry.

3.1.1.4. Lyophilization of Crude Products

1. Transfer the dried peptide to a 100 mL 1-neck round glass flask.
2. Add 4.5 mL H_2O and 0.5 mL glacial acetic acid to the flask in order to dissolve the peptide.
3. Freeze the solution by rotating the round glass flask in a freeze-drying flask at –70 to –80°C.
4. Attach the flask to the lyophilizer overnight (*see* **Note 17**).
5. Weigh the lyophilized peptide and transfer it to small vial and keep it at –20°C.

3.1.2. Peptide Cyclization in Solution

The synthesis of cyclo-RRRWFW is the result of the head-to-tail lactamization of the N and C terminus of the crude peptide acid RRRWFW-OH using HAPyU and DIEA in DMF under dilute condition (*see* **Notes 18–21**).

1. Weigh 1 equiv. of dry peptide acid and transfer it into a 100 mL 1-neck round glass flask with a magnetic bar (*see* **Note 22**) and dissolve it in appropriate amount of dry DMF to reach a peptide solution with concentration of 5×10^{-4} M.
2. Add 1.1 equiv. HAPyU to the solution with stirring (*see* **Note 23**).

3. Let the cyclization reaction run by addition of 3 equiv. DIEA and stir the mixture for 10 min at room temperature.
4. Transfer a 50 μL aliquot of the reaction mixture to an eppendorf vial and dilute it with 100 μL of 0.1% TFA in H_2O to stop the cyclization reaction.
5. Check the progress of the reaction with analytical RP-HPLC and/or ESI-MS (*see* **Note 24**).
6. Make sure that the reaction has completed and try to remove DMF by the rotary evaporator under vacuum condition.
7. Add 100 mL of H_2O/ACN (4/1 [v/v]) solution to the remaining mixture and freeze-dry it.

3.1.3. Purification and Characterization of the Linear and Cyclic Peptides

3.1.3.1. HPLC

Analytical RP-HPLC performed by using a Jasco HPLC system with a diode array detector operating at four different wavelengths ($\lambda = 220, 280, 301, 360$ nm). Runs were carried out on a PolyEncap A 300 (250 × 4.0 mm) column (Bischoff Analysentechnik, Germany). The sample concentration was 1 mg/mL peptide in a solution mixture of 0.1% TFA in 50% ACN in H_2O. The mobile phase A was 0.1% TFA in water, and B was 0.1% TFA in 80% ACN and 20% water (v/v). The retention time (t_R) of the peptides was determined using a linear gradient of 5–95% of mobile phase B over 40 min at room temperature.

The purification of crude peptides performed by RP-HPLC using a Shimadzu LC-10A system (Japan) operating at $\lambda =$ 220 nm using the same eluents with 10 mL/min flow rate. Afterward, the purified peptides are subjected to final lyophilization (*see* **Note 25**).

3.1.3.2. Mass Spectrometry

ESI-MS was used for further characterization of the peptides.

1. Prepare a stock solution of the peptide of 1 mg/mL concentration in a mixture of 0.1% TFA in 50% ACN in water (v/v).
2. Dilute the sample 10 times in a 1.5-mL eppendorf vial with ultrapure water (Milli-Q Element System, Millipore, Molsheim, France).
3. Centrifuge the vial to get rid of any small particles.
4. Transfer 100 μL of the diluted sample into an appropriate vial for ESI-MS. Set the injection volume of sample at 5 μL with a flow rate of 200 μL/min operating at $\lambda = 220$ nm.

3.2. Thermodynamic Characterization Using ITC

3.2.1. Binding Model for Analysis of Data

One simple model to describe thermodynamic profiles of lipid–peptide interaction is the Langmuir adsorption isotherm. In this model, it is assumed that there is limited number of binding sites in a lipid membrane. When the binding sites are energetically equivalent and are made up of *n* single lipid molecules, it is so-called "one set of sites" binding model. This model actually describes the frequently used "chemical equilibrium" term. This model is particularly useful in describing the binding thermodynamics when the number of available binding sites per lipid vesicles is limited, e.g., the interactions between large proteins and the membrane because of steric reasons. However, unlike protein–ligand interactions, the binding of a peptide to a lipid membrane (or a micelle) is best described in terms of a partition equilibrium rather than a stoichiometric binding process. In other words, the solute (i.e., the peptide) can choose between two different phases, namely the aqueous phase (buffer) and the membrane phase (*see* **Note 26**).

The theory underlying the "One Set of Sites" binding model for a 1:1 stoichiometric interaction is explained below. The following binding equilibriums describe the interaction between lipid and peptide:

$$L{+}P \rightleftarrows L.P \qquad [1]$$

$$K_a = \frac{[L.P]}{[L][P]} \qquad [2]$$

where L is lipid, P is peptide, $L.P$ is the lipid-bound peptide, and K_a is the binding constant. The total concentration of lipid and peptide can be represented by Eqs. [3] and [4]

$$[L]t = [L] + [L.P] \qquad [3]$$

$$[P]t = [P] + [L.P] \qquad [4]$$

and the total heat content of the solution (Q) in the calorimeter cell with a volume (V) is defined by Eq. [5],

$$Q = n\theta[P]t\Delta H^{\circ}V \qquad [5]$$

where n is the number of interaction sites in the lipid membrane, θ is the moles of lipids bound per moles of total peptide as shown in Eq. [6], ΔH^0 is the heat of binding of the lipid to the peptide.

$$\begin{aligned}\theta &= \frac{[L.P]}{[P]t} = \frac{[L.P]}{[P]+[L.P]} \\ &= \frac{K_a[L][P]}{[P]+K_a[L][P]} = \frac{K_a[L]}{1+K_a[L]}\end{aligned} \qquad [6]$$

According to Eq. [6] when there is n number of interaction sites $[L.P] = n\theta[P]t$, and thus, $[L] = [L]t - n\theta[P]t$. On substituting $[L]$ into the Eq. [6] θ is solved as follow:

$$\theta = \frac{1}{2}\left[\left(\frac{[L]t}{n[P]t} + \frac{1}{nK_a[P]t} + 1\right) - \sqrt{\left(\frac{[L]t}{n[P]t} + \frac{1}{nK_a[P]t} + 1\right)^2 - \frac{4[L]t}{n[P]t}}\right] \quad [7]$$

As a consequence, Q can be extracted by combination of the Eqs. [5] and [7]

$$Q = \frac{n[P]t\Delta H^{\circ}V}{2}\left[\left(\frac{[L]t}{n[P]t} + \frac{1}{nKa[P]t} + 1\right) - \sqrt{\left(\frac{[L]t}{n[P]t} + \frac{1}{nKa[P]t} + 1\right)^2 - \frac{4[L]t}{n[P]t}}\right] \quad [8]$$

Based on this equation, the total heat of content (Q) is the function of ΔH^0, n, and K_a.

In an ITC thermogram, the data are plotted as the change in the heat content for each injection (Q_i) of a definite volume of lipid into the calorimeter cell containing a known volume of peptide solution versus the molar ratio of $\frac{[L]}{[P]t}$ (**Fig. 7.3**). With each injection of the lipid vesicle in the titration experiment, some liquid is expelled into the calorimetrically inert access tube by a known volume (ΔV_i). Therefore, the change in the heat released from the ith injection is defined by the Eq. [9]

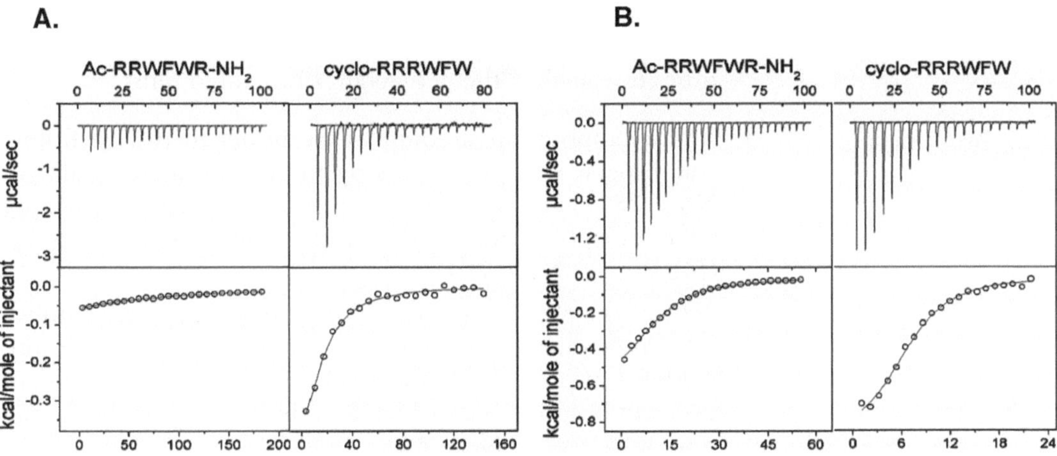

Fig. 7.3. The raw ITC data (power vs. time) and the integrated data (heat vs. molar ratio) for the injection of (**a**) POPC (40 mM) and (**b**) POPC/POPG (3/1 [mol/mol]) (20 mM) SUVs into the calorimeter cell containing linear or cyclic peptide solution (40 μM) at 37°C. *Non-filled circles* give the experimental data and the *solid line* is the result of data fitting based on the non-linear least square analysis with the "one set of sites" binding model. The binding constant (K_a) for linear and cyclic peptide are $K_{\mathrm{POPC,\,linear}} = 119\ \mathrm{M}^{-1}$, $K_{\mathrm{POPC/POPG\,(3/1),\,linear}} = 3.49 \times 10^3\ \mathrm{M}^{-1}$, $K_{\mathrm{POPC,\,cyclic}} = 1.7 \times 10^3\ \mathrm{M}^{-1}$, $K_{\mathrm{POPC/POPG\,(3/1),\,cyclic}} = 1.79 \times 10^4\ \mathrm{M}^{-1}$.

$$\Delta Q_i = Q_i + \frac{\Delta V_i}{V}[\frac{Q_i + Q_i - 1}{2}] - Q_i - 1 \qquad [9]$$

The standard free energy of Gibbs (ΔG^0) for the interaction between the peptide and lipid is related to the binding constant by the Eq. [10]. According to the standard thermodynamic expression shown in Eq. [11], ΔG^0 of binding is related to the contribution of the corresponding enthalpy and entropy (ΔS^0) changes.

$$\Delta G^\circ = -RT\,\mathrm{In}[K_\mathrm{a}({}^1\!/_M)] \qquad [10]$$

$$\Delta G^\circ = \Delta H^\circ - T\Delta S^\circ \qquad [11]$$

Therefore, one can determine the complete thermodynamic profiles of a binding at micromolar level and even smaller by applying the ITC measurement.

3.2.2. Vesicle Preparation

3.2.2.1. Lipid Preparation

1. Weigh a sufficient number of round-bottomed glass tubes with their lids and note them in a lab book before addition of lipid.
2. Add an appropriate volume of lipids dissolved in chloroform to the 10 mL round-bottomed glass tubes using volumetric glass pipette (*see* **Note 27**).
3. Remove the solvent by using the rotary evaporator and finally use a high vacuum pump (~10^{-2} to 10^{-3} mbar) overnight and remove traces of chloroform.
4. Put argon over the lipid-containing tubes. Close the tubes with the lids tightly to prevent oxidation of phospholipids. Weigh the closed vials and subtract the weights of the empty tubes to calculate the net weight of the lipid films.

3.2.2.2. Preparation of SUVs

1. Add an appropriate amount of the buffer to the dried lipid and disperse the lipid by shaking and vortexing for 5 min (*see* **Note 28**). This will result in a turbid suspension containing large and polydisperse vesicles called multilamellar vesicles (MLVs) (*see* **Note 29**).
2. Subject the MLV to ultrasonication in an ice/water bath for 20 min by putting the titanium tip of the ultrasonicator into the lipid suspension. As the result of sonication, the suspension will become clear, which is an indication for the formation of SUVs (*see* **Note 30**).
3. Transfer the content of the tube into a 2-mL eppendorf vial and centrifuge it at 7,000 × *g* for 5 min at 10°C to get rid of titanium debris left during the ultrasonication.
4. Transfer the supernatant carefully into another eppendorf vial.

5. Check the size and polydispersity of the vesicles by DLS (*see* **Notes 31** and **32**).

3.2.3. Instrument Setup and Titration Experiment

1. Set the desired temperature for both the calorimeter and the thermovacuum thermostat (e.g., 37°C).
2. Prepare a stock solution of peptide at a concentration of 200 μM in a 2-mL eppendorf vial in the same buffer as used for the preparation of the lipid solution (*see* **Notes 33** and **34**).
3. Prepare a peptide solution (e.g., 40 μM) in an ITC small tube by dilution of the stock solution (*see* **Notes 35** and **36**). Put a small magnet bar in the tube and place it for at least 5 min in the thermovacuum in order to avoid the formation of bubbles in the calorimeter cell during the titration (*see* **Note 37**).
4. Wash the reference and calorimeter cells of the calorimeter with the same buffer used for the experiment for 20 times. Remove the rest of the buffer remained in the cells at the end (*see* **Notes 38** and **39**).
5. Wash out the ITC syringe with buffer and if necessary with 5% Decon solution too.
6. Place an appropriate amount of the lipid solution (e.g., 500 μL) in a small disposable culture tubes borosilicate glass. Put the ITC syringe into it in order to fill the syringe with lipid solution. Make sure that there are no bubbles inside the ITC syringe by using purge–refill cycles for 2–3 times (*see* **Note 40**).
7. Fill the reference cell with the degassed buffer using the long-needle syringe.
8. Fill the calorimeter cell with the degassed peptide solution until it overflows at the top of the cell and avoid the production of bubbles. Then, remove the overflowing solution above the calorimeter cell.
9. Select the instrument setting such as the total number of injections (at least 20), the volume of injection (3– 10 μL) (*see* **Note 41**), the temperature (as mentioned above), reference power (18 μcal/s), initial injection delay (180 s), spacing between injections (300 s) (*see* **Note 42**), syringe lipid concentration (20–40 mM), cell peptide concentration (0.04 mM), and stirring speed (310 rpm) (*see* **Note 43**).
10. Place the ITC syringe into the calorimeter cell and initiate the titration experiment. If the heats of injection are quite satisfactory, each experiment usually takes 3 h.
11. After the end of the titration experiment, take the ITC syringe out and make the calorimeter cell and ITC syringe

empty. Wash out the calorimeter cell and ITC syringe completely (*see* **Note 38**) and prepare the instrument for the next experiment.

3.2.4. Data Analysis and Fitting

The data are collected based on the program supplied by the manufacturer. Baseline correction, peak integration, and its adjustment for data acquisition and analysis are done using VP-ITC ORIGIN 5.0 software as described by the manufacturer. The raw data were corrected for the heat of dilution of cyclo-RRRWFW and Ac-RRWFWR-NH_2 peptides in the buffer (*see* **Note 44**). After assuming the values for K_a and ΔH^0, the data fitting is done with the "One Set of Sites" binding model for a 1:1 stoichiometric interaction as described before. According to this model, the data analysis and curve fitting by non-linear least square analysis can be done by VP-ITC ORIGIN 5.0 software. The best fits of the experimental data of the titration of linear and cyclic peptides with lipid vesicles are shown in **Fig. 7.3**. The fitting program finally gives the values for the binding constant and enthalpy. The entropy and free energy of Gibbs of the system are calculated using Eqs. [10] and [11].

4. Notes

1. HAPyU is the pyrrolidinium derivative of commercially available *N*-[(dimethylamino)-*1H*-1,2,3-triazolo [4,5-*b*] pyridine-1-ylmethylene]-*N*-methylmethanaminium hexafluorophosphate *N*-oxide (HATU) (**2** in **Fig. 7.2**) (Sigma-Aldrich, Germany) coupling reagent for peptide synthesis. It is of intermediate stability compared to HATU and can be stored at 0°C. It is not hygroscopic and irritating to eyes, the respiratory system, and the skin. It was prepared in our laboratory (10) from 1-hydroxy-7-azabenzotriazole potassium (KOAt) (**1** in **Fig. 7.2**) salt (Fluka, Germany) and commercially available chloro-*N,N,N′,N′*-bis(tetramethylene)-formamidinium hexafluorophosphate (Sigma-Aldrich, Germany) using Knorr's method (11). The molecular mass is 432.31 g/mol.
2. POPC is a zwitterionic phospholipid which is frequently used to mimic the uncharged outer lipid leaflet of mammalian plasma membranes. To mimic the negatively charged inner membranes of Gram-negative bacteria, a mixture of POPC/POPG (3/1 mol/mol) is used.
3. We usually use phospholipids dissolved at concentrations of 20 mg/mL in chloroform. Glass pasture pipettes should be used to handle the solution. The amount of lipid

transferred to the vials has to be determined gravimetrically after overnight high vacuum evaporation.

4. To prevent any microbial contamination, the buffer should be sterile-filtered and stored at 4°C.

5. The heat of protonation and deprotonation of buffer occurring during the interaction between peptide and lipid vesicles can contribute to the heat of binding (ΔH^0). This can be eliminated by using phosphate buffer. In case of a buffer like tris(hydroxymethyl) aminomethane (TRIS) with a high heat of protonation, it may cause uncertainty for the experimental signals when the heat of protonation of buffer is comparable to that of binding of the peptides to lipid vesicles.

6. One major obstacle for the cyclization of fully protected peptides is their insolubility. One particular example is RRRWFW sequence with two hydrophobic Trp and one Phe residues and three adjacent Arg residues fully protected with bulky and hydrophobic Pbf group.

7. Head-to-tail cyclization of a linear sequence with Arg residue at C-terminal results in δ-lactam formation instead of the desired cyclic peptide.

8. A few resin beads are placed in a small glass tube and washed with EtOH. Solution A (100 μL) together with 100 μL of solution B was added via a Hamilton syringe into the glass tube containing the beads. The mixture was incubated at 110°C for 5 min. The appearance of a yellow color indicates the complete amino acid coupling. If blue color appears, it is the indication of the presence of free amino groups on the beads. Thus, the coupling procedure has to be repeated. Handling the solution A needs special care as it is very poisonous. Working under a hood and using lab coat, gloves, and glasses are demanded.

9. Other solid supports might be used for the synthesis of peptide acids such as hydroxymethylbenzoic acid (HMBA)-resin, 4-hydroxymethylphenoxyacetyl-4′-methylbenzyhydrylamine (HMP)-resin, 4-hydroxymethyl-3-methyloxyphenoxybutryic acid (HMPB)-resin, Wang resin, super acid sensitive resin (SASRIN), or TentaGel S AC. The coupling of the first amino acid to these solid supports is usually difficult and sometimes there are risks of low loading of the first amino acid and epimerization of the anchored amino acid chiral center. In contrary, the coupling to (2′-chloro)-chlorotrityl polystyrene resin occurs in a short time and is free of epimerization as it is done without any activated species. However, all the resins mentioned above show the same swelling properties in solvents

such as DMF or DCM which are usually used for peptide synthesis.

10. One advantage of using (2′-chloro)-chlorotrityl polystyrene resin is the prevention of diketopiperazine formation. The steric constraints of the 2′-chloro group on the resins prevent this problem after anchoring of the second amino acid.
11. The trityl resin is highly moisture sensitive and, thus, the reagents and equipments for the synthesis should be dried before use to prevent hydrolysis of active chloro functional group.
12. As the anchoring of the first amino group via esterification to the resin takes place slowly, the remaining reactive chloride group should be capped. This is particularly important to prevent the synthesis of any side products after loading of the second amino acid.
13. Fill the UV cell with 2.7 mL of 20% piperidine in DMF as a reference solution. Set the UV spectrophotometer at 301 nm. Place the cell into the spectrophotometer and set the instrument to zero. Weigh 5 mg of resin in a 2-mL eppendorf vial, add 1 mL of 20% piperidine in DMF, and agitate the suspension for 20 min. Let the resin settle down. Transfer 100 μL of the supernatant to a 1-mL eppendorf vial pre-filled with 900 μL of 20% piperidine in DMF. Transfer 300 μL of this solution to the UV cell, mix, and measure the absorption for three times. Take the average value and calculate the loading of the first amino acid on the resin according to the Beer–Lambert law

$$\text{Loading}\left(\text{mmol}/\text{g}\right) = \frac{2 \times 10^4 A}{\varepsilon b}$$

where A is absorption, b is width of the UV cell in centimeters, and $\varepsilon_{301} = 6{,}000\ \text{M}^{-1}.\text{cm}^{-1}$.
14. The Pbf protecting group displays a slow deprotection kinetic with TFA compared to *t*-Boc. Thus, a longer reaction time is necessary to fully remove the bulky Pbf side chain protecting group, especially here in which three Pbf groups are present.
15. Under these conditions, Trp is still carboxylated and this prevents the transfer of the Pbf group to the indole moiety due to the electron withdrawing property of the carboxylate functionality. The presence of a little amount of water in the cleavage cocktail will facilitate the formation of sulfonated Trp side product.
16. If the Trp residue is used as Fmoc-Trp-OH derivative (without the protecting group), the quality of the crude

products is significantly influenced by the application of different composition of scavengers. In this case, the application of reagent K is highly recommended. Reagent K is a cocktail as follows: TFA/thioanisole/water/phenol/EDT (82.5/5/5/5/2.5 v/v). It should be pointed out that EDT has a terrible smell and requires careful handling. It is recommended to put all staffs contaminated with EDT such as glass flask, disposal tips, and magnet bar in a dish containing a solution of 10% H_2O_2 and 5% KOH overnight and wash it afterward.

17. At this stage, the carboxylate functionality attached to the Trp indole group during the process of cleavage will be removed.
18. Usually, head-to-tail cyclization of peptides containing seven or more amino acid residues takes place without any problem. The cyclization reaction of penta- and hexapeptide is hampered especially in the absence of turn structure inducing amino acids such as *N*-methyl or D-amino acids. One example is the linear sequence RRRWFW-OH which is discussed in this protocol. In this case, the reaction is either slow or happens with the epimerization of amino acid chiral center at the C terminus.
19. The risk of epimerization and slow rate reactions can be minimized by using HOAt-derived coupling reagents such as HAPyU. In some cases, the reaction proceeds with more than 90% of the desired cyclic product (8).
20. When the linear sequence adopts a conformation which facilitate efficient ring closure, highly concentrated linear precursor can be used (9).
21. As the synthesis of short sequence RRRWFW-OH is easy and done without formation of any deficient sequences, the crude peptide acid may be used for the cyclization reaction without further purification.
22. The lyophilized peptides might contain acetate or TFA salts even after lyophilization for a long time, especially when peptides contain basic amino acid residues such as arginine or lysine. These amounts of residual acetate and TFA salts are usually difficult to quantify.
23. In case of using excess amount of HAPyU, the modification of the amino group present in the sequence to guanidino-type side products is probable (12).
24. Analytical RP-HPLC and ESI-MS can give indications for the progress of cyclization reaction. The retention time of the cyclic peptide usually increases compared to the corresponding peptide acid. In addition, the loss of 18 units in the molecular mass of the linear peptide

represents the head-to-tail lactamization. (ESI-MS (*m/z*) $[M+H]^+ = 988.548$).

25. If TFA salt is expected to cause problems either for biological experiments or for thermodynamic studies, it is highly recommended to replace the TFA salt with chloride ion by using a solution of 0.1 N HCl. To do so, dissolve the lyophilized peptide after final purification in 0.1 N HCl solution and freeze-dry it again.

26. The interaction between a peptide and a lipid membrane surface can be described by a two-step model. In the first step, the interaction is initiated by the electrostatic contribution. The second step follows with the hydrophobic interaction causing penetration of the peptide into the hydrophobic core of the lipid bilayer, which will be accomplished by a conformational change of the peptide. In the presence of two immiscible phases such as an aqueous solution (buffer) and the membrane, the membrane-binding peptides which usually carry positive charges distribute between two phases based on their affinity to the phases. This affinity can be described as a partition coefficient and presented as a thermodynamic equilibrium constant. The thermodynamic profiles of the lipid–peptide interaction can be deduced according to this equilibrium.

    The overall binding constant so-called apparent binding constant (K_{app}) includes both hydrophobic and electrostatic contributions for lipid–peptide interaction and is defined as

$$K_{app} = \frac{C_b}{\gamma L C_f} \qquad [13]$$

where γ is the bounds of the lipid accessibility factor (1.0 for membrane-permeant solutes, 0.5 or 0.6 for membrane-impermeant solutes interacting with large unilamellar vesicles or SUVs, respectively), C_f is bulk concentration of peptide (unbound peptide), C_b is the concentration of peptide bound to the membrane surface, and L is the total lipid concentration.

    Depending on the attractive or repulsive nature of the electrostatic adsorption of the peptide on lipid, it will either increase or decrease the peptide interfacial concentration compared to that of bulk solution if the lipid membrane is naturally negative like POPG or neutral like POPC, respectively. If the ratio of lipid/peptide concentration is too low (as usual here in titration of peptide solution with lipid vesicles), this electrostatic contribution cannot be neglected. In the partition equilibrium model, K_{app} is corrected for non-specific electrostatic effects to describe the partitioning of the peptide between the lipid–water interface and the membrane. This equilibrium is so-called the intrinsic binding constant (K^0) and described as in Eq. [14]

$$K^0 = \frac{C_\mathrm{b}}{\gamma L C_\mathrm{M}} \qquad [14]$$

where C_M is the interfacial concentration of peptide. The interfacial concentration of the peptide can be related to that of bulk solution according to the Boltzmann equation described in Eq. [15]:

$$C_\mathrm{M} = C_\mathrm{f} \exp(-z F_0 \Psi_0 / RT) \qquad [15]$$

where z is the effective charge number of peptide, F_0 is the Faraday constant, Ψ_0 is the membrane surface potential, R is gas constant, and T is the thermodynamic temperature.

In consequence, K^0 is divided into two parts as shown in Eq. [16]

$$K^0 = K_\mathrm{app} \exp(z F_0 \Psi_0 / RT) \qquad [16]$$

The value for the Ψ_0 can be calculated from the Gouy-Chapman theory using Eq. [17] based on the membrane surface charge density (σ):

$$\sigma^2 = 2{,}000 \varepsilon_0 \varepsilon_r RT \sum_i C_i [\exp(-z_i F_0 \Psi_0 / RT) - 1] \qquad [17]$$

where ε_0 and ε_r are the electric permittivity of free space and the dielectric constant of water, respectively, z_i is the valence of ith species, and C_i is the concentration of the ith dissociated electrolyte in the solution.

Finally, σ is obtained according to the theoretical calculation described by Seelig (13), which is corrected for the effect of counterion binding due to the association of Na^+ to the POPG headgroups shown in Eq. [18]

$$\sigma = \frac{Z_P e R_b}{(A_L + R_b A_P)(1 + K_{Na} C_{Na} \exp(F_0 \Psi_0 / RT))} \qquad [18]$$

$$R_b = \frac{C_{P,b}}{\gamma C_L} \qquad [19]$$

where e_0 is the elementary charge (1.602×10^{-9} C), A_l is the membrane area occupied by one lipid headgroup ($0.68\ \mathrm{nm}^2$), A_p is the membrane area occupied by the peptide (A_p is assumed zero when the peptide adsorbs superficially), $C_{p,b}$ is the concentration of bound peptide, C_l is the concentration of lipid, K_{Na} is the binding constant of the sodium to the negatively charged lipid headgroup (0.6 M^{-1}) with a concentration of C_{Na}.

Figure 7.4 shows the data analysis of the interaction between the linear and cyclic peptides with POPC and POPC/POPG (3/1 [mol/mol]) SUVs according to this model.

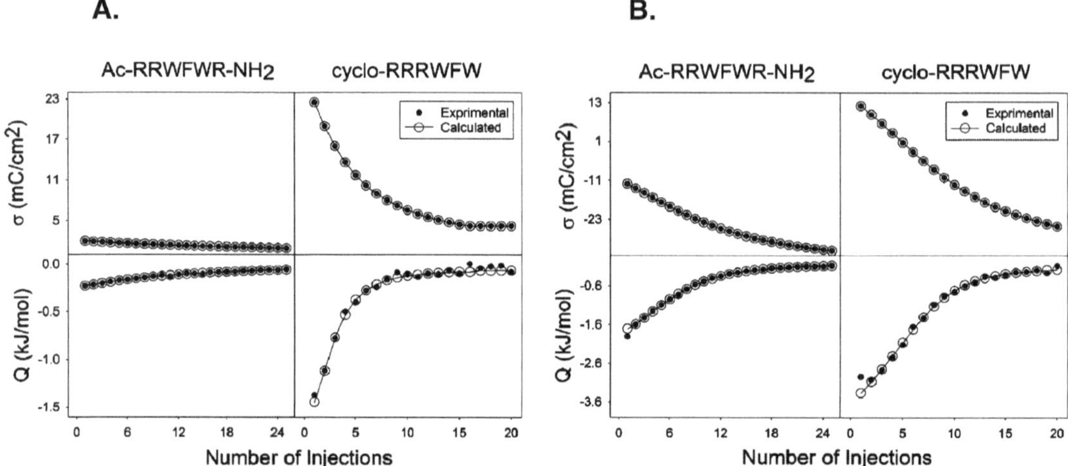

Fig. 7.4. The calculated surface charge density of lipid and the integrated heat of release versus number of injections for the injection of (**a**) POPC and (**b**) POPC/POPG (3/1 [mol/mol]) SUVs into the calorimeter cell containing linear or cyclic peptide solution at 37°C. The intrinsic binding constant (K^0) was derived taking into consideration the electrostatic peptide–bilayer interaction with the Gouy-Chapman theory by using the Excel Solver add-in to fit K^0 (14, 15). K^0 values for the linear and cyclic peptide are $K_{\text{POPC, linear}} = 240\ M^{-1}$, $K_{\text{POPC/POPG (3/1), linear}} = 1.2 \times 10^3\ M^{-1}$, $K_{\text{POPC, cyclic}} = 6.7 \times 10^3\ M^{-1}$, $K_{\text{POPC/POPG (3/1), cyclic}} = 1.2 \times 10^4\ M^{-1}$.

27. To have the vesicles composed of different lipid species (e.g., POPC/POPG [3/1 mol/mol]), the different lipids should be dried and weighed one by one in the same vial.
28. The lipid concentration used in ITC experiments depends on the affinity of the peptides for the lipid membranes. The higher the affinity, the lower the minimal lipid concentration. We normally use lipid concentrations at 20–40 mM. For high-affinity peptides, it may be possible to reduce the concentration.
29. MLVs can be stored under argon at –20°C for several months. The volume of the lipid suspension should not be less than 1 mL.
30. SUVs are usually less stable. They have to be stored at room temperature under argon and should be used no longer than 3 days. Keeping them at low temperature will result in irreversibly agglomeration.

31. The average diameter of SUVs is around 30 nm.
32. From a more practical point of view, it is important to note that most proteins and peptides bind to SUVs in an enthalpy-driven way, whereas their interaction with LUVs is mostly dominated by entropy. This means that binding to SUVs can often be easily followed by ITC, whereas binding of the same compounds to LUVs is more difficult to monitor calorimetrically.
33. It is recommended that the purity of the lyophilized peptides is > 95% by analytical HPLC. Small contaminations cause errors in the concentrations and uncertainty in the thermodynamic binding parameters.
34. It is highly recommended to prepare both lipid and peptide solutions in an identical buffer (pH, salt concentration, kind of buffer). Otherwise, the heat of mixing of two different systems will influence the signals of the binding reaction.
35. The calorimeter cell has a volume of 1.3998 mL. To fill up the cell, a larger volume of the sample (~2.5 mL) is required.
36. The concentration of peptide and lipid should be high enough to ensure complete binding of the injected peptide and to get a reasonable signal.
37. The presence of bubbles in both calorimeter cell and ITC syringe will cause baseline instabilities.
38. Try to wash out the calorimeter cell and/or ITC syringe with 5% Decon followed by at least 30 times washing with the buffer in case you want to run your experiment with another lipid mixture.
39. Try to avoid the cell drying. Otherwise, this can lead to irreversible damage.
40. To have a complete binding isotherm, such as enthalpy, entropy, and free energy of Gibbs, peptide solution has to be placed in the calorimeter cell and lipid vesicles in the ITC syringe. Since, the peptide is titrated with lipid vesicles, it starts with all of the peptide in the aqueous phase, which will then gradually transfer into the membrane phase as the lipid concentration increases in the calorimeter cell. Thus, the ITC peaks will become smaller and smaller because less and less peptide is available in the aqueous phase. This allows not only calculating enthalpy of lipid–peptide interaction, but also the partition coefficient, that is, the preference of the peptide to partition into the membrane phase. If, by contrast, small amounts of the peptide are injected into the lipid vesicle suspension, all signals will be identical.

Upon each injection, the added peptide will bind completely to the lipid phase thus allowing only the determination of the enthalpy of binding.

41. The ITC signal depends on the concentration of lipid and peptide and the injected volume. Try to modify the injection volume depending on the changes in the heat of the lipid–peptide binding; however, a volume less than 3 μL is not recommended. Due to the loss of lipid during the mounting of the syringe and equilibration stage before the actual titration experiment, we usually set the first injection volume to 3 μL and do not take the value of its heat of release into account in the data analysis.
42. After each injection, the ITC signal comes back to the signal of baseline. The enthalpy of lipid–peptide binding is measured as the integral of each individual peak. The time between the injections should be long enough to allow the signal returns to the baseline level before the next injection.
43. The speed can be modified to ensure complete mixing of the sample.
44. For the measurement of the heat of dilution, a separate experiment is usually done by injecting the lipid vesicles solution into buffer. Alternatively, we usually consider the average heat of the last injections in a real titration experiment as the effect of heat of dilution.

Acknowledgments

Dr. Sandro Keller is gratefully acknowledged with ITC. Dr. Margitta Dathe (FMP) and Professor Michael Bienert (FMP) are thanked for their helpful comments on the manuscript.

References

1. Hancock, R. E. W. and Sahl, H. -G. (2006) Antimicrobial and host-defense peptides as new anti-infective therapeutic strategies. *Nat. Biotechnol.* **24**, 1551–1557.
2. Chan, D. I., Prenner, E. J., and Vogel, H. J. (2006) Tryptophan- and arginine-rich antimicrobial peptides: structures and mechanisms of action. *Biochim. Biophys. Acta* **1758**, 1184–1202.
3. Blondelle, S. E. and Houghten, R. A. (1996) Novel antimicrobial compounds identified using synthetic combinatorial library technology. *Trends Biotechnol.* **14**, 60–65.
4. Dathe, M., Nikolenko, H., Klose, J., and Bienert, M. (2004) Cyclization increases the antimicrobial activity and selectivity of arginine- and tryptophan-containing hexapeptides. *Biochemistry* **43**, 9140–9150.
5. Wessolowski, A., Bienert, M., and Dathe, M. (2004) Antimicrobial activity of arginine- and tryptophan-rich hexapeptides: the effects of aromatic clusters, D-amino acid substitution and cyclization. *J. Pept. Res.* **64**, 159–169.
6. Appelt, C., Wessolowski, A., Soderhall, J. A., Dathe, M., and Schmieder, P. (2005) Structure of the antimicrobial, cationic hexapeptide cyclo (RRWWRF) and its

analogues in solution and bound to detergent micelles. *Chem. Bio. Chem.* **6**, 1654–1662.

7. Seelig, J. (1997) Titration of lipid-peptide interactions. *Biochim. Biophys. Acta* **1331**, 103–116.
8. Ehrlich, A., Heyne, H. -U., Winter, R., Beyermann, M., Haber, H., Carpino, L. A., and Bienert, M. (1996) Cyclization of *all*-L-pentapeptides by means of 1-hydroxy-7-azabenzotriazole-derived uronium and phosphonium reagents. *J. Org. Chem.* **61**, 8831–8838.
9. Ehrlich, A., Rothemund, S., Brudel, M., Beyermann, M., Carpino, L. A., and Bienert, M. (1993) Synthesis of cyclic peptides via efficient new coupling reagents. *Tetrahedron Lett.* **34**, 4781–4784.
10. Bienert, M., Henklein, P., Beyermann, M., and Carpino, L. A. (2002) Uronium/guanidinium salts. In *Synthesis of Peptides and Peptidomimetics* (4th ed). Goodman M., Felix A., Moroder L., Toniolo C. (Eds.), pp. 555–580, Houben–Weyl Methods of Organic Chemistry. Stuttgart: Georg Thieme, Vol. E22a, Chapter 3.8.
11. Knorr, R., Trzeciak, A., Bannwarth, W., and Gillessen, D. (1989) New coupling reagents in peptide chemistry. *Tetrahedron Lett.* **30**, 1927–1930.
12. Albericio, F., Bofill, J. M., El-Faham, A., and Kates, S. A. (1998) Use of onium salt-based coupling reagents in peptide synthesis. *J. Org. Chem.* **63**, 9678–9683.
13. Thomas, P. G. and Seelig, J. (1993) Binding of the calcium antagonist flunarizine to phosphatidylcholine bilayers: charge effects and thermodynamics. *Biochem. J.* **291**, 397–402.
14. Keller, S., Heerklotz, H., and Blume, A. (2006) Monitoring lipid membrane translocatlon of sodium dodecyl sulfate by isothermal titration calorimetry. *J. Am. Chem. Soc.* **128**, 1279–1286.
15. Keller, S., Heerklotz, H., Jahnke, N., and Blume, A. (2006) Thermodynamics of lipid membrane solubilization by sodium dodecyl sulfate. *Biophys J.* **90**, 4509–4521.

Chapter 8

Synthesis of Antimicrobial Peptides Using the SPOT Technique

Dirk F.H. Winkler and Kai Hilpert

Abstract

Developing new lead structures for drugs against multiresistant bacteria is an urgent need for modern medicine. Antimicrobial peptides are a class of drugs that can be used to discover such structures. In order to support development of this research, a fast, easy, and inexpensive method to synthesize peptides is necessary. The SPOT synthesis has the potential to produce the required peptide arrays, synthesizing up to 8,000 peptides, peptide mixtures, or other organic compounds on cellulose or other planar surfaces in a positionally addressable and multiple manner. Protocols for the preparation of cellulose membranes and the SPOT synthesis as well as cleavage of peptides from the support are described.

Key words: SPOT synthesis, peptide array, screening, antimicrobial peptides, cellulose membranes.

1. Introduction

We are presently facing an era of resistant and multiresistant bacteria, which undermines the success of modern medicine to prevent and treat bacterial infection. As a consequence, there is a great effort to discover new substances for the development of drugs that can treat infections arising from multiresistant bacteria. One promising class of substances is peptides. Some naturally occurring peptides were discovered to possess antimicrobial properties, which gave rise to the term antimicrobial peptides (AMPs). These peptides are active against both Gram-positive and Gram-negative bacteria, fungi, viruses, and parasites (1). In recent years, it has become obvious that AMPs also have an effect on the immune system of mammals (2). For example, they can influence wound

A. Giuliani, A.C. Rinaldi (eds.), *Antimicrobial Peptides*, Methods in Molecular Biology 618,
DOI 10.1007/978-1-60761-594-1_8, © Springer Science+Business Media, LLC 2010

healing, angiogenesis, tissue repair, and the release of cytokines (3). Therefore, these antimicrobial peptides are now also called host defense peptides (HDP). However, it was shown that at least one AMP does stimulate the metastatic phenotype in breast cancer (4).

An easy method to synthesize AMPs for an activity screen, preferably with sequences smaller than about 20 amino acids, is the SPOT technology (*see* **Table 8.1**). The synthesis requires cellulose membranes set up like filter paper, resulting in a method that is inexpensive and easy to establish in virtually any laboratory (18). Using the SPOT technique, it is possible to synthesize and to screen up to 8,000 peptides, peptide mixtures, or other organic compounds on a letter-size membrane (19 cm × 28 cm) (19–21).

Table 8.1
Recommended literature

Description	References
First description of the SPOT technique in detail	(5)
Detailed review regarding synthesis and application of SPOT synthesis	(6)
Protocols for SPOT synthesis using active esters	(7)
Automated SPOT synthesis using HOBt/DIC activation	(8)
Comparison of OPfp and HOBt/DIC activation	(9)
Investigations on the quality of SPOT syntheses	(10)
Characterization of some useful types of filter paper and description of some linkers	(11)
Review about using SPOT synthesis for studying protein–protein interactions	(12)
SPOT synthesis and screening techniques	(13)
SPOT synthesis and screening on living cells	(14)
Peptide libraries synthesized by SPOT technology to develop new antimicrobials	(15)
Investigation of antimicrobial peptides using SPOT technology	(16)
Study of surface-bound antimicrobial peptides on SPOT membranes	(17)

Peptides produced by the SPOT technology can be cleaved from the membrane and directly used in screening for antimicrobial activity (22). Furthermore, the peptides can also be synthesized on modified membranes and their ability to kill bacteria even when tethered to a surface (17).

Here, we describe a protocol to synthesize peptides on cellulose membranes and then to subsequently cleave the peptides from their support medium. In addition, we also describe the preparation of CAPE and TOTD membranes, which are both suitable for the investigations of tethered peptides.

2. Materials

2.1. Preparation of Cellulose Membranes

1. Solvents: *N*,*N*′-dimethylformamide (DMF), methanol (MeOH), ethanol (EtOH), *N*-methylpyrrolidone (NMP), diethylether (DEE), dichloromethane (methylene chloride, DCM), 1,4-dioxane. The quality of solvents for washing steps should be of at least ACS or synthesis grade. Solvents for dissolving reagents must be amine and water free (*see* **Note 1**).
2. Homemade membranes for cleavable peptides are prepared from filter paper Whatman 50, Whatman 540, or Chrl (Whatman, Maidstone, UK) (23, 24) (*see* **Note 2**).
3. Amine functionalization of the filter paper by esterification using amino acids: diisopropylcarbodiimide (DIPC, DIC; Fluka), *N*-methylimidazole (NMI; Sigma), and Fmoc-β-alanine (EMD Biosciences) (*see* **Note 3**).
4. CAPE (cellulose-amino-hydroxypropyl ether) membranes: 70% perchloric acid (Alfa Aesar), 2-(bromomethyl)oxirane (epibromohydrine, Fluka), 1,3-diaminopropane (Alfa Aesar), and sodium methylate (sodium methoxide; Fluka).
5. TOTD (trioxa-tridecanediamine) membranes: 70% perchloric acid (Alfa Aesar), epibromohydrine (Fluka), 4,7,10-trioxa-1,13-tridecanediamine (Fluka), and sodium methylate (sodium methoxide; Fluka).
6. Staining solution: 0.002% bromophenol blue (BPB; Sigma) in MeOH (20 mg in 1 L).

2.2. SPOT Synthesis of Macroarrays

1. Coupling reagents: diisopropylcarbodiimide (DIPC, DIC; Fluka) and *N*-hydroxybenzotriazole (HOBt; EMD Biosciences). Coupling reagents are only necessary when preactivated amino-acid derivatives are not used (*see* **Note 4**).
2. Nonpreactivated amino acids with protection groups according to the Fmoc-protection strategy (25, 26) (EMD Biosciences and GL Biochem). Preactivated amino-acid derivatives with protection groups according the Fmoc-protection strategy are usually pentafluorophenyl esters (OPfp ester; GL Biochem, EMD Biosciences and Bachem) (27) (*see* **Note 4**).
3. Capping solution: 2% acetic anhydride (Sigma) in DMF. 2% Ethyl-diisopropylamine (DIPEA, DIEA; Sigma) can be

added to deprotonate amino groups and buffer the acetic acid generated during reaction.

4. Deprotection solution A: 90% trifluoroacetic acid (TFA; VWR) (v/v), 5% dist. water (v/v), 3% triisopropylsilane or triisobutylsilane (TIPS or TIBS; Fluka) (v/v), 1% phenol (Sigma) (w/v).
5. Deprotection solution B: 50% TFA (v/v), 3% TIPS or TIBS (v/v), 2% dist. water (v/v), 1% phenol (w/v), 44% DCM (v/v).

2.3. Cleavage from Membrane as Free Peptide Amides

1. Ammonia gas (Air Liquide).

3. Methods

If not noted different elsewhere, all washing, incubation, and reaction steps may be performed using a rocking shaker.

3.1. Preparation of Cellulose Membranes

For the basic cellulose matrix, cut a section of filter paper large enough to accommodate all peptide spots, including controls (*see* **Figs. 8.1 and 8.2**). The distance between two spots should be at least 2.8 mm if 0.1 μL of coupling solution were used, and at least 8 mm for 1 μL spotting volume.

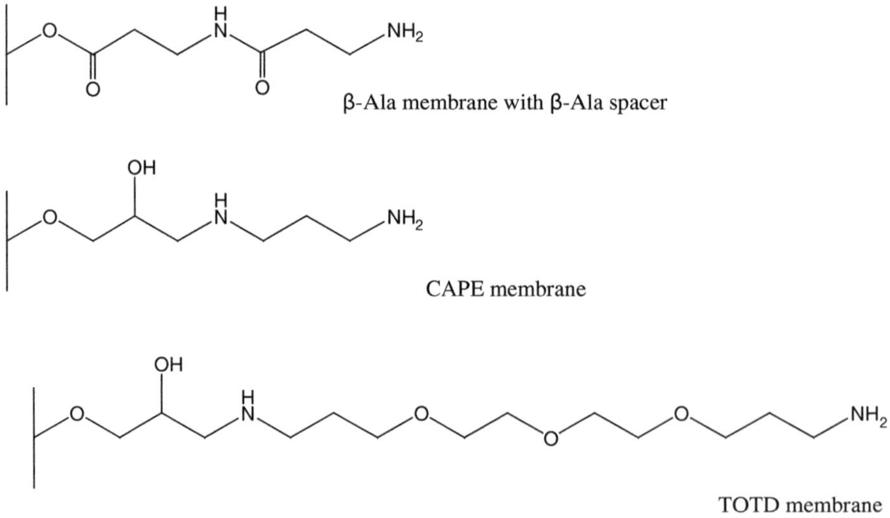

Fig. 8.1. Structure of different linker types described in this publication.

3.1.1. Preparation of Esterified Membranes

1. Amine functionalization by esterification of filter paper: For functionalization of a membrane with a size of

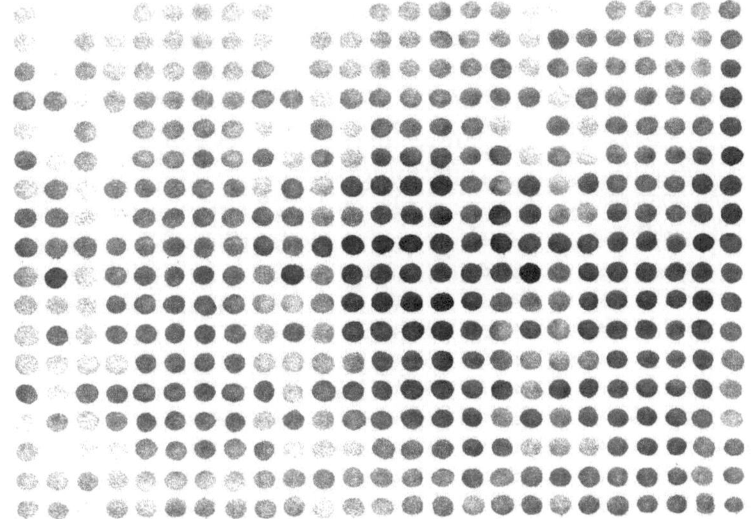

Fig. 8.2. Copy of a spot membrane of about a letter size accommodating 450 large spots.

10 cm × 12 cm (size of a 96-well plate) dissolve 0.64 g Fmoc-β-alanine in 10 mL amine-free DMF. Add 374 μL DIC and 317 μL NMI. Mix well before transferring this solution into a chemically resistant box with lid. For larger membranes, use proportionally more reagent solution. Place the filter paper of the corresponding size in the box, avoiding air bubbles under the paper, and ensure that the surface of the membrane is slightly covered by the solution. Treat the membrane with the reaction mixture in the closed box for at least 2 h, or overnight (*see* **Note 2**). Shaking is not necessary.

After the treatment, wash the membrane three times with DMF for at least 30 s each. The membrane can be stored at –80°C for several months until needed (*see* **Note 5**). For storage, wash the modified membrane at least twice with MeOH or EtOH and air dry in a fume hood or using a hair dryer without heat.

2. Fmoc-deprotection: Treat the membrane twice with 20% piperidine in DMF for at least 5 min each. Wash the membrane at least three times with DMF, followed by washing at least twice with MeOH or EtOH. Duration of each washing should be at least 30 s.

3. Staining (optional) (28) (*see* **Note 6**): Treat the membrane with staining solution for at least 2 min until the filter paper shows a homogeneous blue color. If the staining is insufficient, repeat the treatment with fresh staining solution. After staining, wash the membrane at least twice with MeOH or EtOH, until the wash solution remains colorless.

4. Air dry the membrane in a fume hood or with a hair dryer without heat (*see* **Note 7**). The membrane is now ready for the coupling cycle.

3.1.2. Preparation of Cellulose-Amino-Hydroxypropyl Ether (CAPE) Membranes

Since the membranes can be stored for some time without significant loss of activity, it is efficient to prepare a large membrane (e.g., 18 cm × 28 cm) and then cut it into several smaller membranes if needed. Otherwise, use proportionally less reagent solutions (29).

1. Wash the membrane with about 50 mL of a mixture of 50 mL MeOH and 1 mL 70% aqueous perchloric acid for several minutes. Air dry the membrane.
2. Treat the membrane with 50 mL of a mixture of 5 mL epibromohydrine and 500 μL 70% aqueous perchloric acid in 1,4-dioxane. Keep the box closed. After 3 h, wash the membrane once with EtOH or MeOH for about 15 min.
3. Let the activated membrane react overnight with about 60 mL of a 50% solution of 1,3-diaminopropane in DMF. The following day, wash the membrane three times with DMF, twice with water, and then three times with MeOH.
4. After treatment with a methanolic suspension of 5 M sodium methylate, wash the membrane with MeOH (three times), water (five times), and again MeOH (three times). After air drying the membrane is ready to use.

   The membrane can be stored at –20°C for several months until needed. For storage, wash the modified membrane at least twice with MeOH or EtOH and air dry it in the air stream of a fume hood, or using a hair dryer without heat.
5. Staining (optional) (28) (*see* **Note 6**): Wash the membrane at least three times with MeOH or EtOH for at least 30 s each. Treat the membrane with staining solution for at least 2 min until the filter paper shows a homogeneous blue color (*see* **Note 8**). If staining is insufficient, renew the staining solution. After staining, wash the membrane at least twice with MeOH or EtOH, until the wash solution remains colorless and then air dry the membrane.

3.1.3. Preparation of Trioxa-Tridecanediamine (TOTD) Membranes

The preparation of TOTD membranes follows the same protocol as for the CAPE membranes, with the exception that instead of 1,3-diaminopropane it will use 4,7,10-trioxa-1,13-tridecanediamine at Step 3 (30).

This membrane is also stable and can be stored at –20°C for several months until needed. For storage, wash the modified membrane at least twice with MeOH or EtOH and air dry it in the air stream of a fume hood, or using a hair dryer without heat.

3.2. SPOT Synthesis of Macroarrays

Peptide synthesis using the SPOT technique can be carried out manually or automated. Manual SPOT synthesis is convenient only for a relatively small number of peptides and a large pipetting volume (>0.5 μL). For all other cases, it is recommended to perform the synthesis semi- or fully automated. For synthesis of small spots with a pipetting volume of about 0.1 μL spot synthesizers from Intavis AG (Köln, Germany) are very useful (8).

If we describe working with amino acids, it always includes the use of other organic building blocks (e.g., PNA monomers, peptoidic elements, heterocycles) (20, 31) those can be used under SPOT synthesis conditions. Here, we describe only the basic procedures for SPOT synthesis of linear peptides. Synthesis of modified peptides, for example, cyclization or side-chain modifications, is described elsewhere in the literature (18).

3.2.1. Preparation of Coupling Solutions

For the preparation of coupling solution containing activated amino acids mostly two different methods were used.

First method uses pre-activated Fmoc-protected amino-acid derivatives (e.g., pentafluorophenyl esters). In this approach, only one reagent is necessary, which makes the preparation of coupling solutions very simple and reduces the likelihood of mistakes during the preparation. A disadvantage is the higher price of the amino-acid derivatives. But due to the small amount of activated amino acids used for the synthesis, the absolute cost does not increase significantly. Another disadvantage of this method lies in the fact that activated esters are only commercially available for standard amino acids. For nonstandard amino acids, their pre-activated derivatives would need to be synthesized (27), or the second method may be used instead.

The second method involves the use of in situ-activated amino acids. Activation of the amino acids is carried out by adding an activator and a coupling reagent to the nonactivated Fmoc-protected amino-acid derivative. This method is more time consuming, but can be applied for all building blocks available for use in Fmoc solid-phase peptide synthesis.

Method 1: Preparation of 0.3 M solutions with pre-activated amino-acid derivatives. Dissolve most of the amino-acid derivatives in amine-free NMP at a concentration of 0.3 M. Due to poor solubility, the derivative of serine must be dissolved in amine-free DMF. Except for the arginine derivative, the solutions of pre-activated amino acids can be used for at least 1 week if the stock solution is stored at or below –20°C. Each day, before starting the synthesis cycle, discard the amino-acid solutions of the previous day and replace them with fresh amounts from the stock solutions (*see* **Note 4**). Due to the instability of dissolved pre-activated arginine derivatives, the solution must be prepared fresh every day.

Method 2: Preparation of solutions with in situ-activated amino acids or other building blocks. Prepare a 0.9 M solution of HOBt in amine-free NMP. Dissolve the Fmoc-amino acids or desired protected building blocks with the HOBt solution to a concentration of 0.45 M. Except for the arginine derivatives, these solutions can be stored at –20°C for at least a week. Use new portions of the prepared amino-acid/HOBt solutions every day (*see* **Note 4**). To these solutions, add a freshly prepared mixture of 20% DIC in NMP at a ratio of 3:1 (e.g., for 100 μL in situ-activated amino-acid solution mix 75 μL HOBt/amino-acid solution with 25 μL 20% DIC/NMP) (*see* **Note 9**).

3.2.2. SPOT Synthesis of the Peptide Array

1. Definition of the synthesis pattern: Place the dry modified membrane on an inert surface (e.g., stainless steel or polypropylene). Deliver the required volume of activated Fmoc-β-alanine/DMSO solution to all spot positions (*see* **Note 10**). Repeat the delivery after 20 min.
2. Blocking free amino groups on nonspot areas: In order to block the free amino groups on areas surrounding the spot as well as free amino groups on the spot areas those formed no bond to the delivered amino acid, place the membrane face down in a box filled with an appropriate amount of capping solution. Ensure that there are no air bubbles under the membrane. Do not shake! Remove the liquid after about 5 min and add an appropriate amount of a mixture of capping solution with 2% DIPEA. Place the membrane in this solution and allow it to react for about 20 min.
3. Removal of Fmoc-protecting group: Wash with DMF four times for at least 30 s each. Treat twice with 20% piperidine/DMF for 5 min each. Wash again with DMF four times for at least 30 s each. Wash at least twice with MeOH or EtOH.
4. Staining (optional) (28) (*see* **Note 6**): Treat the membrane with staining solution in a box while shaking. If the staining solution changes its color to blue very rapidly, renew the solution. Perform the staining until the spots are dyed sufficiently (*see* **Note 11**). Wash the membrane at least twice with MeOH or EtOH until the wash solution remains colorless.
5. Dry the membrane in the air stream of a fume hood, or by using a hair dryer without heat (*see* **Note 7**). The membrane is now ready for the coupling cycle.
6. Amino-acid coupling: The stepwise build up of peptides starts from the C-terminus. Using the desired volume, deliver the prepared, activated amino-acid solutions to the

corresponding positions on the membrane (*see* **Note 12**). After each delivery cycle, allow the reaction to occur for at least 20 min. To achieve a high coupling yield, it is recommended to repeat the spotting of the activated amino acids at least once.

7. Capping (blocking unreacted free amino groups): Place the membrane face down in a box filled with an appropriate amount of capping solution. Do not shake! Allow the reaction to occur for about 5 min. Remove the liquid. Add an appropriate amount of a mixture of capping solution with 2% DIPEA. Place the membrane in this solution and allow at least 5 min reaction time.
8. Building up the peptide chain: Except for the last coupling cycle, repeat Steps 3–7. For the last coupling cycle, repeat the above steps without capping and staining!
9. Removal of last Fmoc-protecting group: Wash with DMF four times for at least 30 s each. Treat twice with 20% piperidine/DMF for 5 min each. Wash again four times with DMF followed by three times washing with DCM for at least 30 s each.
10. Final side-chain deprotection: Treat the membrane with at least 25 mL of deprotection solution A. For larger membranes, use proportionally more deprotection solutions A and B. The membrane must always be well covered by the deprotection solutions. Keep the box tightly closed. Do not shake. After 30 min, pour off the solution very carefully. Wash the membrane several times with DCM for at least 1 min each; then treat with at least 25 mL of deprotection solution B for 3 h in the closed box without shaking. Pour off the solution very carefully (*see* **Note 13**). Wash the membrane at least five times with water for at least 1 min each until the pH value of the washing solution is about 7 (*see* **Note 14**). Wash the membrane four times with MeOH or EtOH for at least 1 min each. Dry the membrane in the air stream of a fume hood or with a hair dryer without heat.

3.3. Cleavage from Membrane as Free Peptide Amides

In order to investigate antimicrobial activity of the peptides, it may be necessary to use peptide solutions. In this case, it is necessary to release the peptides from the cellulose membrane. The method described here involves the exposure of the entire dry membrane or the punched-out spots to ammonia gas (*see* **Fig. 8.3**). The strongly basic environment breaks the ester bond between the peptides and cellulose by forming a C-terminal amide (*see* **Note 15**). The release of peptides with ammonia gas may work only with homemade membranes. Commercially available membranes often have other than ester bonds between functional linker and cellulose. For those membranes, the use of additional

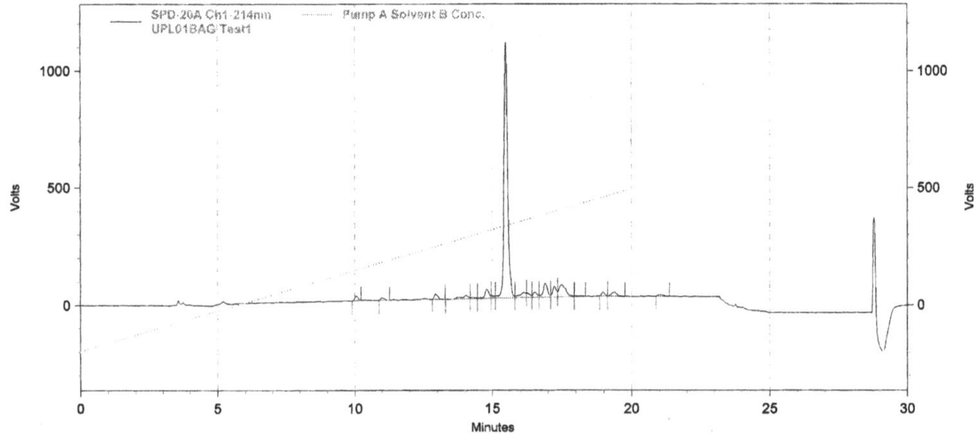

Fig. 8.3. HPLC image of a Bactenecin-related peptide synthesized via SPOT synthesis and cleaved from the membrane using ammonia.

linkers may be necessary (e.g., thioester (29, 32), HMB linker (33), Rink linker (34, 35)).

1. Place the dry membrane or the punched-out spots in a glass desiccator (*see* **Note 16**).
2. Set the desiccator under vacuum.
3. Fill the desiccator with ammonia gas.
4. Repeat Steps 2 and 3 twice.
5. Let it react overnight.
6. On the following day, open the desiccator under a fume hood (Attention! Ammonia gas is highly corrosive and irritant!). Let the gas dissipate over at least 30 min while in the fume hood.
7. If not done before, punch out the spots. Transfer the discs into wells of microtiter plates (MTPs) or into vials in order to dissolve them with the sample buffer (*see* **Note 17**).

4. Notes

1. Due to possible degradation under the influence of light, organic solvents, with the exception of MeOH and EtOH, should be stored in the dark.
2. Ready-to-use PEG membranes are available from AIMS Scientific (Braunschweig, Germany). Those membranes are more stable against high TFA concentration and may be treated only with deprotection solution I for 3.5 h.

3. Due to the flexibility and linear structure of the molecule, β-alanine is commonly used for amine functionalization of the filter paper by esterification. Other amino acids can be used (36), but there is a higher risk of losing functionality due to the inherently lower stability of the ester bond between amino acid and cellulose.
4. Reagents must be protected from moisture. To avoid condensation from air humidity, reagent bottles stored in the fridge or freezer should be unopened and warmed up to room temperature for approx. 30 min before use.
5. For longer storage, a loss of loading is possible. Storage of the membranes at –80°C is recommended. The membranes can also be stored at –20°C or +4°C; however, there is a higher risk of losing activity over a long storage period. Before storage, wash the membrane thoroughly. During storage, any traces of amines could lead to a total loss of activity within a short time! In order to proceed with the synthesis after storage, treat the membrane once with DMF for 20 min.
6. The staining has no influence in the synthesis process, but is recommended as test of the completion of the previous coupling step and the presence of free amino groups. A spot or membrane without coupled amino acid and free amino groups would remain white after staining since the amino groups are blocked by the capping step.
7. For faster drying after the washing steps, it is possible to wash the membrane additionally twice with DEE.
8. In contrast to esterified membranes, the dyed background of CAPE and TOTD membranes may disappear only slowly over several coupling cycles. Nevertheless, it does not affect the quality of the synthesis.
9. In order to avoid problems associated with the formation of hardly soluble urea, shake the in situ-activated amino-acid solutions for about 30 min; then centrifuge the mixtures and transfer the supernatant into vials of the synthesis rack.
10. Because of its flexibility and linear structure, β-alanine is commonly used for the definition of the spot pattern. It works as a spacer to achieve a distance of the amino functions to the cellulose matrix; however, the distance to the cellulose using TOTD or commercially available PEG membranes (37) is already larger than when β-alanine functionalized or CAPE membranes are used (*see* **Fig. 8.1**). In both cases, it is unnecessary to use a spacer. The definition of spot pattern can be performed by spotting the activated derivatives of the first C-terminal amino acids of the peptide sequences.

11. Do not stain the spots too strongly. If the amount of absorbed BPB is too high, some of this dye may be incorporated into the peptide. In that case, removal of the color is difficult and may affect detection after incubation. Due to varying acidity of the coupled amino acids and the built-up peptide chain, differences in the intensity of the staining of the spots are normal (*see* **Fig. 8.2**). For example, spots where aspartic acid or cysteine is the last coupled amino acids may show no blue color, whereas arginine and lysine show a strong staining.
12. In order to completely cover the area of large spots, it is recommended to use at least 20% more amino-acid solution volume than in Step 1 of **Section 3.2.2**.
13. After TFA treatment, the membrane may become very soft. Do not try to lift the membrane out until it becomes harder and less likely to break apart during the washing steps.
14. It is possible to substitute some water-washing steps by washing with PBS or TBS. Nevertheless, after buffer washing the traces of buffer must be removed by washing at least twice with water.
15. If it is necessary to yield a free carboxy terminal, do not treat the membrane with ammonia gas. Instead, treat it with an aqueous basic solution such as ammonium hydroxide or sodium hydroxide solutions (38).

    Another method to release the peptides is the formation of a C-terminal diketopiperazine. In this case, couple Boc-Lys(Fmoc)-OH instead of the β-alanine spacer (Fmoc-β-Ala-OH), followed by coupling of Fmoc-proline as the first coupled amino acid. After TFA treatment, the spots can be punched out and transferred onto a MTP or vials (*see* **Note 10**). The peptides will be released by overnight treatment with basic aqueous buffers (pH $\geq$ 7.5) (5).
16. Many desiccators made from plastics are not inert to ammonia gas; therefore, glass desiccators are strongly recommended.
17. In order to reduce possible contamination with side products from the edge of the spot, the diameter of the punched out membrane discs should be smaller than the spot diameter.

References

1. Jenssen, H., Hamill, P., and Hancock, R. E. W. (2006) Peptide antimicrobial agents. *Clin. Microbiol. Rev.* **19**, 491–511.
2. Sørensen, O. E., Borregaard, N., and Cole, A. M. (2008) Antimicrobial peptides in innate immune responses. *Contrib. Microbiol.* **15**, 61–77.
3. Zhang, L. and Falla, T. J. (2006) Antimicrobial peptides: Therapeutic potential. *Expert Opin. Pharmacother.* 7, 653–663.

4. Weber, G., Chamorro, C. I., Granath, F., Liljegren, A., Zreika, S., Saidak, Z., Sandstedt, B., Rotstein, S., Mentaverri, R., Sanchez, F., Pivarcsi, A., and Stahle, M. (2009) Human antimicrobial protein hCAP18/LL-37 promotes a metastatic phenotype in breast cancer. *Breast Cancer Res.* **11**, R6. (Epub ahead of print).
5. Frank, R. (1992) Spot-synthesis: An easy technique for the positionally addressable, parallel chemical synthesis on a membrane support. *Tetrahedron* **48**, 9217–9232.
6. Hilpert, K., Winkler, D. F. H., and Hancock, R. E. W. (2007) Cellulose-bound peptide arrays: Preparation and applications. *Biotechnol. Genet. Eng. Rev.* **24**, 31–106.
7. Kramer, A. and Schneider-Mergener, J. (1998) Synthesis and application of peptide libraries bound to continuous cellulose membranes. *Methods Mol. Biol.* **87**, 25–39.
8. Gausepohl, H. and Behn, C. (2002) Automated synthesis of solid-phase bound peptides. In *Peptide Arrays on Membrane Support*. J. Koch and M. Mahler (Eds.), pp. 55–68. Berlin, Heidelberg: Springer-Verlag.
9. Molina, F., Laune, D., Gougat, C., Pau, B., and Granier, C. (1996) Improved performances of spot multiple peptide synthesis. *Peptide Res.* **9**, 151–155.
10. Kramer, A., Reineke, U., Dong, L., Hoffmann, B., Hoffmüller, U., Winkler, D., Volkmer-Engert, R., and Schneider-Mergener, J. (1999) Spot-synthesis: observations and optimizations. *J. Peptide Res.* **54**, 319–327.
11. Frank, R., Hoffmann, S., Overwin, H., Behn, C., and Gausepohl, H. (1996) Easy preparation of synthetic peptide repertoires for immunological studies utilizing SPOT synthesis. In *Peptides in Immunology*. C. H. Schneider (Ed.), pp. 197–204. New York: John Wiley & Sons, Ltd.
12. Beutling, U., Städing, K., Stradal, T., and Frank, R. (2008) Large-scale analysis of protein–protein interactions using cellulose-bound peptide arrays. *Adv. Biochem. Eng. Biotechnol.* **110**, 115–152.
13. Winkler, D. F. H. and Campbell, W. D. (2008) The spot technique: Synthesis and screening of peptide macroarrays on cellulose membranes. *Methods Mol. Biol.* **494**, 47–70.
14. Otvos, L., Jr., Pease, A. M., Bokonyi, K., Giles-Davis, W., Rogers, M. E., Hintz, P. A., Hoffman, R., and Ertl, H. C. J. (2000) In situ stimulation of a T helper cell hybridoma with a cellulose-bound peptide antigen. *J. Immunol. Methods* **233**, 95–1051.
15. Cherkasov, A., Hilpert, K., Jenssen, H., Fjell, C. D., Waldbrook, M., Mullaly, S. C., Volkmer, R., and Hancock, R. E. W. (2009) Use of artificial intelligence in the design of small peptide antibiotics effective against a broad spectrum of highly antibiotic-resistant superbugs. *ACS Chem. Biol.* **4**, 65–74.
16. Hilpert, K., Elliott, M. R., Volkmer-Engert, R., Henklein, P., Donini, O., Zhou, Q., Winkler, D. F. H., and Hancock, R. E. W. (2006) Sequence requirements and an optimization strategy for short antimicrobial peptides. *Chem. Biol.* **13**, 1101–1107.
17. Hilpert, K., Elliott, M., Jenssen, H., Kindrachuk, J., Fjell, C. D., Körner, J., Winkler, D. F. H., Weaver, L. L., Henklein, P., Ulrich, A. S., Chiang, S. H., Farmer, S. W., Pante, N., Volkmer, R., and Hancock, R. E. W. (2009) Screening and characterization of surface-tethered cationic peptides for antimicrobial activity. *Chem. Biol.* **16**, 58–69.
18. Hilpert, K., Winkler, D. F. H., and Hancock, R. E. W. (2007) Peptide arrays on cellulose support: SPOT synthesis – a time and cost efficient method for synthesis of large numbers of peptides in a parallel and addressable fashion. *Nat. Protoc.* **2**, 1333–1349.
19. Schneider-Mergener, J., Kramer, A., and Reineke, U. (1996) Peptide libraries bound to continuous cellulose membranes: Tools to study molecular recognition. In *Combinatorial Libraries: Synthesis, Screening and Application Potential*. R. Cortese (Ed.), pp. 53–68. Berlin: Walter de Gruyter & Co.
20. Weiler, J., Gausepohl, H., Hauser, N., Jensen, O. N., and Hoheisel, J. D. (1997) Hybridisation based DNA screening on peptide nucleic acid (PNA) oligomer arrays. *Nucleic Acids Res.* **25**, 2792–2799.
21. Bowman, M. D., Jacobson, M. M., and Blackwell, H. E. (2006) Discovery of fluorescent cyanopyridine and deazalumazine dyes using small molecule macroarrays. *Org. Lett.* **8**, 1645–1648.
22. Hilpert, K. and Hancock, R. E. W. (2007) Use of luminescent bacteria for rapid screening and characterization of short cationic antimicrobial peptides synthesized on cellulose using peptide array technology. *Nat. Protoc.* **2**, 1652–1660.
23. Martens, W., Greiser-Wilke, I., Harder, T. C., Dittmar, K., Frank, R., Örvell, C., Moenning, V., and Liess, B. (1995) Spot synthesis of overlapping peptides on paper membrane supports enables the identification of linear monoclonal antibody binding determinants on morbillivirus phosphoproteins. *Vet. Microbiol.* **44**, 289–298.
24. Santona, A., Carta, F., Fraghi, P., and Turrini, F. (2002) Mapping antigenic sites of

an immunodominant surface lipoprotein of *Mycoplasma agalactiae*, AvgC, with the use of synthetic peptides. *Infect. Immun.* **70**, 171–176.

25. Fields, G. B. and Noble, R. L. (1990) Solid phase synthesis utilizing 9-fluorenylmethoxycarbonyl amino acids. *Int. J. Pept. Protein Res.* **35**, 161–214.
26. Zander, N. and Gausepohl, H. (2002) Chemistry of Fmoc peptide synthesis on membranes. In *Peptide Arrays on Membrane Support.* J. Koch and M. Mahler (Eds.), pp. 23–39. Berlin, Heidelberg: Springer-Verlag.
27. Atherton, E. and Sheppard, R. C. (1989) Activated esters of Fmoc-amino acids. In *Solid Phase Peptide Synthesis – A Practical Approach.* A. Giuliani, A. C. Rinaldi (eds.), pp. 76–78. Oxford, UK: IRL Press at Oxford University Press.
28. Krchnak, V., Wehland, J., Plessmann, U., Dodemont, H., Gerke, V., and Weber, W. (1988) Noninvasive continuous monitoring of solid phase peptide synthesis by acid–base indicator. *Collect. Czech. Chem. Commun.* **53**, 2542–2548.
29. Licha, K., Bhargava, S., Rheinlander, C., Becker, A., Schneider-Mergener, J., and Volkmer-Engert, R. (2000) Highly parallel nano-synthesis of cleavable peptide-dye conjugates on cellulose membranes. *Tetrahedron Lett.* **41**, 1711–1715.
30. Ast, T., Heine, N., Germeroth, L., Schneider-Mergener, J., and Wenschuh, H. (1999) Efficient assembly of peptomers on continuous surfaces. *Tetrahedron Lett.* **40**, 4317–4318.
31. Heine, N., Ast, T., Schneider-Mergener, J., Reineke, U., Germeroth, L., and Wenschuh, H. (2003) Synthesis and screening of peptoid arrays on cellulose membranes. *Tetrahedron* **59**, 9919–9930.
32. Boisguerin, P., Leben, R., Ay, B., Radziwill, G., Moelling, K., Dong, L., and Volkmer-Engert, R. (2004) An improved method for the synthesis of cellulose membrane-bound peptides with free C termini is useful for the PDZ domain binding studies. *Chem. Biol.* **11**, 449–459.
33. Volkmer-Engert, R., Hoffmann, B., and Schneider-Mergener, J. (1997) Stable attachment of the HMB-linker to continuous cellulose membranes for parallel solid phase spot synthesis. *Tetrahedron Lett.* **38**, 1029–1032.
34. Scharn, D., Wenschuh, H., Reineke, U., Schneider-Mergener, J., and Germeroth, L. (2000) Spatially addressed synthesis of amino- and amino-oxy-substituted 1,3,5-triazine arrays on polymeric membranes. *J. Comb. Chem.* **2**, 361–369.
35. Rau, H. K., DeJonge, N., and Haehnel, W. (2000) Combinatorial synthesis of four-helix bundle hemoproteins for tuning of cofactor properties. *Angew. Chem. Int. Ed.* **39**, 250–253.
36. Kamradt, T. and Volkmer-Engert, R. (2004) Cross-reactivity of T lymphocytes in infection and autoimmunity. *Mol. Divers.* **8**, 271–280.
37. Zander, N. (2004) New planar substrates for the in situ synthesis of peptide arrays. *Mol. Divers.* **8**, 189–195.
38. Bhargava, S., Licha, K., Knaute, T., Ebert, B., Becker, A., Grötzinger, C., Hessenius, C., Wiedemann, B., Schneider-Mergener, J., and Volkmer-Engert, R. (2002) A complete substitutional analysis of VIP for better tumor imaging properties. *J. Mol. Recog.* **15**, 145–153.

Chapter 9

High-Throughput Screening for Antimicrobial Peptides Using the SPOT Technique

Kai Hilpert

Abstract

The SPOT technique provides a fast, cost-efficient, and highly parallel method to synthesize peptide arrays on cellulose. Peptides synthesized on cellulose can be easily cleaved from the support and used directly in a screening assay for antimicrobial activity. Depending on the equipment, the synthesis and the screening can be performed in a medium- or high-throughput manner. High-sensitivity screening is achieved using a bacterial strain (e.g., *Pseudomonas aeruginosa* H1001) in which a luminescence-encoding gene cassette has been introduced. The intensity of light produced is directly dependent on the energy level of the bacteria. This screening supports the development of new drugs against multidrug-resistant bacteria.

Key words: Screening, antimicrobial peptides, bacteria, yeast, human pathogens, *Pseudomonas aeruginosa*, luminescence, SPOT, synthesis, peptide array.

1. Introduction

With the introduction of sulfonamides in the 1930s, for the treatment of bacterial infections, conventional Western medicine was revolutionized: epidemics caused by bacterial infections were no longer a threat to human kind. In addition, many new medical procedures, including complicated operations such as implants and transplants, became possible when used in conjunction with antibiotics to prevent infection. What we now acknowledge as a standard in modern medicine cannot be achieved without antibiotics, or at the very least not without an unacceptably high risk of infection for the patient. This need for antibiotics is reflected also in its reported usage: in the year 2007 in Germany alone, 250–300 tons of antibiotics were used in human medicine

A. Giuliani, A.C. Rinaldi (eds.), *Antimicrobial Peptides*, Methods in Molecular Biology 618,
DOI 10.1007/978-1-60761-594-1_9, © Springer Science+Business Media, LLC 2010

and 750 tons in veterinary medicine (1). Unfortunately, we now face a new problem: often only a short time after introducing a new antibiotic to the market, resistant bacteria are observed. Today, modern medicine is undermined by multidrug-resistant bacteria that can withstand the simultaneous application of several different antibiotics, including such examples as methicillin-resistant *Staphylococcus aureus* (MRSA), vancomycin-resistant *Enterococci* (VRE), and multidrug-resistant *Pseudomonas aeruginosa* (2). Without immediate investigation into new substances and treatment strategies, the potential impact of infection by these difficult-to-treat pathogenic bacteria may be devastating.

Combinatory chemistry and biological libraries such as phage display are extensively used in efforts to discover new antimicrobial substances (3). Antimicrobial peptides (AMPs), also called host defense peptides, are a class of substances with the potential to treat multidrug-resistant bacteria. Certain naturally occurring AMPs have been selected and intensively studied. For a long time, it was assumed that all AMPs act on the cell membrane to bring about bacterial death; however, it became clear that some of the AMPs seemed to have internal targets and did not damage the membrane (4).

The normal approach for studying AMPs was to investigate the antimicrobial activity of naturally occurring peptides against Gram-positive and Gram-negative bacteria, as well as any hemolytic effects and structure–activity relationships. Using this information, a few variants of natural peptides with substitutions were designed and synthesized using standard peptide synthesis. Owing to intensive time, manpower, and cost investments with this standard methodology, only a very few of these variants could be produced and therefore screening hundreds or thousands of peptides for antimicrobial activity is very difficult. In addition, it is also desirable to find short peptide sequences to help to reduce the cost of the new drug. Most of the natural AMPs are longer than 12 amino acids and are too long to be ideal candidates for cost-effective drugs. Peptide synthesis on cellulose can provide an inexpensive, easy-to–use, and time-saving strategy to screen for short peptides with the desired antimicrobial activity. For example, we showed before that screening 1,600 peptides yielding enough information to develop bioinformatics tools, which we then used successfully to screen in silico for highly active peptides against multidrug-resistant bacteria (5, 6).

Peptide synthesis by SPOT technology is described in detail in 7 and in this book (8). Around 1,000 peptides can be synthesized on one sheet of cellulose. For each peptide, nine concentration series can be produced and used for tests against nine different bacteria or different isolates of one species. Using two fully automated robots (Intavis AG, Köln, Germany), around 1,800,000 measurements can be performed in 1 year.

Here we describe the process of screening soluble peptides for antimicrobial activity. A genetically modified *P. aeruginosa* strain is used to perform a sensitive and easily measurable tool in a high-throughput manner. A bacterial luciferase gene cassette was introduced via a transposon into *P. aeruginosa* (9). The cassette *lux-CDABE* is comprised of the luciferase itself, as well as enzymes that produce the required fatty aldehyde substrate. The luciferase produces light when reduced flavin mononucleotide ($FMNH_2$) and molecular oxygen is provided by the bacteria. Any event that disturbs the energy balance of the bacteria will also disturb the amount of available $FMNH_2$, which is then registered as a change in light production. The light can be measured sensitively and in a high-throughput manner using a luminescence microtiter plate reader. The method described here already was used for performing substitution analyses of variants of the peptide bactenicin, as well as for a scrambled peptide library and semi-random peptide libraries (5, 10, 11). In principle, this screening can also be performed using unmodified bacteria and standard live/dead staining; however, sensitivity might be strongly reduced.

2. Materials

1. Bromophenol blue (BPB; Sigma).
2. 96-well polypropylene microtiter plates (Corning Inc.).
3. Microseal "F" Foil (BIORAD).
4. 2.1% (wt/vol) Difco Müller–Hinton (MH) broth sterilized by autoclaving (Becton-Dickson).
5. 2.1% (wt/vol) Difco MH broth and 1.5% agar (Becton-Dickson), sterilized by autoclaving, transferred into Petri dishes (Fisher Scientific).
6. 96-well plates suitable for luminescence (Nunc).
7. 12-channel multipipettor (VWR).
8. Bact. I: 5,000 μL MH medium containing 50 μL of a H1001 overnight culture.
9. Bact. II: 50 mL sterile 100 mM Tris–HCl buffer, pH 7.3, containing 20 mM glucose and 1920 μL H1001 culture ($OD_{600} = 0.35$).

3. Methods

When planning to synthesize peptides for screening purposes, be aware that for ten peptides, one positive and one negative control peptide also need to be synthesized. Performing this screening procedure requires two points to be addressed at the stage of synthesizing peptides using SPOT technology. The first point is that at the last round of the coupling cycle, when the peptides are stained with bromophenol blue, the membrane must be labeled using a pencil to indicate the position of each peptide. For example, a grid can be drawn onto the membrane so that each intersection represents the center of a peptide spot (**Fig. 9.1a**). The second point, which is critical to successful screening, is that the cellulose membrane must be carefully washed several times with a neutral pH buffer before the peptides are cleaved from the membrane with ammonia gas. When planning the experiment, make sure to address these two points.

This protocol is written for peptide spots with a density of 0.5–0.8 mmol/cm^2, and a spot area of approximately 0.25 cm^2

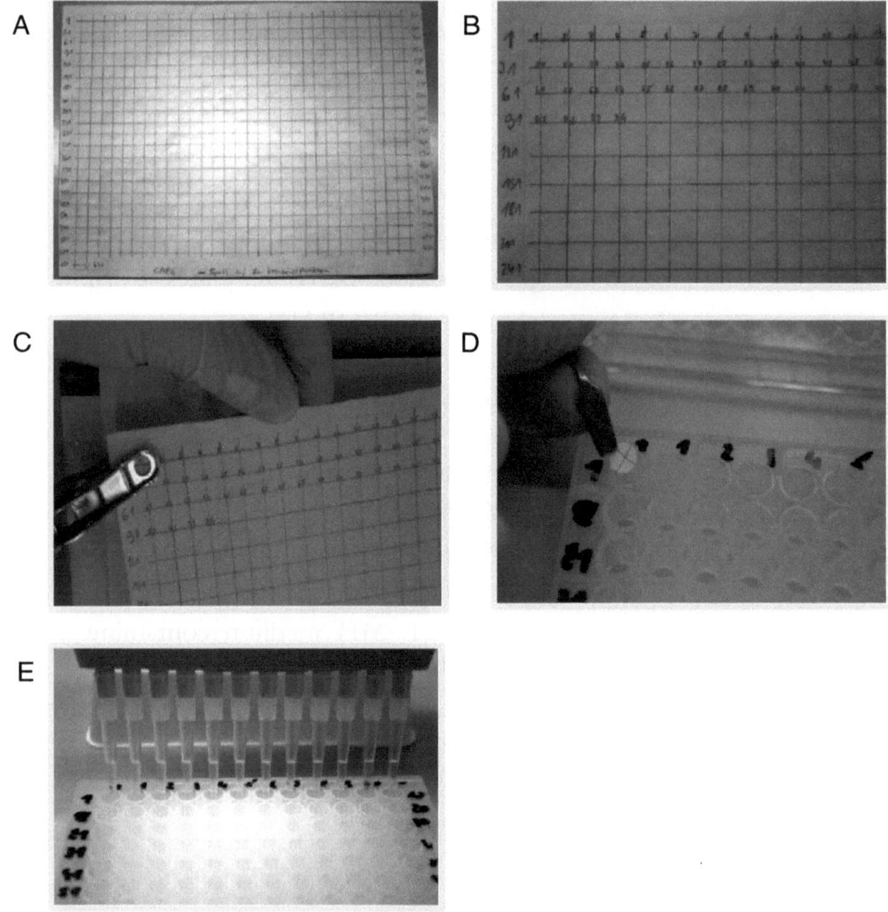

Fig. 9.1. Images showing various steps in the screening procedure: (**a**) Labeled cellulose membrane after synthesis, (**b**) peptides 1–94, showing all spots labeled directly on the membrane, (**c**) extract peptide spot 1 using a single hole puncher, (**d**) transferring the cellulose disk into a well of a 96-well polypropylene microtiter plate, and (**e**) solubilizing the peptides by adding 200 μL of autoclaved distilled water.

In addition, the antimicrobial activity of the peptide can be investigated in a range of the MIC of inactive (<250 g/mL) to around 0.5 μg/mL. Adapt volumes accordingly. The volumes of bacterial solutions provided here are designed to screen 80 test peptides and 16 control peptides. When screening a greater or lesser number of peptides, adapt the volume accordingly. When screening large numbers of peptides per day, a second tube of Bact. I may also be required (*see* **Section 3.4**).

3.1. Labeling Cellulose Membrane and Plates

1. When dealing with cellulose membranes, always wear gloves (*see* **Note 1**). Use a sharp pencil to individually label all peptide spots using consecutive numbers (**Fig. 9.1b**; *see* **Note 2**).
2. Label polypropylene plates with "M1," "M2," and so forth. In each 96-well plate, 80 peptides and 16 control peptides can be accommodated. Column 1 can be used for the positive controls and column 12 for the negative controls. It is crucial to very carefully document the position of each peptide. An example is given in **Fig. 9.2**.

Plate/Position	1	2	3	4	5	6	7	8	9	10	11	12
Plate 1	pos	wt	p1A	p1C	p1D	p1E	p1F	p1G	p1H	p1I	p1K	neg
Plate 2	pos	p1L	p1M	p1N	p1I	p1Q	p1R	p1S	p1T	p1V	p1W	neg
Plate 3	pos	p1Y	wt	p2A	p2C	p2D	p2E	p2F	p2G	p2H	p2I	neg
Plate 4	pos	p2K	p2L	p2M	p2N	p2P	p2Q	p2R	p2S	p2T	p2V	neg
Plate 5	pos	p2W	p2Y	wt	p3A	p3C	p3D	p3E	p3F	p3G	p3H	neg
Plate 6	pos	p3I	p3K	p3L	p3M	p3N	p3P	p3Q	p3R	p3S	p3T	neg
Plate 7	pos	p3V	p3W	p3Y	wt	p4A	p4C	p4D	p4E	p4F	p4G	neg
Plate 8	pos	p4H	p4I	p4K	p4L	p4M	p4N	p4P	p4Q	p4R	p4S	neg
Plate 9	pos	p4T	p4V	p4W	p4Y	wt	p5A	p5C	p5D	p5E	p5F	neg
Plate 10	pos	p5G	p5H	p5I	p5K	p5L	p5M	p5N	p5P	p5Q	p5R	neg
Plate 11	pos	p5S	p5T	p5V	p5W	p5Y	wt	p6A	p6C	p6D	p6E	neg
Plate 12	pos	p6F	p6G	p6H	p6I	p6K	p6L	p6M	p6N	p6P	p6Q	neg
Plate13	pos	p6R	p6S	p6T	p6V	p6W	p6Y	wt	p7A	p7C	p7D	neg
Plate 14	pos	p7E	p7F	p7G	p7H	p7I	p7K	p7L	p7M	p7N	p7P	neg
Plate 15	pos	p7Q	p7R	p7S	p7T	p7V	p7W	p7Y	wt	p8A	p8C	neg
Plate 16	pos	p8D	p8E	p8F	p8G	p8H	p8I	p8K	p8L	p8M	p8N	neg
Plate 17	pos	p8P	p8Q	p8R	p8S	p8T	p8V	p8W	p8Y	wt	p9A	neg
Plate 18	pos	p9C	P9D	p9E	p9F	p9G	p9H	p9I	p9K	p9L	p9M	neg
Plate 19	pos	p9N	p9P	p9Q	p9R	p9S	p9T	p9V	p9W	p9Y	wt	neg
Plate 20	pos	p10A	p10C	p10D	p10E	p10F	p10G	p10H	p10I	p10K	p10L	neg
Plate 21	pos	p10M	p10N	p10P	p10Q	p10R	p10S	p10T	p10V	p10W	p10Y	neg
Plate 22	pos	wt	p11A	p11C	p11D	p11E	p11F	p11G	p11H	p11I	p11K	neg
Plate 23	pos	p11L	p11M	p11N	p11P	p11Q	p11R	p11S	p11T	p11V	p11W	neg
Plate 24	pos	p11Y	wt	p12A	p12C	p12D	p12E	p12F	p12G	p12H	p12I	neg
Plate 25	pos	p12K	p12L	p12M	p12N	p12P	p12Q	p12R	p12S	p12T	p12V	neg
Plate 26	pos	p12W	p12Y	wt	p13A	p13C	p13D	p13E	p13F	p13G	p13H	neg
Plate 27	pos	p13I	p13K	p13L	p13M	p13N	p13P	p13Q	p13R	p13S	p13T	neg
Plate 28	pos	p13V	p13W	p13Y								neg

Fig. 9.2. Sample layout for documenting peptide positions within the 96-well plate. "Pos" stands for positive control, "neg" for negative control, and "wt" for wild type. This example shows a substitution analysis for a 13-mer peptide that was synthesized. The substitution is labeled at each position (e.g., p1L means that at position 1 a leucine (L) was substituted against the wild-type sequence).

3.2. Transfer of the Peptide Spots and Solubilizing of the Peptides

1. Use a single hole punch to carefully punch out each peptide spot. Use tweezers to transfer the cellulose disc into the corresponding well of the 96-well polypropylene microtiter plate (**Fig. 9.1c** and **d**; *see* **Note 3**). After filling one row of the plate, add 200 μL autoclaved, distilled water into each well, which contains a cellulose disc (**Fig. 9.1e**). If low activity is expected, add only 50–100 μL of water. Continue this procedure until all peptides are transferred.
2. When one plate is complete, replace the lid and seal the plate with parafilm. Do this for all plates.
3. Put all plates on a shaker (suitable for 96-well plates) and shake the plates overnight at room temperature.
4. After the overnight solubilization, the peptides can be used directly for the assay or, alternatively, the plates can be sealed with foil and stored at –20°C for several months.

3.3. Preparation of the Bacteria Solution

1. Use a sterile loop to scratch off some frozen bacteria solution (*P. aeruginosa* strain H1001, cryo stock) and streak on a Petri dish containing Müller–Hinton (MH) medium and 1.5% agar. This procedure results in single colonies.
2. Incubate the plate overnight at 37°C.
3. Pick one single colony from the agar plate (Point 1) with a sterile tip. Place this tip into a sterile tube containing 3 mL of Müller–Hinton (MH) medium to transfer the bacterial colony.
4. Incubate the tube overnight at 37°C and 225 rpm on a shaker.
5. In preparation for the next day, make a 500 mL buffer solution of 100 mM Tris–HCl, pH 7.3, and autoclave. In addition, prepare 50 mL of a 2 M glucose solution and sterile filter the solution (0.2 μm pore size).

3.4. Measuring Luminescence

1. Add 50 μL of the H1001 overnight culture into a sterile tube containing 4,950 μL MH medium. Label the tube "Bact. I."
2. Incubate this tube at 37°C and 225 rpm on a shaker. Repeatedly measure the OD_{600} value of the culture until a value of about 0.35 is reached.
3. Prepare 48 mL of a sterile 100 mM Tris–HCl buffer, pH 7.3, containing 20 mM glucose.
4. Add 1,920 μL bacterial culture ($OD_{600} = 0.35$) into the 48 mL buffer/glucose solution.
5. Gently vortex the tube and label this "Bact. II."

6. Add 50 μL of Bact. II into each well of row A of a 96-well microtiter plate that is suitable for luminescence measurements. Label this plate "Test."
7. Measure the luminescence of row A. If the luminescence signal is detectable, proceed to point 8; if not, stop here, carefully analyze for contamination, including the bacterial strain, as well as all sterile media and buffers.
8. Label all required 96-well plates suitable for luminescence.
9. In each luminescence plate, add 80 μL of Bact. II into row A of each luminescence plate (use a 12-channel multipipettor for Steps 9–12).
10. In each luminescence plate, add 50 μL of Bact. II into rows B–H.
11. From the 96-well plates containing the peptides (labeled M1, M2, ...), transfer 20 μL of each well from one row into row A of the corresponding luminescence plate. For example, 20 μL from row A, plate M1 would be transferred into row A of the first luminescence plate, then 20 μL of row B, plate M1 will be transferred into row A of the second luminescence plate, and so on.
12. With each luminescence plate, perform a stepwise dilution using 50 μL, with thorough mixing after each dilution. To illustrate, use 50 μL from row A, transfer to row B and mix thoroughly. Row B should now contain a total of 100 μL in each well. Next, take 50 μL from row B and transfer it to row C; mix thoroughly. Continue this process until row G is reached. At row G, after mixing, take out 50 μL and discard. Repeat procedure until all wells of each luminescence plate have been diluted in this manner.
13. Seal all peptide master plates (M1, M2, ...) with foil and store them at –20°C.
14. Measure the luminescence of each plate containing bacteria, buffer, glucose, and peptides with a plate reader.
15. Incubate the plates at 37°C for 4 h.
16. Repeat the luminescence measurement.
17. In each well of all plates containing bacteria, add 50 μL of MH media.
18. Incubate all plates at 37°C overnight.
19. Measure the OD_{600} of all plates or alternatively perform a live/dead staining.

3.5. Data Analysis

The luminescence data associated with each plate are normally presented as a table, depending on the software of the reader (e.g., *see* **Fig. 9.3**). Detailed instructions for data analysis are pro-

Rel. Peptide concentration	Row/ Column	1	2	3	4	5	6	7	8	9	10	11	12
1	**A**	0.0013	0.0002	0.0006	0.0048	11.54	12.33	14.7	0.0007	14.98	17.31	0.0004	11.01
0.5	**B**	0.0013	0.0007	0.0007	0.0037	10.42	10.3	10.41	0.0013	11.1	11.01	0.0081	10.93
0.25	**C**	0.0154	0.1633	0.1958	0.1089	10.58	11.26	10.92	0.0071	10.7	10.7	0.0192	10.68
0.125	**D**	1.225	1.244	1.478	1.089	10.76	11.17	10.91	0.888	10.86	10.98	0.4204	10.48
0.0625	**E**	7.278	4.786	4.888	3.976	10.11	10.12	10.35	4.522	10.27	10.43	2.077	10.28
0.03125	**F**	8.942	8.261	7.203	6.914	11.03	11.01	11.08	7.442	11.5	11.56	4.526	10.56
0.015625	**G**	9.63	9.553	8.523	8.502	11.01	11.26	11.26	8.682	11.7	11.59	8.26	10.87
0	**H**	11.9	12.13	11.42	11.5	12.33	12.35	12.5	11.6	12.69	12.54	12.32	11.39
	Peptide #	**K+**	**P1**	**P2**	**P3**	**P4**	**P5**	**P6**	**P7**	**P8**	**P9**	**P10**	**K-**
	Rating		**0**	**0**	**0**	**–1**	**–1**	**–1**	**0**	**–1**	**–1**	**+1**	

Fig. 9.3. Typical luminescence results of *P. aeruginosa* strain H1001 incubated with 10 peptides and two controls for 4 h at 37°C. In the left column, relative concentrations are presented, starting with "1" as the highest concentration. Rows marked with *gray* represent concentrations similar to the IC_{50} of the positive control. Each peptide is rated "0" where the IC_{50} was within range (i.e., similar activity to the control), "–1" when above (i.e., lower activity than the control) and "1" when below (i.e., higher activity than the control).

vided in the chapter by Ralf Mikut (12). A simple analysis for small data sets can be performed as follows:

1. Verify that the luminescence values of the negative control are as high as expected; this depends on the instrument and the settings used.
2. Use row H to calculate the mean of untreated wells.
3. Verify that luminescence values of the positive control have been reduced as expected.
4. With the positive control, approximate the concentration where a half-maximal inhibition (IC_{50}) occurs.
5. Mark the row that corresponds most closely to this concentration. Mark the lower and higher rows as well (**Fig. 9.2**).
6. Determine if the IC_{50} of each peptide is within the marked rows or not. When it is inside the marked rows, label the peptide "0," to indicate activity similar to the control. When values are above the IC_{50}, then label "–1," indicating lower activity relative to the control; values below the marked rows label "1," to indicate a higher activity than the control (**Fig. 9.2**).
7. Compare luminescence values with results from the overnight incubation (*see* **Note 4**).

4. Notes

1. The cellulose membrane is easily contaminated by bacteria, antimicrobial substances, and skin oils, so always use gloves when handling the membrane.

2. The punched out cellulose spots will be transferred into 96-well microtiter plates. During the punching, transfer is done in the well before adding liquid, the small cellulose discs can jump into other wells or out of the plate via electrostatic interaction; therefore, it is critical that the discs are labeled so that they can be easily identified in such an event.
3. Since peptides are normally positively charged and polystyrene is negatively charged, use only polypropylene plates. In addition, avoid tissue culture-treated plates, as these will nonspecifically bind with the peptides.
4. In rare cases, some peptides may initially inhibit luminescence of the bacteria, but then with longer periods of incubation, the bacteria recover, resulting in a false-positive hit. In order to avoid selecting such peptides, the data from the overnight incubation are also important to consider during data analysis. DO NOT consider a peptide to be "active" when luminescence values are strongly reduced, even at higher concentrations, if at the same time growth inhibition (from the overnight data) is observed only at high concentrations or there is no inhibition at all.

References

1. GERMAP (2008), Antibiotika-Resistenz und-Verbrauch, Verlag: Antiinfectives Intelligence, Gesellschaft für klinisch-mikrobiologische Forschung und Kommunikation mbH, Rheinbach, Germany.
2. Rossolini, G. M. and Mantengoli, E. (2008) Antimicrobial resistance in Europe and its potential impact on empirical therapy. *Clin. Microbiol. Infect.* **14**, 2–8.
3. Mullen, L. M., Nair, S. P., Ward, J. M., Rycroft, A. N., and Henderson, B. (2006) Phage display in the study of infectious diseases. *Trends Microbiol.* **14**, 141–147.
4. Brogden, K. A. (2005) Antimicrobial peptides: pore formers or metabolic inhibitors in bacteria? *Nat. Rev. Microbiol.* **3**, 238–250.
5. Cherkasov, A., Hilpert, K., Jenssen, H., Fjell, C. D., Waldbrook, M., Mullaly, S. C., Volkmer, R., and Hancock, R. E. W. (2008) Use of artificial intelligence in the design of small peptide antibiotics effective against a broad spectrum of highly antibiotic-resistant superbugs. *ACS Chem. Biol.* **4**, 65–74.
6. Fjell, C. D., Jenssen, H., Hilpert, K., Cheung, W., Panté, N., Hancock, R. E. W., and Cherkasov, A. (2009) Identification of novel antibacterial peptides by chemoinformatics and machine learning. *J. Med. Chem.*, **52**(7), 2006–2015.
7. Hilpert, K., Winkler, D. F. H., and Hancock, R. E. W. (2007) Cellulose-bound peptide arrays: preparation and applications. *Biotechnol. Genet. Eng. Rev.* **24**, 31–106.
8. Winkler, D. F. H. and Hilpert, K. (2009) Synthesis of antimicrobial peptides using the SPOT technique. *Methods Mol. Med.*, in press.
9. Lewenza, S., Falsafi, R. K., Winsor, G., Gooderham, W. J., McPhee, J. B., Brinkman, F. S. L., and Hancock, R. E. W. (2005) Construction of a mini-Tn*5-luxCDABE* mutant library in *Pseudomonas aeruginosa* PAO1: a tool for identifying differentially regulated genes. *Genome Res.* **15**, 583–589.
10. Hilpert, K., Volkmer-Engert, R., Walter, T., and Hancock, R. E. W. (2005) High-throughput generation of small antibacterial peptides with improved activity. *Nat. Biotechnol.* **23**, 1008–1012.
11. Hilpert, K., Elliott, M. R., Volkmer-Engert, R., Henklein, P., Donini, O., Zhou, Q., Winkler, D. F. H., and Hancock, R. E. W. (2006) Sequence requirements and an optimization strategy for short antimicrobial peptides. *Chem. Biol.* **13**, 1101–1117.
12. Mikut, R. (2009) Computer-based analysis, visualization, and interpretation of antimicrobial peptide activities. *Methods Mol. Biol.* **618**, 287–299.

Section II

Analysis, Properties, and Mechanisms of Antimicrobial Peptides

Chapter 10

Antimicrobial Peptides: The LPS Connection

Andrea Giuliani, Giovanna Pirri, and Andrea C. Rinaldi

Abstract

An expanding body of evidence is rendering manifest that many cationic antimicrobial peptides are endowed with different properties and activities, well beyond their direct action on microbes. One of the most interesting and potentially important research avenue on the alternative use of antimicrobial peptides grounds on their affinity toward lipopolysaccharide (LPS), the endotoxin, responsible for the systemic inflammatory response syndrome (SIRS) and related, often fatal, disorders that can follow Gram-negative infections. Indeed, not only do several antimicrobial peptides, such as cathelicidins, display an ability to strongly bind LPS and break its aggregates, but they have also been demonstrated to suppress LPS-induced pro-inflammatory responses in vitro and to protect from sepsis in animal models. Although many aspects still need to be carefully evaluated – some of which are highlighted here – a mix of antimicrobial, LPS-sequestering/neutralization, and immunomodulatory features make cationic peptides, and especially synthetic or semi-synthetic amphiphilic compounds built on their scheme, attractive candidates for novel drugs to be administered in antisepsis therapies. These therapies will probably hinge either on compounds able to intervene at multiple points in the sepsis cascade or on the combination of two or more immunomodulators.

Key words: Antimicrobial peptides, synthetic peptides, LPS, endotoxin, binding, neutralization, sepsis, septic shock, Gram-negative.

1. Introduction

Host microbial invasion may have different forms and outcomes. One of the most easy-to-encounter and at the same time harder-to-contrast infection-related condition is sepsis. This can be broadly described as a systemic response to infection, mounted by the host to combat and limit infection, but that sometimes derails to an uncontrolled immunological response that may lead to multiple organ dysfunction and failure, septic shock (1). Severe

A. Giuliani, A.C. Rinaldi (eds.), *Antimicrobial Peptides*, Methods in Molecular Biology 618,
DOI 10.1007/978-1-60761-594-1_10, © Springer Science+Business Media, LLC 2010

sepsis is by no means a rare condition, and its incidence has even grown during the last decades. Despite progresses in critical care settings and practice, and infectious disease research, severe sepsis' outcome remains often sad. Indeed, mortality rates exceed 60% in some cases, making it a major cause of death in intensive care units worldwide (1). In the United States only, where about 750,000 people are struck by severe sepsis each year, the estimated direct costs of care amounted at $16.7 billion annually (2). However, this is just a fraction of the burden imposed by sepsis, as 70–80% of costs are believed to arise mainly from productivity losses due to mortality (3).

Although major wall components of Gram-positive bacteria, peptidoglycan and lipoteichoic acid, have been associated with the development of sepsis in some instances (4, 5), our understanding of sepsis is indissolubly linked to the action of Gram-negative lipopolysaccharide (LPS), the endotoxin. It is our ability to sense LPS that drives the early innate and subsequent adaptive antibacterial defenses, alerting the host of pathogen invasion and providing us with a shield against infection, but when the inflammatory response gets unlashed septic shock may arise, the toll to pay to survive in a world where microbial life sovereigns and tends to prevail (6, 7). The core LPS signaling pathway starts with the liver-produced LPS-binding protein (LBP), which binds plasma LPS shed from bacteria and conveys it to the surface of macrophages or neutrophils in a form that can be recognized by the membrane-bound protein CD14. The LBP–CD14 complex in turn activates the Toll-like receptor 4 (TLR4) to trigger the biosynthesis of potent inflammatory cytokines such as TNF-α, IL-1, and IL-6. Controlled production and secretion of pro-inflammatory cytokines together with the release of a range of co-stimulatory molecules by macrophages, granulocytes, dendritic cells, or mast cells orchestrates innate and adaptive antibacterial responses, essential events in order to control infection and clearing pathogens (8). Only when massive production of cytokines and inflammatory mediators overcomes host control, the system precipitates toward severe sepsis and septic shock.

Together with the continuous advances in the characterization of antimicrobial peptides' (AMPs) features and properties recorded in the last two decades or so has steadily grown the belief (and hope) that besides classic anti-infective activities, they could also serve as anti-endotoxin drugs, binding and sequestering LPS before it could induce septic shock, thus offering fresh lymph for the development of an effective – and still lacking – pharmacological therapy of severe sepsis. Although comparatively few AMPs have been studied as for their LPS-binding and LPS-neutralizing properties so far, our knowledge in this area has grown significantly. Rather than being an encyclopedic review of the studies on the interactions of AMPs and LPS, this short chapter aims at highlighting our firm points in this interesting research avenue (mentioning, when appropriate, the relevant methodological aspects), while outlining its reasonably concrete perspectives.

2. LPS Structure and Function

The lipopolysaccharide makes a key component of the cell wall of all Gram-negative bacteria, irrespectively of their pathogenic activity toward either humans or animals. The cell wall (or cell envelope) of Gram-negative bacteria is made of an outer membrane, an inner (plasma) membrane, and a peptidoglycan sheet laying in the periplasm, i.e., the space between the two membranes. While the inner face of the outer membrane is composed of phospholipids, the outer face of the outer membrane is composed primarily (~90%) of LPS, with the addition of some phospholipids and various proteins that play a range of functions.

LPS itself shares with phospholipids an amphipathic design (9). The hydrophilic portion consists of a highly variable polysaccharide, made of a core and an outer portion (**Fig. 10.1**). The core polysaccharide usually contains peculiar sugars, such as heptose and 2-keto-3-deoxyoctonoic acid, and its structure variability is consistent within specific bacterial genera. The much bigger outer polysaccharide extends into the extracellular environment, and it is composed of up to 40 repeating units, which may contain up to seven different or identical sugars. This region accounts for the largest part of the LPS structural diversity among bacterial species and strains. The outer polysaccharide – also referred to as the O-antigen of the bacteria – is also a major antigenic determinant, being involved in interactions with antibodies, porins, LPS-binding protein, and in activating the complement system. The hydrophobic, membrane-anchoring region of LPS is named lipid A. Relatively well conserved among bacterial species, lipid A is the most active moiety of LPS, being predominantly responsible for many of the pathophysiological effects associated with infection by Gram-negative bacteria. It is composed of a *bis*-phosphorylated *N*-acetylglucosamine (NAG) dimer with six or seven saturated acyl chains linked through amide and ester bonds (**Fig. 10.1**).

The main role attributed to LPS is that of contributing to the structural integrity of cell wall, establishing at the same time a permeability barrier that prevents penetration of the bacteria by many, potentially toxic substances, including antimicrobial agents. At a closer view, it would be the saccharidic portion of LPS that would act as a hydrophilic shield to impede the passage of hydrophobic molecules. Evidence has also grown that indicates a

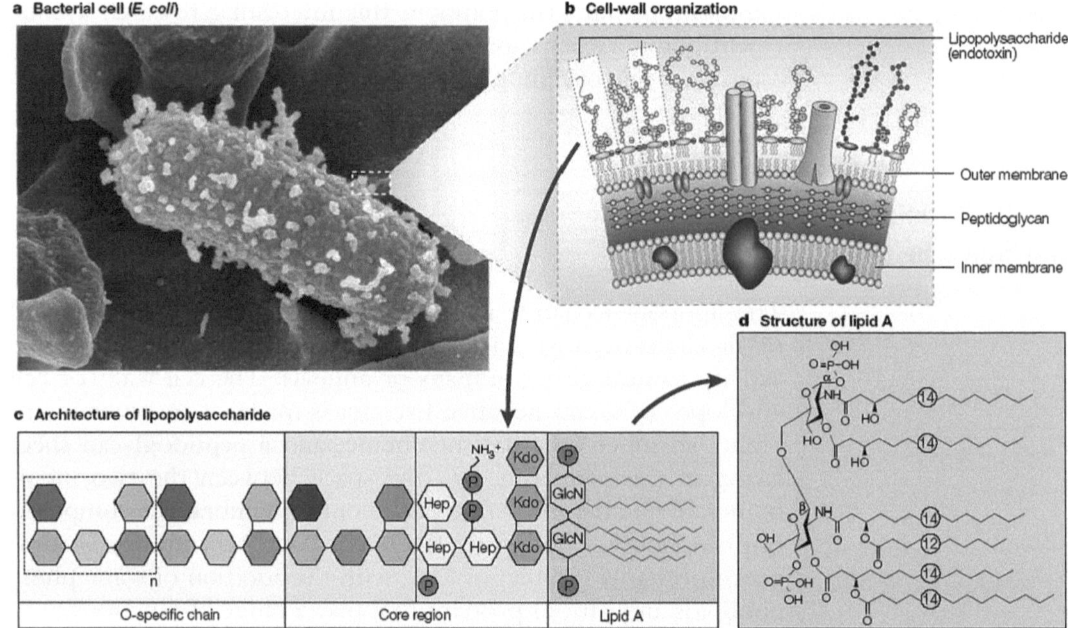

Fig. 10.1. A Gram-negative bacterium. Electron micrograph of *Escherichia coli* (**a**) together with a schematic representation of the location of lipopolysaccharide (LPS; endotoxin) in the bacterial cell wall (**b**) and the architecture of LPS (**c**). Also shown is the primary structure of the toxic center of LPS, the lipid A component (**d**). GlcN, D-glucosamine; Hep, L-glycero-D-manno-heptose; Kdo, 2-keto-3-deoxy-octulosonic acid; P, phosphate. Modified and reprinted from (6), with permission.

direct role of LPS in bacterial resistance to AMPs, exerted essentially through peptide-induced (through gene expression) modification of the LPS structure (reviewed in (10)). Furthermore, the LPS is important in mediating the interactions of the bacterium with its environment, which include colonization and ignition of/escape from immunological responses. The biological functions performed by the different LPS regions have been suggested by the characterization of existing mutants where LPS synthesis is incomplete. So, "rough" mutants lacking the O-antigen are more susceptible to phagocytosis and killing by bactericidal serum factors in vivo, whereas "deep rough" mutants which bear losses in the core region display augmented permeability to hydrophobic compounds. On the other hand, no lipid A mutants can be normally isolated, suggesting that this portion must be essential for bacterium viability.

3. Antisepsis Therapies

Nowadays, treatment of Gram-negative sepsis in critically ill patients is mainly based on the prompt intravenous administration of adequate antimicrobial agents and support of organ dysfunction. As time is of the essence in severe sepsis, an initial empirical broad spectrum antibiotic therapy should be initiated as soon as sepsis is suspected, followed by narrowing the antibiotic therapy based on microbiology and clinical data. In case of patients with severe sepsis and high risk for death, current protocols recommend treatment with drotrecogin alfa (activated) (DrotAA). Approved in 2001 for clinical use, DrotAA is a recombinant (and thus rather expensive) form of human activated protein C and the first drug for which a reduction of mortality, although slight, in severe sepsis has been demonstrated, despite many other investigational drugs have been proposed as treatments for sepsis and were tested in clinical trials (11). Activated protein C is a major physiological anticoagulant and is believed to have anti-thrombotic, pro-fibrinolytic, cytoprotective, anti-apoptotic, and anti-inflammatory effects; protein C levels decline early in patients who develop severe sepsis, and protein C deficiency is now considered a biomarker for this condition (12). The specific mechanism of DrotAA-derived protection from sepsis is not perfectly clear as yet, but it is probably related to the cross-talk between coagulation and inflammation, and to the fact that inhibitors of coagulation might interrupt the cascades of both coagulation and inflammation (e.g., (13)).

As sepsis lies at the crossroad of complex interactions between the invading microbe and the immune, inflammatory, and coagulation responses mounted by the host, many different targets for treatment exist, at least in theory (14, 15). So, pharmaceutical therapies targeting single pro-inflammatory mediators and/or endotoxin are (or were) in development, such as the administration of glucocorticoids, the development of endotoxin-directed monoclonal antibodies and other agents capable of binding and neutralizing LPS, or the inhibition of pro-inflammatory cytokines (e.g., anti-TNF antibodies, IL-1 receptor antagonists). In mice, anti-TLR4 antibodies protected against lethal endotoxemia in experimental models of Gram-negative bacterial sepsis when administered both prophylactically and therapeutically (16). Another supportive therapeutic approach with interesting prospects is offered by extracorporeal blood purification (hemofiltration), which could be effective during severe septic shock because it permits the unselective removal of endotoxin, cytokines, and inflammatory mediators from blood stream (17, 18). Various LPS-based vaccination strategies, that could prevent Gram-negative infection and thus sepsis, are also under development (19). However, despite good preliminary results in animal models, the effects of most new antisepsis therapies in humans have been so far disappointing or their benefits uncertain.

4. AMPs as LPS-Binding Compounds

Although our knowledge of the modes of interactions of AMPs and LPS has grown significantly in the last decade or so, the precise mechanism by which selected peptides might interfere and eventually block endotoxin's biological activities remains elusive. This has certainly to do with the diversity of AMPs object of study – a fact that hampers in several instances the comparison and collage of data obtained on different experimental models – but also with the multifaceted relationships between AMPs and LPS. The outer membrane may be seen in a way as the contact interface between Gram-negative bacteria and host, thus AMPs are forced to come into contact with LPS to exert their direct killing activity. On the other hand, since LPS activates the immune system even if detached from bacteria, giving rise to the stimulatory cascade described above, each step of this signaling and response pathway is a potential opportunity for intervention by AMPs, well known nowadays to act besides direct antimicrobial activity as immunomodulatory and immunostimulatory agents (20). Yosef Rosenfeld and Yechiel Shai have proposed that anti-endotoxin peptides could either bind straight to LPS, making it unavailable to LBP and therefore unable to transfer it to its primary receptor CD14, or the peptide could compete directly with LPS for binding to CD14 (10). In addition, NFκB translocation into the nucleus could also be hit, and peptides could even work by altering inflammatory genes expression (**Fig. 10.2**). Needless to say, a combination of (some of) these actions is also possible.

First, binding LPS. The model for the general mechanism of binding and insertion of peptides into the outer membrane that remains in force is the so-called self-promoted uptake theory (20). Initially, the high affinity of cationic AMPs for anionic LPS drives their interactions with the bacterial outer membrane; docking on the membrane, cationic peptides would bind at sites where divalent cations (Mg^{2+} and Ca^{2+}) cross-bridge adjacent lipopolysaccharide molecules and stabilize the outer membrane architecture. Peptide binding would distort the integrity of the outer membrane and increase its permeability to peptide itself, which once has crossed the outer membrane is bound to interact with the negatively charged surface of the lipid cytoplasmic (inner) membrane, thereupon exerting its cell-killing activity. Observations indicate that contact with the outer membrane induces folding of the peptides into their final membrane-associated form, and that the folded peptides may aggregate into tightly packed "rafts," which in turn would greatly contribute to the disruption of the outer membrane structure (21). Some observations on different peptides have suggested that if AMPs tend to aggregate upon

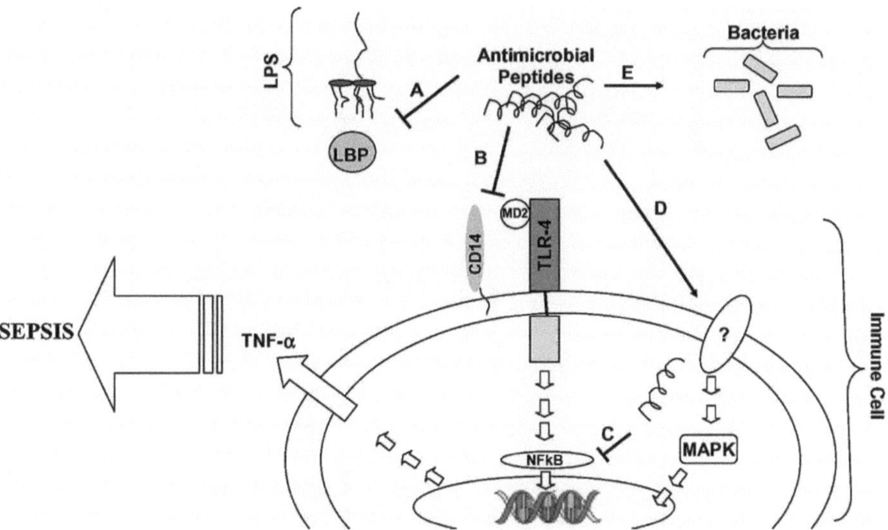

Fig. 10.2. A cartoon outlining the major events in induction of sepsis by bacteria and the points at which cationic peptides are proposed to intervene. Antibiotics can stimulate the release of endotoxin (LPS) which binds to LPS-binding protein (LBP) present in the blood, and transfers it to CD14 receptor on the surface of immune cells. This complex initiates intracellular signaling reactions, which mediate the production of inflammatory cytokines. AMPs inhibit this event by several mechanisms: (**A**) by direct binding to LPS, making it unavailable to LBP; (**B**) by competing with LPS for binding to the TLR signaling complex; (**C**) by inhibiting NFκB translocation into the nucleus; (**D**) by altering inflammatory genes expression through direct triggering of the MAPK pathways; and (**E**) by direct killing of microbes either via disruption of their membranes or by reacting with internal molecules. Reprinted from (10), with permission.

binding to LPS on the outer membrane, the bulky derivative is unable to cross the LPS layer and reach the plasma membrane to kill the bacterium, stressing the protective role of LPS and providing interesting pointers for the design of more potent AMPs (22).

When attention shifts from membrane-bound to isolated LPS, it should be recalled that LPS always forms non-lamellar cubic and/or unilamellar aggregate structures in solution, so AMPs must face this type of structure rather than single LPS molecules. It is generally believed that in order to prevent LPS from unlashing immune over-response, AMPs should be able to tightly bind LPS and to dissociate LPS aggregates, thus sequestering endotoxin and impeding its recognition from sensing systems. Several techniques can indeed be used to measure AMPs' ability to bind LPS, either in solution or in membrane models, as a few specific examples will illustrate.

Recently, Yosef Rosenfeld and colleagues have investigated the LPS-binding properties of a series of synthetic AMPs with the same amino acid composition (K_6L_9, **Table 10.1**) and of their diastereomers, composed of both L- and D-amino acids (23). These authors took advantage of the property of the nitrobenzothiazole (NBD) fluorophore, linked to the peptide, to change the intensity of its fluorescence emission when in a hydrophobic

Table 10.1
Sequences of some of the LPS-binding peptides discussed in the text

Peptide	Source	Sequence	Structure	References
K_6L_9	Synthetic	LKLLKKLLKKLLKLL-NH_2	α-Helical	(23)
Temporin L	*Rana temporaria* (frog skin)	FVQWFSKFLGRIL-NH_2	α-Helical	(24)
S1	*Carcinoscorpius rotundicauda* (horseshoe crab)	GFKLKGMARISCLPNGQWSNFPPKCIRECAMVSS	α-Helical	(27)
S3	*C. rotundicauda* (horseshoe crab)	HAEHKVKIGVEQKYGQFPQGTEVTYTCSGNYFLM	Dimeric, intermolecular S–S bond, cyclic	(27)
Arenicin-1	*Arenicola marina* (lugworm)	RWCVYAYVRVRGVLVRYRRCW	S–S bond, cyclic	(30)
LL-37	Human neutrophils	LLGDFFRKSKEKIGKEFKRIVQRIKDFLRNLVPRTES-NH_2	α-Helical	(34)
CAP11	Guinea pig neutrophils	GLRKKFRKTRKRIQKLGRKIGKTGRKVWKAWREYGQIPYPCRI	Dimeric, intermolecular S–S bond	(35)
CM4	*Bombyx mori* (Chinese silk worm)	RWKIFKKIEKVGQNIRDGIVKAGPAVAVVGQAATI	α-Helical	(38)
PR-39	Pig intestine	RRRPRPPYLPRPRPPPFFPPRLPPRIPPGFPPRFPPRFP	Linear, rich in Pro and Arg	(42)

environment, thus revealing AMP binding to the hydrophobic component of LPS. Extending their observations, the same researchers explored the effect of the peptides on LPS aggregates, by monitoring the changes in the fluorescence of LPS-FITC, which was quenched in the aggregated form but increased once the aggregates dissociated, and used transmission electronic microscopy to visualize the peptide's impact on LPS aggregates (23). Results with the parent, amphipathic-L peptide showed that the peptide binds LPS with a good affinity and exhibits a strong ability to dissociate LPS aggregates, breaking them into smaller particles. Authors concluded that the mechanism of LPS neutralization by these peptides involves strong binding to LPS followed by breaking LPS aggregates to smaller particles that are not available to LBP (23).

Monomolecular films of LPS, alone or interspersed with lipids, have also been increasingly used as suitable model systems to investigate the interactions of AMPs with Gram-negative outer membrane, and as a rapid means to evaluate their LPS-binding properties. Temporin L, an intriguing 13-mer member of the temporin family isolated from the skin of the red European frog *Rana temporaria* (**Table 10.1**), was shown with this technique to efficiently penetrate into *Escherichia coli* LPS monolayers, as demonstrated by the increase in film surface pressure (24). To collect additional information on the Temporin L-LPS modes of interaction, a fluorescent probe displacement method recently developed by Stewart Wood and colleagues was used (25). The fluorescent probe BODIPY TR cadaverine (BC) binds LPS, interacting specifically with its toxic center lipid A, probably via salt bridges with its glycosidic phosphate group, and the binding results in a progressive quenching of fluorescence; BC can then be competitively displaced by compounds displaying an affinity for lipid A, with a proportional dequenching of fluorescence. Temporin L was found to bind to purified *E. coli* LPS and lipid A, inducing a significant displacement of BC (24). This simple and useful method can be seen as an evolution of the classic fluorescence displacement method using dansyl polymyxin B, with the BODIPY fluorophore offering spectral properties that are superior to that of dansyl, in addition to other advantages of fluorescently labeled polyamines (such as cadaverine) with respect to polymyxins (*see* (25)).

Another interesting example of well-characterized LPS-binding AMP is the couple of so-called Sushi peptides (**Table 10.1**). These were derived from the core LPS-binding region of Factor C, a serine protease produced and stored in the amoebocytes of horseshoe crab. In this ancient animal, LPS-sensing through Factor C triggers blood coagulation, a defense mechanism used to trap invading bacteria (26). The extreme sensitivity of Factor C for LPS is the rationale behind the consolidate and widespread use of *Limulus* amoebocyte lysate test to detect trace levels of LPS in solution. Indeed, the N-terminal portion of Factor C contains several LPS-binding regions that span ~60 amino acids, domains named Sushi because of several sulfide bonds that hold the structure into a rolled, unique folding motif. Sushi 1 and Sushi 3 domains were identified as the major LPS-binding regions of Factor C, leading the way to the two derived 34-mer Sushi peptides S1 and S3, spanning the portions of Sushi 1 and Sushi 3 domains where the LPS-binding sequences reside (27). Structurally speaking, both S1 and S3 are believed to exist mostly in a random coil conformation in water, but whereas S1 tends toward an α-helical structure as the environment hydrophobicity increases, S3 adopts a mixture of secondary structures containing β-sheets, with a propensity to dimerize via an intermolecular disulfide bond. The LPS-binding properties of S1 and S3 were analyzed through surface plasmon resonance analysis of peptide interaction with lipid A, revealing a significant (100- to 10,000-fold) reduction in LPS affinity with regard to the recombinant Sushi 1 and 3 domains, a fact ascribed to the lack of a tight structural architecture conferred by disulfide bonds in the case of the peptides (27). Further studies demonstrated that S1 and S3 display detergent-like properties in disrupting LPS aggregates, with S3 being much stronger in this activity with respect to S1, but whereas S1 functions as a monomer, S3 is mainly active in its dimeric form (26). Investigating the action of S3 on LPS micelles, Jeak Ling Ding and colleagues proved that the intermolecular disulfide bonding of S3 resulting in S3 dimers is indispensable for its interaction with LPS and for disruption of LPS micelles (28). These authors suggested that, initially, binding of S1 and S3 to bacterial outer membrane would occur via electrostatic interactions between the positively charged amino acid residues near the N termini of peptides and the negatively charged bisphosphate head groups of the lipid A moiety of LPS, and possibly also of anionic lipids present in the membrane, such as phosphatidylglycerol (POPG) (29). Subsequently, hydrophobic interactions with the acyl chains of LPS/POPG would stabilize peptide binding to the membrane while favoring its insertion, leading to membrane perturbation.

Finally, a few words about Arenicin-1. This is a 21-residue cyclic β-sheeted antimicrobial peptide isolated from the marine lugworm *Arenicola marina* (**Table 10.1**). The group of Jörg Andrä at the Leibniz-Center for Medicine and Biosciences in Borstel, Germany, has studied the binding to, intercalation into, and permeabilization of model membranes composed of phospholipids or LPS extracted from *E. coli*, *Salmonella enterica*, and *Proteus mirabilis* by Arenicin-1 (30). FRET spectroscopy and tryptophan fluorescence confirmed the ability of the peptide to intercalate into LPS aggregates, which was also shown to permeabilize a model outer membrane consisting of LPS on its outer leaflet and phospholipids on its inner leaflet, forming heterogeneous and short-lived lesions. Changes in the zeta-potential of LPS aggregates – a measure for the surface charge of aggregates – upon incubation with Arenicin-1 revealed that besides a primary electrostatic attraction between the negatively charged LPS and the cationic peptide, which led to charge compensation, an additional hydrophobic interaction must take place, followed by a further aggregation or fusion of LPS supramolecular structures, probably as a consequence of peptide-induced LPS charge neutralization (30). The same authors have recently proposed surface acoustic wave biosensors are a sensitive tool for characterizing AMPs' interactions with LPS and lipid membranes, using the human cathelicidin-derived peptide LL32 as a model compound (31). Other examples of LPS-binding peptides, with a mention of used methods, can be found in dedicated reviews (32, 33)

5. Neutralizing LPS In Vitro and In Vivo, the Real Challenge

Following initial binding and interaction, the most desired property of a potential anti-endotoxin peptide is of course to prevent LPS-mediated inflammation and septic shock. Indeed, this has been demonstrated in a number of cases in vitro and, to a much lesser extent, in vivo. Rosenfeld and colleagues have investigated the correlation between antimicrobial activity, LPS-binding properties, and the ability to neutralize LPS of several peptides, including the human cathelicidin LL-37 – a peptide that in addition to antibacterial activity plays numerous roles in inflammatory and tissue remodeling processes (**Table 10.1**) – and magainin – a classic AMP isolated from frog skin. As for LPS neutralization, this was assayed by measuring the inhibition of LPS-mediated cytokine release from macrophages and the peptide competition with LPS for its binding site on the CD14 receptor, in the latter case using flow cytometry (34). All peptides bound LPS in solution, with magainin showing the weaker affinity (but the greatest antimicrobial activity), and all, except magainin, blocked LPS from binding to macrophages, as visualized through fluorescence spectroscopy and confocal microscopy. These data correlated well with the potency of the peptides to reduce the level of mRNA expression of the cytokines IL-6 and TNF-α, as well as to prevent TNF-α secretion from macrophages, with magainin being the most inactive peptide. These data confirmed that there is no direct correlation between the antimicrobial activity of AMPs and their ability to neutralize LPS, and that strong binding of a peptide to LPS is not sufficient to block LPS biological activity. In addition, LL-37 was the only peptide that could compete with LPS on its binding site within the CD14 receptor, indicating that LPS neutralization can be achieved by different peptides following different routes (i.e., dissociation of LPS aggregates and/or competition for CD14 receptor) (34).

More recently, the team led by Isao Nagaoka showed that the cathelicidin peptide CAP11 (**Table 10.1**) inhibits the binding of LPS to macrophages, suppressing the cellular release of high mobility group box-1 (HMGB1), a nuclear protein secreted from mononuclear phagocytes as a consequence of LPS-stimulation and which is believed to play an important role in septic/endotoxin shock as a late-phase mediator, behaving as a trigger of inflammation (35). The same group had previously demonstrated that CAP11 administration can suppress the elevation of serum HMGB1 level in a mouse endotoxin shock model and protect mice from endotoxin lethality, increasing survival; in this model, CAP11 inhibited the binding of LPS to $CD14^{+}$ cells (peritoneal macrophages) and suppressed LPS-induced TNF-α expression by these cells (36). These observations confirm and explain the current interest around cathelicidins and derived, (semi)synthetic analogues as suitable candidates to be developed as anti-sepsis compounds (37).

Besides cathelicidins, other AMPs display interesting LPS-neutralizing activities. CM4, for example, a linear cecropin-like AMP from the Chinese silk worm *Bombyx mori* (**Table 10.1**), inhibited the binding of FITC-conjugated LPS to the $CD14^{+}$ murine macrophage RAW264.7 cells, suppressed LPS-induced TNF-α and IL-6 mRNA expression, and blocked release of TNF-α and NO following LPS challenge (38).

Maria Luisa Mangoni and colleagues have examined the effect of several members of the temporin family of AMPs, alone and when combined with each other, in inhibiting LPS-induced pro-inflammatory cytokines secretion by LPS-activated macrophages (39). Typically, multiple forms of AMPs are synthesized in the skin of a single species of frog (and in other animals) as structurally related members of a family, probably providing the producing organism with a maximum coverage against a wider range of invading microbes (40). In this case, Mangoni and colleagues reported that when Temporin L was mixed with the related Temporin A or B, the anti-endotoxin activity (measured as inhibition of TNF-α secretion by macrophages) was found to be significantly greater than the corresponding additive effect of the single peptides, thus showing synergism. When coupled to the fact the synergism between temporins extends to their antimicrobial activity against a range of bacterial species (including components of the frog's natural flora), these observations would confirm that the contemporary presence of different forms of related AMPs on the same frog specimen is essential in assuring protection from infections, and could also (at least partially) explain why amphibians usually are resistant to the toxic effects of LPS (39, 41). In addition, "these studies should help in the future development of non-single peptide-based therapeutic strategies to kill multidrug-resistant Gram-negative bacteria and/or neutralize the effect of their LPS, subsequently released," remarked the authors (39).

As mentioned above, with respect to the in vitro available evidences, reliable studies on the LPS-neutralizing activities of AMPs in vivo are comparatively scarce, a fact due in large part to the difficulties inherent to setting up a good, reproducible model for septic shock. A few examples, however, do exist, proving that this approach is viable. Melanie Madhani and colleagues have explored the effects of the antibacterial cathelicidin peptide PR-39 – a proline-rich antibacterial peptide first identified in pig intestines and later in neutrophils (**Table 10.1**) – on nitric oxide and liver oxygenation in a mouse model of endotoxaemia (42). PR-39 treatment had beneficial effects on liver in this septic model, mediated by increasing levels of local NO and preserving sinusoidal perfusion and lobule oxygenation; PR-39 induced elevated levels of liver nitric oxide at 6 h following injection of LPS and improved pO_2 across the hepatic tissue bed (42). Other authors tested the effects of LL-37, both used as prophylactic or in therapy, in a rat model of peritoneal contamination and infection (43). In this model, LL-37 therapy improved outcome and increased survival rates after sepsis; in particular, combining LL-37 therapy with hyperthermic preconditioning (exposing mice at 41°C for 1 h, 24 h before sepsis) was successful as compared to controls, a fact attributed to the additive effect of LL-37 and hyperthermia in suppressing the excessive release of pro-inflammatory cytokines (43).

6. Putting AMPs into Context

As outlined above, our comprehension of the AMPs–LPS interactions has grown significantly, although it is far from being complete or even satisfactory. Our current tenets are (i) binding and insertion of peptides into the bacterial outer membrane takes place via the "self-promoted uptake" model; (ii) crude binding of cationic AMPs to LPS released from bacteria involves a mix of electrostatic and hydrophobic interactions; these can lead to breaking LPS aggregates into smaller particles that are not available to LBP, explaining inhibition of LPS-induced activation of immune cells by cationic peptides (in alternative, the mechanism of LPS neutralization could involve re-aggregation of the LPS from an unilamellar/cubic structure into a multilamellar structure, which does not activate the signal cascade, *see* (33));

(iii) several AMPs, such as cathelicidins, can suppress harmful pro-inflammatory responses, neutralizing LPS effects in vitro and in vivo; this is likely done by acting at different points of intervention and following various mechanisms.

Given this scenario, are there any real possibilities that AMPs could be developed in the (near?) future into anti-endotoxin drugs? No simple answer is available at the moment. Sepsis is and remains an ultra-challenging field for drug development. Despite the fact that LPS represents the most popular target for sepsis treatment, indeed, numerous clinical trials featuring the use of antibodies directed against LPS failed (44). Lack of efficacy in late-stage trials led to the discontinuation of Centocor's nebacumab and Xoma & Pfizer's edobacomab, two anti-endotoxin monoclonal antibodies, while in 1997 MorphoSys's decided to suspend all in-house development related to endotoxin-binding peptides for the potential treatment of septic shock (45). In recent years, Novartis developed PMX-622, a dextran 70 conjugate of polymyxin B, a cationic cyclic polypeptide antibiotic that binds strongly to the lipid A portion of Gram-negative bacterial LPS. This derivate was designed to reduce the toxicity of polymyxin B without interfering with its LPS-neutralizing capacity (46, 47). However, despite results of a Phase IIa trial in patients with severe Gram-negative infection and SIRS showing reduction in circulating LPS, IL-6 and IL-10 in patients receiving PMX-622 in late 1999, the product failed to demonstrate efficacy in patients with severe sepsis; studies, however, continue (48). So, even if preclinical data may indicate, at least in experimental animal models, that blocking or neutralizing LPS will prevent death from sepsis, the failure of the clinical trials leads to avoid the safe conclusion that blocking endotoxin might have the same outcome in humans. The various dynamics of pathophysiological responses during sepsis are different in humans and mice, and insights from animal studies cannot be transferred in a straightforward manner from the lab to the clinical setting (49). Moreover, it needs to be taken into account the limitations associated with any endotoxin-neutralizing agent, since its effects will not be manifest in all cases of sepsis, but rather such a drug would only be useful in the context of an infection caused by Gram-negative pathogens.

Nevertheless, if a pharmaceutical therapeutic approach for sepsis is to be developed, AMPs cannot be ignored at all. They display antimicrobial properties, and so can be used to implement (or even substitute, in some instances) conventional antibiotic therapies to contrast infections, while, at the same time, providing a tool to sequestrate circulating LPS (whose plasma concentration often increases because of antibiotic therapy itself) and neutralize its effects. At present, this is more theory than practice, and many problems remain to be solved, including the poor bioavailability and poor proteolytic stability typical of AMPs, and the need to harness their immunostimulatory and immunomodulatory properties before they can be administered systemically.

What to do next? A first, necessary step would be searching for new LPS-binding peptides. This can be done following two main directions, i.e. looking closer at natural AMPs and/or recurring to synthetic peptides. Several hundreds naturally occurring AMPs have been described now, and probably thousands more exist to be discovered yet. A naïve search (performed on March 2009) of the APD2 database (http://aps.unmc.edu/AP/main.php),[1] which now contains information for almost 1,400 AMPs, using specific "molecular targets" codes renders some 21 entries for LPS-interacting peptides, and this is most likely just a tiny fraction of those really existing (50). Even proteins other than AMPs can give a hint. So, the Sushi peptides are not the only example of peptides with the potential to bind and neutralize endotoxin that were derived from natural LPS-binding proteins. Two overlapping 15-mer peptides, corresponding to the residues 91–108 of human LBP, were, for example, found to inhibit the *Limulus* test and to block release of TNF-α after LPS challenge in vitro (in peripheral blood mononuclear cells) and in vivo (in mice) (51). Another 27-mer peptide was derived from the bactericidal permeability increasing protein (BPI), a part of the human innate immune system which binds LPS and has potent killing activity against Gram-negative bacteria, showing the ability of inhibiting LPS-induced TNF-α secretion by macrophages and, when preincubated with LPS, protecting from subsequent endotoxin challenge (52). Other examples on this trail are the various peptides derived from *Limulus* anti-LPS factor (LALF), from human lactoferrin, and from other proteins (26, 33).

Synthetic analogues and peptidomimetics that could mimic the features of AMPs are already actively pursued, since it is increasingly believed that it is possible to endow AMPs with improved antimicrobial and pharmacokinetic properties, coupled to a lower cost of manufacturing (53). The same considerations can be extended to the anti-endotoxin activities of AMPs. It is thus quite safe to forecast that the following years will see a burst of studies aimed at translating what was understood on what makes a given peptide able to bind and neutralize LPS into an engineered drug lead candidate with some chances to enter the clinical treatment of sepsis (33). Natural, synthetic, or semisynthetic lipopeptides, consisting of a hydrophilic cyclic peptide portion attached to a fatty acid chain which facilitates insertion

[1] This is just one of several existing, web-based databases dedicated to AMPs, of either natural or synthetic origin; all of them are currently linked to APD2.

into the LPS/lipid bilayer of bacterial membranes, are another viable alternative (54). Besides peptides, smaller molecules are being evaluated as endotoxin antagonists, molecules that bear a cationic amphiphilic structure in common that can bind to the anionic and amphipathic nature of lipid A, with lipopolyamines at the forefront (e.g., (55)).

We believe future approaches are likely to focus either on the identification of compounds able to intervene at multiple points in the sepsis cascade or on the combination of two or more immunomodulators. In this context, the properties and capacities of several antimicrobial peptides to modulate the cascade of the immune response in isolated cells and in various animal models of disease could prove to be the winning bet. In conclusion, presumably no anti-endotoxin "magic bullet" will be derived from AMPs or analogues, but rather a small panel of compounds that will contribute to shaping the next generation, modern therapeutic approach to severe sepsis and septic shock, grounded on a deeper understanding of the cross-talk between innate and adaptive immunities. That alone, would be a great success.

References

1. Baudouin, S. V. (Ed.) (2007) *Sepsis*. Springer-Verlag: New York.
2. Angus, D. C., Linde-Zwirble, W. T., Lidicker, J., Clermont, G., Carcillo, J., and Pinsky, M. R. (2001) Epidemiology of severe sepsis in the United States: analysis of incidence, outcome, and associated costs of care. *Crit. Care Med.* **29**, 1303–1310.
3. Burchardi, H. and Schneider, H. (2004) Economic aspects of severe sepsis: a review of intensive care unit costs, cost of illness and cost effectiveness of therapy. *Pharmacoeconomics* **22**, 793–813.
4. Wang, J. E., Dahle, M. K., McDonald, M., Foster, S. J., Aasen, A. O., and Thiemermann, C. (2003) Peptidoglycan and lipoteichoic acid in gram-positive bacterial sepsis: receptors, signal transduction, biological effects, and synergism. *Shock* **20**, 402–414.
5. Kimbrell, M. R., Warshakoon, H., Cromer, J. R., Malladi, S., Hood, J. D., Balakrishna, R., Scholdberg, T. A., and David, S. A. (2008) Comparison of the immunostimulatory and proinflammatory activities of candidate Gram-positive endotoxins, lipoteichoic acid, peptidoglycan, and lipopeptides, in murine and human cells. *Immunol. Lett.* **118**, 132–141.
6. Beutler, B. and Rietschel, E. Th. (2003) Innate immune sensing and its roots: the story of endotoxin. *Nat. Rev. Immunol.* **3**, 169–176.
7. Freudenberg, M. A., Tchaptchet, S., Keck, S., Fejer, G., Huber, M., Schültze, N., Beutler, B., and Galanos, C. (2008) Lipopolysaccharide sensing an important factor in the innate immune response to Gram-negative bacterial infections: benefits and hazards of LPS hypersensitivity. *Immunobiol.* **213**, 193–203.
8. Beutler, B., Hoebe, K., Du, X., and Ulevitch, R. J. (2003) How we detect microbes and respond to them: the Toll-like receptors and their transducers. *J. Leukoc. Biol.* **74**, 479–485.
9. Raetz, R. H. and Whitfield, C. (2002) Lipopolysaccharide endotoxins. *Annu. Rev. Biochem.* **71**, 635–700.
10. Rosenfeld, Y. and Shai, Y. (2006) Lipopolysaccharide (endotoxin)-host defense antibacterial peptides interactions: role in bacterial resistance and prevention of sepsis. *Biochim. Biophys. Acta* **1758**, 1513–1522.
11. Hosac, A. M. (2002) Drotrecogin alfa (activated): the first FDA-approved treatment for severe sepsis. *BUMC Proc.* **15**, 224–227.
12. Shorr, A. F., Nelson, D. R., Wyncoll, D. L. A., Reinhart, K., Brunkhorst, F., Vail, G. M., and Janes, J. (2008) Protein C: a potential biomarker in severe sepsis and a possible tool for monitoring treatment with drotrecogin alfa (activated). *Crit. Care* **12**, R45.
13. de Pont, A. C., Bakhtiari, K., Hutten, B. A., de Jonge, E., Vroom, M. B., Meijers, J. C., Büller, H. R., and Levi, M. (2005)

Recombinant human activated protein C resets thrombin generation in patients with severe sepsis – a case control study. *Crit. Care* **9**, R490–R497.

14. Russell, J. A. (2006) Management of sepsis. *N. Engl. J. Med.* **355**, 1699–1713.
15. Claessens, Y.-E. and Dhainaut, J.-F. (2007) Diagnosis and treatment of severe sepsis. *Crit. Care* **11**(Suppl. 5), S2.
16. Roger, T., Froidevaux, C., Le Roy, D., Reymond, M. K., Chanson, A.-L., Mauri, D., Kim Burns, K., Riederer, B. M., Akira, S., and Calandra, T. (2009) Protection from lethal Gram-negative bacterial sepsis by targeting Toll-like receptor 4. *Proc. Natl. Acad. Sci. USA* **106**, 2348–2352.
17. Venkataraman, R., Subramanian, S., and Kellum, J. A. (2003) Clinical review: extracorporeal blood purification in severe sepsis. *Crit. Care* 7, 139–145.
18. Nalesso, F. (2005) Plasma filtration adsorption dialysis (PFAD): a new technology for blood purification. *Int. J. Artif. Organs* **28**, 731–738.
19. Nagy, G. and Pál, T. (2008) Lipopolysaccharide: a tool and target in enterobacterial vaccine development. *Biol. Chem.* **389**, 513–520.
20. Hancock, R. E. W. (2001) Cationic peptides: effectors in innate immunity and novel antimicrobials. *Lancet Infect. Dis.* **1**, 156–164.
21. Chapple, D. S., Hussain, R., Joannou, C. L., Hancock, R. E. W., Odell, E., Evans, R. W., and Siligardi, G. (2004) Structure and association of human lactoferrin peptides with *Escherichia coli* lipopolysaccharide. *Antimicrob. Agents Chemother.* **48**, 2190–2198.
22. Papo, N. and Shai, Y. (2005) A molecular mechanism for lipopolysaccharide protection of Gram-negative bacteria from antimicrobial peptides. *J. Biol. Chem.* **280**, 10378–10387.
23. Rosenfeld, Y., Sahl, H.-G., and Shai, Y. (2008) Parameters involved in antimicrobial and endotoxin detoxification activities of antimicrobial peptides. *Biochemistry* **47**, 6468–6478.
24. Giacometti, A., Cirioni, O., Ghiselli, R., Mocchegiani, F., Orlando, F., Silvestri, C., Bozzi, A., Di Giulio, A., Luzi, C., Mangoni, M. L., Barra, D., Saba, V., Scalise, G., and Rinaldi, A. C. (2006) Interaction of temporin L with lipopolysaccharide in vitro and in experimental rat models of septic shock caused by Gram-negative bacteria. *Antimicrob. Agents Chemother.* **50**, 2478–2486.
25. Wood, S. J., Miller, K. A., and David, S. A. (2004) Anti-endotoxin agents. 1. Development of fluorescent probe displacement method optimized for the rapid identification of lipopolysaccharide-binding agents. *Combin. Chem. High Throughput Screen.* **7**, 239–249.
26. Ding, J. L., Li, P., and Ho, B. (2008) The Sushi peptides: structural characterization and mode of action against Gram-negative bacteria. *Cell. Mol. Life Sci.* **65**, 1202–1219.
27. Tan, N. S., Ng, M. L., Yau, Y. H., Chong, P. K., Ho, B., and Ding, J. L. (2000) Definition of endotoxin binding sites in horseshoe crab factor C recombinant sushi proteins and neutralization of endotoxin by sushi peptides. *FASEB J.* **14**, 1801–1813.
28. Li, P., Wohland, T., Ho, B., and Ding, J. L. (2004) Perturbation of lipopolysaccharide (LPS) micelles by Sushi 3 (S3) antimicrobial peptide. The importance of an intermolecular disulfide bond in S3 dimer for binding, disruption, and neutralization of LPS. *J. Biol. Chem.* **279**, 50150–50156.
29. Li, P., Sun, M., Wohland, T., Yang, D., Ho, B., and Ding, J. L. (2006) The molecular mechanisms that govern the specificity of Sushi peptides for gram negative bacterial membrane lipids. *Biochemistry* **45**, 10554–10562.
30. Andrä, J., Hammer, M. U., Grötzinger, J., Jakovkin, I., Lindner, B., Vollmer, E., Fedders, H., Leippe, M., and Gutsmann, T. (2009) Significance of the cyclic structure and of arginine residues for the antibacterial activity of arenicin-1 and its interaction with phospholipid and lipopolysaccharide model membranes. *Biol. Chem.* **390**, 337–349.
31. Andrä, J., Böhling, A., Gronewold, T. M., Schlecht, U., Perpeet, M., and Gutsmann, T. (2008) Surface acoustic wave biosensor as a tool to study the interaction of antimicrobial peptides with phospholipid and lipopolysaccharide model membranes. *Langmuir* **24**, 9148–9153.
32. Jerala, R. and Porro, M. (2004) Endotoxin neutralizing peptides. *Curr. Top. Med. Chem.* **4**, 1173–1184.
33. Andrä, J., Gutsmann, T., Garidel, P., and Brandenburg, K. (2006) Mechanisms of endotoxin neutralization by synthetic cationic compounds. *J. Endotoxin Res.* **12**, 261–277.
34. Rosenfeld, Y., Papo, N., and Shai, Y. (2006) Endotoxin (lipopolysaccharide) neutralization by innate immunity host-defense peptides: peptide properties and plausible modes of action. *J. Biol. Chem.* **281**, 1636–1643.
35. Shibusawa, K., Murakami, T., Yomogida, S., Tamura, H., and Nagaoka, I. (2009) Antimicrobial cathelicidin peptide CAP11 suppresses HMGB1 release from

lipopolysaccharide-stimulated mononuclear phagocytes via the prevention of necrotic cell death. *Intern. J. Mol. Med.* **23**, 341–346.
36. Nagaoka, I., Hirota, S., Niyonsaba, F., Hirata, M., Adachi, Y., Tamura, H., and Heumann, D. (2001) Cathelicidin family of antibacterial peptides CAP18 and CAP11 inhibit the expression of TNF-α by blocking the binding of LPS to CD14(+) cells. *J. Immunol.* **167**, 3329–3338.
37. Mookherjee, N., Rehaume, L. M., and Hancock, R. E. (2007) Cathelicidins and functional analogues as antisepsis molecules. *Expert Opin. Ther. Targets* **11**, 993–1004.
38. Lin, Q. P., Zhou, L. F., Li, N. N., Chen, Y. Q., Li, B. C., Cai, Y. F., and Zhang, S. Q. (2008) Lipopolysaccharide neutralization by the antibacterial peptide CM4. *Eur. J. Pharmacol.* **596**, 160–165.
39. Mangoni, M. L., Epand, R. F., Rosenfeld, Y., Peleg, A., Barra, D., Epand, R. M., and Shai, Y. (2008) Lipopolysaccharide, a key molecule involved in the synergism between temporins in inhibiting bacterial growth and in endotoxin neutralization. *J. Biol. Chem.* **283**, 22907–22917.
40. Rinaldi, A. C. (2002) Antimicrobial peptides from amphibian skin: an expanding scenario. *Curr. Opin. Chem. Biol.* **6**, 799–804.
41. Rosenfeld, Y., Barra, D., Simmaco, M., Shai, Y., and Mangoni, M. L. (2006b) A synergism between temporins toward Gram-negative bacteria overcomes resistance imposed by the lipopolysaccharide protective layer. *J. Biol. Chem.* **281**, 28565–28574.
42. Madhani, M., Barchowsky, A., Klei, L., Ross, C. R., Jackson, S. K., Harold, M., Swartz, H. M., and James, P. E. (2002) Antibacterial peptide PR-39 affects local nitric oxide and preserves tissue oxygenation in the liver during septic shock. *Biochim. Biophys. Acta* **1588**, 232–240.
43. Torossian, A., Gurschi, E., Bals, R., Vassiliou, T., Wulf, H. F., and Bauhofer, A. (2007) Effects of the antimicrobial peptide LL-37 and hyperthermic preconditioning in septic rats. *Anesthesiology* **107**, 437–441.
44. Cohen, J. (1999) Adjunctive therapy in sepsis: a critical analysis of the clinical trial programme. *Br. Med. Bull.* **55**, 212–225.
45. Datamonitor (2006) *Stakeholder Opinions: Sepsis. Under Reaction to An Overreaction.* Datamonitor: London.
46. Bucklin, S. E., Lake, P., Lögdberg, L., and Morrison, D. C. (1995) Therapeutic efficacy of a polymyxin B-dextran 70 conjugate in experimental model of endotoxemia. *Antimicrob. Agents Chemother.* **39**, 1462–1466.
47. Fuchs, P. C., Barry, A. L., and Brown, S. D. (1998) PMX-622 (polymyxin B-dextran 70) does not alter in vitro activities of 11 antimicrobial agents. *Antimicrob. Agents Chemother.* **42**, 2765–2767.
48. Lake, P., DeLeo, J., Cerasoli, F., Logdberg, L., Weetall, M., and Handley, D. (2004) Pharmacodynamic evaluation of the neutralization of endotoxin by PMX622 in mice. *Antimicrob. Agents Chemother.* **48**, 2987–2992.
49. Rittirsch, D., Hoesel, L. M., and Ward, P. A. (2007) The disconnect between animal models of sepsis and human sepsis. *J. Leukoc. Biol.* **81**, 137–143.
50. Wang, G., Li, X., and Wang, Z. (2009) APD2: the updated antimicrobial peptide database and its application in peptide design. *Nucleic Acids Res.* **37**(database issue), D933–D937.
51. Taylor, A. H., Heavner, G., Nedelman, M., Sherris, D., Brunt, E., Knight, D., and Ghrayeb, J. (1995) Lipopolysaccharide (LPS) neutralizing peptides reveal a lipid A binding site of LPS binding protein. *J. Biol. Chem.* **270**, 17934–17938.
52. Battafaraono, R. J., Dahlberg, P. S., Ratz, C. A., Johnston, J. W., Gray, B. H., Haseman, J. R., Mayo, K. H., and Dunn, D. L. (1995) Peptide derivatives of three distinct lipopolysaccharide binding proteins inhibit lipopolysaccharide-induced tumor necrosis factor-alpha secretion in vitro. *Surgery* **118**, 318–324.
53. Giuliani, A., Pirri, G., Bozzi, A., Di Giulio, A., Aschi, M., and Rinaldi, A. C. (2008) Antimicrobial peptides: natural templates for synthetic membrane-active compounds. *Cell. Mol. Life Sci.* **65**, 2450–2460.
54. Pirri, G., Giuliani, A., Nicoletto, F. S., Pizzuto, L., and Rinaldi, A. C. (2009) Lipopeptides as antiinfectives: a practical perspective. *CEJB* **4**, 258–273.
55. Wu, W., Sil, D., Szostak, M. L., Malladi, S. S., Warshakoon, H. J., Kimbrell, M. R., Cromer, J. R., and David, S. A. (2009) Structure–activity relationships of lipopolysaccharide sequestration in guanylhydrazone-bearing lipopolyamines. *Bioorg. Med. Chem.* **17**, 709–715.

Chapter 11

Binding and Permeabilization of Model Membranes by Amphipathic Peptides

Paulo F. Almeida and Antje Pokorny

Abstract

Measurement of binding and activity of antimicrobial and cytolytic amphipathic peptides on membranes is essential to understanding their function and cell specificity. The use of model systems has provided a wealth of information on the interactions of amphipathic peptides with membranes. Binding of peptides to membranes can be monitored by measuring Förster resonance energy transfer from a Trp residue on the peptide to a lipid fluorophore incorporated in the membrane. Especially for peptides that perturb or disrupt the membrane, it is advantageous to perform these measurements as a function of time, rather than in steady state. The activity of these amphipathic peptides toward model membranes is usually measured by dye efflux kinetics. One of those methods, based on self-quenching of carboxyfluorescein, is described here, together with a discussion of caveats and pitfalls of the corresponding analysis and interpretation.

Key words: Membrane-active peptides, antimicrobial peptides, binding kinetics, FRET, dye release, carboxyfluorescein efflux, cell-penetrating peptides.

1. Introduction

Our understanding of the function and mechanism of antimicrobial and cytolytic peptides has benefited enormously from studies in model membranes. Lipid vesicles have been most commonly used, but supported bilayers, micelles, and other membrane-mimetic systems have also been employed. These model membranes can be prepared easily and reproducibly, and permit the variation of the lipid composition, which is difficult in a biological membrane. The effects of different lipids can therefore be studied in a very convenient way. The investigation of peptide function

A. Giuliani, A.C. Rinaldi (eds.), *Antimicrobial Peptides*, Methods in Molecular Biology 618,
DOI 10.1007/978-1-60761-594-1_11, © Springer Science+Business Media, LLC 2010

usually requires knowledge of at least two pieces of information. How well does the peptide bind to, and how active is it toward the membrane?

Binding of amphipathic peptides to membranes can be measured by various methods, including isothermal titration calorimetry and circular dichroism. If the peptide contains a Trp residue fluorescence spectroscopy is especially convenient and sensitive. In particular, Förster resonance energy transfer (FRET) (1) between the Trp residue and a lipid fluorescent probe incorporated in the membrane in small amounts (about 1–2 mol% of the lipid) is described in this article. The fluorescent lipid probe is chosen so that it can accept energy transfer from the Trp residue when the peptide is bound to the membrane. The Trp is excited at 280 nm and would normally emit at about 320–350 nm, depending on its environment, but if the lipid probe is within a small distance, FRET occurs instead, and fluorescence emission is then observed from the lipid probe, at a higher wavelength. Dansyl lipid fluorophores have been used as acceptors for Trp (1), but dansyl has the disadvantage that it has a very small extinction coefficient. We have recently synthesized a suitable probe and used it to measure the kinetics of binding of amphipathic peptides to membranes (2). The fluorophore proper, 7-methoxycoumarin-3-carboxylic acid (7MC) (**Fig. 11.1**), is attached to the headgroup of 1-palmitoyl-2-oleoylphosphatidylethanolamine (POPE) to produce a lipid fluorophore (7MC–POPE) that absorbs maximally at $\lambda = 348$ nm (2). If this probe is within about 26 Å from the Trp, it can accept its excitation energy and emits fluorescence with a maximum at $\lambda = 396$ nm (**Fig. 11.2**). This provides a fluorescence signal that can easily be monitored as a function of time (**Fig. 11.3a**). Peptide binding to lipid vesicles can also be measured by steady-state fluorescence using the same type of vesicles and probes. There are, however, many advantages in measuring kinetics rather than equilibrium of peptide binding. In steady-state fluorescence measurements, either the lipid is added to peptide or peptide is added to lipid. In the first case, the lipid vesicle suspension is added to the peptide solution until saturation is eventually reached. Thus, what is measured is the partitioning of the peptide from water to the membrane. However, in the lower range of lipid concentrations, the peptide/lipid ratio is high, and the peptide may severely disrupt the vesicles.

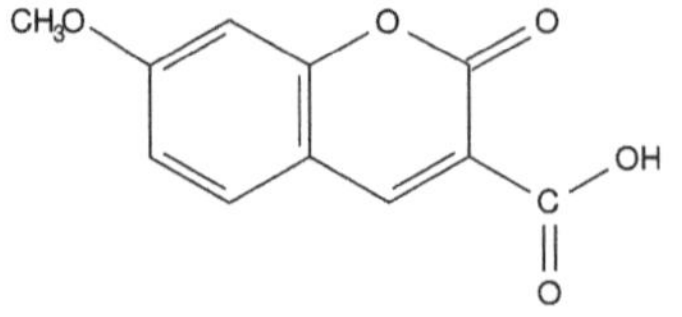

Fig. 11.1. Structure of 7-methoxycoumarin-3-carboxylic acid (7MC).

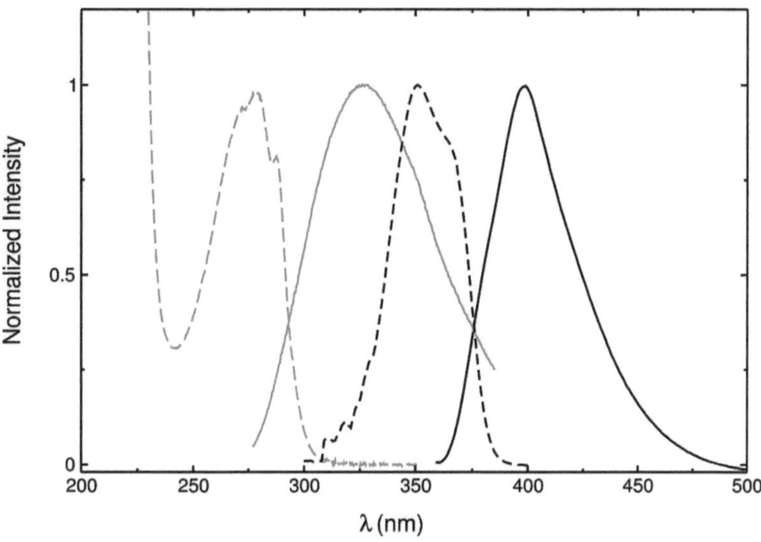

Fig. 11.2. Overlap of the fluorescence emission or Trp in a peptide with absorption of 7MC–POPE, a condition for FRET. The absorption spectrum of Trp is shown by the *dashed gray line* and its emission spectrum by the *solid gray line*. The absorption of 7MC is shown by the *dashed black line* and its emission by the *solid black line*.

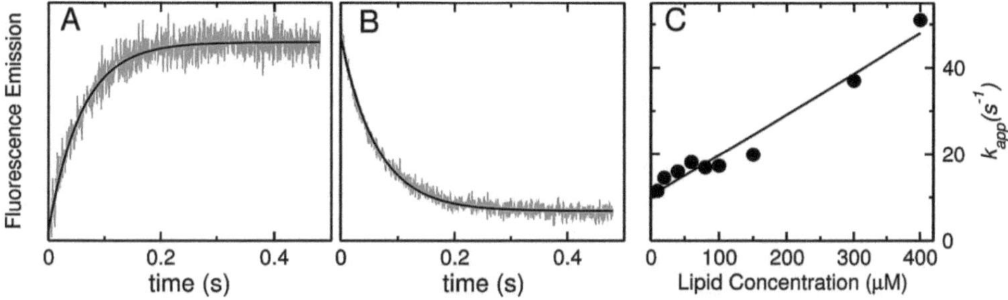

Fig. 11.3. Binding kinetics of the amphipathic peptide tp10W (which contains a Trp residue) to POPC vesicles labeled with 1 mol% 7MC–POPE (20). (**a**) Association and (**b**) dissociation kinetics of tp10W. The *gray traces* are the experimental curves and the *solid lines* are exponential fits. (**c**) Plot of the apparent rate constant (k_{app}), obtained from the fits of the association kinetics, against the lipid concentration to obtain the on- and off-rate constants.

Moreover, obtaining an equilibrium reading of the fluorescence may be a problem because these amphipathic peptides often perturb the vesicles, inducing other fluorescence changes that are not primarily caused by binding. In the second case, the peptide is added to the lipid suspension, avoiding very high peptide/lipid ratios. But conceptually this experiment is poorly defined. What is measured in this case is the transfer of a lipid patch from contact with water to contact with peptide. Further, the fluorescence of the donor (Trp) increases linearly during the experiment. A correction is possible, by dividing the fluorescence by the donor (peptide) concentration, but the error introduced can be large, if the increase due to the donor concentration is much larger than the increase in FRET. To obtain the free ligand concentration it is necessary to assume a size for the membrane patch bound by the peptide. The kinetics experiment, on the other hand, can usually be performed at sufficiently low peptide or high lipid concentration for this is not to be a significant problem. Further, it is a differential measurement, which is always advantageous.

Large unilamellar vesicles (LUV) (3) are the best choice to measure binding and activity of amphipathic peptides. They are the least strained among the most readily produced types of phospholipid vesicles. LUVs are produced by extrusion of a suspension of multilamellar vesicles (MLV) through 0.1 μm polycarbonate filters. This is a standard technique that has been shown to produce a suspension of unilamellar vesicles of about 100 nm diameter. LUVs can be made reproducibly, in a homogeneous size population, and in various sizes, if needed.

The activity of antimicrobial and cytolytic peptides on lipid vesicles can be measured by the rate of dye efflux they induce from lipid vesicles (4, 5). We have used this method extensively with several peptides (2, 6–9). The most commonly used dyes are calcein and carboxyfluorescein (CF) (**Fig. 11.4**), which can be encapsulated at high, self-quenching concentrations in the lumen of lipid vesicles (10). When the peptides interact with the vesicles, these fluorophores are released to the external solution, where they are diluted and, therefore, highly fluorescent. The increase in fluorescence that occurs upon peptide addition is monitored as a function of time, yielding a measure of the peptide activity (**Fig. 11.5**). The efflux kinetics curves can be analyzed to extract more quantitative information. However, that analysis is beyond the scope of this chapter. We will only mention the methods and corresponding references. The first method was developed in Schwarz's laboratory (11–13) and uses an analytical approach to extract pore events from the efflux curves. It does not deal with reversible aggregation and handles fractional recovery in an operational manner. The second method was developed in our labo-

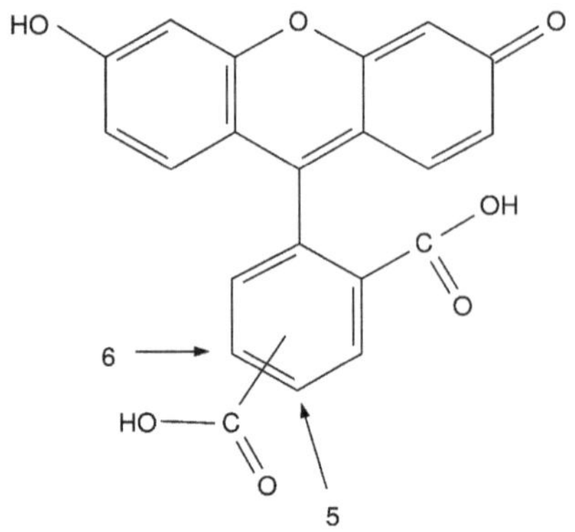

HO
O
O
OH
C
O
6
HO—C
O
5

Fig. 11.4. Structure of carboxyfluorescein (CF). A mixture of the 5- and 6-isomers is usually used.

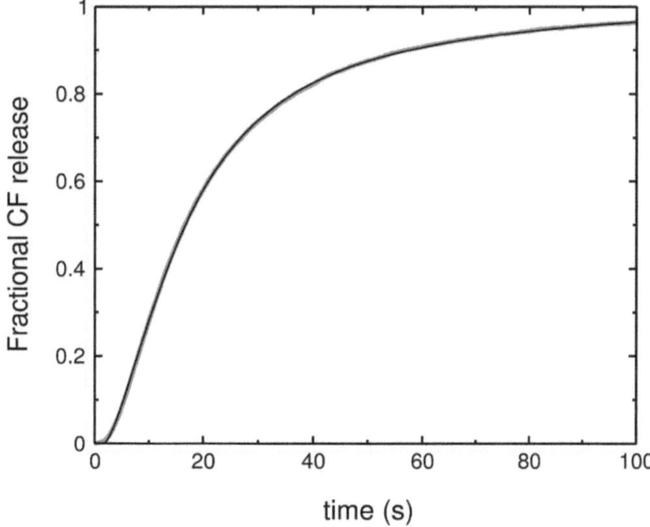

Fig. 11.5. Carboxyfluorescein release induced by δ-lysin from POPC vesicles (6). The *gray trace* is the experimental data and the *thin black line* is a three-exponential fit to the data using the equation $F(t) = 1 - \alpha_1 \exp(-t/\tau_1) - \alpha_2 \exp(-t/\tau_2) - \alpha_3 \exp(-t/\tau_3)$. The curves are only shown up to 100 s so that the initial inflection is visible, but the data extend to 200 s. The fit parameters are $\alpha_1 = -0.462$, $\alpha_2 = 1.120$, $\alpha_3 = 0.342$, $\tau_1 = 3.84$ s, $\tau_2 = 11.8$ s, and $\tau_3 = 43.7$ s. The mean lifetime, calculated from these amplitudes and the individual τ_i, is $\tau = 26.4$ s, exactly the same as obtained by direct integration using equation [1] (*see* **Section 3.8**). If the fit is performed only for the first 100 s, $\tau = 25.9$ s is obtained, which is still a very accurate representation ($t > 3\tau$). But if only the first 50 s ($t < 3\tau$) are fit, $\tau = 38$ s, which is incorrect. For comparison, estimation of the mean lifetime from the half-time with $\tau = t_{1/2}/\ln 2$ yields $\tau = 23.5$ s.

ratory (2, 6–9) and uses a numerical approach, whereby theoretical efflux curves are generated from a set of differential equations that represent a particular model, which are then numerically integrated and directly fit to the experimental data. The method deals with partial recovery and non-linear effects, such as aggregation, but is more involved and time consuming. Short of these two methods, for which the reader is referred to the original literature, there is a simple type of analysis that can be performed (14). It does not provide information on the mechanism of the peptide, but it has the advantage that is model independent and provides a quantitative measure of the activity. This latter, simpler method will be described here.

2. Materials

2.1. Synthesis of 7MC–POPE

1. Lipids (Avanti Polar Lipids, Alabaster, AL, USA): 1-palmitoyl-2-oleoyl-*sn*-glycero-3-phosphoethanolamine (POPE).

2. 7-Methoxycoumarin-3-carboxylic acid-succinimidyl ester (7MC-SE) (Molecular Probes/Invitrogen, Carlsbad, CA, USA).
3. Molecular sieves (Sigma-Aldrich, S. Louis, MO, USA).
4. Potassium carbonate (K_2CO_3).
5. Organic solvents (HPLC grade): chloroform ($CHCl_3$), dichloromethane (CH_2Cl_2), methanol (MeOH), dimethylformamide (DMF).
6. Zinzade solution: sodium molybdate dihydrate ($Na_2MoO_4{\cdot}2H_2O$), 6.85 g, hydrazine sulfate, 400 mg, sulfuric acid (H_2SO_4), 250 mL.

2.2. Lipid Large Unilamellar Vesicles (LUVs)

1. Standard buffer: 20 mM MOPS, pH 7.5, 100 mM KCl, 0.1 mM EGTA, 0.02% sodium azide.
2. Carboxyfluorescein buffer: 20 mM MOPS, pH 7.5, 50 mM carboxyfluorescein (Molecular Probes/Invitrogen), 0.1 mM EGTA, 0.02% sodium azide.
3. Phospholipids (Avanti Polar Lipids) in chloroform solutions: 1-palmitoyl-2-oleoyl-*sn*-glycero-3-phosphocholine (POPC), 1-palmitoyl-2-oleoyl-*sn*-glycero-3-phosphoglycerol (POPG), 1-palmitoyl-2-oleoyl-*sn*-glycero-3-phosphoserine (POPS).

3. Methods

3.1. Synthesis of 7MC–POPE

The synthesis of 7MC–POPC follows the general method of Vaz & Halmann (15), modified by Gregory et al. (2). The version described here is simpler and more effective.

1. Dry the molecular sieves and K_2CO_3 for over 48 h in an oven at 80°C (*see* **Note 1**).
2. Dry $CHCl_3$, MeOH, and DMF by pouring them over molecular sieves and leaving in a closed screw-cap tube for 24 h. The volume of the solvent should be only about 2–4 mL larger than the bed volume occupied by the molecular sieves.
3. Lipid (POPE) should be in 10 mol% excess over the probe, that is, a ratio L/P = 1.1/1. Typically 5–10 mg of 7MC-SE is used. The numbers given here assume 10 mg of 7MC-SE (31.5 μmol).
4. Pipet the desired amount of POPE (34.65 μmol) into a round-bottom flask and rotoevaporate at 60°C. Redissolve in a minimal amount of dry $CHCl_3$ (about 700 μL) and place in a reaction tube. Use a simple screw-cap tube wrapped in aluminum foil to avoid exposure of the

fluorophores to light. Place a small magnetic bar into the tube to stir the solution during the reaction.

5. Crush up the required amount of K_2CO_3 (in 10 mol% excess over the probe, about 2.5 mg) and add to the reaction tube.
6. Dissolve 10 mg 7MC-SE in about 0.2 mL dry DMF in a separate test tube.
7. Add the 7MC-SE solution dropwise to the reaction tube in the dark.
8. Allow the reaction to stir for about 5 h, monitoring its progress by thin layer chromatography (TLC).
9. The TLC is eluted in a small chamber with CH_2Cl_2/MeOH/H_2O 65:25:4 (v/v/v). The 7MC-SE and 7MC–POPE are detected by their fluorescence under a hand-held UV light. POPE is detected by the Zinzade reagent (*see* **Sections 2.1/3.2** and (16)), which stains the phosphate of the headgroup blue (*see* **Note 2**).
10. Upon completion of the reaction, the product is isolated by preparative TLC (Analtech, Newark, DE, USA). Spot the product along a straight line about 1 cm to the bottom of the plate and elute in a large chamber with CH_2Cl_2/MeOH 3.5:1 (v/v). Let the plate dry, then elute it again in the same solvent (prepare the solvent anew) for better separation of bands. Keep the plate exposed to minimal light during the entire procedure. Again, detect the product using UV light and Zinzade's reagent, carefully protecting all of the plate except for a small strip on one side (the Zinzade reagent destroys the probe).
11. Scrape off product band and deposit it in about 10 mL of CH_2Cl_2/MeOH 2:1 (v/v). Filter off silica through a glass Buchner funnel with a fritted disk.
12. Remove the solvent in a rotary evaporator.
13. Redissolve in 4 mL dry $CHCl_3$.
14. Determine the 7MC–POPE concentration by UV-Vis absorption spectroscopy, in a $CHCl_3$ solution. 7MC–POPE has an absorbance maximum at $\lambda = 348$ nm and an emission maximum at $\lambda = 396$ nm. The extinction coefficient of 7MC-SE is $\varepsilon = 2.6 \times 10^4$ $M^{-1}cm^{-1}$ and that of 7MC (carboxylic acid) is $\varepsilon = 2.0 \times 10^4$ $M^{-1}cm^{-1}$ (16). We estimated a similar value, $\varepsilon = 2.5 \times 10^4$ $M^{-1}cm^{-1}$ for 7MC–POPE.

3.2. Preparation and Use of the Zinzade Solution

This is a summary of the method described by Kates (17).

1. Dissolve 6.85 g of sodium molybdate (dihydrate) and 400 mg of hydrazine sulfate in 100 mL of deionized water.

2. Place the molybdate solution on ice and slowly add 250 mL of ice-cold, concentrated sulfuric acid. Allow to cool.
3. Dilute to 1 L with 600 mL of deionized water. This solution can be kept for several months.
4. Place some of the solution in a glass spray bottle. This can be operated with a small rubber bulb or with house air.
5. In a fume hood, place the TLC plate in a cardboard box to reduce contamination of surrounding with sulfuric acid. Spray the TLC plate (partially protected if appropriate) with the Zinzade solution at a distance of about 30 cm. Do not over-wet the TLC.
6. Remove the plate from the cardboard box with forceps and place on a hot plate. Observe the appearance of blue spots corresponding to phosphate groups (phospholipids). Then turn the hot plate to about 200°C and allow the TLC plate to ash completely. This eliminates the excess sulfuric acid and leaves the carbon imprint of the compounds on the plate, revealing all non-volatile organic compounds.

3.3. Preparation of Lipid Large Unilamellar Vesicles (LUVs)

1. These instructions are to prepare 1 mL of a 20 mM suspension of LUVs. The volume can be increased as desired for vesicles that do not contain encapsulated dye, but for CF-containing vesicles it should not exceed 2 mL because that will impair the column separation of vesicles from the external dye.
2. Using a Hamilton glass syringe, pipet 20 μmol of lipid (POPC, for example) in chloroform solution, into a round-bottom flask. If the volume is small add about 2 mL of chloroform. To measure peptide binding, include 1–2 mol% 7MC–POPE relative to the total amount of lipid. In this case the vesicles are hydrated with standard buffer. Pre-warm the water bath of the rotary evaporator to 60°C. Evaporate on a high rotation speed setting to obtain a thin lipid film on the wall of the flask. Keep under vacuum for 4 h.
3. Hydrate the lipid film by adding 2 mL of standard buffer or CF buffer (*see* **Sections 2.2** and **3.4**), as desired (*see* **Note 3**). Vortex for about 5 min, making sure to place the vortex plate directly under the film and rotating the flask to get the lipid off the wall of the flask. This procedure produces a turbid suspension of multilamellar vesicles (MLVs).
4. If this is a preparation of vesicles with encapsulated CF, the extent of dye encapsulation can be enhanced by subjecting the MLVs to five freeze–thaw cycles (*see* **Note 4**).
5. Extrude the MLV suspension 10 times with an extruder (Lipex Biomembranes, Vancouver, Canada) through two stacked polycarbonate filters of 0.1 μm pore size (e.g.,

Nuclepore, from Whatman, Florham, NJ, USA) at a temperature above the T_m of the lipid. This produces LUVs of about 100 nm in diameter.

6. If this is a preparation of vesicles with encapsulated CF, the hydration is done with CF buffer in Step 3. In this case, the CF in the external medium must be separated from the vesicles. This separation is achieved with a small (30 cm) column of Sephadex G-25 (Sigma-Aldrich) equilibrated with standard buffer. Collect 2 mL fractions. The vesicles will elute close to the void volume of the column, and will not be fluorescent because of self-quenching, but are easily detected by the eye because of turbidity. In the end, the CF that remains adsorbed to the column can be eluted with 50 mM KOH.
7. Determine the lipid concentrations using the standard method of Bartlett (18) or one of its modified versions (6).

3.4. Preparation of Carboxyfluorescein Buffer

1. CF is only soluble in aqueous solutions when it is fully deprotonated. Therefore, the pH needs to be 7.5 or higher (if this is a problem, use calcein instead of CF).
2. Prepare a buffer identical to the standard buffer but without any KCl (*see* **Note 5**).
3. Mix the CF powder with the buffer and titrate the solution with concentrated (about 1 M) KOH under the pH meter, until all the CF is dissolved and the pH is 7.5 (*see* **Note 6**). Then bring up to the desired volume with buffer (without KCl).
4. Filter the CF buffer through a syringe filter with 0.2 μm pore size (Whatman 25 mm GD/X, NYL) (*see* **Note 7**).
5. Store the CF buffer in the refrigerator.

3.5. Carboxyfluorescein Efflux Kinetics

1. The efflux of CF as a function of time, induced by peptide interaction with the lipid vesicles, is measured in a stopped-flow fluorimeter (*see* **Note 8**). We have used the SX-18MV instrument from Applied Photophysics (Leatherhead, UK), but other companies make alternative, high-quality instruments (Bio-Logic, Claix, France and KinTek, Austin, USA, for example).
2. Identical volumes of 1–2 μM peptide solutions are mixed with vesicle suspensions of 50–1,000 μM in lipid, in the stopped-flow system, to produce final concentrations that are one-half of the original ones. The excitation is at $\lambda =$ 470 nm (the CF maximum is actually at 490 nm) and the emission at 520 nm. In the absence of a second monochromator, the emission can be recorded through a long-pass filter (e.g., OG 530, Edmund Industrial Optics, Barrington, NJ, USA).

3. The final CF release has to be normalized to the maximum possible release from the vesicles. This maximum fluorescence is obtained by adding the detergent Triton X-100 to the vesicles, to a final detergent concentration of 1% (of the aqueous solution). All dye is released under these conditions (*see* **Note 9**).

3.6. Measurement of Peptide Binding to LUVs

1. Prepare a peptide solution of 1–2 μM and a series of about 8–10 vesicle suspensions with lipid concentrations varying from 50 to 800 μM. After mixing in the stopped-flow system, these concentrations will be halved.
2. The kinetics of association of peptide with LUVs are recorded in a stopped-flow fluorimeter (e.g., SX.18MV from Applied Photophysics) measuring the increase in FRET from the Trp of the peptide to 7MC–POPE. Trp is excited at 280 nm and the emission of 7MC is measured at 396 nm or through a GG-385 cut-off filter (Edmund Industrial Optics) (*see* **Note 10**).
3. Record the fluorescence emission of 7MC–POPE, which increases as a function of time, as more peptide binds to the vesicles (**Fig. 11.3a**) (*see* **Note 11**).
4. To obtain the on- and off-rate constants (*see* analysis below at **Section 3.7**) perform this experiment as a function of lipid concentration, from about 25 to 400 μM (final lipid concentrations, after mixing). Record about 10 repeats (at least triplicates) for each curve. If the signal-to-noise ratio is poor, it is necessary to acquire more traces and average of about 10–20 to improve the signal/noise ratio (*see* **Note 12**).

3.7. Analysis of Peptide Association Kinetics

1. Fit the experimental binding kinetic curves with a single exponential rising function (**Fig. 11.3a**). Extract the value of the apparent rate constant k_{app}, which is related to the molecular on- and off-rate constants for binding to the vesicles by $k_{app} = k_{off} + k_{on}$ [L], where [L] is the lipid concentration.
2. Since the experiments were performed for various lipid concentrations, now plot k_{app} against the lipid concentration. This yields a straight line, with slope k_{on} and y-intercept k_{off} (**Fig. 11.3c**).
3. The equilibrium dissociation constant is obtained by $K_D = k_{off}/k_{on}$.

3.8. Analysis of Dye Efflux Kinetics

1. The most straightforward method is to calculate the mean lifetime (τ) of efflux numerically (14). First, differentiate the CF efflux curve (**Fig. 11.5**), $F(t)$, numerically to obtain its derivative $F\prime(t)$. This function is the probability density

function. Second, evaluate τ by integration between time zero and the completion time of the curve, according to equation [1] (*see* **Note 13**):

$$\tau = \int tF'(t)dt / \int F'(t)dt$$

2. A simpler, albeit less satisfactory, alternative to the calculation of the mean lifetime by equation [1] is to fit the CF efflux curves with the smallest number of exponential functions needed to obtain a very good fit (**Fig. 11.5**). Typically 2 or 3, but rarely more than 3 exponentials are required. The lifetime is then obtained by an amplitude-weighted average of the characteristic time constants (τ_i) of each exponential. For example, for 2 exponentials, bring the origin of the experimental $F(t)$ to zero, and fit the curve with the function $f(t) = A_o\ [1-\alpha_1 \exp(-t/\tau_1)-\alpha_2 \exp(-t/\tau_2)]$, with $\alpha_1 + \alpha_2 = 1$, and where A_o is a global amplitude factor. The lifetime is then given by $\tau = \alpha_1\tau_1 + \alpha_1\tau_2$. The apparent rate constant for efflux is the reciprocal of τ (*see* **Note 14**).

4. Notes

1. Exclusion of water both from solvents and reactants is the essential condition for a successful synthesis. Hence, the need to dry all solvents and K_2CO_3. K_2CO_3 is the base in the reaction, which is necessary to keep the nucleophile (the amino group of the ethanolamine headgroup) in the deprotonated state ($-NH_2$).
2. The 7MC–POPE spot should be the only spot on the TLC that is detected both by fluorescence and the Zinzade solution. Approximate R_fs are POPE, 0.53; 7MC–POPE, 0.76; and 7MC-SE, 0.92.
3. The buffer must be at a temperature higher than the phase transition temperature of the lipid (T_m). For POPC T_m is –4°C, so room temperature is appropriate, but if saturated lipids are used, care must be taken to ensure that the buffer temperature is higher than T_m.
4. The freeze–thaw procedure also helps eliminate small vesicles, which fuse with larger ones, and could otherwise result in artifacts in the peptide-induced dye release kinetics.
5. The osmolarity of the CF buffer solution will be equivalent to the osmolarity of the standard buffer, as the contribution

of CF and its counter ions will be equivalent to that of 100 mM KCl.

6. Be careful not to overtitrate, as this will cause an imbalance in the osmotic strength compared to standard buffer.
7. This step is essential to remove any undissolved material, which will clog the filters during extrusion.
8. If the efflux is slow, typically longer than 1,000 s for completion, the curves can be recorded in a steady-state spectrofluorimeter, by adding the peptide solution to the vesicle suspension in the cuvette. The initial value must be measured prior to peptide addition, and the Triton X-100 release level is determined in the same sample by adding detergent, to a concentration of 1%, at the end of the experiment. If the efflux occurs in a time longer than 1 s but shorter than about 20 s, measurement in the cuvette is very inaccurate. In the absence of a stopped-flow instrument, it is possible to use a rapid kinetics accessory (e.g., RX2000 from Applied Photophysics) that is mounted on the steady-state fluorimeter.
9. The Triton X-100 release level can also be determined separately, in a steady-state spectrofluorimeter [we have used the 8100 SLM-Aminco (Urbana, IL, USA) instrument, upgraded by ISS (Urbana, IL, USA), or the Jobin-Yvon (Edison, NJ, USA) instruments Fluorolog and Fluoromax], by measuring the initial fluorescence, before adding detergent, and the final fluorescence, after detergent addition. The initial fluorescence value is necessary to convert the measurements in the two experiments (stopped flow and steady state) to a common scale.
10. FRET between an intrinsic Trp residue of the peptide and 7MC–POPE incorporated in the lipid membrane is used to monitor peptide binding. If the peptide is bound, instead of emitting fluorescence at 320–350 nm, it transfers its excitation energy to 7MC–POPE, which absorbs maximally at 348 nm. The fluorescence emission maximum of 7MC–POPE occurs at 396 nm. In order for FRET to occur, the Trp residue and the 7MC fluorophore need to be approximately within the Förster distance. For this pair of fluorophores, the Förster distance is 26 Å if the Trp emission maximum is at about 330 nm, but the exact value may vary by 1–2 Å, depending on the wavelength of the Trp emission maximum, which is determined by its physical environment. The distance for transfer from Trp to other coumarin fluorophores, such as 7-hydroxycoumarin (6) and methylcoumarin (19), is similar (25 Å).

11. Binding is monitored typically for less than 1 s, sometimes less than 0.1 s. Many antimicrobial and cytolytic peptides induce other, post-binding changes in the vesicles that also result in fluorescence changes, which should not, however, be mistaken for binding.
12. If binding is very tight, it may be necessary to perform a different experiment to obtain a good estimate of the off-rate constant. In this case, peptide (about 1 μM) is preincubated with vesicles (about 100 μM) that contain the lipid probe (7MC–POPE), which are then mixed in the stopped-flow system with an excess of vesicles (about 1 mM) that do not contain any probe. Binding to the acceptors can be made quasi-irreversible by changing their lipid composition. For example, many antimicrobial peptides are cationic and therefore bind much better to anionic vesicles. In this case, one would prepare acceptor vesicles with a higher content of anionic lipids (such as POPG) than present in the donor vesicles. Peptide will now dissociate from the donor vesicles and will bind to the acceptor vesicles, leading to a decrease in fluorescence of 7MC–POPE from the donors (**Fig. 11.3b**).
13. This analysis of efflux curves does not allow for inferences regarding the peptide mechanism but can be used with confidence to compare the peptide activity as a function of lipid concentration, vesicle, composition, temperature, etc. Calculating the average lifetime is far superior to using initial rates of efflux, which take into account only a small portion of the curve, discarding the majority of the data points. Using initial rates also overlooks the problem that the shapes of curves vary considerably, leading to erroneous results. For example, sigmoidal curves (**Fig. 11.5**), which exhibit an initial lag time, will produce low initial rates of efflux, but often this is more than compensated by a sudden increase in peptide activity after the lag. For the numerical differentiation to be successful, a large number of data points is necessary (about 500 or more), and the curves cannot be too noisy. Ideally, the curve should have reached a plateau; at the very least, it is necessary to collect data for a period = 3τ (**Fig. 11.5**), otherwise the result may be meaningless.
14. To obtain a meaningful fit, the data must be collected for a period of at least 2.5 times the longest τ_i (**Fig. 11.5**). If the curves contain a small initial inflection or sigmoidicity (a lag period), a good fit can still be obtained if one of the amplitudes (α_i) is negative (*see* legend of **Fig. 11.5**). However, for curves with long lags, which occur with some peptides, such as δ-lysin (6), this type of fit would require

a series of exponentials with negative amplitudes, which quickly becomes a complicated task.

Acknowledgments

This work was supported by National Institutes of Health grant No. GM072507. We thank Erin Kilelee for her help during the preparation of the manuscript.

References

1. Vaz, W. L., Kaufmann, K., and Nicksch, A. (1977) Use of energy transfer to assay the association of proteins with lipid membranes. *Anal. Biochem.* **83**, 385–393.
2. Gregory, S. M., Cavenaugh, A., Journigan, V., Pokorny, A., and Almeida, P. F. F. (2008) A Quantitative model for the all-or-none permeabilization of phospholipid vesicles by the antimicrobial peptide cecropin A. *Biophys. J.* **94**, 1667–1680.
3. Hope, M. J., Bally, M. B., Webb, G., and Cullis, P. R. (1985) Production of large unilamellar vesicles by a rapid extrusion procedure. Characterization of size distribution, trapped volume and ability to maintain a membrane potential. *Biochim. Biophys. Acta* **812**, 55–65.
4. Weinstein, J. N., Yoshikami, S., Henkart, P., Blumenthal, R., and Hagins, W. A. (1977) Liposome-cell interaction. Transfer and intracellular release of a trapped fluorescent marker. *Science* **195**, 489–492.
5. Kayalar, C. and Düzgüneçs, N. (1986) Membrane action of colicin El. Detection by the release of carboxyfluorescein and calcein from liposomes. *Biochim. Biophys. Acta* **860**, 51–56.
6. Pokorny, A., Birkbeck, T. H., and Almeida, P. F. F. (2002) Mechanism and kinetics of δ-lysin interaction with phospholipid vesicles. *Biochemistry* **41**, 11044–11056.
7. Pokorny, A. and Almeida, P. F. F. (2004) Kinetics of dye efflux and lipid flip-flop induced by δ-lysin in phosphatidylcholine vesicles and the mechanism of graded release by amphipathic, α-helical peptides. *Biochemistry* **43**, 8846–8857.
8. Yandek, L. E., Pokorny, A., Florén, A., Knoelke, K., Langel, U., and Almeida, P. F. F. (2007) Mechanism of the cell-penetrating peptide transportan 10 permeation of lipid bilayers. *Biophys. J.* **92**, 2434–2444.
9. Gregory, S. M., Pokorny, A., and Almeida, P. F. F. (2009) Magainin 2 revisited: a test of the quantitative model for the all-or-none permeabilization of phospholipid vesicles. *Biophys. J.* **96**, 116–131.
10. Chen, R. F. and Knutson, J. R. (1988) Mechanism of fluorescence concentration quenching of carboxyfluorescein in liposomes: energy transfer to nonfluorescent dimers. *Anal. Biochem.* **172**, 61–77.
11. Schwarz, G. and Robert, C. H. (1990) Pore formation kinetics in membranes, determined from the release of marker molecules out of liposomes or cells. *Biophys. J.* **58**, 577–583.
12. Robert, C. H. and Schwarz, G. (1992) Kinetics of pore-mediated release of marker molecules from liposomes or cells. *Biophys. Chem.* **42**, 291–296.
13. Rex, S. and Schwarz, G. (1998) Quantitative studies of the melittin-induced leakage mechanism of lipid vesicles. *Biochemistry* **37**, 2336–2345.
14. Pokorny, A., Yandek, L. E., Elegbede, A. I., Hinderliter, A., and Almeida, P. F. F. (2006) Temperature and composition dependence of the interaction of δ-lysin with ternary mixtures of sphingomyelin/cholesterol/POPC. *Biophys. J.* **91**, 2184–2197.
15. Vaz, W. L. C. and Hallmann, D. (1983) Experimental evidence against the applicability of the Saffman-Delbrück model to the translational diffusion of lipids in phosphatidylcholine bilayer membranes. *FEBS Lett.* **152**, 287–290.
16. Haugland, R. P. (2002) *Handbook of Fluorescent Probes and Research Products* (9th ed.). Eugene: Molecular Probes.
17. Kates, M. (1972) Techniques in lipidology. In *Laboratory Techniques in Biochemistry and Molecular Biology*. Work, T. S. and Work, E. (Eds.), Amsterdam: North-Holland Publishing Company.

18. Bartlett, G. R. (1959) Phosphorous assay in column chromatography. *J. Biol. Chem.* **234**, 466–468.
19. Wimley, W. C. and White, S. H. (2000) Determining the membrane topology of peptides by fluorescence quenching. *Biochemistry* **39**, 161–170.
20. Yandek, L. E., Pokorny, A., and Almeida., P. F. F. (2008) Small changes in the primary structure of transportan 10 alter the thermodynamics and kinetics of its interaction with phospholipid vesicles. *Biochemistry* **47**, 3051–3060.

Chapter 12

Comparing Bacterial Membrane Interactions of Antimicrobial Peptides and Their Mimics

Nathaniel P. Chongsiriwatana and Annelise E. Barron

Abstract

Interactions with bacterial membranes are integral to the mechanisms of action of all antimicrobial peptides (AMPs), regardless of their final cellular targets. Here, we describe in detail two biophysical techniques that can be used to measure the membrane activities of AMPs and antimicrobial peptidomimetics: (1) a calcein leakage assay to investigate interactions between AMPs/peptidomimetics with large unilamellar vesicles and (2) a potential-sensitive dye-based depolarization assay to investigate interactions with the membranes of live bacteria. By comparing the membrane interactions of AMPs and their mimics, these techniques can provide insights into their extent of mimicry and their antimicrobial mechanisms.

Key words: Calcein leakage, large unilamellar vesicles, membrane depolarization, membrane potential, diSC3–5.

1. Introduction

Many antimicrobial peptidomimetics, such as β-peptides (1, 2), β-peptoids (3), α/β-peptides (4), oligoureas (5), oligo (phenylene ethynylene)s (6), and peptoids (7–9), successfully emulate antimicrobial peptide (AMP) function within a non-natural framework; that is, mimicry of AMP structure generally leads to mimicry of AMP function. However, because the mechanisms of action of AMPs themselves are poorly understood, it is difficult to discern whether antimicrobial peptidomimetics kill bacteria via the same mechanisms as the peptides that they emulate.

A. Giuliani, A.C. Rinaldi (eds.), *Antimicrobial Peptides*, Methods in Molecular Biology 618,
DOI 10.1007/978-1-60761-594-1_12, © Springer Science+Business Media, LLC 2010

Here, we discuss two techniques that can be used to establish, to some degree, the extent of mechanistic analogy between AMPs and their mimics. Since membrane interactions are integral to the mechanisms of all AMPs regardless of their final targets (10, 11), calcein leakage assays (12–17) and bacterial membrane potential measurements (18–26) can be used to probe peptidomimetic-membrane interactions and compare them to like interactions with AMPs.

Fluorescence monitoring of leakage from lipid vesicles is a model system used to evaluate membrane permeabilization by AMPs. This relatively simple model isolates the membrane and eliminates the possibility of interactions with other cellular components, which could confound the interpretation of results. However, leakage results must be seen within a broader context, since such other interactions also may prove integral to the overall mechanism. Typically, large unilamellar vesicles (LUVs) are used for leakage studies, with a variety of dyes entrapped therein, such as calcein (12–17), carboxyfluorescein (27–29), FITC-labeled dextrans (16, 17), and dipicolinic acid, the latter of which forms a fluorescent complex with terbium(III) (27, 28). In a calcein leakage assay, the dye is entrapped inside LUVs at a high, self-quenching concentration of 70 mM. If the membrane is permeabilized, dye leaks out of the vesicles, becomes diluted by the extra-vesicular buffer, and begins to fluoresce. The amount of calcein dequenching is then quantified relative to a 100% leakage control (typically a detergent like Triton X-100).

Rather than measuring the leakage of artificial fluorophores from model vesicles, membrane permeabilization can also be studied by monitoring the translocation of physiological species, like protons or other small ions, across membrane bilayers. As with leakage assays, this can be accomplished with fluorescence spectroscopy using fluorescent dyes that are sensitive to changes in transmembrane potential or intracellular pH (29), such as diSC3–5 (30), BCECF-AM (31), 9-aminoacridine (32), and oxonol VI (33).

Bacteria typically have an inside-negative transmembrane potential that arises from the interplay between charge gradients and concentration gradients across the plasma membrane. This is made possible by the selective permeability of the plasma membrane to certain ions, particularly K^+, Na^+, and Cl^- (34). A number of net-anionic proteins are concentrated in the cytoplasm, whose charge is balanced by positively charged counterions. Were the membrane impermeable to every species, no electrical potential would be generated, since both anions and counterions would be restricted to the cytoplasm, albeit against their concentration gradients. However, bacterial membranes contain a number of protein channels that each accommodate a specific ion, such as the potassium channel KcsA (35). Potassium can freely diffuse through this channel, and so a small number of K^+ ions leak out of the cell down their concentration gradient. In contrast, anionic proteins are held in the cytoplasm, thereby creating a charge imbalance, and the cell is left with an inside-negative voltage across the plasma membrane. Note that many different ions are responsible for the overall membrane potential, although the contributions of potassium and sodium predominate.

DiSC3–5 is perhaps the most commonly used potential-sensitive dye and is particularly useful because it can be employed in conjunction with live bacteria in addition to lipid vesicles (18–26). A cationic cyanine, diSC3–5, accumulates in the electronegative cell interiors, where its fluorescence is self-quenched. Membrane disruption by AMP-mimetic oligomers can cause a loss of the membrane's selective ion permeability, leading to depolarization (dissipation of the membrane potential). As a result, the dye is released from the cell and its subsequent fluorescence can be quantified relative to the action of a molecule, such as melittin, an amphipathic, cationic peptide that is known to depolarize bacteria completely.

2. Materials

2.1. Calcein Leakage Assay

2.1.1. Extrusion of Calcein-Filled LUVs

1. Self-quenching aqueous calcein solution: 70 mM calcein, 10 mM Tris–HCl, pH 7.4 (*see* **Note 1**).
2. Extra-vesicular buffer: 10 mM Tris–HCl, 100 mM NaCl, pH 7.4 (36–39).
3. Lipids [cholesterol, palmitoyloleoylphosphatidylcholine (POPC), palmitoyloleoylphosphatidylethanolamine (PO PE), palmitoyloleoylphosphatidylglycerol (POPG), etc.] (Avanti Polar Lipids, Alabaster, AL, USA) are purchased as lyophilized powders, weighed out, and dissolved in chloroform to a final known stock concentration (typically around 10.0 mg/mL). Store at –20°C for up to 2 weeks; stock solutions should be allowed to reach room temperature before using and should be clear and colorless.
4. LUV extruder (Avanti Polar Lipids) with accompanying short-needle Hamilton syringes, polycarbonate membrane filters with 0.1 μm pores (larger sizes are optional; *see* **Note 2**).
5. Liquid nitrogen.
6. Water bath at 35–40°C.

2.1.2. Separation of Calcein-Filled LUVs from Free Calcein

1. Sephadex G-50, medium grain, swollen by boiling in distilled, deionized water for approx. 1 h. Boiling the slurry accelerates swelling of the gel and degasses the water to prevent the appearance of air bubbles after packing the column. The amount of water in the gel suspension should be adjusted to approx. 75% settled gel. Store at room temperature.
2. Glass column with 1 cm inner diameter, ~30 cm length, fritted glass bottom, and stopcock.

2.1.3. Phosphorus Assay to Determine Lipid Concentration of Calcein-Filled LUVs After Gel Filtration

1. Digestion acid: concentrated H_2SO_4 plus 70% $HClO_4$ (2:1). Store at room temperature; discard if not colorless.
2. Phosphate solutions for standards: 10 and 1 mM aqueous monobasic sodium phosphate (NaH_2PO_4). Store at room temperature.
3. Aqueous solution of 0.5% (w/w) ammonium molybdate tetrahydrate [$(NH_4)_6Mo_7O_{24}{\cdot}4H_2O$] for developing the color reaction. Store at room temperature.
4. 1-Amino-2-naphthol-4-sulfonic acid (ANSA) reagent (commercially available from Fisher Scientific, Pittsburgh, PA, USA): mix 1 g/L ANSA, 37 g/L sodium sulfite (Na_2SO_3), and 62 g/L sodium metabisulfite ($NaHSO_3$) in water. Store at –20°C for up to 2 months.
5. Glass test tubes.
6. Sand bath or test tube heating block.

2.1.4. Measuring Calcein Leakage

1. Rectangular quartz fluorescence cuvette, 1.5 mL volume.
2. Extra-vesicular buffer: 10 mM Tris–HCl, 100 mM NaCl, pH 7.4.
3. Aqueous stock solutions of peptides and/or peptidomimetics.
4. 100% leakage control: 10% (v/v) aqueous Triton X-100 detergent.

2.2. Membrane Depolarization Assay

2.2.1. Preparation of Bacterial Suspension

1. Gram-negative (i.e., *Escherichia coli* ATCC 35218) or Gram-positive (i.e., *Bacillus subtilis* ATCC 6633) bacteria, grown to mid-logarithmic phase in standard growth medium, such as Luria–Bertani (LB) broth or Mueller–Hinton broth (MHB), made according to manufacturer's instructions.
2. Respiration buffer: 5 mM HEPES, 20 mM glucose, pH 7.4. Store in the dark, as HEPES can develop toxic by-products when exposed to light.

3. Dye stock: 30 μM 3,3'-dipropylthiadicarbocyanine iodide (diSC3–5) in DMSO. Store in the dark at 4°C and let thaw at room temperature before using.
4. 0.5 M aqueous EDTA, pH 8.0 (*see* **Note 3**).

2.2.2. Measuring Membrane Depolarization

1. Rectangular quartz fluorescence cuvette, 1.5 mL volume.
2. Aqueous stock solutions of peptides and/or peptidomimetics.
3. 100% depolarization control: aqueous melittin stock solution, at an accurately known concentration of ~10.0 mg/mL.

3. Methods

3.1. Calcein Leakage Assay

3.1.1. Extrusion of Calcein-Filled LUVs

1. In a glass vial, combine lipid stock solutions (in chloroform) in the desired molar ratio to a combined mass of 20–50 mg (*see* **Note 4**).
2. Dry the lipid mixture to obtain a film on the side of the vial by rotating slowly under a stream of nitrogen.
3. Lyophilize the film overnight to remove residual chloroform.
4. Add enough buffered calcein solution (*see* **Section 2.1.1**) to bring the lipids to a concentration of 10–30 mM.
5. Let the lipid film hydrate in a water bath at 35–40°C for 1 h.
6. After hydration, vortex the lipid suspension until no traces of the lipid film remain on the sides of the vial. The suspension now contains multilamellar vesicles (MLVs).
7. Subject the MLV suspension to 5 freeze/thaw cycles (liquid N_2/35–40°C water bath).
8. Assemble the extrusion device according to the manufacturer's instructions, using two stacked polycarbonate filters with 100 nm pores, then load the MLV suspension into one of the syringes.
9. Pass the suspension through the 100 nm pores 15 times (40). The suspension will become considerably more translucent as the MLVs are extruded into large unilamellar vesicles (LUVs). For some lipid compositions, preliminary extrusion through a larger pore size may be necessary (*see* **Note 2**). After the final passage, the LUV suspension should be in the syringe opposite to the starting syringe.

This ensures that as few MLVs as possible are present in the final suspension.

3.1.2. Separation of Calcein-Filled LUVs from Free Calcein

1. Prepare a gel filtration column by pouring the slurry of hydrated Sephadex G-50 into a 1 cm inner diameter glass column. Allow the gel to settle completely before proceeding. The final settled height of the gel inside the column should be 6–10 cm.
2. Open the column stopcock and let the water above the gel drain. Do not let the liquid level fall below the top of the gel, or air will be introduced into the column and it will not permit adequate separation.
3. Equilibrate the column by flowing at least two column heights of extra-vesicular buffer (*see* **Section 2.1.1**) through the column (*see* **Note 5**).
4. To prepare to add the LUV suspension, let the level of buffer above the column fall just to the top of the gel, but not below.
5. Using a long Pasteur pipette, transfer the LUV suspension to the top of the column. Open the stopcock until the liquid level once again falls just to the top of the gel. Moving quickly, carefully add at least two column heights of extra-vesicular buffer to the top of the column and open the stopcock.
6. Watch the effluent closely. The calcein-filled LUVs will elute before the free calcein and appear visibly turbid compared to the free calcein solution.
7. As soon as the turbid LUVs begin to elute, start collecting fractions (~0.3–1 mL each). Recombine the fractions containing the most LUVs (i.e., turbidity) and with least free calcein (i.e., color without turbidity). The resulting purified LUV suspension should be light orange in color and somewhat iridescent. Store at 4°C for up to 1 week, although the performance of the vesicles will degrade over time as the entrapped calcein slowly leaks out of the vesicles into the surrounding buffer (this can be detected by an increase in the fluorescence of the suspension).

3.1.3. Phosphorus Assay to Determine Lipid Concentration of Calcein-Filled LUVs After Gel Filtration

This assay was adapted from Bartlett (41).

1. To create phosphorus standards, carefully pipette 50 μL of 1 mM NaH_2PO_4 and 10, 20, 30, and 40 μL of 10 mM NaH_2PO_4 into separate test tubes. Leave one tube empty. These standards contain 0, 50, 100, 200, 300, and 400 nmol of phosphorus.
2. Estimate the phosphorus concentration of the LUV suspension, accounting for the fact that the lipids will have

been diluted during gel filtration. Since there is 1 phosphorus atom per lipid molecule (except cholesterol), the molar phosphorus concentration is equal to the molar lipid concentration. Using the estimated phosphorus concentration, pipette enough LUV suspension into a test tube to yield ~200 nmol phosphorus. Make triplicates of each unknown at two different volumes to ensure that one set of unknowns will fall on the standard curve. For example, if lipids were extruded at 20 mM, typically three unknown samples of 20 μL LUV suspension and three of 40 μL LUV suspension will yield data that fall on the standard curve.

3. Dry unknown and standard samples in a sand bath or heating block at 100°C for 1 h.
4. To each sample, add three drops of concentrated sulfuric acid/perchloric acid mixture (digestion acid, *see* **2.1.3.**).
5. To digest samples, heat in sand bath or heating block above 250°C until the samples are clear and colorless (approx. 1–2 h).
6. To each sample, add 1.8 mL 0.5% (w/w) aqueous ammonium molybdate and mix (*see* **2.1.3.**).
7. Add 20 μL of ANSA reagent (*see* **2.1.3.**).
8. Let color develop for at least 30 min at room temperature.
9. In a spectrophotometer, read the absorbance of the samples at 660 nm.
10. Plot A_{660} versus nanomoles of P for the standards, fit a line, and use the resulting trend line to calculate the lipid concentration of the LUV suspension (typically ~1–10 mM). If the LUVs contain cholesterol, be sure to account for the fact that cholesterol does not contain phosphorus when calculating the total lipid concentration of the suspension.

3.1.4. Measuring Calcein Leakage

1. Set the fluorometer to measure fluorescence intensity at $\lambda_{excitation} = 490$ nm, $\lambda_{emission} = 520$ nm, and a bandwidth of 1 mm.
2. Pipette 1.5 mL extra-vesicular buffer (*see* **Section 2.1.1**) into a rectangular quartz fluorescence cuvette.
3. Begin recording the fluorescence.
4. Pipette enough of the LUV suspension into the cuvette to have a final concentration of 50 μM lipids and aspirate gently to mix. The signal increases sharply and quickly stabilizes, reflecting the background fluorescence of the LUV mixture; this value is taken as 0% leakage for this particular run.

5. After the background fluorescence stabilizes, pipette the appropriate amount of aqueous peptide or peptidomimetic stock solution into the cuvette to achieve the desired final peptoid or peptide test concentration. Aspirate gently to mix. If leakage occurs, dilution of the calcein by the extra-vesicular buffer causes dequenching and the fluorescence increases.
6. After 5 min, add 5 μL of 10% (v/v) aqueous Triton X-100 and mix to get the maximal fluorescence value (100% leakage) for this particular run. The percent dequenching (percent leakage) after 5 min is given by

$$\%\text{Dequenching} = \frac{(F - F_0)}{(F_T - F_0)} \times 100$$

where F is the fluorescence 5 min after the addition of the peptide/peptidomimetic, F_0 is the initial fluorescence (before peptide addition), and F_T is the total fluorescence (after Triton X-100 addition).
7. Repeat Steps 2–6 for each desired concentration of peptide or peptidomimetic to generate a curve of % dequenching (leakage) after 5 min versus peptide/peptidomimetic concentration.

3.2. Membrane Depolarization Assay

3.2.1. Preparation of Bacterial Suspension

1. Grow bacteria at 37°C in growth medium (i.e., LB broth or MHB) to the mid-logarithmic phase.
2. Centrifuge bacterial culture at 2500×*g* for 10 min at room temperature.
3. Draw off supernatant and discard.
4. Resuspend the bacterial pellet in respiration buffer (*see* **Section 2.2.1**).
5. Add enough bacterial suspension to 25–40 mL fresh respiration buffer to make a dilute suspension with an optical density OD_{590} 0.05.
6. Divide the dilute suspension into 1.5 mL aliquots. Each aliquot should be dyed with diSC3–5 1 hour before measuring depolarization by adding DMSO dye stock to a final dye concentration of 0.2 μM (*see* **Notes 6–8**).
7. For *E. coli* and other Gram-negative bacteria, 0.5 M aqueous EDTA may be added to a final concentration of 0.1–0.2 mM to each aliquot immediately before addition of dye in order to facilitate penetration of the dye through the outer membrane into the bacteria (*see* **Note 9**).

3.2.2. Measuring Membrane Depolarization

1. Set the fluorimeter to measure fluorescence intensity at $\lambda_{emission} = 670$ nm, $\lambda_{excitation} = 622$ nm, and a bandwidth of 1 mm.
2. One hour after adding dye to the suspension aliquot, transfer the aliquot to a 1.5 mL rectangular quartz fluorescence cuvette.
3. After recording the baseline fluorescence, pipette the appropriate amount of aqueous peptide/peptidomimetic stock solution into the cuvette to achieve the desired final test concentration. Aspirate gently to mix. If depolarization occurs, the change in membrane potential causes dequenching of the diSC3–5 and the fluorescence increases.
4. Five minutes after the addition of peptide/peptidomimetic, pipette aqueous melittin stock to a final concentration of 20 μM to obtain the maximal depolarization for this particular run. Percent depolarization is calculated analogously to that in calcein leakage experiments:

$$\%\text{Depolarization} = \frac{(F - F_0)}{(F_M - F_0)} \times 100$$

where F is the fluorescence 5 min after the addition of the peptide/peptidomimetic, F_0 is the initial fluorescence (before peptide addition), and F_M is the total fluorescence (after melittin addition).

5. Repeat Steps 2–4 for each desired concentration of peptide or peptidomimetic to generate a curve of % depolarization after 5 min versus peptide/peptidomimetic concentration.

4. Notes

1. Because each molecule of calcein can donate four protons in solution, calcein will quickly acidify a solution, preventing further dissolution of the compound. To make 25 mL of the buffered 70 mM calcein solution, bring 1.089 g calcein and 0.0303 g Tris–HCl up to 25 mL with deionized water; the calcein will remain undissolved. Add saturated aqueous NaOH dropwise until the EDTA dissolves, then bring the pH back down to 7.4 with concentrated HCl.
2. Some lipid compositions (i.e., POPC/cholesterol) will be much more difficult to extrude than others (i.e., POPE/POPG); that is, such compositions may require much more pressure to extrude, sometimes enough to rupture the polycarbonate membrane filters. For these

compositions, it may help to extrude the lipids first through membranes with larger pores (i.e., 600 or 400 nm) then move to increasingly smaller pores and finishing with 100 nm.

3. Like calcein, EDTA will not dissolve at high concentrations until the pH is raised to 9–10. To make 50 mL of 0.5 M aqueous EDTA, bring 9.306 g EDTA up to 50 mL with deionized water. Add saturated aqueous NaOH dropwise until the EDTA dissolves, then bring the pH back down to 8.0 with concentrated HCl.
4. Two examples of model lipid compositions are POPC/cholesterol (2:1) for erythrocyte-mimetic vesicles (42) and POPE/POPG (7:3) for bacteria-mimetic vesicles (43).
5. If liquid flows through the column too slowly, remove some of the gel.
6. The dye incorporates slowly into the bacteria (over ~30 min) and its fluorescence is quenched in the process.
7. Each aliquot of suspension must be dyed separately because the fluorescence response of diSC3–5 to depolarization varies with the amount of time it has been in contact with the bacteria. Thus, all aliquots of dyed bacterial suspension must be used the same amount of time after addition of dye to ensure consistency of response.
8. Some published diSC3–5 depolarization protocols involve the use of 100 mM KCl to equilibrate the inner and outer K^+ concentrations prior to measurement of depolarization (18, 19, 23–25). However, in our hands, depolarization was observed in response to peptides/peptidomimetics using the above protocol, even without addition of KCl. Indeed, AMPs and their mimics are believed to cause a change in membrane potential via non-specific permeabilization of bacterial membranes; thus, depolarization should occur and be observable regardless of whether or not the inner and outer K^+ concentrations have been equilibrated.
9. For each bacterium used, try running the assay first in the absence of EDTA to determine if an adequate fluorescence response can be observed without its addition.

Acknowledgments

The authors would like to thank Professor Robert MacDonald, Dr. Joshua Rausch, Dr. Jiwon Seo, and Ms. Meera Rao for their assistance in the development of these protocols. This work was supported by a Department of Homeland Security Fellowship and NIH Grants 1 R01 HL67984 and 1 R01 AI072666.

References

1. Seebach, D., Overhand, M., Kühnle, F. N. M., Martinoni, B., Oberer, L., Hommel, U., and Widmer, H. (1996) β-peptides: synthesis by Arndt-Eistert homologation with concomitant peptide coupling. Structure determination by NMR and CD spectroscopy and by X-ray crystallography. Helical secondary structure of a β-hexapeptide in solution and its stability towards pepsin. *Helv. Chim. Acta* **79**, 913–941.
2. Porter, E. A., Weisblum, B., and Gellman, S. H. (2002) Mimicry of host-defense peptides by unnatural oligomers: antimicrobial beta-peptides. *J. Am. Chem. Soc.* **124**, 7324–7330.
3. Hamper, B. C., Kolodziej, S. A., Scates, A. M., Smith, R. G., and Cortez, E. (1998) Solid phase synthesis of β-peptoids: N-substituted β-aminopropionic acid oligomers. *J. Org. Chem.* **63**, 708–718.
4. Epand, R. F., Schmitt, M. A., Gellman, S. H., Sen, A., Auger, M., Hughes, D. W., and Epand, R. M. (2005) Bacterial species selective toxicity of two isomeric alpha/beta-peptides: role of membrane lipids. *Mol. Membr. Biol.* **22**, 457–469.
5. Violette, A., Averlant-Petit, M. C., Semetey, V., Hemmerlin, C., Casimir, R., Graff, R., Marraud, M., Briand, J.-P., Rognan, D., and Guichard, G. (2005) *N,N'*-linked oligoureas as foldamers: chain length requirements for helix formation in protic solvent investigated by circular dichroism, NMR spectroscopy, and molecular dynamics. *J. Am. Chem. Soc.* **127**, 2156–2164.
6. Arnt, L., and Tew, G. N. (2002) New poly(phenyleneethynylene)s with cationic, facially amphipathic structures. *J. Am. Chem. Soc.* **124**, 7664–7665.
7. Patch, J. A., and Barron, A. E. (2003) Helical peptoid mimics of magainin-2 amide. *J. Am. Chem. Soc.* **125**, 12092–12093.
8. Seurynck, S. L., Patch, J. A., and Barron, A. E. (2005) Simple, helical peptoid analogs of lung surfactant protein B. *Chem. Biol.* **12**, 77–88.
9. Statz, A. R., Meagher, R. J., Barron, A. E., and Messersmith, P. B. (2005) New peptidomimetic polymers for antifouling surfaces. *J. Am. Chem. Soc.* **127**, 7972–7973.
10. Brogden, K. A. (2005) Antimicrobial peptides: pore formers or metabolic inhibitors in bacteria? *Nat. Rev. Microbiol.* **3**, 238–250.
11. Hale, J. D. F., and Hancock, R. E. W. (2007) Alternative mechanisms of action of cationic antimicrobial peptides on bacteria. *Expert Rev. Anti Infect. Ther.* **5**, 951–959.
12. Jelokhani-Niaraki, M., Prenner, E. J., Kay, C. M., McElhaney, R. N., and Hodges, R. S. (2002) Conformation and interaction of the cyclic cationic antimicrobial peptides in lipid bilayers. *J. Pept. Res.* **60**, 23–36.
13. Jin, Y., Mozsolits, H., Hammer, J., Zmuda, E., Zhu, F., Zhang, Y., Aguilar, M. I., and Blazyk, J. (2003) Influence of tryptophan on lipid binding of linear amphipathic cationic antimicrobial peptides. *Biochemistry* **42**, 9395–9405.
14. Matsuzaki, K., Murase, O., Tokuda, H., Funakoshi, S., Fujii, N., and Miyajima, K. (1994) Orientational and aggregational states of magainin 2 in phospholipid bilayers. *Biochemistry* **33**, 3342–3349.
15. Raghuraman, H., and Chattopadhyay, A. (2004) Interaction of melittin with membrane cholesterol: a fluorescence approach. *Biophys. J.* **87**, 2419–2432.
16. Tachi, T., Epand, R. F., Epand, R. M., and Matsuzaki, K. (2002) Position-dependent hydrophobicity of the antimicrobial magainin peptide affects the mode of peptide-lipid interactions and selective toxicity. *Biochemistry* **41**, 10723–10731.
17. Vogt, T. C. B., and Bechinger, B. (1999) The interactions of histidine-containing amphipathic helical peptide antibiotics with lipid bilayers. *J. Biol. Chem.* **274**, 29115–29121.
18. Friedrich, C. L., Moyles, D., Beveridge, T. J., and Hancock, R. E. W. (2000) Antibacterial action of structurally diverse cationic peptides on Gram-positive bacteria. *Antimicrob. Agents Chemother.* **44**, 2086–2092.
19. Friedrich, C. L., Rozek, A., Patrzykat, A., and Hancock, R. E. W. (2001) Structure and mechanism of action of an indolicidin peptide derivative with improved activity against Gram-positive bacteria. *J. Biol. Chem.* **276**, 24015–24022.
20. Sal-Man, N., Oren, Z., and Shai, Y. (2002) Preassembly of membrane-active peptides is an important factor in their selectivity toward target cells. *Biochemistry* **41**, 11921–11930.
21. Suzuki, H., Wang, Z.-Y., Yamakoshi, M., Kobayashi, M., and Nozawa, T. (2003) Probing the transmembrane potential of bacterial

cells by voltage-sensitive dyes. *Anal. Sci.* **19**, 1239–1242.

22. Toyomizu, M., Okamoto, K., Akiba, Y., Nakatsu, T., and Konishi, T. (2002) Anacardic acid-mediated changes in membrane potential and pH gradient across liposomal membranes. *Biochim. Biophys. Acta* **1558**, 54–62.
23. Wu, M., Maier, E., Benz, R., and Hancock, R. E. W. (1999) Mechanism of interaction of different classes of cationic antimicrobial peptides with planar bilayers and with the cytoplasmic membrane of *Escherichia coli*. *Biochemistry* **38**, 7235–7242.
24. Zhang, L., Dhillon, P., Yan, H., Farmer, S., and Hancock, R. E. W. (2000) Interactions of bacterial cationic peptide antibiotics with outer and cytoplasmic membranes of *Pseudomonas aeruginosa*. *Antimicrob. Agents Chemother.***44**, 3317–3321.
25. Zhang, L., Scott, M. G., Yan, H., Mayer, L. D., and Hancock, R. E. W. (2000) Interaction of polyphemusin I and structural analogs with bacterial membranes, lipopolysaccharide, and lipid monolayers. *Biochemistry* **39**, 14504–14514.
26. Zhu, W. L., Song, Y. M., Park, Y., Park, K. H., Yang, S.-T., Kim, J. I., Park, I.-S., Hahm, K. S., and Shin, S. Y. (2007) Substitution of the leucine zipper sequence in melittin with peptoid residues affects self-association, cell selectivity, and mode of action. *Biochim. Biophys. Acta* **1768**, 1506–1517.
27. Sainz, B., Jr., Rausch, J. M., Gallaher, W. R., Garry, R. F., and Wimley, W. C. (2005) Identification and characterization of the putative fusion peptide of the severe acute respiratory syndrome-associated coronavirus spike protein. *J. Virol.* **79**, 7195–7206.
28. Rausch, J. M., and Wimley, W. C. (2001) A high-throughput screen for identifying transmembrane pore-forming peptides. *Anal. Biochem.* **293**, 258–263.
29. Waggoner, A. (1976) Optical probes of membrane potential. *J. Membr. Biol.* **27**, 317–334.
30. Letellier, L., and Shechter, E. (1979) Cyanine dye as monitor of membrane potentials in *Escherichia coli* cells and membrane vesicles. *Eur. J. Biochem.* **102**, 441–447.
31. Ozkan, P., and Mutharasan, R. (2002) A rapid method for measuring the intracellular pH using BCECF-AM. *Biochim. Biophys. Acta* **1572**, 143–148.
32. Casadio, R., Di Bernardo, S., Fariselli, P., and Melandri, B. A. (1995) Characterization of 9-aminoacridine interaction with chromatophore membranes and modelling of the probe response to artificially induced transmembrane ΔpH values. *Biochim. Biophys. Acta* **1237**, 23–30.
33. Apell, H.-J., and Bersch, B. (1987) Oxonol VI as an optical indicator for membrane potentials in lipid vesicles. *Biochim. Biophys. Acta* **903**, 480–494.
34. Alberts, B., Johnson, A., Lewis, J., Raff, M., Roberts, K., and Walter, P. (2002) *Molecular biology of the cell.* Garland Science: New York.
35. Noskov, S. Y., and Roux, B. (2006) Ion selectivity in potassium channels. *Biophys. Chem.* **124**, 279–291.
36. Seelig, J. (1997) Titration calorimetry of lipid-peptide interactions. *Biochim. Biophys. Acta* **1331**, 103–116.
37. Wieprecht, T., Apostolov, O., and Seelig, J. (2000) Binding of the antibacterial peptide magainin 2 amide to small and large unilamellar vesicles. *Biophys. Chem.* **85**, 187–198.
38. Wieprecht, T., Beyermann, M., and Seelig, J. (1999) Binding of the antibacterial magainin peptides to electrically neutral membranes: thermodynamics and structure. *Biochemistry* **38**, 10377–10387.
39. Wieprecht, T., and Seelig, J. (2002) Isothermal titration calorimetry for studying interactions between peptides and lipid membranes. In *Peptide-Lipids Interactions.* S. A. Simon and T. J. McInotosh (Eds.), *Current Topics in Membranes*, Vol. 52, pp. 31–56. USA: Elsevier Science.
40. Hope, M. J., Bally, M. B., Webb, G., and Cullis, P. R. (1985) Production of large unilamellar vesicles by a rapid extrusion procedure. Characterization of size distribution, trapped volume and ability to maintain a membrane potential. *Biochim. Biophys. Acta* **812**, 55–65.
41. Bartlett, G. R. (1959) Phosphorus assay in column chromatography. *J. Biol. Chem.* **234**, 466–468.
42. Dodge, J. T., and Phillips, G. B. (1967) Composition of phospholipids and of phospholipid fatty acids and aldehydes in human red cells. *J. Lipid Res.* **8**, 667–675.
43. Kruijff, B. D., Killian, J. A., Rietveld, A. G., and Kusters, R. (1997) Phospholipid structure and *Escherichia coli* membranes. In *Lipid Polymorphism and Membrane Properties.* R. M. Epand (Ed.), *Current Topics in Membranes and Transport*, Vol. 44, pp. 477–515. London:Academic Press.

Chapter 13

Dynamic Transitions of Membrane-Active Peptides

Stephan L. Grage, Sergii Afonin, and Anne S. Ulrich

Abstract

Membrane-active peptides or protein segments play an important role in many biological processes at the cellular interface to the environment. They are involved, e.g., in cellular fusion or host defense, where they can cause not only merging but also the destabilization of cell membranes. Many factors determine how these typically amphipathic peptides interact with the lipid bilayer. For example, the peptide orientation in the membrane determines which parts of the peptide are exposed to the hydrophobic bilayer interior or to the polar lipid/water interface. As another example, oligomerization is required for many activities such as pore formation. Peptides have been often classified according to a single characteristic mode of interaction with the bilayer, but over the years a more versatile picture has emerged. It appears that any single peptide can adopt several different alignments and/or oligomeric states in response to changes in the environment. For instance, many antimicrobial peptides adopt a surface-parallel alignment at low concentration, but they tilt obliquely into or even fully insert transmembrane into the bilayer above a critical peptide-to-lipid ratio, often in the form of oligomeric pores. Similar changes in peptide orientation or oligomeric state have been observed as a function of, e.g., temperature, lipid composition, pH, or induced by a synergistic partner peptide. Such transitions between peptide states can be regarded as the result of a re-adjustment in the balance between peptide–peptide and peptide–lipid interactions, as the environment conditions are changed. Though often studied in model membrane systems, such rich variety of peptide states is even more likely to occur in native biomembranes with their diverse compositions and physicochemical properties. The ability to undergo transitions between different states thus plays a fundamental role for the biological activities of membrane-active peptides.

Key words: Antimicrobial peptides, biological membranes, peptide structure, peptide orientation tilt, re-alignment, barrel-stave pore, toroidal wormhole pore, solid-state nuclear magnetic resonance, oriented circular dichroism, synergy

1. Introduction

Biomembranes create a barrier between cells and their environment, providing control of and protection against any physical and chemical imbalance. They present a natural target for many

A. Giuliani, A.C. Rinaldi (eds.), *Antimicrobial Peptides*, Methods in Molecular Biology 618,
DOI 10.1007/978-1-60761-594-1_13, © Springer Science+Business Media, LLC 2010

kinds of membrane-active peptides, which can destabilize this protective shield and/or translocate into the cytosol. These short peptides typically possess a pronounced amphipathicity which facilitates their integration into the lipid bilayer, and they do not appear to require any receptor proteins for recognition. Several different classes of membrane-active peptides are distinguished mainly on the basis of their predominant biological function, and an intriguing question concerns the selectivity of their interaction with the specific lipid composition of the target membrane. The so-called antimicrobial or host defense peptides are particularly active against bacteria, leading to the death of the intruding microorganism. Antiviral and antifungal peptides may also be categorized as self-defense peptides, yet they are supposed to interact with non-prokaryotic membranes. Quite spectacular are anticancer peptides and some toxins that are lethal against specific mammalian (human) cells. Another related class of fusogenic peptides is capable of promoting fusion between lipid bilayers in vitro and in vivo. More recently, cell penetration has been linked to certain peptide sequences, coined cell-penetrating peptides or protein-transducing domains. Today, the boundaries between these different categories of membrane-active peptides are getting increasingly diffuse, as many peptides have been demonstrated to exhibit several different activities at the same time (1–4).

It is obvious that many of the diverse biological activities are based on similar physicochemical principles of peptide–lipid interactions. To get better insight into the underlying mechanisms of membrane recognition, perturbation, and bilayer translocation, numerous peptides have been extensively studied (5–8). Prominent examples are the α-helical antimicrobial peptides magainin 2 and PGLa from frog skin, the bee venom peptide melittin, and the fungal peptide alamethicin. All of these have been found to permeabilize membranes, albeit in different ways (*see* **Fig. 13.1**). In the "carpet model," the peptides are thought to disturb the integrity of the membrane by binding to its surface at a high concentration that changes the mechanical properties of the bilayer and causes a spontaneous local collapse. A "detergent-like" mechanism has also been proposed, wherein the amphiphilic peptides dissolve the lipid bilayer, yielding micellar or bicellar membrane patches. Finally, peptides can also increase permeability of the membrane by direct pore formation. Such pores can be lined by the peptide as in the "barrel-stave" model, or the peptides can stabilize a pore that is intrinsically formed by lipids according to the "wormhole" model.

The different models of peptide–lipid interactions have offered explanations for the activities of various peptides considered in the individual studies. An underlying assumption has often been that there should be a single mode of action in each case, and just one dominant mode of peptide–lipid interactions

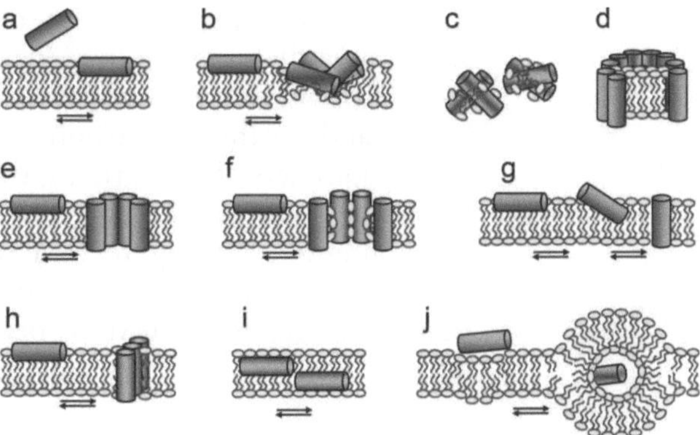

Fig. 13.1. Different modes of membrane interaction suggested for membrane-active peptides: (**a**) binding to the bilayer or water interface; (**b**) "carpet"-like interaction; (**c**) detergent-like action; (**d**) disintegration of the bilayer producing bicelle-like discs; (**e**) transient or long-lived "barrel-stave"; (**f**) "toroidal" or "wormhole" pores; (**g**) insertion into a transmembrane alignment; (**h**) formation of a "slit"; (**i**) diffusion across the membrane; (**j**) modulation of the membrane curvature, induction of non-bilayer lipid phases and/or translocation via an inverted micelle.

intrinsic to each system. The mechanism of action has thus been attributed to the particular state the peptide had been found to adopt. For example, the surface alignment of an antimicrobial peptide has been taken to reflect a carpet-like interaction with the membrane, whereas an upright transmembrane alignment implied a pore mechanism, and motionally averaged molecules would have to be part of a micelle. However, such concept of characteristic states representing any one type of mechanism might be misleading. Accumulating evidence suggests that most peptides can assume several different states, and the transitions between these states need to be considered to explain the activity of each particular peptide system. Antimicrobial peptides, e.g., have been found to change their tilt angle with respect to the membrane as a function of temperature (9, 10), as a function of the peptide-to-lipid ratio (11–15), as a function of lipid composition and bilayer hydration, etc. (16). More remarkably even, fusogenic peptides have been shown to change their membrane-bound conformation and mode of interaction in response to the peptide-to-lipid ratio (17–21) and as a function of pH (22–24). Also the fusion activity itself can be modulated by pH and ionic strength (25), and ionic strength influences the oligomerization and conformation of melittin in solution and modulates its bactericidal activities (26). Some cell-penetrating peptides have been found to aggregate above certain threshold concentrations, which leads to a modulation in the ratio of free peptide vs. aggregated one and influences the effective amount of transduction-competent species (27).

In this review we highlight the diverse behavior of membrane-active peptides undergoing dynamic transitions, in which they change between different states as a response to changes in the bilayer and/or their environment. In the first section several relevant parameters characterizing the different states of membrane-bound peptides will be presented. The second section addresses the conditions leading to structural transitions between different peptide states in model membranes. Finally, we discuss these transitions in the context of peptide–peptide and peptide–lipid interactions.

2. Membrane-Active Peptides in Different States

At first the question arises as to how the different states of a peptide in the membrane can be defined and characterized. One way is to use experimentally accessible parameters that reflect the mode of membrane activity of the peptide. Membrane interactions of peptides have been studied with an enormous range of different techniques. In the following we give a brief overview of the different parameters typically addressed.

2.1. Peptide Conformation

The physicochemical properties of peptides and many aspects of peptide–lipid or peptide–peptide interactions are determined by the three-dimensional structure of the peptide. For instance, a peptide is only able to display its (secondary) amphipathicity when in the right fold. Yet there seems to be no particular structure required for membrane activity. For example, antimicrobial effects can be exerted by α-helical peptides, such as magainin or melittin, but also by β-helical, β-stranded, or β-turn containing peptides, such as gramicidin A, gramicidin S, or protegrin 1, respectively, to name just a few (28). Cell-penetration activity may even be mediated by merely the presence of appropriately spaced guanidinium moieties, irrespective of the peptide (or peptidomimetic) backbone conformation (29). For one of the most well-studied fusogenic peptides, FP23 from HIV gp41, the question whether the structure of its active form is either α-helical, β-stranded, or something else is still under debate (30–33).

Linear peptides such as antimicrobial sequences with typically 10–30 amino acids tend to be highly flexible in aqueous solution, as they often lack stabilizing intramolecular interactions. The structure thus depends strongly on the dielectric properties of the environment, such as a solvent or a lipid bilayer. Therefore, it is not surprising that antimicrobial peptides often adopt their functionally relevant conformation only when they are embedded in the membrane. Hence, a defined secondary structure often indicates binding of the peptide and can thus reflect the adoption of a membrane-active state.

As the environment of the peptide is a crucial determinant of its conformation, reconstitution in a lipid bilayer or a membrane-mimicking environment is a prerequisite for obtaining a functionally relevant structure, which raises the question as to the relevance of X-ray crystallization results. Circular dichroism is well compatible with studies of membrane-active peptides and represents the most useful routine tool to evaluate qualitative secondary structures. For a full three-dimensional structure analysis in a lipid bilayer or a membrane-mimicking environment, NMR spectroscopy plays an essential role. For soluble proteins of moderate size, solution state NMR represents an established route to obtain structures at atomic resolution. This technique is applicable to membrane-associated peptides, provided a suitable membrane mimic can be found. Often organic solvents such as trifluoroethanol (TFE), ethanol, methanol, chloroform, acetonitrile, hexafluoroisopropanol, hexafluoroacetone, or DMSO, as well as their mixtures with water are used to provide a simple model of the hydrophobic membrane environment. However, in particular the use of TFE with α-helical peptides can be problematic, as this solvent enforces the formation of intramolecular hydrogen bonds; hence any observed helix or turn should be interpreted as a propensity for such conformation that is intrinsic to the peptide. DMSO as a solvent, on the other hand, is rather prone to disrupt secondary structures of peptides (since S=O is a stronger hydrogen bond acceptor than C=O). Hence it is most useful to study conformations that are stabilized by very strong intramolecular hydrogen bonds, e.g., in cyclic peptides. Even for such stable folds as that of the cyclic gramicidin S (two to four hydrogen bonds predicted), a slight but significant solvent dependence of the conformation has been observed (34–36). Gramicidin A, representing the first antimicrobial peptide whose three-dimensional membrane-bound structure has been solved, was studied at first using solution NMR, where a severe conformational heterogeneity depending on the membrane-mimetic solvent was observed (37–39). The conformation of gramicidin A dissolved in DMSO, in particular, turned out to be fundamentally different from its structure in flat bilayer membranes, as determined by solid-state NMR (*see* below). Instead of a head-to-head dimer of single-stranded right-handed β-helices, an intertwined left-handed double-helical structure was found, indicating that the environment has a severe impact on the fold and self-assembly.

Instead of organic solvents, detergent micelles or isotropic bicelles provide an environment significantly closer to a lipid bilayer, while still allowing for the motional averaging necessary for solution state NMR. Numerous structures of membrane-active peptides have been determined this way (40–42). Though mimicking the hydrophobic interior of the membrane as well as the polar/apolar interface, micelles are not able to represent correctly the curvature of a membrane, possibly leading to deformed conformations for surfacially bound peptides (43, 44). To avoid such artifacts, isotropic bicelles, composed of a mixture of a detergent-like short-chain lipid and a long-chain lipid, have been used as an improved substitute (45).

Ultimately, one aims at obtaining the peptide structure in a planar lipid bilayer (41, 46). However, the molecular weight of extended membrane systems exceeds by far the limit imposed by the need for fast isotropic tumbling, which is the basis of the high resolution achieved in solution state NMR (47). As an alternative, solid-state NMR has been employed in a number of cases to successfully obtain backbone structures of membrane-active peptides at atomic resolution. The "static" NMR signals of membrane-bound peptides exhibit a pronounced orientation dependence, which gives rise to characteristic line broadening but as a consequence leads to deleterious signal overlap. Two different strategies have been pursued to cope with this strong anisotropy of the NMR interactions (48, 49). Using magic angle spinning, reasonably rapid isotropic motion can be artificially introduced, and high resolution is achieved on the same grounds as in solution state NMR. As an alternative, and at least equally successfully, strategies have been developed to exploit rather than suppress the orientation dependence in uniaxially aligned membrane samples. The first solid-state NMR peptide structure, that of gramicidin A, has been acquired using this approach (50). The β-helix backbone structure and the tryptophan side chain conformations for this peptide have been assembled from orientation-dependent NMR signals of selective ^{15}N-, ^{13}C-, and ^{2}H-labels.

2.2. Alignment of Peptides in the Membrane

What defines the mode of membrane activity is often not only the three-dimensional structure as such, but also how the molecule is positioned with respect to the lipid bilayer. While the backbone conformation defines the amphipathic nature of the peptide, it is the molecular alignment in the membrane which determines whether the different faces of the folded molecule are exposed to the hydrophobic interior of the bilayer, to the apolar/polar interface, or to the aqueous subphase. The actual interaction of the peptide with the lipids and/or with other peptides therefore depends critically on its orientation with respect to the bilayer. As the interactions with the membrane are the underlying cause for membrane activity, the peptide alignment within the membrane is also indicative of the mechanistic mode of action. For many peptides, where amphipathicity divides the molecule into a hydrophilic and hydrophobic face (typically the apolar/polar dividing line may run along the long molecular axis in amphipathic α-helices and β-strands or along the backbone ring in "sided" cyclic peptides) it is the hydrophobic moment vector that characterizes the alignment (51, 52). For the most thoroughly studied cases of amphipathic α-helices, the hydrophobic moment has been found to orient either (roughly) parallel, to be tilted, or to be aligned nearly perpendicular to the membrane normal. Hence the alignment of the entire helix, seen as a conformationally rigid body, can be classified in terms of three distinct states. In the surface-aligned or "S"-state, the polar side of the peptide faces the aqueous phase, while the hydrophobic side points toward the core of the bilayer. In the tilted or "T"-state, the peptide is obliquely immersed in the bilayer with a slanted orientation. Finally, in the inserted or "I"-state, the peptide helix is fully inserted with its apolar/polar dividing line perpendicular to the membrane surface; hence both faces of the amphipathic peptide are positioned within the bilayer slab.

Depending on its orientational state, different faces of the peptide are exposed to the interior of the bilayer or the aqueous subphase. In the S-state the peptide adapts optimally to the apolar/polar interface of the bilayer. As this state represents the natural alignment of any amphipathic membrane-bound peptide, it is the expected and indeed most frequently encountered alignment. This state allows extensive interactions of the peptide with the interface region of the membrane. Such mode of binding may lead to a destabilization of the bilayer if its physicochemical properties such as spontaneous curvature, molecular packing of certain lipids, or lateral stress profile are changed. This way, antimicrobial peptides can increase the membrane permeability in accordance with the "carpet model" (**Fig. 13.1**), and fusogenic peptides can promote curvature-driven stages of the bilayer fusion event. In the I-state, on the other hand, the polar face of the amphiphatic peptide is immersed in the hydrophobic interior of the membrane. A single molecule would not be stable in this alignment. However, by interacting with other peptides to form an oligomeric pore, with the hydrophilic face pointing inside, a disadvantageous exposure of the polar face to the hydrophobic lipid chains can be avoided. Antimicrobial peptides assume this orientational state when acting via the "barrel-stave" or "wormhole" mechanism. Likewise, amphiphilic α-helical cell-penetrating peptides have to adopt an inserted state at least in a transient fashion to be able to cross the membrane.

Knowledge of the peptide orientation in the lipid bilayer thus gives crucial clues about its mode of membrane activity. The molecular alignment can be specified by two angles, the tilt angle τ and the azimuthal rotation ρ (*see* **Fig. 13.2**). It has been recently demonstrated that a distribution of tilt and rotation angles is present rather than a single alignment. It thus depends on the timescale of the experimental method, whether an averaged effective orientation or the distribution is accessible.

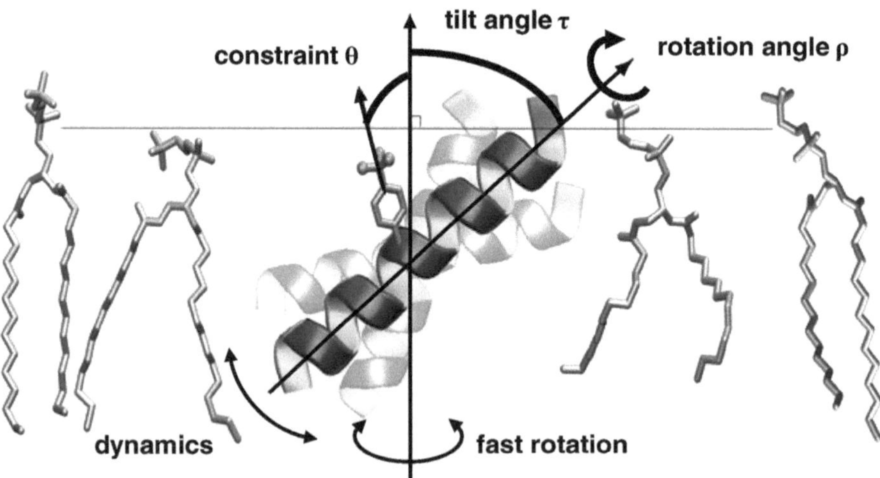

Fig. 13.2. A number of orientational constraints (i.e., a set of local angles θ) have to be measured by solid-state NMR to calculate the peptide conformation (here: helical), its alignment (described by the tilt angle τ and the azimuthal rotation ρ), and its dynamic behavior in the lipid membrane.

Experimentally, the peptide orientation can be obtained either indirectly from the penetration depth of a set of probes introduced into the peptide or directly by using orientation-dependent spectroscopic techniques. Depth measurements have been used in several recent fluorescence studies to address the inclination of a peptide (53–55). Exploiting the quenching of NMR probes by paramagnetic ions in the aqueous or membrane phase, it has also been possible to determine peptide orientation by solid-state NMR (56–58). To measure the alignment state directly, the anisotropic parameters accessible by many spectroscopic techniques can be exploited. Most of these rely on a uniform orientation of the studied peptide in a macroscopically aligned sample. Membranes can be aligned mechanically on glass plates or on other (inert polymeric) solid supports. Alternatively, bicelles formed by a mixture of a short-chain and a long-chain lipid provide a way to orient bilayers magnetically for anisotropic solid-state NMR studies. Optical techniques such as infrared spectroscopy (IR) or circular/linear dichroism on peptides in oriented bilayers provide a complementary way to qualitatively assess the peptide orientation. Particularly for α-helical antimicrobial peptides, oriented circular dichroism (OCD) has been a useful tool in determining their alignment state. The absorption of the CD band around 208 nm depends on the alignment between the electrical field component and the transition dipole moments associated with the peptide bond. An intense OCD signal is obtained if the α-helix is aligned perpendicular to the incident light, whereas the band practically disappears when the peptide long axis lies in the direction of the light propagator.

While the theoretical complexity of OCD or polarized attenuated total reflection infrared spectroscopy (ATR-IR) limits the orientational analysis to rather qualitative conclusions, solid-state NMR with its simple and direct orientation dependencies can provide precise angular information even in atomic detail. To this aim, the anisotropic chemical shift or dipolar couplings of a set of isotope labels in the peptide are measured, and they are combined to yield the τ and ρ angles of the peptide. Several labeling schemes and associated experimental approaches have been applied. The backbone orientation of a given peptide can be directly addressed using ^{15}N or ^{13}C. In particular for uniformly ^{15}N-labeled α-helical peptides, the correlation of ^{15}N chemical shift anisotropy and ^{15}N–^{1}H dipolar coupling in the PISEMA or related experiments has been very successful. The α-helical pattern of the peptide is mapped directly onto a regular wheel-like pattern in the two-dimensional NMR spectrum, whose location in the spectrum yields the value of τ. The position of a single, selectively ^{15}N-labeled residue along the rim of the wheel, on the other hand, can reveal the azimuthal angle ρ of the peptide helix. Side chain labels, too, can be used to determine the peptide alignment, provided that the NMR label on the side chain is rigidly linked to the backbone. Using, for example, $^{2}H_3$-alanine in the GALA approach, a number of peptide tilts have been determined (59, 60). More recently, some highly sensitive ^{19}F labels have also been employed in alignment studies of membrane-active peptides. Special amino acids were designed to provide a direct connection of the ^{19}F label with the peptide backbone, without altering their biological activity in most cases (61–65). The exquisite sensitivity has allowed systematic screening work to study the influence of various conditions on the peptide alignment behavior, rendering solid-state ^{19}F NMR a highly important tool for the discovery of transitions between peptide states.

3. Transitions Between Peptide States

Membrane-active peptides were originally thought to be associated with a specific biological function, as the result of the particular physicochemical properties of the peptide. However, numerous examples in which a peptide has been found to adjust to changes in the environment have become apparent over recent years. Hence it is clear that the behavior of peptides is not only governed by their intrinsic physical properties, but also strongly influenced by their interactions with the target membrane.

As outlined in the first section, the mode of interaction of a peptide with membrane lipid bilayer is reflected in several observable parameters, such as the conformation, the molecular alignment in the membrane, or the oligomeric state. Many different states have been described, such as the S-state, T-state, I-state, monomeric, dimeric, oligomeric, barrel-stave, toroidal wormhole. Notable, a single peptide is likely to be able to adopt several of these states, which are not static but can interchange depending on various conditions. In the following we describe some of the most relevant changes in the environmental parameters that have been found to lead to transitions between states.

3.1. Peptide-to-Lipid Ratio

Perhaps the strongest influence on the state of a peptide is exerted by the peptide-to-lipid ratio (P/L), i.e., the peptide concentration in the membrane. Already in the 1990s, Huang et al. have suggested the presence of a critical concentration, separating two states with different alignments in the membrane (13). Many amphipathic α-helical peptides have been found since to change from a surface-bound S-state at low concentration to a tilted or inserted state at high concentration. For example, the antimicrobial PGLa adopts an angle of $\tau \sim 98°$ (between the helix axis and the membrane normal), i.e., a helix alignment nearly parallel to the bilayer surface, below a threshold of approximately P/L = 1/50 to 1/100 (in DMPC) (11, 15, 16). Above this concentration, the tilt angle changes to $\tau \sim 126°$, and the peptide adopts an oblique T-state orientation (*see* **Fig. 13.3**). A similar behavior was found for other peptides with the same architecture, for example, MSI-103 and MAP. At low concentration, amphipathic α-helical peptides thus bind to the membrane adopting their alignment to the apolar/polar interface. If the peptide concentration exceeds a critical threshold, neighboring peptides will play a role in determining the favorable orientation. At high concentration the peptides then interact not only with the lipid bilayer alone, but also among themselves. A plausible model suggests that they may form

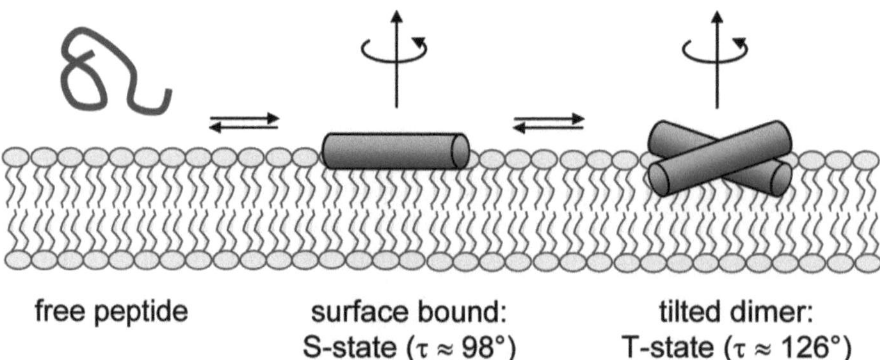

Fig. 13.3. Solid-state ^{19}F- and ^{2}H-NMR revealed the binding, folding, and dimerization of the antimicrobial peptide PGLa, and this was also observed for the related analogues MSI-103 and MAP (11, 12, 15, 16).

dimers, as this would explain the stability of the otherwise less favorable tilted alignment of the amphiphile (11, 16).

While increasing concentration of α-helical antimicrobial peptides leads to a transition from an S-state to a T-state, a complete I-state insertion was observed for cyclic β-sheet/β-turn peptides such as gramicidin S and protegrin B (66, 67). For gramicidin S an insertion in an upright position in the membrane was observed for a peptide-to-lipid ratio above approximately 1/80 (in DMPC). Interestingly, this transition into the I-state was also found to be promoted at temperatures near the lipid phase transition, thus indicating a more complex dependence on different environmental conditions. To account for the immersion of the hydrophilic face of gramicidin S into the hydrophobic interior of the membrane in the I-state, self-assembly as a β-barrel pore was suggested as a stable H-bonded structure (66, 68) (*see* **Fig. 13.4**). A similar equilibrium between a monomeric S-state at low concentration and an insertion to form an oligomeric pore at high concentration was found for alamethicin (14). However, this transition exhibits also a rather different dependence on the lipid phase state (*see* below).

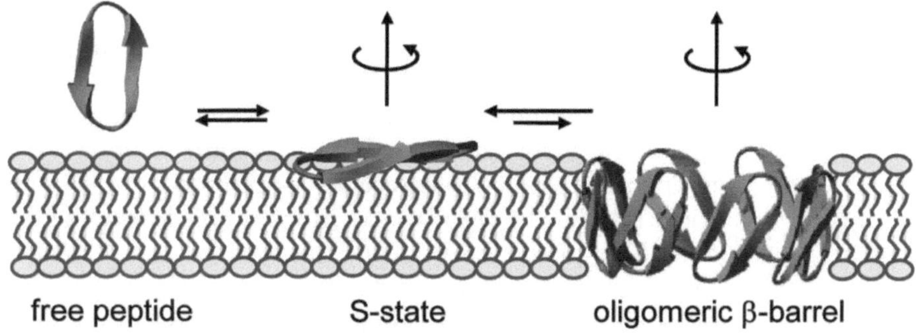

Fig. 13.4. Formation of a β-barrel pore by gramicidin S, a cyclic amphiphilic bacterial peptide.

Very large peptide assemblies in membranes have been visualized by atomic force microscopy (AFM), revealing extended striated domains over several micrometers for gramicidin A or the model peptide WALP reconstituted in solid-supported bilayers (69, 70, 82–84). Though arguably the solid support may influence the physical properties of the bilayer, peptide–peptide, and peptide–lipid interactions have comparable strengths, allowing for a broad range of oligomer sizes and supermolecular architectures in such peptide/lipid systems.

Another type of aggregates frequently encountered with many peptides are fibrillar self-assemblies. Many diseases such as Alzheimer are connected to proteins or peptides assuming a pathogenic, supramolecular form, which often possesses a cross-β-sheet secondary structure. Some of these proteins are themselves postulated to have a membrane-related activity. For example, the Alzheimer Aβ peptide is speculated to form ion channels (71, 72), for which a crowding model has been recently suggested (73). A similar behavior has been suggested for further peptides related to neurodegenerative diseases, such as α-synuclein, and might be a more general phenomenon (74, 75). Following binding to the membrane and bilayer-induced conformational changes, an increase in peptide concentration here leads to the formation of fibrillar aggregates at the bilayer surface. Quite common are such observations where typical α-helical peptides were unexpectedly found to form amyloid fibrils under specific conditions. For example, the fusogenic protein segment B18 from the sea urchin fertilization protein bindin assembles as fibrils (17, 18, 20), and recently also the model amphipathic peptide MAP was encountered in a fibrillar form (76, 77). Only when aggregation is suppressed in MAP by inclusion of a rigid D-amino acid does it exhibit the typical concentration-dependent transition from a monomeric S- to a potentially dimeric T-state, similar to other peptides with the same architecture such as PGLa or MSI-103. It appears that the aggregation state plays a functionally relevant role in the use of MAP and other sequences as cell-penetration peptides (27).

3.2. Temperature Dependence

Concentration was long believed to be the dominant factor determining the state of membrane-active peptides. However, already in the early studies of alamethicin, Vogel, Huang and coworkers found a change in molecular alignment upon varying the temperature, which was recently confirmed by solid-state NMR and EPR (10, 14, 78, 79). At temperatures below the gel/liquid crystalline phase transition of the lipid bilayer, alamethicin was found in the S-state, whereas insertion of this rather hydrophobic sequence occurred at higher temperatures in the fluid phase. More recently, a different type of temperature-dependent re-alignment was encountered for the cyclic gramicidin S and for the α-helical amphipathic peptide PGLa (9, 66). As discussed above, gramicidin S adopts either a monomeric S-state or an inserted oligomeric state in the bilayer, the latter being favored at high peptide concentration. The transition into the I-state, however, was also found to depend on temperature, as it was only observed at temperatures near the gel/lipid phase transition of the bilayer. Furthermore, the temperature range over which the I-state is observed gets significantly wider with increasing peptide concentration. As the lipid phase transition also broadens with higher peptide concentration, the appearance of the I-state of gramicidin S thus seems to be connected to the lipid phase transition (and to the lipid chain length, *see* below). In the case of PGLa, a multi-stage re-alignment has been comprehensively analyzed by ^{2}H and ^{19}F NMR. A concentration-dependent transition from S- to T-state had been described and attributed to the formation of a putative homodimers at high concentration. Complete insertion into an I-state, however, could not be achieved by increasing the PGLa concentration further. Only by adding magainin 2, its synergistic partner peptide, could an I-state be achieved under ambient conditions (*see* below) (16, 80). However, as in the case of gramicidin S, also the temperature was found to be able to trigger the transitions between the S, T, and even the I-state (9) (*see* **Fig. 13.5**). Thus far only observed in the presence of magainin 2, upon lowering the temperature below the lipid phase transition PGLa was found to fully insert on its own in an I-state alignment. While this transition is related to the lipid phase state (acyl chain melting of DMPC at ~24°C), a second temperature-dependent transition was observed at higher temperatures. Above about 45°C, PGLa no longer retains its T-state alignment at high concentration, but instead reverts to the monomeric S-state. This transition is not related to the lipid phase state, but might be attributed to a disintegration of the PGLa dimer.

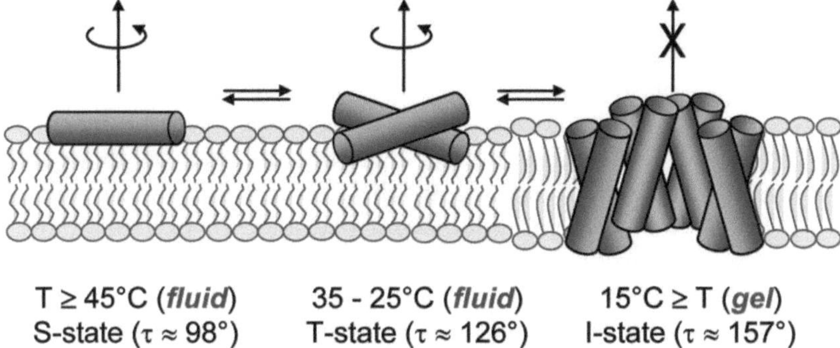

Fig. 13.5. Temperature-dependent re-alignment of PGLa from S-state to T-state in fluid lipid bilayers, and its insertion as an oligomeric transmembrane pore in the gel phase.

Most temperature-dependent transitions of peptide states seem to be related to the melting transition of the lipid chains. Here the change in temperature leads to a different lipid environment, with significantly different physical properties. Many peptide transitions thus can be regarded as an adaptation to the respective membrane in the gel or fluid phase. As will be outlined below, a lipid bilayer in its different states possesses different thicknesses and molecular dynamics, leading to different lipid–lipid and lipid–peptide interactions depending on the bilayer phase state. Hence the balance of the interactions between the different molecules can be shifted, leading to changes in the oligomeric state or the insertion of the peptide upon going from one lipid phase to another. Some peptide states, however, are not associated with either the gel or the fluid phase of the membrane, but they exist only in the vicinity of the lipid phase transition temperature itself (as in the case of gramicidin S). These states appear to be related to local defects in the bilayer or to fluctuations between solid–fluid phase boundaries occurring at the lipid phase transition. Yet the PGLa example demonstrates that there also exist peptidic transitions disconnected from the lipid phase behavior, which are governed solely by peptide–peptide interactions.

3.3. Lipid Composition

Besides peptide–peptide interactions, peptide–lipid interactions are the determinants for the peptide behavior in the membrane environment. Both the lipid phase and the intrinsic properties of the lipids will exert an important influence on the peptide state. Such physicochemical aspects concern the hydrophobic thickness, the degree of chain saturation, headgroup type and charge, and the spontaneous curvature of the lipid bilayer. In biological membranes with their very diverse and complex lipid composition, particular properties can lead to distinctly different peptide activity. For example, cationic antimicrobial peptides are known to target the plasma membrane of the invading bacteria, but leave the host cell membranes intact. This selectivity is attributed to the fact that bacterial membranes tend to be negatively charged, whereas the plasma membrane of eukaryotic cells contains mainly zwitterionic lipids in its outer leaflet as well as a high amount of cholesterol.

The membrane thickness is one of the parameters discussed intensively in the literature as determining the behavior of membrane-bound peptides. In particular, the concept of hydrophobic mismatch suggests that (predominantly hydrophobic) peptides tend to adapt their hydrophobic region to the thickness of the hydrophobic region of the bilayer. This concept was explored using the model peptide WALP and analogues thereof (81). Different possibilities were discussed controversially, namely that the peptide could adapt to a thinner membrane by assuming a tilted orientation, or that instead the annular lipids could adjust to the peptide by stretching their chains and changing the local bilayer curvature. Evidence for both scenarios has been found. For example, AFM studies of WALP and gramicidin A suggest an adaptation of the lipids to the peptide (69, 70, 82–84). On the other hand, tilt angle measurements using solid-state NMR can be interpreted as an adaptation of the peptide orientation to match the bilayer thickness, provided that mobility is considered in the data analysis (85, 86). While hydrophobic mismatch in the case of the WALP leads at most to a change in tilt angle, in the case of alamethicin the above-mentioned studies (78, 79) suggest that this peptide can no longer remain inserted in a membrane of inappropriate thickness and gets expelled (*see* **Fig. 13.6**). Besides the alignment in the membrane, also the mobility of a peptide can be influenced by the membrane thickness, as, for example, observed for $(LA)_n$ model peptides in a solid-state NMR study by Sharpe et al. (87). Here, peptide–peptide interactions were

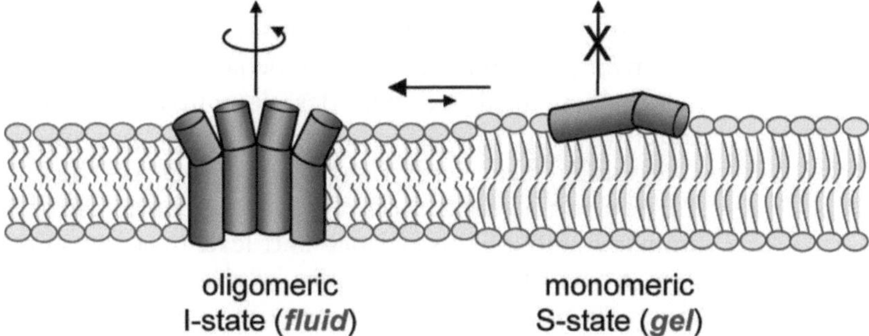

Fig. 13.6. In fluid membranes alamethicin forms a transmembrane barrel-stave type of pore, but it gets expelled from the bilayer at temperatures below the lipid chain melting transition, presumably due to hydrophobic mismatch.

modulated by the hydrophobic mismatch, and as a consequence, a different degree of mobility was observed and interpreted in terms of different oligomeric states.

Despite their small dimensions, which are usually not sufficient to span the lipid bilayer, even cyclic β-sheet peptides have been found to exhibit a dependence on the bilayer thickness. For example, retrocyclin-2 assumes a tilted or surface alignment, depending on whether it is reconstituted in DLPC or POPC bilayers, respectively (88). As both lipids are in the fluid phase, the change in peptide orientation is clearly related to an adaptation to the different thickness of these bilayers. Also gramicidin S was found to exhibit a dependence on the bilayer thickness (89). As discussed already above, the usually surfacially aligned peptide inserts into the membrane close to the phase transition temperature. The temperature range for the I-state not only increases with increasing peptide concentration, but also changes with bilayer thickness. In membranes with increasing thickness, from DLPC to DPPC, the inserted state was found over a decreasing temperature range. Similar to alamethicin, these cyclic β-sheet peptides were thus also found to insert into the thinner membrane, whereas the mismatch in the thicker bilayer rendered insertion less favorable.

While in most cases the lipid chains are responsible for the alignment or oligomeric state of the peptide, the lipid headgroups play an important role in the initial interaction with membrane-active peptides. In particular, the binding of peptides to the membrane is often mediated through electrostatic interactions, and the presence of charged residues influences this initial step of membrane activity. As many antimicrobial peptides are positively charged, negatively charged lipids have thus often found to enable or enhance peptide binding. For example, in aqueous solution PGLa has virtually no affinity to zwitterionic vesicles, but is stimulated in the presence of anionic lipids (16).

3.4. Dependence on pH and Hydration

The amphipathic nature of membrane-active peptides is one of the main determinants of their behavior in the bilayer environment. As seen above, peptides align in membranes such as to match their hydrophobic faces with the bilayer interior, and the hydrophilic faces with the polar regions of the environment. Introducing a charged residue into a peptide can thus lead to a dramatic shift of this balance and lead to a transition from one state to another. In a number of examples, a pH change has been able to trigger such change of the total peptide charge. The fusion peptide of influenza hemagglutinin, for example, has been found to adopt a boomerang-like conformation with a different kink angle at acidic or neutral pH, allowing a deeper insertion at acidic pH and promoting fusion (90). Using histidine-containing LAH_n model peptides, it has also been demonstrated that pH can be used to alter their orientation in the bilayer, by switching a hydrophobic helix into an amphipathic one upon histidine protonation (91).

Hydration has been one of the first parameters found to influence the alignment state of antimicrobial peptides. In studies of alamethicin or melittin in oriented bilayer stacks using OCD, a re-alignment of these helices was observed upon varying the level of sample hydration (92–94). Here, dehydration was found to lower the critical peptide-to-lipid ratio, at which peptide insertion into the bilayer takes place. A fully hydrated interface region of the bilayer thus seems to be a prerequisite for a surface orientation, embedding the peptide according to its amphipathic nature into the polar/apolar interface region. A reduction of the water content then renders an insertion more favorable (16).

3.5. Interactions with Different Peptides

A rare but very interesting change in the alignment state of a peptide can be induced by the specific interaction with another peptide. This situation differs from the mere concentration-dependent effect of increasing the number of membrane-bound molecules. For example, magainin 2 and PGLa are known as a synergistic pair of antimicrobial peptides that occur in the same glands of the frog *Xenopus laevis.* Both of them have been individually shown to undergo a transition from an S-state to a T-state above a critical peptide/lipid ratio, which has been attributed to the formation of homodimers. Interestingly, when combined with magainin 2, PGLa starts to tilt much earlier and it is able to insert fully into the membrane adopting an I-state (16, 80). Correspondingly, in antimicrobial assays a synergistic increase of the activity was observed at a 1/1 molar ratio of PGLa/magainin 2 (95). A likely explanation lies in the formation of synergistically active heterodimers, which have been observed by solution state NMR in detergent micelles (96) (*see* **Fig. 13.7**). Other synergistic pairs of antimicrobial peptides have been described, for example,

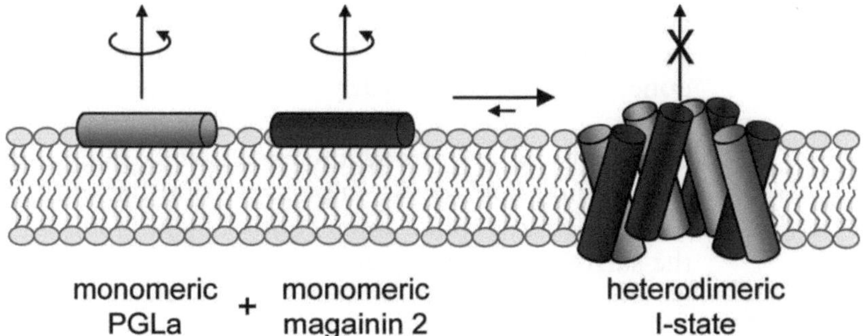

Fig. 13.7. Synergistic formation of a stable heterodimeric transmembrane pore by PGLa and magainin 2.

among temporins, also extracted from frog skins. Combinations of either temporin A or B with temporin L have been found to lead to an increase in antimicrobial activity against Gram-negative bacteria (97). Even though the structural basis is not yet resolved, specific interactions between the partner peptides appear to enable their penetration through the outer bacterial membrane, which in the case of Gram-negative bacteria is made of both lipids and lipopolysaccharide.

4. Transitions: Tipping the Balance

So far we have seen that any single membrane-active peptide can display a variable behavior in the way it is folded, and especially how it is spatially embedded and arranged in the membrane. Transitions between different states can be reversible (e.g., between S-, T-, I-states) or irreversible (conversion of an α-helical peptide into β-amyloid). The different states are usually characterized in separate samples when prepared, e.g., with different P/L ratios, or they are monitored under different conditions, e.g., within a series of experiments at different temperatures. When two states are simultaneously present, it depends on the timescale of interchange and on the experimental method, whether separate populations are detected or a dynamic exchange process leads to time-averaged signals. Both situations have been encountered in solid-state NMR studies of PGLa (9, 11, 12, 16, 80, 85, 86). In view of these intrinsically diverse and dynamic transitions of structures and alignments, which happen in response to changes in the environment, it is clear that the notion of a single mode of membrane interaction may not always be a suitable picture to explain the biological function of a peptide.

In the concluding section we now want to address the possible driving forces for transitions between different peptide states.

Clearly, the system consisting of peptides and lipid has to be treated as a whole, as the mutual interactions between the components are of equal order of magnitude. Nonetheless, it may help to distinguish several scenarios. The behavior of a peptide in the membrane will on the one hand involve interactions with the surrounding lipids as an important contribution. On the other hand, peptide–peptide interactions can be dominant, such that the surrounding lipids play only a minor role. Finally, it may also be conceivable that lipid–lipid interactions are the major player. The important aspect as to what triggers a transition is not necessarily the strength of the interactions involved, but the degree of change when varying the parameter that causes the transition.

Considering the initial attraction of a peptide to the membrane surface, electrostatic interactions between the polar lipid headgroups and charged moieties of the peptide are mainly responsible. Hydrophobic interactions then play a role once the peptide is (partially) immersed in the hydrophobic core of the membrane. Important for the tendency to insert more deeply into the membrane are also the elastic properties of the bilayer, modulated, e.g., by the spontaneous curvature of the lipids, and the spatial conformation of the peptide.

A theoretical framework to explain the distribution over several peptide states has been established by Heimburg and coworkers (98), expanding on theories of Minton and Chatelier and others (99, 100). To describe the behavior of a membrane-bound peptide, they evaluated the adsorption of a peptide to the bilayer, assuming that the peptide can bind in different states of insertion and alignment to the membrane. One possibility is adsorption to the surface, i.e., in an S-state orientation, without inserting into the bilayer. A second possibility is binding to the membrane in an inserted state. Both modes of binding are characterized by different free energies and different requirements of surface area to accommodate the peptide. Surfacially bound peptides were treated as a two-dimensional gas with a finite area requirement for each particle. When inserted, peptide–lipid interactions lead to an "edge tension," describing the energetic cost (or gain) of insertion. The lipid area requirement of an inserted peptide, on the other hand, is zero. Using this model, Heimburg and others were able to predict preferential binding in the S-state at low peptide concentration, whereas at high concentration the inserted state prevails. At low concentration, the energetic costs of insertion prevent the peptide from insertion, hence it remains surfacially bound. With an increasing number of adsorbed peptides, the pressure or chemical potential in the two-dimensional surface gas increases and insertion becomes more favorable. Furthermore, inserted peptides present obstacles for S-state binding, decreasing the area available for potential binding sites for S-state peptides, thereby inhibiting any further S-state binding.

This way, the experimentally observed concentration-dependent change from S-state to an inserted state above a critical peptide to lipid ratio could be well reproduced, as predicted, e.g., by Huang and coworkers (13, 14, 92, 94, 101). Moreover, a pronounced, sharp threshold concentration was obtained when considering oligomeric pores as the inserted state. In this case, higher cooperativity upon insertion, a lower distributional entropy, and a lower area lead to such sharp transitions. Altogether, the pressure of the two-dimensional surface gas formed by the surfacially bound peptides, which would drive insertion, is balanced by the lipid–peptide interactions occurring in the inserted state and resisting insertion. This model is thus able to explain not only concentration-dependent transitions, but also transitions occurring as a consequence of changes in the lipid environment. A temperature variation leading to a change in the lipid phase state, or a different hydrophobic thickness of the membrane leads to a shift of the lipid–peptide interactions, and can therefore shift the concentration threshold to one side or the other. Vice versa, altering the charge distribution of a peptide through, e.g., a pH change will also result in an change of the lipid–peptide interactions.

Also Huang has developed a model to explain the transition from a surface-bound state to an inserted state on the basis of the different free energies of these two states (101–103). Similar to the approach of Heimburg above, he assumes peptide binding to the water/membrane interface in two forms, either in the S-state or in a monodisperse oligomeric inserted state. Again, a two-dimensional gas-like behavior is assumed for the S-state. However, differently from Heimburg, in the model of Huang an increase of the membrane area induced by the S-state peptide, and the resulting elastic energy of the membrane is identified as the major contribution to the energy balance of the system. Because of the attractive binding force to the membrane surface, an increase in the interface area between peptide and bilayer surface is energetically favorable. S-state bound peptide thus leads to a stretching of the bilayer, and hence an increase in the elastic energy contribution acting then against further binding in the S-state. If a critical concentration is exceeded, the balance between elastic forces in the membrane and forces due to the high concentration at the surface is shifted, and the inserted state becomes preferred reducing the elastic energy contribution. Experimental evidence for the model of Huang arises from measurements of the membrane thickness (104). Due to the low compressibility of the membrane, the peptide-induced area increase of the bilayer leads to a thinning of the membrane, which indeed has been observed experimentally in a number of cases.

Some experimentally observed transitions between peptide states, however, are not covered by the picture of a lipid–peptide interaction-based binding to the membrane. For example, the homo- and heterodimers proposed for PGLa and magainin 2 (on their own and as a pair, respectively), must involve also some specific peptide–peptide interactions and/or entropic contributions of the peptide. Also the temperature-dependent transitions manifested for PGLa are not fully explained by alterations in the lipid–peptide interactions. In the first transition from I-state to T-state near the lipid chain melting temperature, the peptide molecules essentially follow the lipid behavior, but a second transition from T-state to an S-state was observed at higher temperature in fluid membranes. The latter might be ascribed to the disintegration of the putative peptide dimers, pointing again to some peptide-originating contributions.

Even lipid–lipid interactions can under particular circumstances be the driving force for peptides to change their state. In particular at the lipid phase transition many different kinds of peptides have been observed to change their alignment state. For example, gramicidin S is found to insert at temperatures near the lipid phase transition temperature, and high peptide concentrations are able to widen the temperature range of the inserted state to the same extent as they broaden the lipid phase transition itself. A possible explanation are the fluctuations inherent to a phase transition, which lead to local defects in the membrane, which in turn may facilitate peptide penetration.

Finally, one might ask critically as to what extent the findings obtained mostly from model membrane systems are able to explain events in actual biological membranes. Could transitions between peptide states be a purely academic exercise, occurring only in synthetic lipid bilayers with their discrete phase transitions and limited compositions? How relevant are the peptide transitions observed at the lipid chain melting temperature, or those peptide states that are found to be stable only in the gel phase? Here, it should be pointed out that the different scenarios encountered in model systems are also found in biomembranes. For example, microdomains possess a more gel-like phase state, and even though a complex membrane has no clearly defined phase transition temperature it may yet be considered as existing within a lipid phase transition over a broad range of ambient temperatures. Hence it is very likely that the diverse behavior of peptides encountered in model systems will also be found in biological systems. Furthermore, any peptide states that can be experimentally observed only when trapped under relatively artificial conditions (e.g., in the gel phase of short-chain lipids) may well occur transiently in native membranes under the impact of concentration gradients or asymmetric membrane binding. Moreover, the tendency of peptides to undergo multiple transitions may well reflect a flexibility that is required to successfully carry out their functions, hence this property appears to be an intrinsic characteristic of membrane-active peptides.

References

1. Kawasaki, H. and Iwamuro, S. (2008) Potential roles of histones in host defense as antimicrobial agents. *Infect. Disord. Drug. Targets* **8**, 195–205.
2. Henriques, S. T., Melo, M. N., and Castanho, M. A. (2006) Cell-penetrating peptides and antimicrobial peptides: how different are they?. *Biochem. J.* **399**, 1–7.
3. Melino, S., Rufini, S., Sette, M., Morero, R., Grottesi, A., Paci, M., and Petruzzelli, R. (1999) Zn^{2+} ions selectively induce antimicrobial salivary peptide histatin-5 to fuse negatively charged vesicles. Identification and characterization of a zinc-binding motif present in the functional domain. *Biochemistry* **38**, 9626–9633.
4. Dos Santos Cabrera, M. P., Arcisio-Miranda, M., da Costa, L. C., de Souza, B. M., Broggio Costa, S. T., Palma, M. S., Ruggiero Neto, J., and Procopio, J. (2009) Interactions of mast cell degranulating peptides with model membranes: A comparative biophysical study. *Arch. Biochem. Biophys.* **486**, 1–11.
5. Zasloff, M. (2002) Antimicrobial peptides of multicellular organisms. *Nature* **415**, 389–395.
6. Matsuzaki, K. (1999) Why and how are peptide-lipid interactions utilized for self-defense? Magainins and tachyplesins as archetypes. *Biochim. Biophys. Acta* **1462**, 1–10.
7. Yang, L., Weiss, T. M., Lehrer, R. I., and Huang, H. W. (2000) Crystallization of antimicrobial pores in membranes: Magainin and protegrin. *Biophys. J.* **79**, 2002–2009.
8. Shai, Y. (1999) Mechanism of the binding, insertion and destabilization of phospholipid bilayer membranes by alpha-helical antimicrobial and cell non-selective membrane-lytic peptides. *Biochim. Biophys. Acta* **1462**, 55–70.
9. Afonin, S., Grage, S. L., Ieronimo, M., Wadhwani, P., and Ulrich, A. S. (2008) Temperature-dependent transmembrane insertion of the amphiphilic peptide PGLa in lipid bilayers observed by solid state ^{19}F NMR spectroscopy. *J. Am. Chem. Soc.* **130**, 16512–16514.
10. Vogel, H. (1987) Comparison of the conformation and orientation of alamethicin and melittin in lipid membranes. *Biochemistry* **26**, 4562–4572.
11. Glaser, R. W., Sachse, C., Dürr, U. H. N., Wadhwani, P., Afonin, S., Strandberg, E., and Ulrich, A. S. (2005) Concentration-dependent re-alignment of the peptide PGLa in lipid membranes observed by solid state ^{19}F-NMR. *Biophys. J.* **88**, 3392–3397.
12. Glaser, R. W., Sachse, C., Dürr, U. H. N., Wadhwani, P., and Ulrich, A. S. (2004) Orientation of the antimicrobial peptide PGLa in lipid membranes determined from ^{19}F-NMR dipolar couplings of 4-CF_3-phenylglycine labels. *J. Mag. Res.* **168**, 153–163.
13. Ludtke, S. J., He, K., Wu, Y., and Huang, H. W. (1994) Cooperative membrane insertion of magainin correlated with its cytolytic activity. *Biochim. Biophys. Acta.* **1190**, 181–184.
14. He, K., Ludtke, S. J., Heller, W. T., and Huang, H. W. (1996) Mechanism of alamethicin insertion into lipid bilayers. *Biophys J.* **71**, 2669–2679.
15. Bürck, J., Roth, S., Wadhwani, P., Afonin, S., Kanithasen, N., Strandberg, E., and Ulrich, A. S. (2008) Conformation and membrane orientation of amphiphilic helical peptides by oriented circular dichroism. *Biophys J.* **95**, 3872–3881.
16. Tremouilhac, P., Strandberg, E., Wadwhani, P., and Ulrich, A. S. (2006) Conditions affecting the re-alignment of the antimicrobial peptide PGLa in membranes as monitored by solid state ^{2}H-NMR. *Biochim. Biophys. Acta* **1758**, 1330–1342.
17. Ulrich, A. S., Tichelaar, W., Förster, G., Zschörnig, O., Weinkauf, S., and Meyer, H. W. (1999) Ultrastructural characterization of peptide-induced membrane fusion and peptide self-assembly in the lipid bilayer. *Biophys. J.* 77, 829–841.
18. Afonin, S., Dürr, U. H. N., Glaser, R. W., and Ulrich, A. S. (2004) "Boomerang"-like insertion of a fusogenic peptide in a lipid membrane revealed by solid state ^{19}F NMR. *Mag. Res. Chem.* **42**, 195–203.
19. Binder, H., Arnold, K., Ulrich, A. S., and Zschörnig, O. (2000) The effect of Zn^{2+} on the secondary structure of a histidine-rich fusogenic peptide and its interaction with lipid membranes. *Biochim. Biophys. Acta* **1468**, 345–358.
20. Glaser, R. W., Grüne, M., Wandelt, C., and Ulrich, A. S. (1999) NMR and CD structural analysis of the fusogenic peptide sequence B18 from the fertilization protein bindin. *Biochemistry* **38**, 2560–2569.
21. Grage, S. L., Afonin, S., Grüne, M., and Ulrich, A. S. (2004) Interaction of the fusogenic peptide B18 in its amyloid-state with DMPC bilayers studied by solid-state ^{2}H- and ^{31}P-NMR and DSC. *Chem. Phys. Lipids* **132**, 65–77.

22. Li, W., Nicol, F., and Szoka, F. C., Jr. (2004) GALA: A designed synthetic pH-responsive amphipathic peptide with applications in drug and gene delivery. *Adv. Drug Deliv.* **56**, 967–985.
23. Tatulian, S. A. and Tamm, L. K. (2000) Secondary structure, orientation, oligomerization, and lipid interactions of the transmembrane domain of influenza hemagglutinin. *Biochemistry* **39**, 496–507.
24. Sammalkorpi, M. and Lazaridis, T. (2007) Configuration of influenza hemagglutinin fusion peptide monomers and oligomers in membranes. *Biochim. Biophys. Acta* **1768**, 30–38.
25. Rapaport, D., Hague, G. R., Pouny, Y., and Shai, Y. (1993) pH- and ionic strength-dependent fusion of phospholipid vesicles induced by pardaxin analogues or by mixtures of charge-reversed peptides. *Biochemistry* **32**, 3291–3297.
26. Raghuraman, H. and Chattopadhyay, A. (2007) Melittin: A membrane active peptide with diverse functions. *Biosci. Rep.* **27**, 189–223.
27. Pujals, S., Fernández-Carneado, J., López-Iglesias, C., Kogan, M. J., and Giralt, E. (2006) Mechanistic aspects of CPP-mediated intracellular drug delivery: Relevance of CPP self-assembly. *Biochim. Biophys. Acta* **1758**, 264–279.
28. Bulet, P., Stöcklin, R., and Menin, L. (2004) Antimicrobial peptides: From invertebrates to vertebrates. *Immunol. Rev.* **198**, 169–184.
29. Futaki, S. (2006) Oligoarginine vectors for intracellular delivery: Design and cellular-uptake mechanisms. *Biopolymers* **84**, 241–249.
30. Castano, S. and Desbat, B. (2005) Structure and orientation study of fusion peptide FP23 of gp41 from HIV-1 alone or inserted into various lipid membrane models (mono-, bi- and multibi-layers) by FT-IR spectroscopies and Brewster angle microscopy. *Biochim. Biophys. Acta* **1715**, 81–95.
31. Qiang, W., Bodner, M. L., and Weliky, D. P. (2008) Solid-state NMR spectroscopy of human immunodeficiency virus fusion peptides associated with host-cell-like membranes: 2D correlation spectra and distance measurements support a fully extended conformation and models for specific antiparallel strand registries. *J. Am. Chem. Soc.* **130**, 5459–5471.
32. Reichert, J., Grasnick, D., Afonin, S., Bürck, J., Wadhwani, P., and Ulrich, A. S. (2007) A critical evaluation of the conformational requirements of fusogenic peptides in membranes. *Eur. J. Biophys.* **36**, 405–413.
33. Grasnick, D., Sternberg, U., Strandberg, E., Wadhwani, P., and Ulrich, A. S. (2009) Irregular structure of the HIV fusion peptide in membranes demonstrated by NMR and MD simulations. In preparation
34. Huang, D. H., Walter, R., Glickson, J. D., and Krishna, N. R. (1981) Solution conformation of gramicidin S: An intramolecular nuclear Overhauser effect study. *Proc. Natl. Acad. Sci.* **78**, 672–675.
35. Jones, C. R., Kuo, M., and Gibbons, W. A. (1979) Multiple solution conformations and internal rotations of the decapeptide gramicidin S. *J. Biol. Chem.* **25**, 10307–10312.
36. Rackovsky, S. and Scheraga, H. A. (1980) Intermolecular anti-parallel β-sheet: Comparison of predicted and observed conformations of gramicidin S. *Proc. Natl. Acad. Sci.* **77**, 6965–6967.
37. Bystrov, V. F., Arseniev, A. S., Barsukov, I. L., and Lomize, A. L. (1986) 2D NMR of single and double stranded helices of gramicidin A in micelles and solutions. *Bull. Magn. Reson.* **8**, 84–94.
38. Zhang, Z., Pascal, S. M., and Cross, T. A. (1992) A conformational rearrangement in gramicidin A: From a double-stranded left-handed to a single-stranded right-handed helix. *Biochemistry* **31**, 8822–8828.
39. Kelkar, D. A. and Chattopadhyay, A. (2007) The gramicidin ion channel: A model membrane protein. *Biochim. Biophys. Acta* **1768**, 2011–2025.
40. Fernández, C. and Wüthrich, K. (2003) NMR solution structure determination of membrane proteins reconstituted in detergent micelles. *FEBS Lett.* **555**, 144–150.
41. Wang, G. (2008) NMR of membrane-associated peptides and proteins. *Curr. Protein Pept. Sci.* **9**, 50–69.
42. Sanders, C. R. and Sönnichsen, F. (2006) Solution NMR of membrane proteins: Practice and challenges. *Magn. Reson. Chem.* **44**(Spec. issue), S24–S40.
43. Galanth, C., Abbassi, F., Lequin, O., Ayala-Sanmartin, J., Ladram, A., Nicolas, P., and Amiche, M. (2009) Mechanism of antibacterial action of dermaseptin B2: Interplay between helix-hinge-helix structure and membrane curvature strain. *Biochemistry* **48**, 313–327.
44. Chou, J. J., Kaufman, J. D., Stahl, S. J., Wingfield, P. T., and Bax, A. (2002) Micelle-induced curvature in a water-insoluble HIV-1 Env peptide revealed by NMR dipolar coupling measurement in stretched polyacrylamide gel. *J. Am. Chem. Soc.* **124**, 2450–2451.

45. Prosser, R. S., Evanics, F., Kitevski, J. L., and Al-Abdul-Wahid, M. S. (2006) Current applications of bicelles in NMR studies of membrane-associated amphiphiles and proteins. *Biochemistry* **45**, 8453–8465.
46. Strandberg, E. and Ulrich, A. S. (2004) NMR methods for studying membrane-active antimicrobial peptides. *Concepts Magn. Res.* **23A**, 89–120.
47. Tugarinov, V., Hwang, P. M., and Kay, L. E. (2004) Nuclear magnetic resonance spectroscopy of high-molecular-weight proteins. *Annu. Rev. Biochem.* **73**, 107–146.
48. Gong, X. M., Franzin, C. M., Thai, K., Yu, J., and Marassi, F. M. (2007) Nuclear magnetic resonance structural studies of membrane proteins in micelles and bilayers. *Methods Mol. Biol.* **400**, 515–529.
49. Nielsen, N., Malmendal, A., and Vosegaard, T. (2004) Techniques and applications of NMR to membrane proteins. *Mol. Membr. Biol.* **21**, 129–141.
50. Ketchem, R. R., Hu, W., and Cross, T. A. (1993) High resolution conformation of gramicidin A in a lipid bilayer by solid state NMR. *Science* **261**, 1457–1460.
51. Eisenberg, D., Weiss, R. M., and Terwilliger, T. C. (1982) The helical hydrophobic moment – a measure of the amphiphilicity of a helix. *Nature* **299**, 371–374.
52. Dathe, M. and Wieprecht, T. (1999) Structural features of helical antimicrobial peptides: Their potential to modulate activity on model membranes and biological cells. *Biochim. Biophys. Acta* **1462**, 71–87.
53. Ho, D. and Merrill, A. R. (2009) Evidence for the amphipathic nature and tilted topology of helices 4 and 5 in the closed state of the Colicin E1 channel. *Biochemistry.* **48**, 1369–1380.
54. Orioni, B., Bocchinfuso, G., Kim, J. Y., Palleschi, A., Grande, G., Bobone, S., Park, Y., Kim, J. I., Hahm, K. S., and Stella, L. (2009) Membrane perturbation by the antimicrobial peptide PMAP-23: A fluorescence and molecular dynamics study. *Biochim. Biophys. Acta Biomembranes* **1788**, 1523–1533.
55. Raghuraman, H. and Chattopadhyay, A. (2007) Orientation and dynamics of melittin in membranes of varying composition utilizing NBD fluorescence. *Biophys. J.* **92**, 1271–1283.
56. Gröbner, G., Glaubitz, C., and Watts, A. (1999) Probing membrane surfaces and the location of membrane-embedded peptides by ^{13}C MAS NMR using lanthanide ions. *J. Magn. Reson.* **141**, 335–339.
57. Buffy, J. J., Hong, T., Yamaguchi, S., Waring, A. J., Lehrer, R. I., and Hong, M. (2003) Solid-state NMR investigation of the depth of insertion of protegrin-1 in lipid bilayers using paramagnetic Mn^{2+}. *Biophys. J.* **85**, 2363–2373.
58. Prosser, R. S., Luchette, P. A., Westerman, P. W., Rozek, A., and Hancock, R. E. (2001) Determination of membrane immersion depth with O_2: A high-pressure ^{19}F NMR study. *Biophys. J.* **80**, 1406–1416.
59. van der Wel, P. C. A., Strandberg, E., Killian, J. A., and Koeppe, R. E. I. I. (2002) Geometry and intrinsic tilt of a tryptophan-anchored transmembrane α-helix determined by ^{2}H NMR. *Biophys. J.* **83**, 1479–1488.
60. Strandberg, E., Özdirekcan, S., Rijkers, D. T. S., van der Wel, P. C. A., Koeppe, R. E., II, Liskamp, R. M., and Killian, J. A. (2004) Tilt angles of transmembrane model peptides in oriented and non-oriented lipid bilayers as determined by ^{2}H solid-state NMR. *Biophys. J.* **86**, 3709–3721.
61. Ulrich, A. S. (2007) Solid state ^{19}F-NMR analysis of oriented biomembranes. In *Modern Magnetic Resonance.* Webb, G. A., (ed.), pp. 261–268. Dordrecht, The Netherlands: Springer.
62. Mykhailiuk, P. K., Afonin, S., Palamarchuk, G. V., Shishkin, O. V., Ulrich, A. S., and Komarov, I. V. (2008) Synthesis of trifluoromethyl-substituted proline analogues as ^{19}F NMR labels for peptides in the polyproline II conformation. *Angewandte Chem. Int. Ed.* **47**, 5765–5767.
63. Mykhailiuk, P. K., Afonin, S., Ulrich, A. S., and Komarov, I. V. (2008) A convenient route to trifluoromethyl-substituted cyclopropane derivatives. *Synthesis* **11**, 1757–1760.
64. Afonin, S., Mikhailiuk, P. K., Komarov, I. V., and Ulrich, A. S. (2007) Evaluating the amino acid CF_3-bicyclopentylglycine as a new label for solid-state ^{19}F-NMR structure analysis of membrane-bound peptides. *J. Pept. Sci.* **13**, 614–623.
65. Mikhailiuk, P. K., Afonin, S., Chernega, A. N., Rusanov, E. B., Platonov, M. O., Dubinina, G. G., Berditsch, M., Ulrich, A. S., and Komarov, I. V. (2006) Conformationally rigid trifluoromethyl-substituted α-amino acid designed for peptide structure analysis by solid state ^{19}F-NMR spectroscopy. *Angew. Chem. Int. Ed.* **45**, 5659–5661.
66. Afonin, S., Dürr, U. H. N., Wadhwani, P., Salgado, J., and Ulrich, A. S. (2008) Solid state NMR structure analysis of the antimicrobial peptide gramicidin S in lipid membranes: Concentration-dependent re-

alignment and self-assembly as a β-barrel. *Topics Curr. Chem.* **273**, 139–154.
67. Mani, R., Cady, S. D., Tang, M., Waring, A. J., Lehrer, R. I., and Hong, M. (2006) Membrane-dependent oligomeric structure and pore formation of β-hairpin antimicrobial peptide in lipid bilayers from solid-state NMR. *Proc. Natl. Acad. Sci.* **103**, 16242–16247.
68. Llamas-Saiz, A. L., Grotenbreg, G. M., Overhand, M., and van Raaij, M. J. (2007) Double-stranded helical twisted β-sheet channels in crystals of gramicidin S grown in the presence of trifluoroacetic and hydrochloric acids. *Acta Crystallogr. D, Biol. Crystallogr.* **63**, 401–407.
69. Ivanova, V. P., Makarov, I. M., Schäffer, T. E., and Heimburg, T. (2003) Analyzing heat capacity profiles of peptide-containing membranes: Cluster formation of gramicidin A. *Biophys. J.* **84**, 2427–2439.
70. Sparr, E., Ganchev, D. N., Snel, M. M., Ridder, A. N., Kroon-Batenburg, L. M., Chupin, V., Rijkers, D. T., Killian, J. A., and de Kruijff, B. (2005) Molecular organization in striated domains induced by transmembrane α-helical peptides in dipalmitoyl phosphatidylcholine bilayers. *Biochemistry* **44**, 2–10.
71. Rhee, S. K., Quist, A. P., and Lal, R. (1998) Amyloid β-protein (1–42) forms calcium-permeable Zn^{2+} sensitive channels. *J. Biol. Chem.* **273**, 13379–13382.
72. Arispe, N., Rojas, E., and Pollard, H. B. (1993) Giant multilevel cation channels formed by Alzheimer disease amyloid β-protein (AβP[1–40]) in bilayer membranes. *Proc. Natl. Acad. Sci.* **90**, 10573–10577.
73. Bokvist, M. and Gröbner, G. (2007) Misfolding of amyloidogenic proteins at membrane surfaces: The impact of macromolecular crowding. *J. Am. Chem. Soc.* **129**, 14848–14849.
74. Selkoe, D. J. (2004) Cell biology of protein misfolding: The examples of Alzheimer's and Parkinson's diseases. *Natl. Cell Biol.* **6**, 1054–1061.
75. Aisenbrey, C., Borowik, T., Byström, R., Bokvist, M., Lindström, F., Misiak, H., Sani, M. A., and Gröbner, G. (2008) How is protein aggregation in amyloidogenic diseases modulated by biological membranes?. *Eur. Biophys. J.* **37**, 247–255.
76. Wadhwani, P., Bürck, J., Strandberg, E., Mink, C., Afonin, S., and Ulrich, A. S. (2008) Using a sterically restrictive amino acid as a ^{19}F NMR label to monitor and to control peptide aggregation in membranes. *J. Am. Chem. Soc.* **130**, 16515–16517.
77. Kerth, A., Erbe, A., Dathe, M., and Blume, A. (2004) Infrared reflection absorption spectroscopy of amphipathic model peptides at the air/water interface. *Biophys. J.* **86**, 3750–3758.
78. Maisch, D. (2008) Synthese und Strukturuntersuchungen des membranaktiven Peptaibols Alamethicin mittels Festkörper-^{19}F-NMR, PhD thesis, University of Karlsruhe, Karlsruhe, Germany.
79. Salnikov, E. S., Zotti, M. D., Formaggio, F., Li, X., Toniolo, C., O'Neil, J. D., Raap, J., Dzuba, S. A., and Bechinger, B. (2009) Alamethicin topology in phospholipid membranes by oriented solid-state NMR and EPR spectroscopies: A comparison. *J. Phys. Chem. B* **113**, 3034–3042.
80. Strandberg, E., Tremouilhac, P., Wadhwani, P., and Ulrich, A. S. (2009) Synergistic transmembrane insertion of the heterodimeric PGLa/magainin 2 complex studied by solid-state NMR. *Biochim. Biophys. Acta Biomembranes* **1788**, 1667–1679.
81. de Planque, M. R. and Killian, J. A. (2003) Protein-lipid interactions studied with designed transmembrane peptides: Role of hydrophobic matching and interfacial anchoring. *Mol. Membr. Biol.* **20**, 271–284.
82. Mou, J., Czajkowsky, D. M., and Shao, Z. (1996) Gramicidin A aggregation in supported gel state phosphatidylcholine bilayers. *Biochemistry* **35**, 3222–3226.
83. de Kruijff, B., Killian, J. A., Ganchev, D. N., Rinia, H. A., and Sparr, E. (2006) Striated domains: Self-organizing ordered assemblies of transmembrane alpha-helical peptides and lipids in bilayers. *Biol. Chem.* **387**, 235–241.
84. Rinia, H. A., Boots, J. W., Rijkers, D. T., Kik, R. A., Snel, M. M., Demel, R. A., Killian, J. A., van der Eerden, J. P., and de Kruijff, B. (2002) Domain formation in phosphatidylcholine bilayers containing transmembrane peptides: Specific effects of flanking residues. *Biochemistry* **41**, 2814–2824.
85. Esteban-Martín, S., Strandberg, E., Fuertes, G., Ulrich, A. S., and Salgado, J. (2009) Influence of whole-body dynamics on ^{15}N PISEMA NMR spectra of membrane proteins: A theoretical analysis. *Biophys. J.* **96**, 3233–3241.
86. Strandberg, E., Esteban-Martín, S., Salgado, J., and Ulrich, A. S. (2009) Orientation and dynamics of peptides in membranes calculated from ^{2}H-NMR data. *Biophys. J.* **96**, 3223–3232.
87. Sharpe, S., Barber, K. R., Grant, C. W., Goodyear, D., and Morrow, M. R. (2002)

Organization of model helical peptides in lipid bilayers: Insight into the behavior of single-span protein transmembrane domains. *Biophys. J.* **83**, 345–358.

88. Tang, M., Waring, A. J., Lehrer, R. I., and Hong, M. (2006) Orientation of a β-hairpin antimicrobial peptide in lipid bilayers from two-dimensional dipolar chemical-shift correlation NMR. *Biophys. J.* **90**, 3616–3624.
89. Ulrich, A. S., Wadhwani, P., Dürr, U. H. N., Afonin, S., Glaser, R. W., Strandberg, E., Tremouilhac, P., Sachse, C., Berditchevskaia, M., and Grage, S. L. (2005) Solid-state ^{19}F-nuclear magnetic resonance analysis of membrane-active peptides. In *NMR Spectroscopy of Biological Solids*. Ramamoorthy, A., ed., pp. 215–236. Boca Raton, FL, USA: Francis & Taylor.
90. Han, X., Bushweller, J. H., Cafiso, D. S., and Tamm, L. K. (2001) Membrane structure and fusion-triggering conformational change of the fusion domain from influenza hemagglutinin. *Nat. Struct. Biol.* **8**, 715–720.
91. Bechinger, B. (1996) Towards membrane protein design: pH-sensitive topology of histidine-containing polypeptides. *J. Mol. Biol.* **263**, 768–775.
92. Huang, H. W. and Wu, Y. (1991) Lipid-alamethicin interactions influence alamethicin orientation. *Biophys. J.* **60**, 1079–1087.
93. Heller, W. T., He, K., Ludtke, S. J., Harroun, T. A., and Huang, H. W. (1997) Effect of changing the size of lipid headgroup on peptide insertion into membranes. *Biophys. J.* **73**, 239–244.
94. Yang, L., Harroun, T. A., Weiss, T. M., Ding, L., and Huang, H. W. (2001) Barrel-stave model or toroidal model? A case study on melittin pores. *Biophys. J.* **81**, 1475–1485.
95. Westerhoff, H. V., Zasloff, M., Rosner, J. L., Hendler, R. W., de Waal, A., Vaz Gomes, A., Jongsma, P. M., Riethorst, A., and Juretić, D. (1995) Functional synergism of the magainins PGLa and magainin-2 in *Escherichia coli*, tumor cells and liposomes. *Eur. J. Biochem.* **228**, 257–264.
96. Haney, E. F., Hunter, H. N., Matsuzaki, K., and Vogel, H. J. (2009) Solution NMR studies of amphibian antimicrobial peptides: Linking structure to function?. *Biochim. Biophys. Acta Biomembranes* **1788**, 1639–1655.
97. Mangoni, M. L. and Shai, Y. (2009) Temporins and their synergism against Gram-negative bacteria and in lipopolysaccharide detoxification. *Biochim. Biophys. Acta Biomembranes* **1788**, 1610–1619.
98. Zuckermann, M. J. and Heimburg, T. (2001) Insertion and pore formation driven by adsorption of proteins onto lipid bilayer membrane-water interfaces. *Biophys. J.* **81**, 2458–2472.
99. Chatelier, R. C. and Minton, A. P. (1996) Adsorption of globular proteins on locally planar surfaces: Models for the effect of excluded surface area and aggregation of adsorbed protein on adsorption equilibria. *Biophys. J.* **71**, 2367–2374.
100. Minton, A. P. (1999) Adsorption of globular proteins on locally planar surfaces. II. Models for the effect of multiple adsorbate conformations on adsorption equilibria and kinetics. *Biophys. J.* **76**, 176–187.
101. Huang, H. W. (2009) Free energies of molecular bound states in lipid bilayers: Lethal concentrations of antimicrobial peptides. *Biophys. J.* **96**, 3263–3272.
102. Huang, H. W. (2006) Molecular mechanism of antimicrobial peptides: The origin of cooperativity. *Biochim. Biophys. Acta.* **1758**, 1292–1302.
103. Lee, M. T., Chen, F. Y., and Huang, H. W. (2004) Energetics of pore formation induced by membrane active peptides. *Biochemistry* **43**, 3590–3599.
104. Ludtke, S., He, K., and Huang, H. W. (1995) Membrane thinning caused by magainin 2. *Biochemistry* **34**, 16764–16769.

Chapter 14

Solid-State NMR Investigations of Membrane-Associated Antimicrobial Peptides

Christopher Aisenbrey, Philippe Bertani, and Burkhard Bechinger

Abstract

Solid-state NMR and other biophysical investigations have revealed many mechanistic details about the interactions of antimicrobial peptides with membranes. These studies have shaped our view on how these peptides cause the killing of bacteria, fungi, or tumour cells and how they permeabilize model membranes. As a result, we better understand the biological activities of these peptides and we are now able to design new and better sequences. Here we present some of the tools that have allowed these solid-state NMR investigations, including detailed protocols on how to reconstitute the peptides into oriented or non-oriented membranes as well as simple set-up procedures for ^{2}H as well as proton-decoupled ^{31}P or ^{15}N solid-state NMR measurements. Static and magic angle spinning experiments are described. Where adequate, the special requirements for or limitations of some of the measurements are discussed. Solid-state NMR spectra of both lipids and peptides have been recorded, and through the ensemble of measurements a detailed picture of these complex peptide–lipid supramolecular systems has finally emerged.

Key words: Membrane reconstitution, oriented bilayer, helix topology, amphipathic peptide, surface alignment, transmembrane orientation, membrane protein structure, peptide–lipid interactions, pore formation, channel, magic angle spinning.

1. Introduction

The investigation ofantimicrobial peptidesby biophysical methods has much enlightened our understanding of the underlying mechanisms of how they interact with membranes, cause membrane permeability increases, and selectively kill bacteria (1). For example, proton-decoupled ^{15}N solid-state NMR spectra of magainin 2 provided the first indication that these amphipathic

A. Giuliani, A.C. Rinaldi (eds.), *Antimicrobial Peptides*, Methods in Molecular Biology 618, DOI 10.1007/978-1-60761-594-1_14, © Springer Science+Business Media, LLC 2010

antimicrobial sequences orient parallel to the membrane surface rather than form pores by association into transmembrane helical bundles (2). For other peptides, the issue was unambigously clarified by designing antimicrobial sequences and analysing their behaviour using the same technique (3) in combination with biological assays (4). Biophysical studies have revealed that selected peptides can interact with membranes in a very dynamic and diverse manner, and have shaped our modern view on these systems (reviewed and developed in (1, 5)). To this scope, several solid-state NMR approaches are available each of them with a particular strength.

In contrast to solution NMR spectroscopy, where fast molecular tumbling ensures isotropic averaging of the nuclear interactions, in static solid or semi-solid samples the chemical shift, dipolar interactions, and quadrupolar splittings are all dependent on the molecular alignment relative to the magnetic field direction. Therefore, the anisotropy of nuclear interactions is a predominant feature of the NMR spectra also in the case of membrane-associated peptides. On the one hand, fast spinning of the sample around the magic angle (several 1,000 rotations per second at an angle of 54.7° with respect to the magnetic field) results in the elimination of many of these anisotropic features, leading to spectra that resemble much those obtained in solution. These so-called magic angle spinning (MAS) solid-state NMR spectra, therefore, reveal small changes in the isotropic chemical shifts and thereby the chemical environments of the nucleus under observation. For example, proton-decoupled ^{31}P MAS solid-state NMR has been used to follow the selective association of antimicrobial peptides with one component of a mixed membrane (6).

Further advantage can be taken from the orientation dependence of NMR interactions in the solid or semi-solid state. The chemical shifts (e.g. ^{15}N, ^{13}C, ^{1}H), quadrupolar interactions (e.g. ^{2}H and ^{14}N), or dipolar couplings (e.g. ^{15}N–^{13}C and ^{1}H–^{15}N) all depend on the alignment of molecules (and/or bonds) relative to the magnetic field direction. A valuable approach has therefore been to reconstitute peptides and proteins into macroscopically oriented membranes, to insert them in the magnetic field of the NMR spectrometer with the bilayer normal parallel to the magnetic field direction and to record the solid-state NMR spectra. In this manner the ^{15}N chemical shift, the deuterium quadrupolar splitting, and many other parameters provide orientational constraints and thereby the secondary structure and the alignment of the peptides (7). The sample preparation and the NMR experiments relevant to this approach will be described in **Sections 2.2, 3.2,** and **3.3** (**Fig. 14.1**).

Furthermore, non-oriented samples, where all directions of the membrane are present and visible in the spectra, result in

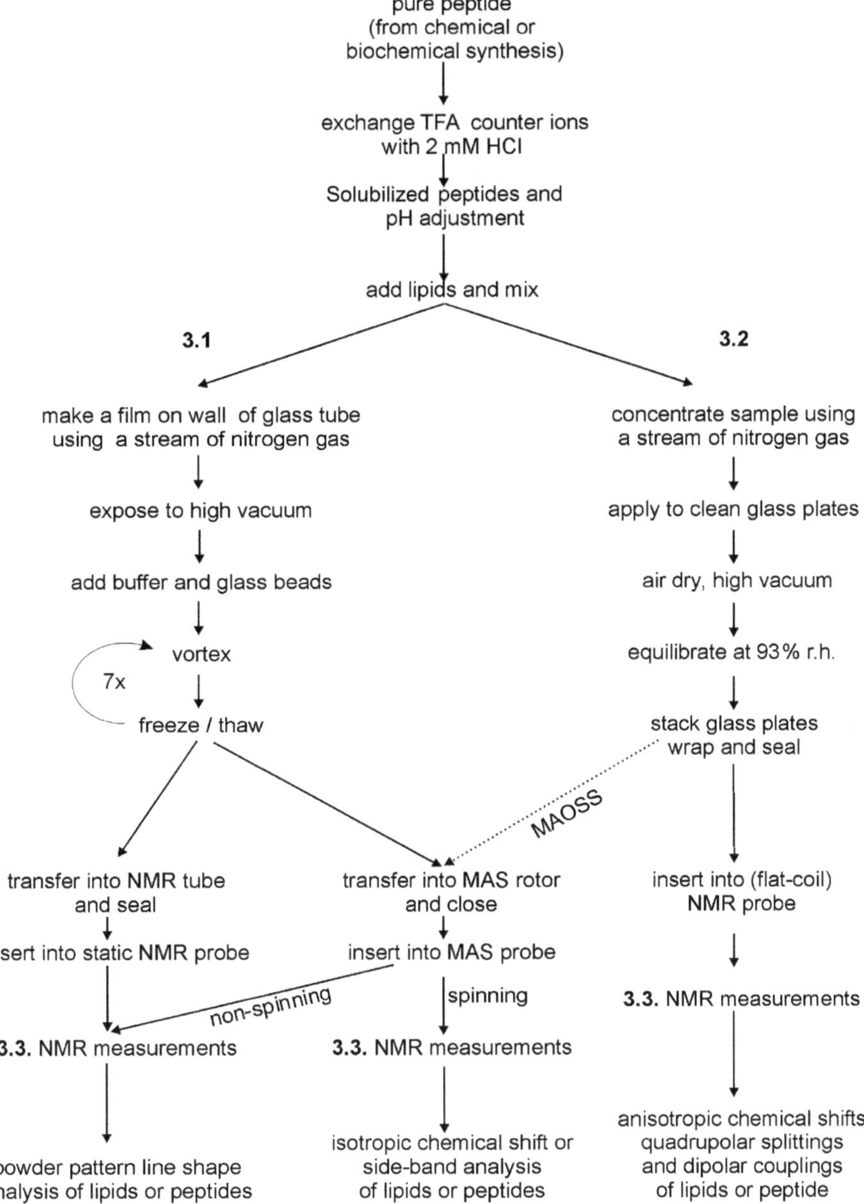

Fig. 14.1. Flow chart for the preparation of samples and the consecutive NMR measurements. The *hatched line* makes reference to the possibility of investigating oriented samples under MAS conditions (not discussed in the text, *see* (34, 35)). Typical sample sizes for the analysis of powder pattern line shapes are 10 mg for ^{31}P or ^{2}H spectra of lipids and 40 mg for ^{15}N or $^{2}H_3$ NMR spectroscopy of peptides, respectively; for the isotropic chemical shift analysis by MAS solid-state NMR 5 mg of lipids or 10 mg of peptide, respectively; and for oriented samples $\geq$10 mg of lipid and $\geq$5 mg of single-site labelled peptide, respectively. Numbers (**3.1, 3.2,** and so on) refer to main text sections.

broad spectral line shapes which provide valuable information on the motional averaging of the labelled sites, i.e. local motions (8) as well as lateral and rotational diffusion of the molecule as a whole (9, 10). This approach has a long-standing record in monitoring the macroscopic phase properties of membranes, by analysing the very different ^{31}P NMR line shapes of bilayers, hexagonal, micellar, or bicellar phases (11, 12). A more recent analysis of the resulting peptide spectra indicates that these powder patterns represent the alignment of the peptide relative to the membrane normal provided that fast rotation around this axis occurs (9, 13, 14). Samples of peptides associated with membranes made of phospholipids deuterated fatty acyl chains have also been used to analyse the peptide penetration depth, the selective interaction of cationic peptides with anionic lipids in mixed membranes, and the resulting membrane disruptive properties of the sequences (6, 15). The sample preparation and methodology for these NMR approaches will be discussed in **Sections 2.1, 3.1,** and **3.3** (**Fig. 14.1**).

The theoretical background of solid-state NMR spectroscopy of oriented samples shall not be presented here but a simple correlation exists for α-helical peptides that have been labelled with ^{15}N at one of the amide bonds and which have been reconstituted into oriented or non-oriented membranes (9, 14). In short, it can be shown that for samples that have been uniaxially oriented with the membrane normal parallel to the magnetic field (*see* **Sections 3.2** and **3.3.3**), transmembrane α-helical peptides exhibit ^{15}N resonances in the 200 ppm range whereas they resonate at <100 ppm when aligned parallel to the membrane surface (9) (**Fig. 14.3c**). Furthermore, the deuterated alanine methyl group exhibits several favourable properties that make this nucleus amenable to structural investigation by ^{2}H solid-state NMR spectroscopy. In this case, the deuterium quadrupolar splitting measured in an aligned sample reflects the orientation of the alanine C_α–C_β bond relative to the membrane normal (Θ), and with the C_α carbons being an integral part of the peptide backbone, also of the peptide as a whole. The quadrupolar splitting is proportional to $(3\cos^2\Theta-1)$ with a maximum of about 80 kHz. This parameter provides valuable complementary information to ^{15}N chemical shift measurements obtained from the same helix (16, 17), but there is no simple correlation between the helix tilt angle and the quadrupolar splitting. Other parameters that can be used for structural analysis are ^{13}C chemical shift or the dipolar couplings between nuclei (9, 18).

Clearly the reconstitution of the peptides into phospholipids bilayers is a critical step before the analysis by NMR spectroscopic methods can take place. Whereas reconstitution into non-oriented membranes is described in **Section 3.1**, the preparation of oriented membranes is detailed in **Section 3.2**. Here we present a protocol for the reconstitution of magainin 2 or similar amphipathic peptides into phosphatidylcholine membranes at a peptide-to-lipid ratio typically equal to 1:50, and **Notes 1** and **2** will give indications on how to proceed with other peptides or other lipids. Furthermore, the main protocol reported below describes a procedure where a homogenous peptide/lipids mixture is prepared in organic solvent before membrane formation. However, **Note 6** is dedicated to the addition of the peptide dissolved in aqueous solution to the lipids, which in some cases provides different results (12).

It should be mentioned that by being non-invasive NMR spectroscopy is also relatively insensitive when compared to other types of spectroscopy. It is therefore worthwhile to spend a few considerations on sensitivity and the requirements of the different approaches for isotopic labelling. One of the most straightforward measurements is the analysis of the macroscopic phase properties of the membranes by ^{31}P solid-state NMR as the ^{31}P nucleus is present in every phospholipid at 100% natural abundance and exhibits a relatively high sensitivity (e.g. (12)). Also the deuterium spectrum of about 10 mg of ^{2}H-labelled lipids can be acquired within minutes (e.g. (6, 15)). However, recording the NMR spectra of peptides reconstituted into membranes renders the situation more difficult as the peptide is diluted in the membrane and in water, even though signal enhancement techniques such as cross-polarization are used. In addition, in oriented membranes much of the space within the coil volume is occupied by glass plates, and the spins typically observed (^{15}N, ^{2}H) are relatively insensitive when compared to ^{1}H or ^{31}P. Therefore, a typical sample requires 5–15 mg of peptide, labelled with ^{15}N, ^{13}C, and/or ^{2}H (at one or a few sites) in 100–200 mg of lipid. These peptide sequences are typically prepared by chemical solid-phase peptides synthesis and purified by reverse-phase HPLC. Alternatively, solid-state NMR spectra have been obtained from about 1 mg of uniformly labelled peptide prepared by bacterial over expression (19) or by other biochemical methods (20).

2. Materials

2.1. Peptide Reconstitution into Non-oriented Membranes

1. Disposable glass tubes ≥10 mL, glass pipettes with long tip for organic solvents, 250 μL Hamilton syringe, glass beads 2–3 mm in diameter (e.g. Aldrich, St. Louis, MO, USA), Teflon tape, parafilm.
2. HCl (min. 37% analytical reagent Riedel-de Haen, St. Louis, MO, USA), 2,2,2-trifluoroethanol (TFE; ReagentPlus® >99%, Aldrich), Milli-Q water and isopropanol (industrial grade)/dry ice mixture. For other peptides and lipids

(*see* **Notes 1** and **2**), organic solvents of analytical grade are used, such as hexafluoroisopropanol (HFIP, 99+%, Aldrich), dichloromethane (DCM normapur^{TM} analytical reagent stabilized with 0.1% ethanol, Prolabo/VWR, West Chester, PA, USA), methanol (HPLC grade >99.8%, Fluka, St. Louis, MO, USA), formic acid (purum >98%, Fluka), and acetic acid (Ultra $\geq$ 99.5%, Fluka).

3. 1-Palmitoyl-2-oleoyl-*sn*-glycero-3-phosphocholine (POPC; Avanti Polar Lipids, Birmingham, AL, USA) (*see* **Note 2**).
4. Nitrogen gas (bottled or prepared by evaporation of liquid nitrogen).
5. >1.2 mL buffer solution: 150 mM NaCL, 20 mM Tris–HCl, pH 7.4 (*see* **Note 3**).
6. NMR sample tubes that fit the NMR coil and probes. For ^{31}P NMR, Ultra Clear centrifuge tubes (Beckman, Paolo Alto, CA, USA) with home-made Teflon lids have been used in our settings. The 5 mm diameter tubes (175 μL volume) fit into a 7 mm MAS rotor after shortening (*see* **Note 4**). The next available size is 8 mm which also fits some NMR coils (400 μL volume). Depending on the polymer composition, the signal intensity, and the detailed experimental parameters, these plastic tubes result in background signals that can interfere with, e.g., ^{13}C or ^{15}N NMR measurements; thus, glass tubes are a better choice (e.g. Verrerie Villeurbannaise, Villeurbanne, France). For MAS or static experiments it is also possible to directly transfer the sample into MAS rotors. Take care when spinning as the lipids act like lubricants that can easily cause loosening of the rotor lid concomitantly with a rotor crash and the loss of the sample. Special inserts or lids are available to reduce these problems (e.g. Wilmad-Labglass, Buena, NJ, USA or high-resolution MAS inserts from Bruker, Rheinstetten, Germany).
7. Access to a solid-state NMR spectrometer (at least two-channels X{^{1}H} except for ^{2}H where a single X-channel frequency is applied) with the corresponding probes (solenoidal coil or equivalent): static and/or MAS probes for ^{31}P{^{1}H}, ^{15}N{^{1}H}, ^{13}C{^{1}H}, and/or ^{2}H. To investigate the lipid macroscopic phase properties good temperature control is desirable (±1 K). For magic angle spinning solid-state NMR experiments MAS probes and accessories are a prerequisite.

2.2. Peptide Reconstitution into Oriented Membranes

1. Disposable glass tubes ≥10 mL, glass pipettes with long tip for organic solvents, 10 μL glass capillaries, 250 μL Hamilton syringe, glass Petri dishes with lid (diameter about

10 cm), watchmaker forceps, one with pointed and one with flat tips (e.g. Outils Dumont SA, Montignez, Switzerland), Teflon tape, plastic foil that goes with an impulse sealer of the type used in kitchens (e.g. TEW Electrical Heating Equipment, Taipei, Taiwan).

2. Solvents and reagents of analytical grade (*see* Step 2 of **Section 2.1**): HCl, 2,2,2-trifluoro ethanol and Milli-Q water. For other peptides and lipids, organic solvents such as hexafluoroisopropanol, dichloromethane, methanol, formic acid, and naphthalene (puriss. for scintillation $\geq$99% Fluka) can be used (*see* **Notes 1** and **2**).
3. 1-Palmitoyl-2-oleoyl-*sn*-glycero-3-phosphocholine (POPC; Avanti Polar Lipids) (*see* **Note 2**).
4. A few mL each of the following: 1 N NaOH (e.g. normapur^{TM} analytical reagent >85.9% Prolabo/VWR); 200 mM Tris–HCl, pH 7.4; saturated solution of KNO_3 (for analysis, min 99%, e.g. Merck, Darmstadt, Germany) in a closed chamber (*see* **Note 5**).
5. pH indicator strips covering the range 0–9 in intervals of 0.5.
6. Nitrogen gas (bottled or prepared by evaporation of liquid nitrogen).
7. Ultra thin glass plates that fit your NMR coil (thickness 00 or 0 from Marienfeld, Lauda-Könishofen, Germany). Before preparing the sample test if a stack of glass plates sealed and wrapped in plastic film fits the coil (*see* Step 8 of **Section 3.2** and **Fig. 14.2d**). Custom-made glass plates of dimensions 11 $\times$ 20, 9 $\times$ 20 or 8.5 $\times$ 20 mm^2 have been used in our settings. If you have the choice, 9 $\times$ 20 is about half the prize when compared to the 8.5 $\times$ 20 size.
8. Access to a solid-state NMR spectrometer (at least two-channels except for ^{2}H) with the corresponding probes: static ^{31}P{^{1}H}, ^{15}N{^{1}H}, ^{13}C{^{1}H}, ^{2}H. For better sensitivity during the investigation of peptides in membranes, flat coil-specific NMR probes have been developed (21, 22) and are commercially available from several manufacturers (e.g. Bruker, Rheinstetten, Germany; Varian, Paolo Alto, CA, USA; Doty Scientific, Columbia, SC, USA), although for the investigation of lipids by ^{31}P{^{1}H} or ^{2}H solid-state NMR smaller quantities of sample are required (e.g. a ^{31}P NMR signal can already be obtained from 10 mg phospholipids oriented between a pair of glass plates) and conventional solenoidal coil shapes have also been used.

3. Methods

3.1. Reconstitution of Antimicrobial Peptides into Non-oriented Samples for ^{31}P or ^{2}H NMR Spectroscopy of Lipids (Concentration Series of 0–6 mol% Magainin 2 in POPC) (see **Notes 6** *and* **7***)*

1. In order to remove the TFA counter ions that are usually associated with peptides after cleavage from resins and/or after purification by HPLC, dissolve the peptide at a concentration of 1 mg/mL in 2 mM HCl and lyophilize (23). The cycle of solubilization in 2 mM HCl and freeze-drying should be repeated at least three times (*see* **Note 8**).
2. Prepare a stock solution of about 10 mg peptide in 2 mL TFE/100 μL H_2O (5 mg/mL). When handling organic solvents use glass ware and avoid contact with plastic and rubber as these may dissolve and/or release organic molecules (*see* **Note 1**).
3. Prepare a stock solution of 100 mg of POPC in 2 mL TFE (50 mg/mL) (*see* **Note 2**).
4. Mix lipid and peptide solution in glass tubes (or small round bottom flasks) that should have enough volume to accommodate ≥2 mL of solution, position the glass tubes in contact with a warm water bath, and remove the bulk solvent with a gentle stream of nitrogen (*see* **Note 9**) in order to deposit a lipid film on the walls of the glass tube, and thereafter expose the film to high vacuum overnight (**Table 14.1**).

Table 14.1
Example of how stock solutions are mixed to prepare a series of phospholipid membranes with increasing peptide-to-lipid ratio

Peptide (mol%)	50 mg/mL lipid stock (μL) (MW 760)	5 mg/mL peptide stock (μL) (MW 3,000)	Water (μL)	TFE (μL)
1	400	156	57	727
2	400	313	49	578
3	400	470	42	428
6	400	940	0	0
0 (last to be pipetted)	400	0	19	921

5. Hydrate the film with 200 μL of buffer (*see* **Note 3**) and close the tube with parafilm. Leave for about an hour or longer followed by vortexing. The addition of 3–4 glass beads of about 2 mm in diameter helps in transforming the film into a suspension of multilamellar vesicles (*see* **Note 10**).
6. To equilibrate well the samples submit to ≥7 freeze (using dry ice in isopropanol)/thaw (using lukewarm water)/vortex cycles (*see* **Note 11**).

7. Transfer into sample tubes that fit the NMR coil (*see* Step 6 of **Section 2.1**) by using a micropipette. The membranes are sticky, therefore it is necessary to increase the opening of the tip by cutting off the very end. For the same reason it is recommended to transfer several small quantities which avoids the wetting of a large inner surface of the tip. A final centrifugation step allows one to collect the membrane suspension in the bottom of the tube, e.g. by placing the NMR tube in an microcentrifuge tube (e.g. Eppendorf tube) and by slowly spinning in a corresponding centrifuge (e.g. Eppendorf centrifuge at 1,000–2,000 rpm). The inner length of the tubes and/or the positioning of the lids should be adjusted to confine all samples in the centre of the NMR coil. The position in the coil can be stabilized with Teflon tape. For MAS experiments the suspension can be transferred directly into a zirconium MAS rotor (*see* **Section 3.3.4**).

3.2. Reconstitution of Antimicrobial Peptides into Oriented Membranes (POPC) for ^{15}N, ^{13}C, and/or ^{2}H Solid-State NMR Measurements of the Peptides (at a Lipid/Peptide Ratio of 50)

1. In order to remove the TFA counter ions that are usually associated with peptides after cleavage from resins and/or after purification by HPLC, dissolve the peptide at a concentration of 1 mg/mL in 2 mM HCl and lyophilize (23). The cycle of solubilization in 2 mM HCl and freeze-drying should be repeated at least three times (*see* **Note 8**).
2. Dissolve about 12 mg peptide in 2 mL TFE in a glass tube which leaves sufficient space for subsequent manipulations (e.g. 10 mL).
3. Adjust pH to approximately neutrality by the successive additions of small amounts of 1 N NaOH (5–10 μL at a time). The presence of 70 μL of 200 mM Tris–HCl, pH 7.4, helps during the titration (but *see* **Note 3**). The pH is controlled after each addition of base by removing small quantities of peptide solution using a 10 μL capillary and the application of a small droplet of this solution onto pH indicator paper. Once the organic solvent has evaporated a drop of Milli-Q water is added to read out the pH from the colour change (*see* **Note 12**).
4. Add 150 mg of POPC to the solution (*see* **Note 13**).
5. Concentrate the solution until it turns viscous (to about 1 mL) by using a gentle stream of nitrogen (*see* **Note 9**), in order to be able to deposit small droplets that stick to the glass plates in the next step.
6. Carefully place about 30 clean ultra-thin microscope cover glasses side by side on two or three glass Petri dishes. Use fine watchmaker forceps to manipulate the glass plates. Apply equal quantities of the previous solution as small elongated droplets onto the central areas of all but one glass

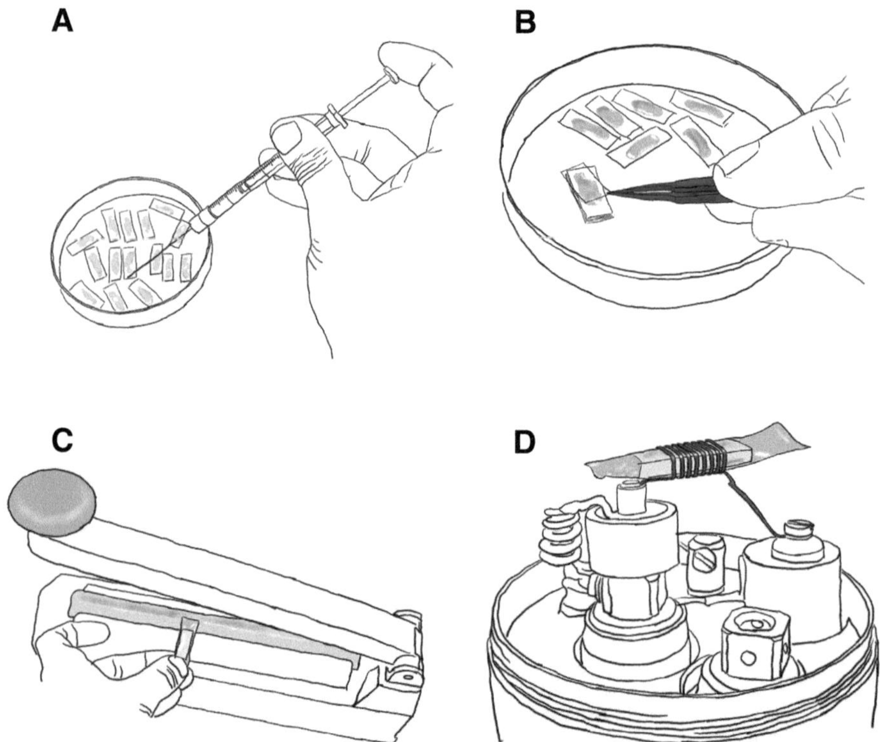

Fig. 14.2. The final steps of the preparation of oriented membrane samples are illustrated. (**a**) Application of viscous lipid/peptide solutions onto ultra-thin cover plates. (**b**) After elimination of the organic solvent and rehydration the gas plates are stacked on top of each other. (**c**) The sample is sealed into plastic foil and inserted into a flat coil solid-state NMR probe for measurement (**d**).

plates using a Hamilton syringe (**Fig. 14.2a**). Keep one glass plate on the side to be used during the stacking of the plates (during Step 8). Let the sample dry in air until most of the solvent has evaporated. It is possible to apply several layers onto the same regions when necessary. After the sample has been equally distributed and the bulk solvent has evaporated in air, expose the sample to high vacuum overnight. In our own laboratory we have connected a desiccator to the lyophilizer to allow the drying of many samples simultaneously in their horizontal position.

7. Thereafter the samples are placed in a closed chamber at 93% relative humidity (rh) and left to equilibrate for typically about 3 days. This can be achieved by establishing a closed atmosphere in contact with a KNO_3 saturated solution (*see* **Note 5**).
8. Once the lipid films become translucent and soft (*see* **Note 5**) the glass plates are stacked on top of each other, with one empty glass plate used to cover the last top surface (**Fig. 14.2b**). Membrane alignment can be improved

by shearing with delicate pressure and small movements. The stack is stabilized by wrapping with Teflon tape and the whole sample sealed in a plastic wrapping using an impulse sealer (**Fig. 14.2c**). The volume of the sample should fit tightly into the NMR coil (*see* Step 8 of **Section 2.2** and **Fig. 14.2d**).

3.3. NMR Measurements

In this paragraph the set-up procedure for three types of measurements is described. However, it is assumed that the beginner receives a hands-on introduction into the use of a solid-state NMR spectrometer, as well as the computer and pulse programmes. The latter should be installed and tested by an experienced user. Beware of the dangers associated with the high-field magnet and the high operating voltage of the electronic equipment. Some simple set-up procedures and pulse sequences such as standard ^{1}H acquisition, Hartmann–Hahn matching, or cross-polarization are also explained in an applications-oriented manner in reference (24).

3.3.1. ^{2}H Solid-State NMR Spectra (Oriented or Non-oriented)

1. Insert a sample of D_2O into the centre of the NMR coil. Close the probe head and insert the probe into the NMR magnet. With the amplifiers silent connect the ^{2}H (X-channel) electronic circuit of the probe to the console. During operation this line carries incoming radio frequency (RF) pulses from the console to the probe and outgoing signal from the sample via the preamplifier (that is tuned to the ^{2}H Larmor frequency) to the detector.
2. Tune and match the probe for the ^{2}H carrier frequency.
3. Using a simple pulse acquisition sequence does vary the pulse length to determine the maximal and the zero intensities (90° and 180° pulses, respectively). Adjust the carrier frequency to have the signal close to the centre of your spectral window and all the other parameters according to **Table 14.2**. Adjust the receiver gain to avoid saturation and good sensitivity at the same time. Using a short pulse, adjust and save the phases to have a pure positive signal intensity (Lorentzian). Calibrate the chemical shift scale by setting the maximum of the D_2O signal to 0. Adjust the power of the pulses to have a 90° pulse in the range of ≤ 5 μs. Once these parameters have been determined for a given spectrometer and probe, they provide a valuable starting point for future set-up procedures, which should then require only minor re-adjustments.
4. Using the ^{2}H power levels determined in Step 3 adjust the durations of a 90_x–τ–90_y–τ acquisition pulse programme (25), which is presented schematically in **Fig. 14.3a**. Adjust the 90° pulse as determined and all the other parameters

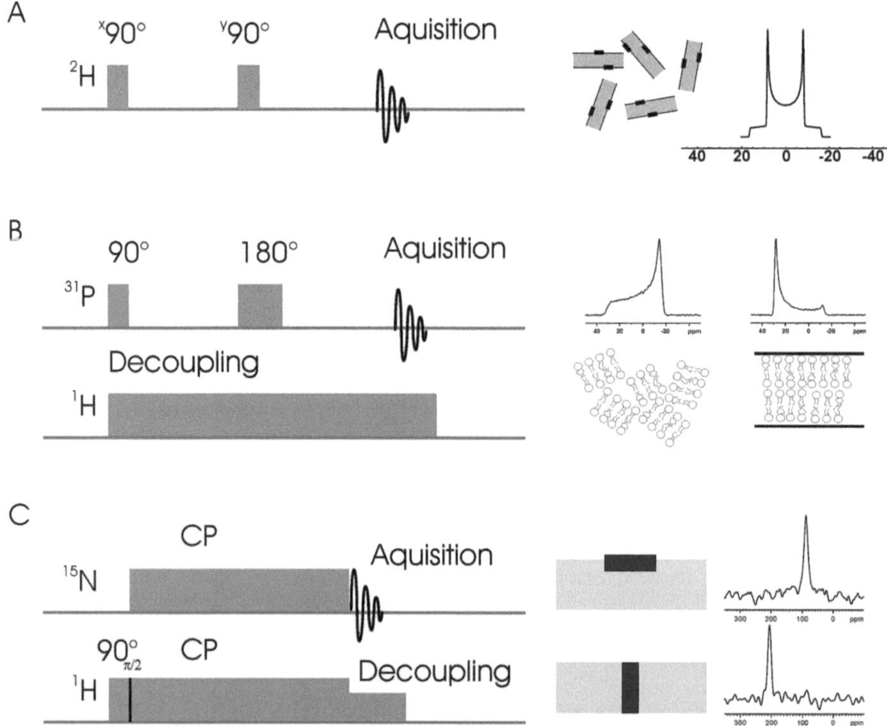

Fig. 14.3. Solid-state NMR pulse sequences and typical spectra of lipids and peptides: (**a**) Deuterium quadrupolar echo sequence where the phase of the pulse and receiver are cycled in the following manner (25): First 90°: x -y -x y x -y -x y; second 90°: -y -x y x y x -y -x; acquisition: y x -y -x y x -y -x (25). To the right of the pulse sequence are shown membrane fragments that are oriented randomly and in which are incorporated helical peptides. Furthermore, a motionally averaged 2H powder pattern line shape of peptides labelled with $^2H_3$-alanine is shown. (**b**) ^{31}P Hahn-echo sequence with proton decoupling (26). Typical spectra of unoriented and (partially) oriented samples are shown to the right together with an illustration of the phospholipid alignment. The phase cycles of the pulses and the receiver are (26): 90°: x y -x -y x y -x -y -x -y x y -x -y x y; 180°: y x y x -y -x -y -x x y x y -x -y -x -y; acquisition: x y -x -y. (**C**) ^{1}H-^{15}N cross-polarization experiment. The 90°-^{1}H pulse is phase shifted by π/2 relative to the phase of the subsequent ^{1}H CP contact pulse. Oriented membranes carrying in-plane oriented (*top*) or transmembrane helical peptides (*bottom*) are schematically illustrated and typical ^{15}N spectra resulting from such configurations are shown to the *right*.

according to **Table 14.2**. Using the reference sample the performance of the pulse programme can be tested by adding a full-phase cycle (eight scans). Check that the oscillations (free induction decay, FID) recorded from each repetition add to the overall signal intensity, and for the absence of artefacts in its beginning (probe ringing may require longer τ) and of spikes (cleaning the coil may help). With some spectrometers, an installed lock unit may interfere with the signal. This problem can be solved by reducing/turning off the lock power.

Stably insert the membrane sample into the centre of the NMR coil (*see* **Note 4**). Reconnect the probe, tune, and match (Steps 1 and 2) for every new sample and after equilibrating to a new temperature. Use the parameters determined before

(**Table 14.2**). Readjust the receiver gain for the lipid sample to obtain the maximal sensitivity without saturating the receiver. The signal should become visible within minutes (lipids) to a few hours (peptides), and it should be possible to obtain a spectrum with reasonable signal-to-noise ratio in about 30–60 min (lipids) or overnight (peptides). Note that the signal from naturally abundant deuterium in water shows up first and can be reduced by using deuterium-depleted water for sample hydration.

3.3.2. Proton-Decoupled ^{31}P Solid-State NMR Spectra (Oriented or Non-oriented)

1. Insert a sample of 85% H_3PO_4 into the centre of the NMR coil. Close probe head and insert the probe into the NMR magnet. With the amplifiers silent connect the ^{31}P (X) and the ^{1}H electronic circuits of the probes to the respective electrical connections to allow during the following steps passage of incoming RF pulses from the console and outgoing signal from the sample via filters and the preamplifier that covers the ^{31}P Larmor frequency.
2. Tune and match the probe (^{31}P and ^{1}H electronic circuits).
3. Using a simple pulse acquisition sequence vary the pulse length to determine the maximal and the zero intensities (90° and 180° pulses, respectively). Using an isotropic liquid during this set-up procedure renders ^{1}H decoupling unnecessary, and it can even heat and damage the sample and/or probe head as long acquisition times are used to fully record the free induction decay. Adjust the carrier frequency to have the signal close to the centre of your spectral window and all the other parameters according to **Table 14.2**. Adjust the receiver gain to avoid saturation and good sensitivity at the same time. Using a short pulse, adjust and save the phases to have a purely positive signal intensity. Calibrate the chemical shift scale with the maximum of the phosphoric acid signal at 0. Adjust the power of the pulses to have a 90° pulse in the range of 5 μs. Once these parameters have been determined for a given spectrometer and probe, they provide a valuable starting point for future set-up procedures, which should then require only minor re-adjustments.
4. Change the spectrometer settings to acquire the ^{1}H signal. Adjust the power levels of the ^{1}H channel to have a 90° pulse length $\leq$7 μs and the ^{1}H carrier frequency using a procedure analogous to the previous step. This step is required only once in a while to measure or control the ^{1}H decoupling field strength and the ^{1}H frequency.
5. Using the ^{31}P and ^{1}H power levels determined in Steps 3 and 4 adjust the durations of the $90°–\tau–180°–\tau$ (acquisition with ^{1}H decoupling) pulse programme (26) graphically represented in **Fig. 14.3b**. Adjust the 90° and 180° pulses as determined and all the other parameters according to **Table 14.2**. Using the reference sample the performance of

Table 14.2
Typical solid-state NMR acquisition parameters for experiments on static samples. Use a copy of the table to keep track of your own values

Section 3.3.1			Section 3.3.2		
Experiment and substance	**Pulse acqu. ^2H_2O**	**^2H echo lipid/$^2H_3$-alanyl**	**Pulse acqu. 85% $^1H_3PO_4$**	**Pulse acqu. 85% $H_3{}^{31}PO_4$**	**^{31}P echo liquid disordered membranes**
Power level/90° pulse	___/___ (≤6 μs)	Idem	___/___ (≤7 μs)	___/___ (≤6 μs)	Idem
180 pulse length	n.a.	n.a.		____μs	Idem
Recycle delay (*see* **Note 14**)	1 s	0.3 s/1.5 s	2 s	2 s	2 s
τ delay	n.a.	40 μs	n.a.	n.a.	40 μs
Spectral width	10 kHz	100 kHz	40 kHz	30 kHz	40 kHz
Number of scans (typical values)	1	5,000/50,000	1	1	1,024
Acquisition time	51 ms	6 ms	200 ms	32 ms	10 ms
Receiver settings	________	______	________	________	________
Reference frequency	0 Hz @ ________Hz	Idem	max @ ________Hz	0 Hz @ ________Hz	Idem

Section 3.3.3		
Experiment and substance	**^1H–^{15}N cross-polarization $^{15}NH_4Cl$ powder**	**^1H–^{15} N cross-polarization peptide**
^1H power level and 90 pulse length	___/___ (≤ 8 μs)	Idem
Recycle delay (*see* **Note 14**)	5 s	3 s
Cross-polarization contact time	1 ms	1 ms
Spectral width	20 kHz	40 kHz
Number of scans (typical values)	4	20,000
Acquisition time	6 ms	3 ms
Receiver settings	________	______
Reference frequency	39.3 ppm @ ________Hz	Idem

n.a. – not applicable.

the pulse programme can be tested by adding a full-phase cycle (16 scans). Check the FID for being additive from scan to scan, and for the absence of artefacts in its beginning (probe ringing may require longer τ) and of spikes (reduction of the decoupling power and/or cleaning the coil may help).

6. Stably insert the sample into the centre of the NMR coil (*see* **Note 4**). Reconnect the probe, tune, and match (Steps 1 and 2) for each new sample and after equilibrating to a new temperature. Use the same parameters as before (**Table 14.2**) but reduce the recycle delay to 1.5–3 s. Readjust the receiver gain for the lipid sample for maximal sensitivity without saturating the receiver. The signal should become visible within minutes, and it should be possible to obtain a spectrum with good signal-to-noise ratio in about 30 min (for a sample size of 10 mg of phospholipid).

3.3.3. Proton-Decoupled ^{15}N Cross-Polarization Solid-State NMR Spectra (Oriented or Non-oriented)

1. Using a reference sample of $^{15}NH_4Cl$ follow the Steps 1 and 2 in **Section 3.3.2** in an analogous manner using the ^{15}N and ^{1}H electronic circuits. An analogous procedure to the one described here can be applied to set up $^{15}N/^{1}H$ cross-polarization experiments using adamantane.
2. Adjust the power levels of the ^{1}H channel to have a 90° pulse length ≤ 7 μs and the ^{1}H carrier frequency using a simple pulse acquisition sequence analogous to Step 3 of **Section 3.3.2** (*see* **Note 15**).
3. Switch to the acquisition of the ^{15}N signal. It is recommended to use a cross-polarization sequence (24) with the possibility to vary the power levels of the 90° ^{1}H pulse, the ^{1}H and ^{15}N contact pulses and the ^{1}H decoupling independently (**Fig. 14.3c**). Keep the ^{1}H power level fixed as determined in Step 2 for all intervals (90° pulse, contact time, and decoupling during acquisition), set the length of the 90° ^{1}H pulse as determined, choose a contact time of $\leq$ 5 ms, and vary the power level of the ^{15}N channel to obtain the maximal intensity (*see* **Note 15**). Once you obtain signal adjust the ^{15}N carrier frequency to have the signal close to the centre of your spectral window, correct, and save the phases for a positive ^{15}N signal. In case major corrections of the frequencies had to be made check and re-adjust the power levels for maximal signal intensity. Use the parameters according to **Table 14.2**. Adjust the receiver gain to avoid saturation and permit good sensitivity at the same time. Calibrate the chemical shift scale with the maximum of the ammonium peak at 39.3 ppm. Once these parameters have been determined for a given spectrometer and probe, they provide a valuable starting point for future set-up procedures, which

should then require only minor corrections. By constantly comparing the noise level and the signal intensity with previous settings using the same reference sample, it is possible to keep track of the performance of the probe and the spectrometer.

4. Using the reference sample the performance of the pulse programme can be tested by adding a full-phase cycle (typically, eight scans). Check the growing FID for being additive from scan to scan.
5. Stably insert the sample into the centre of the NMR coil, tune, and match the ^{15}N and ^{1}H channels for each new sample.
6. Switch to the acquisition of ^{1}H. Record the ^{1}H NMR spectrum of the sample and adjust the ^{1}H power levels to obtain the same 90°/180° pulse length as determined in Step 2 and used during Step 3. Depending on the content of water and salt significant increases in power may be necessary (typically by 1–3 dB) to obtain the same probe performance (i.e. effective B_1 field).
7. Switch to the cross-polarization pulse programme and switch the spectrometer to the acquisition of ^{15}N. Adjust the following parameters by keeping the previous ^{15}N settings, but by using the new ^{1}H power levels: 90° ^{1}H pulse as determined in Steps 2 and 6, contact time 0.8–1 s, spectral width 40 kHz, and an acquisition time of about 3 ms **(Table 14.2)**. Adjust the receiver gain to avoid saturation and permit good sensitivity at the same time. Use the chemical shift calibration of Step 3. Check the first acquisitions (mostly noise) for the absence of artefacts in its beginning and spikes throughout the FID (reduction of the decoupling power and/or cleaning the coil may help). During acquisition the sample should be cooled with a stream of air, which can be humidified by bubbling through a gas wash bottle provided that acquisition is at ≥ room temperature (if the temperature of the water bath is above that of the connecting lines and probe water condensation will occur at the colder sites). It is recommended to check the tuning of the ^{1}H channel after a few hours of accumulation as water loss or redistribution within the sample can occur due to RF heating.

3.3.4. Magic Angle Spinning Experiments

3.3.4.1. Transferring Crystalline or Nano-crystalline Solids into the MAS Rotor

1. Powder the sample finely with mortar and pestle.
2. Transfer the sample into the NMR rotor (e.g. zirconium) using the appropriate funnel in several steps and compress between the steps. To ensure stable spinning, a long life of the probe head stator and sensitivity the sample should be homogeneously filled and positioned in the middle of the

coil. If there is not enough sample available the use of inserts allows one to centre the sample in the middle of the rotor.

3. Close the sample with the appropriate cap (e.g. Kel F; however, if the rotor or cap material creates background signals other material should be chosen) and make sure that the cap is properly inserted and fits tightly the rotor.

3.3.4.2. Transferring Wet or Soft Solids (Membrane Samples) into the MAS Rotor

1. Transfer the sample with a spatula into the rotor in several steps. Pellet the sample into the rotor in between steps using a microfuge (e.g. Eppendorf centrifuge). Special home-made devices can be prepared to directly centrifuge the sample into the rotor.
2. Close the rotor with the appropriate cap (*see* Step 3 of **Section 3.3.4.1**). Use lids or specially designed inserts for wet samples to avoid the accidental opening of the rotor during rotation (*see* Step 6 of **Section 2.1**).

3.3.4.3. Setting up Proton-Decoupled ^{13}C or ^{15}N CP/MAS Spectra

1. Set the magic angle following the manufacturers' instructions. Usually this is done by maximizing the sideband intensities of KBr or $NaNO_3$ spectra or the rotational echoes (spikes) on the FID. Once this calibration has been performed it is usually stable over long time periods.
2. Insert the reference sample (natural abundance adamantane for ^{13}C, $^{15}NH_4Cl$ for ^{15}N, natural abundance 85% phosphoric acid or phosphoserine for ^{31}P). The changes in the proton wobbling curve can be used to follow the insertion process.
3. Start spinning following manufacturer's instructions. Start slowly to allow the sample to pack into the rotor under centrifugal forces. The ^{1}H wobbling curve and MAS spinning frequency can be used to monitor spinning stability.
4. Set the spinning speed at the desired value respecting probe head and rotor specifications.
5. Tune and match both channels of the probe head.
6. Follow Steps 2–7 of **Section 3.3.3** using typical values reported in **Table 14.3** to set up cross-polarization experiments using adamantane or $^{15}NH_4Cl$ (see **Note 15**). Follow Steps 1–4 of **Section 3.3.2** and the values reported in **Table 14.3** for the set-up of ^{31}P NMR spectra using 85% H_3PO_4 under non-spinning conditions. Under MAS conditions the spectra of the lipids are acquired using a simple 90° (acquisition with ^{1}H decoupling) sequence.

Table 14.3
Typical solid-state NMR acquisition parameters for experiments under MAS conditions. Use a copy of the table to keep track of your own values

Section 3.3.4

Experiment and substance	**^{1}H–^{15}N cross-polarization $^{15}NH_4Cl$ powder**	**^{1}H–^{15}N cross-polarization peptide**
^{1}H power level and 90 pulse length	___/___ (≤ 8 μs)	Idem
Recycle delay (*see* **Note 14**)	5 s	5 s
Cross-polarization contact time	1 ms	1 ms
Spectral width	40 kHz	40 kHz
Number of scans (typical values)	4	1,024
Acquisition time	300 ms, reduce decoupling power!!!!	30 ms
Receiver settings	___________	________
Reference frequency	39.3 ppm @ __________Hz	Idem
Experiment and substance	**^{1}H–^{13}C cross-polarization adamantane powder**	**^{1}H–^{13}C cross-polarization peptide**
^{1}H power level and 90 pulse length	___/___ (≤8 μs)	Idem
Recycle delay (*see* **Note 14**)	5 s	5 s
Cross-polarization contact time	1 ms	1 ms
Spectral width	40 kHz	40 kHz
Number of scans (typical values)	4	1,024
Acquisition time	300 ms, reduce decoupling power!!!!	30 ms
Receiver settings	___________	________
Reference frequency	38.5 and 29.5 ppm @ __________Hz	Idem

Table 14.3 (continued)

Experiment and substance	^{31}P simple acquisition H_3PO_4	^{1}H-decoupled ^{31}P phospholipid sample
^{1}H ppwer level and 90 pulse length	___/___ (≤ 8 μs)	Idem
Recycle delay (*see* **Note 14**)	10 s	3 s
Cross-polarization contact time	n.a.	n.a.
Spectral width	20 kHz	25 kHz
Number of scans (typical values)	4	1,024
Acquisition time	300 ms, reduce decoupling power!!!!	100 ms
Receiver settings	_________	_______
Reference frequency	0 ppm @ _________Hz	_______

n.a. – not applicable.

4. Notes

1. It is pivotal to find a solvent system where peptides, lipids, and other constituents are homogenously mixed. For amphipathic peptides this can usually be achieved in binary and ternary mixtures of DCM, TFE, and water (e.g. 1.2/1.8/0.3), DCM/methanol or in 2 mL HFIP or formic acid (18). The solubility of cationic sequences is usually larger in acidic environments, and solubilization can be accelerated by vigorous mixing or sonication. The use of simple systems made up of only one (or two) volatile solvent is advantageous. In multi-component solvent mixtures the more volatile solvents, such as DCM, evaporate earlier and phase separations may occur. In these instances this solvent can be re-added during the drying process. Also, in some instances it has been observed that suspensions form during the drying process, which in most cases was without detrimental effect on the quality of the sample alignment (*see* **Section 3.2**).
2. Often it is desired to test the effect of membrane lipid composition on the peptide–lipid interactions. On one hand POPC represents well the phospholipid composition of many biological membranes, as the PC head group is abundant in the bilayers of higher organisms. The composition of these membranes is furthermore mirrored by a saturated fatty acyl chain at position 1 and an unsaturated chain at position 2 and POPC represents well their average hydrophobic thickness. Furthermore, the negative charges of eukaryotic membranes are often carried by PS, and these membranes are also rich in cholesterol. On the other hand, a simple bacterial membrane mimic could be PE/PG 3:1. Notably, however, many of these lipid components exhibit low solubility in alcohols. Therefore, dichloromethane (or chloroform) provides an alternative for most long-chain lipids (cf., for example, the catalogue information by Avanti Polar Lipids).
3. A wide variety of buffers can be used including Tris, HEPES, or citrate buffer. However, the interference of the buffer with the NMR experiments should be considered. For example, phosphate buffers result in large isotropic peaks hampering the analysis of ^{31}P NMR powder patterns, water contains a significant amount of deuterium (which can be reduced by using deuterium-depleted water), and other buffers cause background signals in ^{13}C NMR experiments. Furthermore, high salt

concentrations interfere with the tuning of NMR probes and the radio frequency irradiation during NMR spectroscopy. Also, extreme pH values cause the hydrolysis of ester bonds and thereby lipid degradation. Long-term stability of lipids can be assured by working at pH 6–7 (27), reduced hydration and/or storage in the freezer. When oriented samples are prepared at <100% rh it should be further considered that buffer and salt are concentrated into the small intermembrane space where the residual water molecules interact with the adjacent membranes. We have noted that the presence of buffer and salts often degrades the quality of the membrane alignment (e.g. 28).

4. When a MAS probe is used for static NMR measurements the samples can be exchanged and transferred into the NMR coils using the sample lift or other fast-loading devices. The sample positioning within the coil results in a visible shift of the tuning parameters (observe wobbling curve or reflected power during loading process). No matters as convenient, the efficiency of the probe is reduced due to losses in filling factor and the inclination of the coil at 54.7° relative to the magnetic field direction. This has not been a problem for proton-decoupled ^{31}P NMR investigations (12), but would certainly hamper the acquisition of spectra from peptides which are considerably more demanding in their NMR set-up procedures.

5. The equilibration period for hydration ranges from a few hours up to a few days. Hydration can be accelerated by passing through a vapour steam or by adding microlitre amounts of water per glass plate. Diverse equilibration conditions are obtained by contact of the atmosphere with saturated solutions of LiCl (15% rh), K_2CO_3 (44% rh), NaBr (57% rh), NaCl (75% rh), KCl (86% rh), KNO_3 (93% rh), or pure water (100% rh) (29). Extended equilibration and storage at 100% rh should be avoided as this promotes microbial growth. Hydration chambers can be easily prepared by filling a plastic box (e.g. lunch box, tupperware) with salt and some water. An inverted Petri dish serves as a platform to avoid the direct contact of the salt solution with the sample.

6. Here a protocol is described in which peptide and lipids are mixed in organic solvent before the membranes are formed in aqueous solution. It is also possible to prepare samples in which the peptide is added in aqueous buffer during the formation of vesicles. In these cases the protocol given in **Section 3.1** has to be modified in the following manner: omit Step 2 as well as the addition of peptide and additional solvent as reported in Step 4. The peptide is then

dissolved in the buffer at the desired quantities and added to the lipid films as in Step 5.

7. The same protocol can be used to prepare a non-oriented sample of peptides labelled with ^{15}N, $^{2}H_3$, and/or ^{13}C (14). To overcome the problems of sensitivity the size per sample has to be up-scaled to about 40 mg of peptide (if labelled selectively at one site per isotope) and 400 mg of lipid per sample.
8. Higher HCl concentrations have been used but there is a danger of structural alterations of the peptide (23, 30). Alternatively it is possible to use HPLC, ion exchange or deprotonation/reprotonation cycles (31). In some technical notes exchange of TFA by dissolving the peptides in glacial acetic acid or 10% acetic acid combined with lyophilization cycles is also suggested. It is important to note that the counter ions contribute to the peptide molecular weight. These are, e.g. acetate (MW 59), TFA (MW 113), Na^+ (MW 23), and/or Cl^- (MW 36) and, if not considered, can make a significant difference in the resulting peptide concentrations or peptide-to-lipid molar ratios.
9. It is convenient to direct the nitrogen stream through a Pasteur pipette directly to the surface of the solution. Carefully adjust the flow of nitrogen as turbulences may cause spray dispersion of the liquid and the loss of sample. The glass pipettes can be exchanged easily to avoid cross-contamination from one preparation to the next.
10. To ensure full and more homogenous hydration larger amounts of buffer (e.g. 5 mL) have been added and the resulting membranes have been pelleted by centrifugation at 21,000*g* for 20 min (6). In such cases membrane association should be strong enough to assure that almost all peptide is associated with the membrane pellet. Alternatively, the peptide concentration remaining in solution can be determined, for example, by optical or NMR spectroscopy, compared to the total amount of peptide and lipid added, and the membrane partitioning constant calculated.
11. For freezing the lipid samples liquid nitrogen is also used. However, by doing so we have observed the occurrence of isotropic signal intensities in ^{31}P NMR spectra of pure lipids which may be related to the speed of freezing and the corresponding ice/glass formation.
12. The sample pH can be controlled by scratching a small quantity of the final sample and applying it onto wet pH paper. The method described in Step 3 of **Section 3.2**

works rather well as closely related titration curves have been obtained when the same peptide–lipid mixtures have been overlaid with defined buffer solutions for ATR-FTIR measurements (3, 32).

13. It should be noted that it is more difficult to re-hydrate PE, PG, and other membranes when compared to bilayers composed of PC or PS. In order to obtain well-oriented membranes, where hydration is an issue, a 1:1 molar ratio of naphthalene:lipid–peptide has been added during the preparation (33). Naphthalene (MW 128) is soluble in DCM at 100 mg/mL but also in some alcohols (e.g. methanol/ethanol at 77 mg/mL). After high vacuum drying (for several days, to remove all naphthalene) the resulting open spaces help in the hydration of these membranes.
14. The recycle delays indicated are values that allow for convenience during the set-up of experiments, in some instances in steady state rather than equilibrium conditions. At about 10 T (^{1}H @ 300–500 MHz) and ambient temperatures the T_1 relaxation times of $^1H_3PO_4 \approx 150$ ms for ^{1}H and $\approx$ 300 ms for ^{31}P, $^2H_2O \approx 400$ ms, ^{1}H in adamantane (MAS) $\approx$ 1.6 s, ^{1}H in solid $^{15}NH_4Cl$ (cross-polarization) $\approx$ 600–1,000 ms, ^{1}H in membrane sample (cross-polarization) $\approx$ 300 ms. To avoid saturation of the signal the recycle delay should be $\geq 5\ T_1$, but the interval should also be long enough to avoid extensive sample heating which can be a problem particularly during cross-polarization of biological samples.
15. Once the approximate settings of ^{1}H power and pulse length have been obtained with the probe and a similar type of samples is being used, Step 2 can be omitted as these parameters can be optimized in further experiments by first adjusting the Hartmann–Hahn matching conditions (Step 3) and then by variation of the ^{1}H power used during the cross-polarization interval (**Fig. 14.3c**). At the Hartmann–Hahn matching condition of static samples the X and the ^{1}H spins precess around the B_1 fields at the same speed (i.e. $\gamma_X B_{1X} = \gamma_H B_{1H}$). As the precession frequency is calculated from the 90° pulse length (t_p) by $\nu_1 = 1/4t_p$ this also means that the t_p intervals are identical for both nuclei. Matching angle spinning changes the Hartmann–Hahn condition to $\gamma_X B_{1X} - \gamma_H B_{1H} = 2\pi\nu_r$, where ν_r is the spinning frequency.

Acknowledgments

We are thankful to Vaincre la Mucoviscidose (TG0101), the Association pour la Recherche sur le Cancer (No. 3100), the Agence Nationale pour la Recherche, the RMN Grand-Est Network of the Ministry of Recherche, the International Center for Frontier Research in Chemistry, and the European Union (MCRTN 33439-Biocontrol) for supporting our projects. The Institute for Supramolecular Chemistry of the University of Strasbourg (ISIS) is acknowledged for hosting the laboratory.

References

1. Bechinger, B. (1999) The structure, dynamics and orientation of antimicrobial peptides in membranes by solid-state NMR spectroscopy. *Biochim. Biophys. Acta* **1462**, 157–183.
2. Bechinger, B., Kim, Y., Chirlian, L. E., Gesell, J., Neumann, J.-M., Montal, M., Tomich, J., Zasloff, M., and Opella, S. J. (1991) Orientations of amphipathic helical peptides in membrane bilayers determined by solid-state NMR spectroscopy. *J. Biomol. NMR* **1**, 167–173.
3. Bechinger, B. (1996) Towards membrane protein design: pH dependent topology of histidine-containing polypeptides. *J. Mol. Biol.* **263**, 768–775.
4. Vogt, T. C. B. and Bechinger, B. (1999) The interactions of histidine-containing amphipathic helical peptide antibiotics with lipid bilayers: the effects of charges and pH. *J. Biol. Chem.* **274**, 29115–29121.
5. Bechinger, B. (2009) Rationalizing the membrane interactions of cationic amphipathic antimicrobial peptides by their molecular shape. *Curr. Opin. Colloid Interface Sci.* 14, 349–355.
6. Mason, A. J., Martinez, A., Glaubitz, C., Danos, O., Kichler, A., and Bechinger, B. (2006) The antibiotic and DNA-transfecting peptide LAH4 selectively associates with, and disorders, anionic lipids in mixed membranes. *FASEB J.* **20**, 320–322.
7. Bechinger, B., Zasloff, M., and Opella, S. J. (1993) Structure and orientation of the antibiotic peptide magainin in membranes by solid-state NMR spectroscopy. *Protein Sci.* **2**, 2077–2084.
8. Bechinger, B., Zasloff, M., and Opella, S. J. (1998) Structure and dynamics of the antibiotic peptide PGLa in membranes by multidimensional solution and solid-state NMR spectroscopy. *Biophys. J.* **74**, 981–987.
9. Bechinger, B. and Sizun, C. (2003) Alignment and structural analysis of membrane polypeptides by 15 N and 31P solid-state NMR spectroscopy. *Concepts Magn. Reson. A* **18A**, 130–145.
10. Aisenbrey, C. and Bechinger, B. (2004) Investigations of peptide rotational diffusion in aligned membranes by ^{2}H and ^{15}N solid-state NMR spectroscopy. *J. Am. Chem. Soc.* **126**, 16676–16683.
11. Cullis, P. R. and De Kruijff, B. (1979) Lipid polymorphism and the functional roles of lipids in biological membranes. *Biochim. Biophys. Acta* **559**, 399–420.
12. Bechinger, B. (2005) Detergent-like properties of magainin antibiotic peptides: a ^{31}P solid-state NMR study. *Biochim. Biophys. Acta* **1712**, 101–108.
13. Cady, S. D., Goodman, C., Tatko, C. D., DeGrado, W. F., and Hong, M. (2007) Determining the orientation of uniaxially rotating membrane proteins using unoriented samples: a ^{2}H, ^{13}C, and ^{15}N solid-state NMR investigation of the dynamics and orientation of a transmembrane helical bundle. *J. Am. Chem. Soc.* **129**, 5719–5729.
14. Prongidi-Fix, L., Bertani, P., and Bechinger, B. (2007) The membrane alignment of helical peptides from non-oriented ^{15}N chemical shift solid-state NMR spectroscopy. *J. Am. Chem. Soc.* **129**, 8430–8431.
15. Salnikov, E. S., Mason, A. J., and Bechinger, B. (2009) Membrane order perturbation in the presence of antimicrobial peptides by ^{2}H solid-state NMR spectroscopy. *Biochimie* **91**, 734–743.
16. Aisenbrey, C. and Bechinger, B. (2004) Tilt and rotational pitch angles of membrane-

inserted polypeptides from combined 15 N and 2H solid-state NMR spectroscopy. *Biochemistry* **43**, 10502–10512.

17. Aisenbrey, C., Sizun, C., Koch, J., Herget, M., Abele, U., Bechinger, B., and Tampe, R. (2006) Structure and dynamics of membrane-associated ICP47, a viral inhibitor of the MHC I antigen-processing machinery. *J. Biol. Chem.* **281**, 30365–30372.
18. Bechinger, B., Bertani, P., Werten, S., Mendonca de Moraes, C., Aisenbrey, C., Mason, A. J., Perrone, B., Prudhon, M., Sudheendra, U. S., and Vidovic, V. (2009) The structural and topological analysis of membrane polypeptides by oriented solid-state NMR spectroscopy: sample preparation and theory. In *Membrane-Active Peptides: Methods and Results on Structure and Function*. M. Castanho (Ed.), in press. La Jolla: International University Line.
19. Vidovic, V., Prongidi-Fix, L., Bechinger, B., and Werten, S. (2009) A versatile and highly efficient method for the production of antimicrobial peptides in *Escherichia coli*. *J. Pept. Sci.* **15**, 278–284.
20. Salnikov, E. S., Friedrich, H., Li, X., Bertani, P., Hertweck, C., Reissmann, S., O'Neil, J. D. J., Rapp, J., and Bechinger, B. (2009) Structure and alignment of the membrane-associated peptaibols ampullosporin A and alamethcin by ^{15}N and ^{31}P solid-state NMR spectroscopy. *Biophys. J.* **96**, 86–100.
21. Bechinger, B. and Opella, S. J. (1991) Flat-coil probe for NMR spectroscopy of oriented membrane samples. *J. Magn. Reson.* **95**, 585–588.
22. Nielsen, N. C., Daugaard, P., Langer, V., Thomsen, J. K., Nielsen, S., Sorensen, O. W., and Jakobsen, H. J. (1995) A flat-coil NMR probe with hydration control of oriented phospholipid-bilayer samples. *J. Biomol. NMR* **5**, 311–314.
23. Andrushchenko, V. V., Vogel, H. J., and Prenner, E. J. (2007) Optimization of the hydrochloric acid concentration used for trifluoroacetate removal from synthetic peptides. *J. Pept. Sci.* **13**, 37–43.
24. Berger, S. and Braun, S. (2004) *200 and More Basic NMR Experiments: A Practical Course*. Weinheim: Wiley-VCH Verlag.
25. Davis, J. H., Jeffrey, K. R., Bloom, M., Valic, M. I., and Higgs, T. P. (1976) Quadrupolar echo deuteron magnetic resonance spectroscopy in ordered hydrocarbon chains. *Chem. Phys. Lett.* **42**, 390–394.
26. Rance, M. and Byrd, R. A. (1983) Obtaining high-fidelity spin-1/2 powder spectra in anisotropic media: phase-cycled Hahn echo spectroscopy. *J. Magn. Res.* **52**, 221–240.
27. Ottiger, M. and Bax, A. (1998) Characterization of magnetically oriented phospholipid micelles for measurement of dipolar couplings in macromolecules. *J. Biomol. NMR* **12**, 361–372.
28. Salnikov, E. S., Friedrich, H., Li, X., Bertani, P., Reissmann, S., Hertweck, C., O'Neil, J. D., Raap, J., and Bechinger, B. (2009) Structure and alignment of the membrane-associated peptaibols ampullosporin A and alamethicin by oriented ^{15}N and ^{31}P solid-state NMR spectroscopy. *Biophys. J.* **96**, 86–100.
29. O'Brian, F. E. M. (1948) The control of humidity by saturated salt solutions. *J. Sci. Instr.* **25**, 73–76.
30. Sani, M. A., Loudet, C., Grobner, G., and Dufourc, E. J. (2007) Pro-apoptotic bax-alpha1 synthesis and evidence for beta-sheet to alpha-helix conformational change as triggered by negatively charged lipid membranes. *J. Pept. Sci.* **13**, 100–106.
31. Roux, S., Zekri, E., Rousseau, B., Paternostre, M., Cintrat, J. C., and Fay, N. (2008) Elimination and exchange of trifluoroacetate counter-ion from cationic peptides: a critical evaluation of different approaches. *J. Pept. Sci.* **14**, 354–359.
32. Bechinger, B., Ruysschaert, J. M., and Goormaghtigh, E. (1999) Membrane helix orientation from linear dichroism of infrared attenuated total reflection spectra. *Biophys. J.* **76**, 552–563.
33. Hallock, K. J., Henzler, W. K., Lee, D. K., and Ramamoorthy, A. (2002) An innovative procedure using a sublimable solid to align lipid bilayers for solid-state NMR studies. *Biophys. J.* **82**, 2499–2503.
34. Glaubitz, C. and Watts, A. (1998) Magic angle oriented sample spinning (MAOSS) – A new approach toward biomembrane studies. *J. Magn. Reson.* **130**, 305–316.
35. Sizun, C. and Bechinger, B. (2002) Bilayer samples for fast or slow magic angle oriented sample spinning solid-state NMR spectroscopy. *J. Am. Chem. Soc.* **124**, 1146–1147.

Chapter 15

Use of Atomic Force Microscopy as a Tool to Understand the Action of Antimicrobial Peptides on Bacteria

Ang Li, Bow Ho, Jeak Ling Ding, and Chwee Teck Lim

Abstract

Atomic force microscopy (AFM) has been extensively used to image the three-dimensional surface morphology of a broad range of biological samples, including Gram-negative bacteria, imaged in the presence of antimicrobial peptides (AMPs). Although this technique provides high molecular resolution, it only requires minimum sample treatment and can even be performed in liquid and at varying temperatures while keeping the bacterial cells viable. In this chapter, we describe an easy, fast, and yet effective method for preparing AMP-treated Gram-negative bacteria samples for AFM imaging. The results obtained using this method show a series of morphological changes of Gram-negative bacteria upon treatment with selected AMPs, thus providing vivid insights into the mechanisms of how AMPs perturb and destroy Gram-negative bacteria in a stepwise manner. Technical details for performing AFM so as to obtain reliable and high-resolution images will also be discussed, together with some possible artifacts and troubleshooting.

Key words: Atomic force microscopy, Gram-negative bacteria, bacteria surface morphology, antimicrobial peptides, lipopolysaccharide, Sushi peptides.

1. Introduction

1.1. Atomic Force Microscopy: A Powerful Tool

Facing the rising threats of infection by antibiotic-resistant bacteria and the Gram-negative bacterial lipopolysaccharide (LPS)-induced septic shock, antimicrobial peptides (AMPs) have been recognized as prospective treatment agents (1–3). Recent studies suggest that certain cationic peptides and proteins display antimicrobial effects but prevent LPS-induced septic shock through binding to LPS and neutralizing its lethal effects without adverse effect toward mammalian cells. Among these peptides are the

A. Giuliani, A.C. Rinaldi (eds.), *Antimicrobial Peptides*, Methods in Molecular Biology 618,
DOI 10.1007/978-1-60761-594-1_15, © Springer Science+Business Media, LLC 2010

Sushi peptides, S1 and S3, derived from the horseshoe crab Factor C, a serine protease that recognizes and binds Gram-negative bacteria and LPS with high affinity (4–7). Although it is typically accepted that AMP molecular charge and electrostatic interactions play critical roles in triggering perturbation of Gram-negative bacteria membrane, the exact mechanism which finally causes lysis of the bacteria is still unknown. In order for Sushi peptides and other AMPs to be developed into potential drugs to prevent septic shock and to combat increasing bacterial resistance toward traditional antibiotics, it is critical to understand their mechanisms of action.

Ever since the invention of the atomic force microscope (AFM) in 1986 by Binnig and co-workers in IBM Research Laboratory (8), AFM has been extensively applied in many fields of science. Not only can it image the cell surfaces and biomolecules with nanometer resolution, it can also be applied to measure molecular interaction forces at the pico-newton scale. AFM has several advantages that make it particularly attractive to biologists. First, it can yield high-resolution, three-dimensional visualization of biological materials under a wide variety of conditions. As there is no need for staining or coating and no special requirement on the sample conductivity property, the biological samples can be easily prepared and examined. Second, it enables the assessment of material properties and biophysical characterization of biological materials due to its ability to measure forces at the molecular level. As such, AFM imaging and AFM-based material property characterization have been established as powerful techniques that enable in vitro studies of highly dynamic biological processes. This approach has greatly revolutionized our knowledge over a broad range of interesting biological subjects including the action of AMPs against Gram-negative bacteria (9).

The basic principle of operation of the AFM is shown in **Fig. 15.1**. AFM scans the surface of a sample by "feeling" rather than "looking." The core of the technique is a sharp tiny tip mounted at the end of a flexible cantilever which scans the sample surface in a raster scanning fashion. Precise lateral and vertical displacement of the sample with respect to the probe is achieved by a computer-controlled piezoelectric stage holding the sample, or conversely, the cantilever holder. During scanning, the interaction force between the tip and sample surface induces deflection of the cantilever which is recorded by a laser beam focused on the back of the cantilever and reflected toward a sensitive quadrant photodiode detector. A feedback loop system monitors the signal on the photodiode and adjusts the scanner position to achieve a constant interaction between the tip and sample (**Fig. 15.1**). The amount of feedback signal, measured at each scanning point of a 2D matrix, is displayed as representative image of the sample topography.

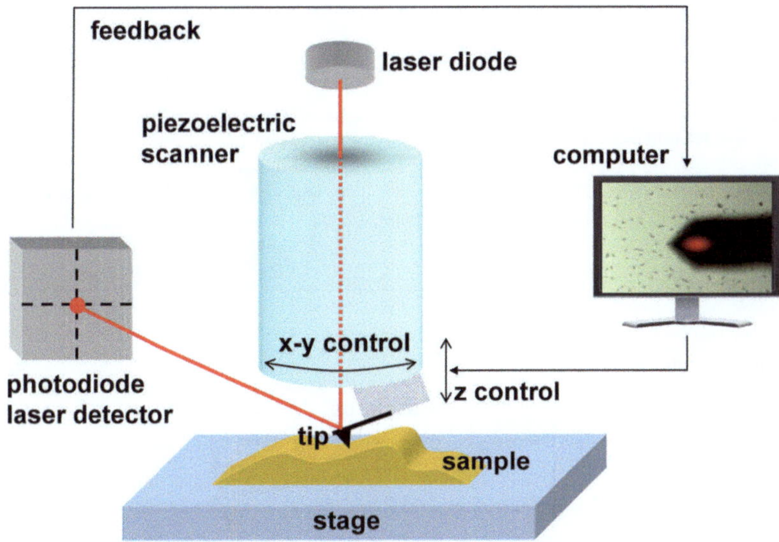

Fig. 15.1. Schematic drawing of the basic components and working principle of an AFM.

1.2. Imaging Modes

There are many imaging modes available when using AFM. Each mode has its advantages and disadvantages and some modes are only useful for a particular application. For imaging AMP-treated bacteria, the two most important and widely used imaging modes are discussed below.

1.2.1. Contact Mode

Here, the tip and sample remain in close contact as the scanning proceeds. The deflection of the cantilever (Δd in **Fig. 15.2A**) is sensed and maintained at a constant predefined level by the feedback control loop. At each scanning point, if the measured deflection is changed from the desired value due to topographical changes on the sample, the feedback control system adjusts the scanner position and moves the sample or tip in the appropriate direction to restore the preset value of deflection. The recorded scanner position is used as a measure of the real height of the features on the sample surface, whereas the change of cantilever deflection can be used to generate the error signal image or deflection image which sometimes gives a better 2D presentation of fine details of topography. The height image can be rendered into a 3D surface plot and the quantitative dimension information can be measured.

1.2.2. Tapping Mode

In order to minimize the lateral contact force, tapping mode (also called intermittent contact mode) was developed (10). This potent technique allows high-resolution topographic imaging of sample surfaces that are easily deformed, loosely attached to their substrate, or difficult to image by other AFM techniques. In the tapping mode, the cantilever is acoustically oscillated near

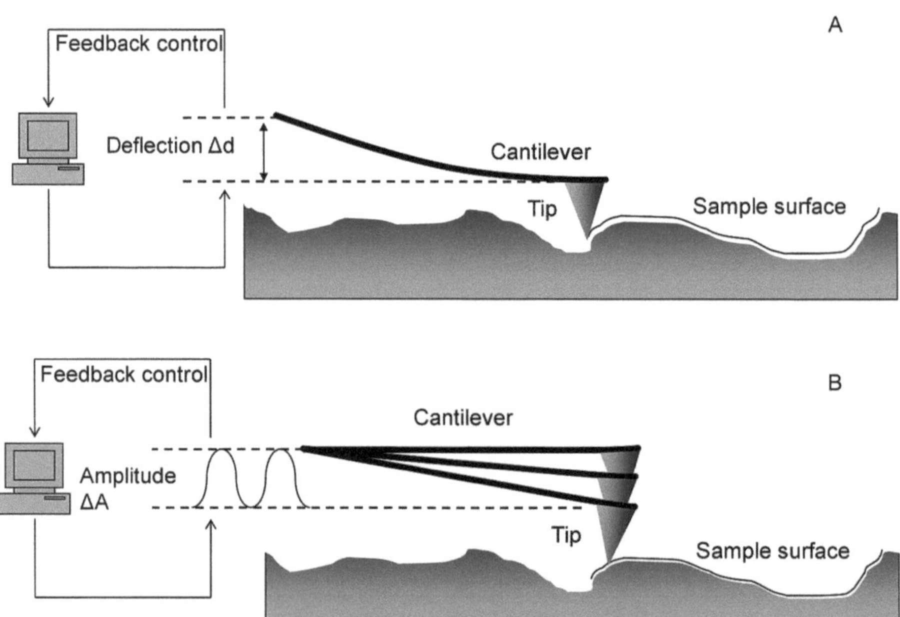

Fig. 15.2. Working principles of the contact mode and tapping mode. (**A**) Contact mode: the deflection of the cantilever is maintained as constant while the tip contacts with and scans over the sample surface. The scanner position represents the surface topography. (**B**) Tapping mode: the cantilever is oscillated near its resonant frequency. The image is generated from scanner position by keeping the damped oscillation amplitude constant during scanning.

its resonance frequency (at the order of kilohertz) by a piezoactuator in contact with cantilever supporting chip. Starting from the free oscillation amplitude (tens to hundreds of nanometers), when the cantilever approaches and hits the sample, the oscillation amplitude is dampened and the approach is stopped when the amplitude reaches a predefined value. During scanning, the tip "taps" the surface for a very short time in each oscillation period (**Fig. 15.2B**) and the oscillation amplitude (ΔA in **Fig. 15.2B**) is monitored and controlled as a constant and feedback on changes in ΔA is used to identify and measure surface features. In this manner, the tip still contacts the sample but for a very short period of time, so that lateral forces are dramatically reduced. Thus, when imaging poorly immobilized or soft samples, tapping mode would be a better choice than contact mode. One additional advantage with tapping mode is that besides displaying height image and amplitude image (resembling deflection image of contact mode), another mode of contrast called phase imaging is also available. This works by measuring the phase difference between the oscillations of the driving piezoactuator and the detected oscillations of cantilever. When the tip is in contact with the sample, the phase of oscillation is disturbed and is no longer precisely in the same step with the phase of the driving oscillation. This is because the tip transfers a small amount of energy to the sample when the two touch dynamically, and the amount of energy transferred is dependent on the material properties of the sample such as viscoelasticity and adhesiveness. Although the information in phase image cannot be interpreted to quantify the material properties, it is sometimes useful to obtain extra contrast and reveal "hidden" features that are invisible in the topographical images (11).

There have been many successful application examples of using AFM to study the morphological and structural changes of microbes induced by AMPs and antibiotics treatments (9, 12–15). In this chapter we present a simple and reliable method for preparing and imaging AMP-treated Gram-negative bacteria. Unfixed samples are prepared in ultrapure water and vacuum dried and imaged by tapping mode methodology. The protocol can be extended to in-fluid real-time imaging and force spectroscopy study if a suitable instrument is provided (16).

2. Materials

Sample preparation is as important – if not even more important than – as operating the AFM instrument for the successful recording of high-resolution and reliable data of normal and AMP-treated Gram-negative bacteria samples. Here, we outline a simple and yet efficient method to prepare unfixed dried Gram-negative bacteria samples for AFM observation. This approach avoids over-handling of the bacteria and maintains them in as native a condition as possible in order to prevent any potential misinterpretation due to artifacts. It ensures that the antimicrobial effects of S1 and S3 – and/or of other AMPs – on the bacterial morphology are efficiently imaged.

2.1. Bacterial Culture and Peptide Treatment

1. *Escherichia coli* ATCC 25922.
2. *Pseudomonas aeruginosa* ATCC 27853.
3. S3 peptide (Genemed Synthesis, San Francisco, CA, USA) (*see* **Note 1**).
4. Nutrient broth (Oxoid Ltd, Basingstoke, UK).
5. Shaker incubator (New Brunswick, NJ, USA).
6. LAL reagent water (Cambrex Bio Science, Walkersville, MD, USA).
7. Pyrogen- and nuclease-free ultrapure water (Millipore, Billerica, MA, USA).
8. Poly-L-lysine-coated microscope slides (Electron Microscopy Sciences, Hatfield, PA, USA).

9. Slide holder (Electron Microscopy Sciences Hatfield, PA, USA).
10. Nalgene® vacuum desiccator (Sigma-Aldrich, St. Louis, MO, USA).
11. Vacuum pump (KNF Neuberger, Freiburg-Munzingen, Germany).
12. Biological safety cabinet (Thermo Fisher Scientific, Waltham, MA, USA).

2.2. AFM Instrumentation

1. Dimension 3100 Atomic force microscope and Nanoscope III controller (Veeco Instrument, Santa Barbara, CA, USA).
2. Antivibration table and acoustic isolation hood (Veeco Instrument).
3. PPP-NCLR cantilever (Nanosensors, Neuchatel, Switzerland) (*see* **Note 2**).
4. Dehumidifier (*see* **Note 3**).
5. Ultraviolet ozone cleaner (Veeco Instrument).
6. TGT1 tip characterization grating (NT-MDT Co., Moscow, Russia).

3. Methods

The following section describes how *E. coli* and *P. aeruginosa* are cultured, harvested, and suspended in ultrapure water, then treated with S3 peptide and imaged by using tapping mode AFM.

3.1. Sample Preparation for AFM Imaging

1. Take the 1 mM S3 peptide (or other AMP) stock solution out from the freezer and place it in a beaker of ice to allow it to thaw slowly. Then dilute peptide from 1 mM to the desired concentrations in pyrogen-free LAL reagent water in sterile depyrogenated borosilicate tube and immerse the tube containing the peptide in a beaker of ice.
2. Prepare an overnight bacterial culture by incubating 5 mL sterile nutrient broth and an inoculum of 300 μL bacterial culture, and shake overnight at 230 rpm at 37°C in a shaker incubator. The concentration of the bacterial inoculum is calculated by correlating with the average optical density, OD_{600} (the inoculum is usually estimated to be around 10^8 cfu/mL) (*see* **Note 4**).
3. Harvest overnight culture by centrifugation at 4,500 ×*g* for 5 min at 4°C.
4. Wash the overnight bacterial culture in ultrapure water four times at 4,500 ×*g* for 5 min each and resuspend the cell pellet in ultrapure water to a final concentration of 10^6 cfu/mL.

5. Mix equal volume of cells and peptide in a depyrogenized borosilicate tube to a desirable final concentration for S3 or other AMP (for example, 1 μM which may refer to the minimum inhibitory concentration (MIC)/minimum bacterial concentration (MBC) in the case of S3). Then seal the tube with parafilm and incubate the mixture at room temperature for a certain period of time (for example, 5 min). As a control, equal volume of bacterial cells and ultrapure water are mixed and incubated at room temperature for the same time length of time.
6. Peel off the parafilm from the tube after incubation and spread 50 μL cell-peptide mixture on several pre-marked spots on the surface of poly-L-lysine pre-coated microscope slides and leave the slides in the biological safety cabinet for 1 min to allow the treated cells to attach to the surface.
7. Dip the glass slide gently in a 50 mL falcon tube filled with ultrapure water for several times to wash away the unattached cells and place the slides vertically on a piece of kimwipe tissue to remove extra liquid (*see* **Note 5**).
8. Put the slides in a slide holder and leave in a vacuum desiccator overnight to obtain completely dried sample for AFM imaging (*see* **Note 6**).

3.2. Equipment Setup and Sample Mounting

1. Place the cantilever in the ultraviolet ozone cleaner and clean for 30 min.
2. Load cantilever onto cantilever holder and mount cantilever holder onto the scanner following the standard instruction.
3. Align the laser at the back of the tip position, maximize sum of laser signal by adjusting along lengthwise of the cantilever (normal value for sum is around 2 for noncoated cantilever and around 6 for coated cantilever) (*see* **Note 7**).
4. Centralize the signal spot on the photodiode detector. The deviation from the center should not exceed 0.5 V.
5. Activate the stage motor and find the focus for both tip and sample stage using the attached optical accessories.
6. Tune to find the resonance frequency of the tip through autotune function (*see* **Note 8**).
7. Leave the whole set on for half an hour to allow the system to thermally equilibrate.

3.3. AFM Tip Check

1. Put tip-check sample onto the stage and activate vacuum to hold the sample stationary.
2. Initialize scan parameters as follows

Field	Scan size	Scan rate	i-gain	p-gain	Set point	Resolution	Channels
Value	5 μm	1 Hz	0.5–1	1–2	Default	256×256	Height and amplitude

3. Engage the tip. The motor and piezoscanner will take turns to move and extend till the amplitude of cantilever reaches the preset set point and then scanning starts.
4. Tune the parameters carefully to get sharp images (*see* **Note 9**).
5. Capture an image and look at 3D reconstruction mode at varied view points to check the tip shape and defects, if any (*see* **Fig. 15.3**) (*see* **Note 10**).

3.4. Imaging of Normal and Treated Bacteria

1. Change the tip-check sample to the glass slide with the thoroughly dried bacteria samples, locate the spots where previously marked, zoom in the optical mode to the maximum

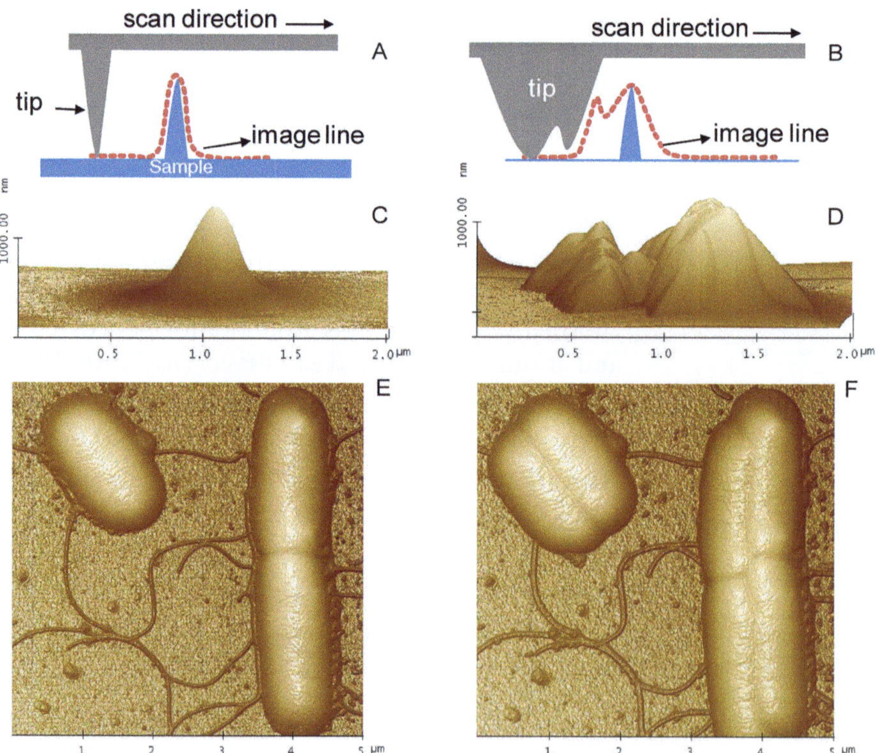

Fig. 15.3. Tip geometry-induced image artifacts. Schematic comparison between true imaging (**A**) and tip imaging (**B**). (**C**) A good tip imaging. (**D**) A contaminated, bad tip imaging. (**E**) Original bacteria imaging using the tip shown in (**C**). (**F**) Faulty double tip imaging using the tip shown in (**D**).

magnification (1,000×), and look for the bacteria which are shown as black dots.

2. Set parameters as follows

Field	Scan size	Scan rate	i-gain	p-gain	Set point	Resolution	Channels
Value	10 μm	0.5 Hz	0.5–1	1–2	Default	128×128	Height and amplitude

3. Repeat Steps 3 and 4 in **Section 3.3**, adjust offset to bring the cells to the center of scan screen, change the resolution to 512×512 (scan lines and data per line), close the isolation hood and capture the image.
4. Zoom in to a smaller scale to scan the fine structures on the single cell, or disengage and change to another site to image other cells after the image is captured (*see* **Note 11**).
5. Process the captured images with the original software, (Nanoscope V5.30r3, in our case). Only planefit and first-order flattening are applied. Three-dimensional reconstruction can be directly made from the height channel (*see* **Fig. 15.4**).
6. Unload the sample and keep in a humidity-controlled cabinet. The samples can be stored for months without any detectable changes.

4. Notes

1. The peptides used in this study were synthesized and purified to > 95% peptide purity. The S3 34'mer sequence N-HAEHKVKIGVEQKYGQFPQGTEVTYTCSGNYFLM-C, with a molecular weight of 3,892, corresponds to residues 268–301 of Factor C and Sushi 3 domain (GenBankTM/EBI accession number S77063).
2. Many types of cantilevers and tips are available on the market. Most of them are microfabricated from silicon or silicon nitride. Generally speaking, silicon tips are usually cone shaped and silicon nitride tips are pyramid shaped. A reasonable sharp tip (tip radius < 10 nm) with small open angle and high resonance frequency (200–300 kHz) is found suitable for bacteria imaging in air.
3. One of the significant issues of imaging in air is the influence of a thin layer of water present on the surface of both tip and sample, which gives rise to the so-called meniscus effect and the resultant strong adhesion force can trap the

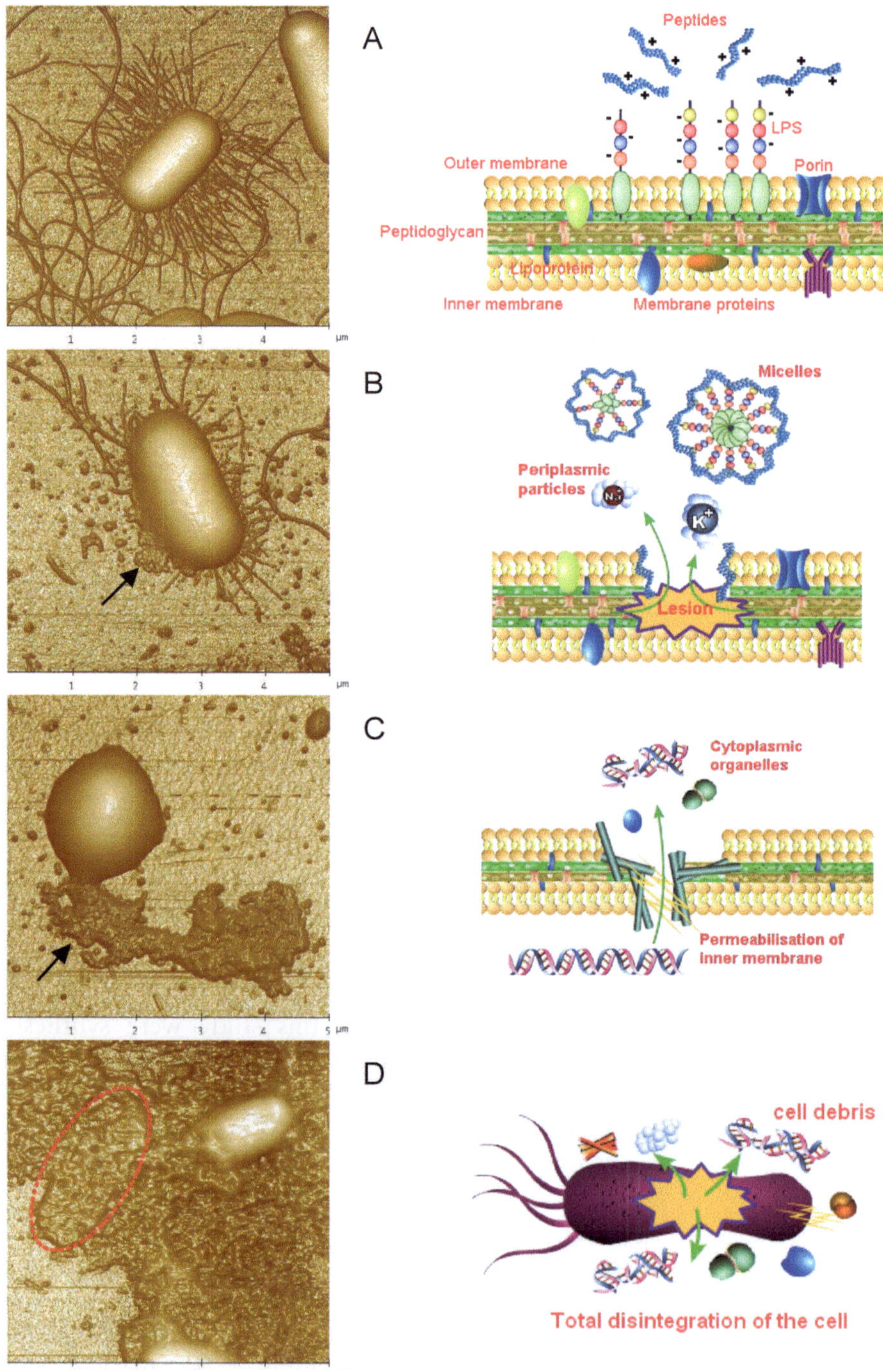

Fig. 15.4. Three-dimensional AFM height images and the corresponding cartoons of proposed S3–Gram-negative bacteria interactions. (**A**) Control normal cell. (**B**) Low-concentration treatment with S3 (*arrow indicates* the formation of micelle of LPS–S3 complex). (**C**) Medium-concentration treatment with S3 (*arrow indicates* the leakage of intracellular materials). (**D**) High-concentration treatment with S3 (*dotted red line circles* a totally lysed cell surrounded by masses of intracellular materials and cell debris). Specimen was prepared as described in **Section 3.2**

tip on the surface and cause unstable imaging. Thus, controlling the environment humidity below 50–60% will be helpful.

4. It is useful to prepare a standard curve by plotting the OD_{600} of different dilutions of the test culture against the colony forming units per mL. Correlate the average OD_{600} of the overnight test culture with the standard curve to determine the culture cell number.
5. If the sample gets dried up in the culture solution, the salt crystals and impurities from the solution will severely contaminate the sample surface. Thus, make sure no salt-containing solutions are left on the glass slide surface. Moreover, no fixation is needed since the cell wall structure prevents the cell from collapsing during the drying process and fixation may introduce artifact on the ultrastructure. However, keeping cells in ultrapure water exerts osmotic pressure on the cells and will eventually lyse the cells. So try to minimize the duration of incubation in ultrapure water and carefully compare with the control (same incubation time in ultrapure water but no peptide treatment) for proper interpretation. If a specific bacterial culture – such as some Gram-positive bacterial species – cannot be kept in ultrapure water, other preparation methods may need to be adapted and used.
6. A completely dried sample is critical for obtaining reliable AFM imaging, but fast drying methods such as heating may not be optimized for this application as they make the cells dehydrate too quickly and can cause the cell wall to collapse. Drying under nitrogen flow may achieve the same results as that of a vacuum desiccator.
7. If laser is correctly aligned but no signal appears on the photodiode screen, either there is a problem with the reflective side of the AFM cantilever or the signal falls out of the detectable area of photodiode. In such a case, adjust the two screws for photodiode detector positioning to get the signal back or, in some types of fixed photodiode, adjust the mirror in between the tip and photodiode to get the signal back to the photodiode screen.
8. If a warning of exceeding safety driving amplitude range appears, check the sweeping range to make sure it is within the estimated resonance frequency indicated by manufacturer (normally, set the range between 100 and 500 kHz). Also, there could be a laser misalignment problem where the laser was aligned onto the chip instead of cantilever, or a tip positioning problem where tip is not fixed well onto the tip holder. In such a case, try to realign the laser or

reload the tip onto the tip holder. If the "autotune" function is not available in the software, one can manually tune the cantilever under oscilloscope mode. Refer to the specification of the cantilever from the manufacturer and find the right peak from oscilloscope mode with target amplitude set to 1 V. Choose the working frequency at about 10–20% shift left to the resonance peak and set point at 70–80% of the target amplitude. Tuning check should be performed after several images are captured to avoid any contamination-induced resonance peak shift and the subsequent imperfect performance.

9. Monitor the cross section trace lines in height channel and adjust gains and set point step by step to achieve closely matched traction and retraction curves. Gains should be maximized to track effectively the surface features while set point should be set at a high level (tapping force is kept low) to avoid sample destruction and distorted imaging arising from high tapping force.
10. This step is very important to exclude any tip-induced artifact in imaging. As it can be seen in **Fig. 15.3**, a bad tip may give rise to a faulty image without user's realization.
11. Small-scale imaging may require readjustment of parameters (smaller gains and faster scan speed) for optimization. Check different sample sites to confirm the uniformity of the observed structures.

Acknowledgments

We thank Mr. Han Chong Ng for the preparation of the bacterial cultures and Ms. Pei Yun Lee for the assistance on AFM imaging and discussion. This work was funded by the Singapore-MIT Alliance (SMA2) and MoE Tier 2 grant (J.L. Ding and B. Ho).

References

1. Gordon, Y. J., Romanowski, E. G., and McDermott, A. M. (2005) A review of antimicrobial peptides and their therapeutic potential as anti-infective drugs. *Curr. Eye Res.* **30**, 505–515.
2. Sang, Y. and Blecha, F. (2008) Antimicrobial peptides and bacteriocins: alternatives to traditional antibiotics. *Anim. Health Res. Rev.* **9**, 227–235.
3. Zasloff, M. (2002) Antimicrobial peptides of multicellular organisms. *Nature* **415**, 389–395.
4. Ding, J. L., Li, P., and Ho, B. (2008) The Sushi peptides: structural characterization and mode of action against Gram-negative bacteria. *Cell. Mol. Life Sci.* **65**, 1202–1219.
5. Tan, N. S., Ng, M. L., Yau, Y. H., Chong, P. K., Ho, B., and Ding, J. L. (2000) Definition of endotoxin binding sites in horseshoe crab factor C recombinant sushi proteins and neutralization of endotoxin by sushi peptides. *FASEB J.* **14**, 1801–1813.
6. Li, P., Wohland, T., Ho, B., and Ding, J. L. (2004) Perturbation of lipopolysaccharide

(LPS) micelles by Sushi 3 (S3) antimicrobial peptide. The importance of an intermolecular disulfide bond in S3 dimer for binding, disruption, and neutralization of LPS. *J. Biol. Chem.* **279**, 50150–50156.

7. Li, P., Sun, M., Wohland, T., Ho, B., and Ding, J. L. (2006) The molecular mechanism of interaction between sushi peptide and *Pseudomonas* endotoxin. *Cell. Mol. Immunol.* **3**, 21–28.
8. Binnig, G., Quate, C. F., and Gerber, C. (1986) Atomic force microscope. *Phys.l Rev. Lett.* **56**, 930–933.
9. Li, A., Lee, P. Y., Ho, B., Ding, J. L., and Lim, C. T. (2007) Atomic force microscopy study of the antimicrobial action of Sushi peptides on Gram negative bacteria. *Biochim. Biophys. Acta.* **1768**, 411–418.
10. Zhong, Q., Inniss, D., Kjoller, K., and Elings, V. B. (1993) Fractured polymer/silica fiber surface studied by tapping mode atomic force microscopy. *Surf. Sci. Lett.* **290**, 688–692.
11. Suo, Z., Yang, X., Avci, R., Kellerman, L., Pascual, D. W., Fries, M., and Steele, A. (2007) HEPES-stabilized encapsulation of *Salmonella typhimurium*. *Langmuir* **23**, 1365–1374.
12. da Silva, A., Jr. and Teschke, O. (2003) Effects of the antimicrobial peptide PGLa on live *Escherichia coli*. *Biochim. Biophys. Acta.* **1643**, 95–103.
13. Meincken, M., Holroyd, D. L., and Rautenbach, M. (2005) Atomic force microscopy study of the effect of antimicrobial peptides on the cell envelope of *Escherichia coli*. *Antimicrob. Agents Chemother.* **49**, 4085–4092.
14. da Silva Junior, A. and Teschke, O. (2005) Dynamics of the antimicrobial peptide PGLa action on *Escherichia coli* monitored by atomic force microscopy. *World J. Microbiol. Biotechnol.* **21**, 1103–1110.
15. Braga, P. C. and Ricci, D. (1998) Atomic force microscopy: application to investigation of Escherichia coli morphology before and after exposure to cefodizime. *Antimicrob. Agents Chemother.* **42**, 18–22.
16. Dufrene, Y. F. (2008) Atomic force microscopy and chemical force microscopy of microbial cells. *Nat. Protocols* **3**, 1132–1138.

Chapter 16

Fluorescence and Electron Microscopy Methods for Exploring Antimicrobial Peptides Mode(s) of Action

Ludovica Marcellini, Maria Giammatteo, Pierpaolo Aimola, and Maria Luisa Mangoni

Abstract

Due to the increasing resistance of microbial pathogens to the available drugs, the identification of new antimicrobial agents with a new mechanism of action is urgently needed. In this context, cationic antimicrobial peptides (AMPs) are considered promising candidates. Although there is evidence that, in contrast to conventional antibiotics, microbial membranes are the principal target of a large number of AMPs, thus making it difficult for the pathogen to acquire resistance, their mode(s) of action is not yet completely clear. Intense research is currently devoted to understand the effect(s) of AMPs on intact cells, either at sub-lethal or at lethal peptide concentrations, and fluorescence/electron microscopy techniques represent a valid tool to get insight into the damage caused by these molecules on the morphology and membrane structure of the target cell. We here present an overview of some microscopic methodologies to address this issue.

Key words: Antimicrobial peptides, mode(s) of action, fluorescence microscopy, scanning electron microscopy, transmission electron microscopy, membrane permeation.

1. Introduction

The discovery of new anti-infective formulations has become a pressing necessity, because of the growing emergence of pathogenic microorganisms that have acquired resistance to a wide range of commercially used antibiotics. Natural antimicrobial peptides (AMPs), which are produced by almost all living forms, act as the first line of defense against noxious invaders (1–5), and represent attractive molecules for the design and

A. Giuliani, A.C. Rinaldi (eds.), *Antimicrobial Peptides*, Methods in Molecular Biology 618, DOI 10.1007/978-1-60761-594-1_16, © Springer Science+Business Media, LLC 2010

manufacture of new peptide-based antibiotic therapies (6). Studies aimed at elucidating the effects caused by AMPs on the permeability of the target cell membrane and the correlation between membrane perturbation/surface structure and microbial viability will certainly contribute to broaden our knowledge on the mode(s) of action of AMPs on intact cells. For this purpose, a useful tool is represented by fluorescence and electron microscopy techniques.

1.1. Fluorescence Microscopy (Triple-Staining Method)

Fluorescence microscopy allows to image cells and/or intracellular components, by labeling with specific fluorophores. We developed a triple-staining method to visualize the effects of a short-sized AMP from frog skin (namely, temporin L; *see* 7) on the viability and membrane integrity of individual *Escherichia coli* cells. This method is based on the following fluorochromes:

(i) the vital dye 5-cyano-2,3-ditolyl tetrazolium chloride (CTC), to stain metabolically active microorganisms. CTC is a monotetrazolium redox dye, which is colorless and nonfluorescent unless it is exposed to an electron-transport chain. Indeed, in respiring cells, CTC is readily reduced to a red fluorescent formazan salt that accumulates intracellularly (8, 9);

(ii) the DNA-binding dye 4,6-diamidino-2-phenylindole (DAPI), which stains all bacterial cells irrespective of viability. DAPI binds to a double-stranded DNA, forming a stable complex which gives a blue fluorescence when excited by UV light;

(iii) the low molecular mass (389.4 Da) green fluorescent probe fluorescein isothiocyanate (FITC), which is unable to traverse the cytoplasmic membrane of intact cells unless this is damaged by a permeabilizing agent. This probe is used to assess alterations in the membrane permeability.

The combination of the double CTC–DAPI staining with FITC allows to simultaneously identify in a single preparation the total number of bacteria (cells with a blue fluorescence, due to DAPI binding to DNA), the bacteria with an altered membrane permeability (cells with a bright green fluorescence, due to FITC uptake), and the metabolically active bacteria (cells with a red fluorescence, due to formation of CTC–formazan crystals) (7).

Interestingly, this technique permits to screen peptides or other molecules that are able to modify the membrane permeability of microbial cells without affecting their viability (as detected by the emission of both green and red fluorescence). In addition, the triple-staining method permits to identify bacteria in which the killing process and/or membrane permeation causes DNA leakage (revealed by absence of blue fluorescence).

1.2. Scanning and Transmission Electron Microscopy

With reference to electron microscopy, both scanning electron microscopy (SEM) and transmission electron microscopy (TEM) represent valid techniques to provide information on the effects caused by AMPs on microorganisms, at higher resolution than optical microscopy. In particular, SEM allows to investigate changes in the morphology of the sample's surface (e.g., surface of the microbial cell), whereas TEM allows to shed light into the effects induced by AMPs on cellular components at the ultrastructural level (e.g., membrane damage).

2. Materials

2.1. Triple-Staining Procedure

2.1.1. Bacterial Culture

1. Luria-Bertani (LB) broth: 1% (w/v) bacto tryptone, 0.5% (w/v) NaCl, and 0.5% (w/v) yeast extract (Oxoid, LTD, Basingstoke, UK) in distilled water. The final pH has to be adjusted to 7.4 with 1 N sodium hydroxide (NaOH). Using a graduated cylinder, dispense aliquots into glass bottles, autoclave them at 120°C for 20 min, and store the sterilized medium at room temperature (rt).
2. 10 mM sodium phosphate buffer, pH 7.4 (NaPB): stock solution of 100 mM NaPB is prepared by mixing 77.4 mL of 1 M Na_2HPO_4 with 22.6 mL of 1 M NaH_2PO_4 while stirring. Both monobasic and dibasic sodium phosphates are dissolved in distilled water (*see* **Note 1**). Working buffer is prepared by diluting the stock solution 1:10 in distilled water. Sterilize by filtration (0.45-μm filters).
3. 24-well polystyrene Costar plates (Corning, Acton, MA, USA).
4. Spectrophotometer (e.g., UV-1700 Pharma Spec from Shimadzu, Kyoto, Japan).
5. Thermomixer (e.g., Eppendorf, Barkhausenweg, Germany).

2.1.2. Poly(l-lysine)-Coated Cover Glasses

1. Microscope cover glasses (12 mm diameter).
2. Petri dishes (10 cm diameter).
3. Alkaline wash solution: 10% (w/v) NaOH + 60% (v/v) ethanol (*see* **Note 2**). Use distilled water to prepare both NaOH and ethanol solutions before mixing.
4. Poly(L-lysine) 70,000–150,000 Da (Sigma, St. Louis MO, USA) (0.1 mg/mL in distilled water).
5. Pure ethanol.

2.1.3. Buffers and Fluorescent Probes

1. Phosphate-buffered saline (PBS): 10 × stock solution contains 1.37 M NaCl, 27 mM KCl, 100 mM Na_2HPO_4, 20 mM KH_2PO_4 in distilled water. Adjust pH to 7.4 with HCl if necessary. Autoclave at 120°C for 20 min before storage at rt. Working buffer is prepared by diluting one part of the stock solution with nine parts of distilled water.
2. CTC solution: dissolve 6 mg CTC (Sigma) in 1 mL distilled water. Sterilize by filtration. Working solution is prepared by diluting the stock solution 1:4 in sterile 1 × PBS to obtain a 5 mM CTC solution.
3. DAPI solution: dissolve 10 mg DAPI (Sigma) in 1 mL distilled water. Sterilize by filtration. Working solution is prepared by diluting the stock solution 1:1,000 in sterile 1 × PBS to obtain a 10 μg/mL DAPI solution.
4. FITC solution: dissolve 10 mg FITC (Sigma) in 1 mL acetone. Sterilize by filtration. Dilute the stock solution to 6 μg/mL in sterile NaPB (*see* **Note 3**).

2.1.4. Mowiol-DABCO Antifade-Mounting Medium

1. In a 50 mL Falcon tube, mix 6 mL glycerol (Sigma), 2.4 g mowiol 4-88 (Fluka, Buchs, Switzerland), and 6 mL of distilled water.
2. Vortex well and shake at rt for 2 h to dissolve mowiol.
3. Add 12 mL of 0.2 M Tris–HCl, pH 8.5, and incubate at 50°C with occasional mixing until the mowiol dissolves (this will take approximately 3 h).
4. Filter through 0.45 μm filters and store aliquots of 100–500 μL in glass vials at –20°C (*see* **Note 4**).
5. On the day of use, add 0.6% (w/v) 1,4-diazobicyclo-(2, 2, 2)-octane (DABCO) and vortex for about 30 min (*see* **Note 5**).

2.2. Scanning Electron Microscopy

2.2.1. Bacterial Culture and Poly(l-lysine)-Coated Cover Glasses

(*See* **Sections 2.1.1** and **2.1.2**)

2.2.2. Fixative Solutions

1. 100 mM potassium phosphate buffer, pH 7.4 (KPB): stock solution is prepared by mixing 80.2 mL of 1 M K_2HPO_4 and 19.8 mL of 1 M KH_2PO_4 while stirring. Both monobasic and dibasic potassium phosphates are dissolved in distilled water. Sterilize by filtration (0.45-μm filters).
2. 2.5% (v/v) glutaraldehyde solution in KPB (*see* **Note 6**).

2.2.3. Dehydration

1. Graded ethanol solution series: prepare 30, 50, 70, 80, 96, 100% (v/v) ethanol solutions in distilled water.

2.2.4. Critical Point Dryer (e.g., Critical Point Dryer CPD020 from Bal-Tec, Balzers, Liechtenstein)

2.2.5. Sample Mounting

1. Leit-C conducting carbon cement (Agar Scientific, Stansted, UK) (*see* **Note 7**).
2. SEM Pin Stubs.

2.2.5. Gold Coating

1. Sputter coater device (e.g., Sputter SCD 040, Bal-Tec, Balzers, Liechtenstein).

2.3. Transmission Electron Microscopy

2.3.1. Bacterial Culture and Poly(l-lysine)-Coated Cover Glasses

(*See* **Sections 2.1.1** and **2.1.2**)

2.3.2. Buffers and Fixative Solutions

1. 0.1 M cacodylate buffer, pH 7.4 (CB): dissolve 2.14 g of sodium cacodylate trihydrate ($Na(CH_3)_2AsO_2$ $3H_20$) in approximately 60 mL of distilled water. Adjust the pH to 7.4 with 0.1 M HCl (approximately 2.7 mL), then add distilled water up to a final volume of 100 mL (*see* **Note 8**).
2. 2.5% (v/v) glutaraldehyde solution in CB: add 1 mL of 25% glutaraldehyde stock solution to 9 mL of CB. Preferably use fresh solutions (*see* **Note 6**).
3. 1% (v/v) osmium tetroxide in CB: clean a glass ampoule containing 0.1 g osmium tetroxide, then score it with diamond pencil, and place it in a dark glass bottle. Break the ampoule by vigorously shaking the bottle and add 10 mL of CB. Leave at rt, in the dark, for at least 24 h for a complete dissolution and finally filter the fixative solution with filter paper to eliminate glass fragments deriving from ampoule breakage (*see* **Note 9**).

2.3.3. Dehydration and Resin Embedding

1. Graded ethanol solution series: prepare 30, 50, 70, 96, and 100% (v/v) ethanol solutions in distilled water.
2. Propylene oxide (1,2-epoxypropane) (*see* **Note 10**).
3. Durcupan® ACM Fluka, epoxy resin (Fluka): the embedding medium is prepared by combining four different components in the following proportions: component A, M (epoxy resin), 30 mL; component B (hardener 964), 30 mL; component C (accelerator 960), 1.2 mL; component D (containing plasticizer), 0.6 mL. Pour into a disposable beaker, mix thoroughly with a glass rod, then with a

magnetic stirrer. The resulting embedding mixture can be stored in a refrigerator for short periods, if tightly sealed (*see* **Note 11**).

2.3.4. Semi-thin and Ultrathin Section Staining

1. 1% (w/v) toluidine blue in 1% (w/v) borax solution: dissolve 1 g of borax (sodium tetraborate) in 100 mL of distilled water, then add 1 g of toluidine blue. Filter with filter paper and store at rt.
2. 5% (w/v) uranyl acetate in 70% (v/v) ethanol: dissolve 0.25 g of uranyl acetate in 5 mL of 70% (v/v) ethanol in distilled water. Filter before usage and keep the solution in the dark (*see* **Note 12**).
3. Reynold's lead citrate: mix 1.33 g of lead nitrate ($Pb(NO_3)_2$) and 1.76 g of trisodium citrate ($Na_3(C_6H_5O_7){\cdot}2H_2O$) with 30 mL distilled water (freshly prepared and carbonate-free) in a stoppered volumetric flask, inverting at intervals of about 10 min. Leave it standing for 30 min, then add 8 mL of 1 N NaOH and distilled water (both freshly prepared and carbonate-free) up to a final volume of 50 mL. Mix again, by inversion, until the solution becomes clear. The final pH should be about 12. Store at 4°C (*see* **Note 13**).

2.3.5. Negative Staining

1. 2% (w/v) phosphotungstic acid solution in water. Add 0.1 g of sodium phosphotungstate powder (PTA, Fluka) to 10 mL of distilled water in a well-cleaned glass vial. Adjust the pH to 7.2–7.4 with 0.1 N NaOH (*see* **Note 14**).
2. Carbon–formvar-coated EM grids (300–400 mesh).

3. Methods

3.1. Fluorescence Microscopy

Combining fluorescence and transmitted light signal is a particularly powerful technique for illustrating details of a cell layer that may not be fully fluorescent. The choice of labels depends upon the availability of filters and fluorochromes.

3.1.1. Bacterial Culture

1. Bacteria are inoculated in 10 mL LB broth and grown at 37°C (use an incubator with shaking at 150 rpm), until an OD_{590} 0.8 is reached.
2. Bacteria are harvested by centrifugation, washed twice in NaPB, and resuspended in the same buffer.
3. Approximately 2×10^7 cells (in 50 μL) are transferred into a 1.5-mL eppendorf tube, where the peptide is added at the desired concentration. A control sample, without peptide, is also included.

4. Samples are incubated in a thermomixer, with shaking at 600 rpm, at 37°C for 15 min.

3.1.2. Preparation of Poly(l-lysine)-Coated Cover Glasses

1. Clean cover glasses by immersion in alkaline wash solution for at least 2 h. For this purpose, cover glasses are picked up one by one with tweezers and placed into a plastic petri dish filled with the alkaline solution.
2. Remove the alkaline solution with a pipette and rinse glasses five times with distilled water.
3. Remove water and add pure ethanol.
4. Leave cover glasses immersed for 30 min (*see* **Note 15**).
5. Take out each single cover glass with sterile tweezers and flame it quickly (*see* **Note 16**).
6. Transfer each sterile cover glass into a well of a 24-wells polystyrene plate.
7. Coat each glass with 1 mL of the poly(L-lysine) solution.
8. Incubate at rt for 2 h.
9. Remove the poly(L-lysine) solution with a pipette and wash glasses with sterile water.
10. Let cover glasses drying at rt for at least 3 days before usage (*see* **Note 17**).

3.1.3. Preparation of Samples for Triple-Staining (CTC–DAPI–FITC)

1. After peptide treatment, 900 μL of CTC (5 mM in PBS 1X) are added to each eppendorf tube.
2. Incubate samples in a thermomixer, at 37°C for 90 min, with agitation (150 rpm).
3. Pour the cell suspension of each sample onto a poly (L-lysine)-coated cover glass, previously placed in the well of a 24-wells polystyrene plate.
4. Incubate the plate at 30°C for 45 min to allow the adhesion of bacteria.
5. Withdraw the CTC solution and wash with 800 μL of NaPB to remove non-attached cells.
6. Repeat washing four times (*see* **Note 18**).
7. Add 1 mL of DAPI solution (10 μg/mL in PBS) to each well and incubate for 30 min at 30°C.
8. Remove DAPI solution and rinse cover glasses with 800 μL of NaPB. Repeat washing three times.
9. Add 1 mL of FITC solution (6 μg/mL in NaPB) to each well.
10. Incubate the plate at 30°C for 30 min.
11. Withdraw the FITC solution and rinse cover glasses with 800 μL of NaPB.
12. Repeat washing four times.

3.1.4. Sample Mounting

1. With a Pasteur pipette, put 1 or 2 drops of antifade-mounting medium on a microscope glass slide.
2. Place the cover glass (upside down) over the mounting medium.
3. Dry slides horizontally at rt overnight, well covered.
4. Observe the sample with a fluorescence microscope. Excitation at 359 nm induces DAPI fluorescence (blue emission, 461 nm); excitation at 490 nm induces FITC fluorescence (green emission, 520 nm); excitation at 450 nm induces CTC–formazan fluorescence (red emission, 630 nm). Examples are shown in **Fig. 16.1**.

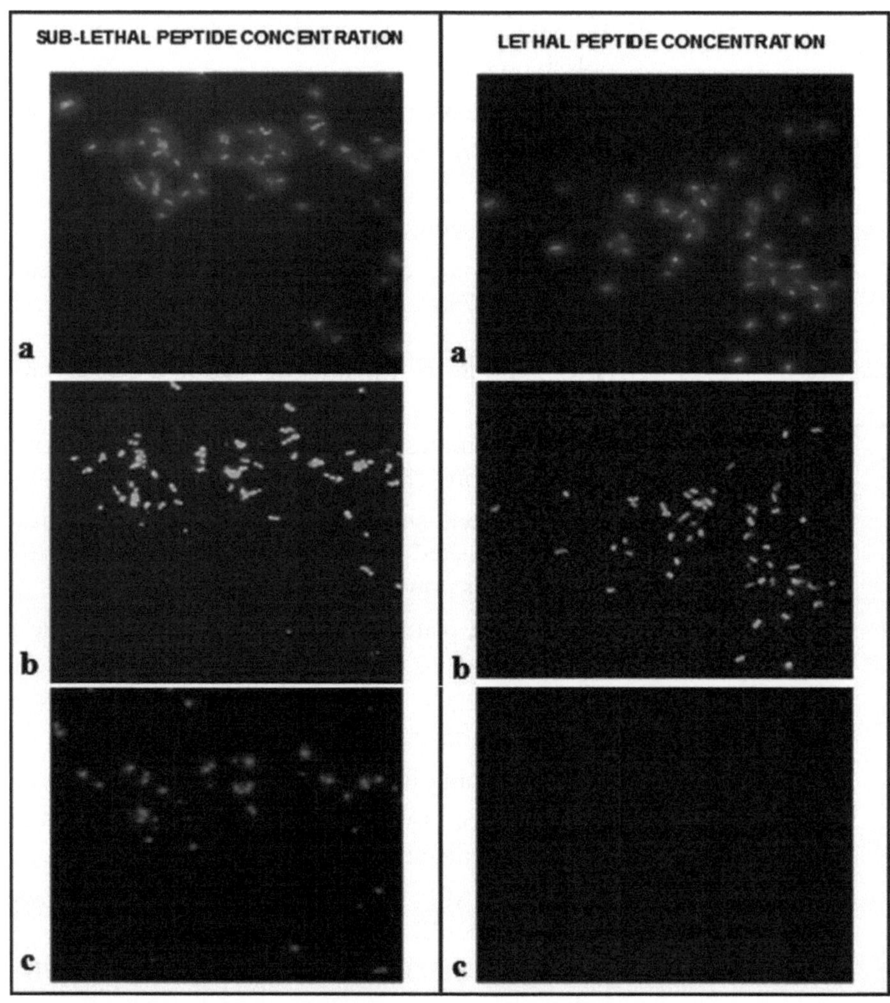

Fig. 16.1. Detection of total, viable, and membrane-perturbed bacteria using triple staining with fluorescent probes. *E. coli* cells were grown and treated with temporin L at sub-lethal (2 μM) and lethal (50 μM) concentrations. Bacteria were then immobilized on poly(L-lysine)-coated glass slides and incubated with CTC, DAPI and FITC as described in the protocol. Panels **(a)**, **(b)**, and **(c)** illustrate staining with DAPI, FITC, and CTC, respectively. (Reproduced with permission, from Mangoni et al. (7). © the Biochemical Society.).

3.2. Electron Microscopy

For electron microscopy, samples must be completely dry, because both SEM and TEM work in vacuum. However, to avoid disruption of biological samples, as a result of the loss of their own water, you can preserve and stabilize the sample's structure by using different fixatives that cross-link molecules and trap them together as a stable structure. After fixation, samples are dried in graded ethanol solution series. In the case of SEM, this procedure is followed by a physical (critical point drying, CPD) (*see* **Note 19**) or chemical dehydration process (*see* **Note 20**).

3.2.1. Scanning Electron Microscopy

In the case of SEM, dehydrated samples must be electrically conductive. For this purpose, they are coated with a thin film of electrically conductive material, usually gold.

3.2.1.1. Adhesion of Bacteria to Poly(l-lysine)-Coated Cover Glasses

1. Exponentially growing bacteria in LB medium (*see* **Section 3.1.1**) are harvested by centrifugation, washed twice in NaPB, and resuspended in the same buffer. Approximately 2×10^7 cells (in 50 μL) are transferred into a 1.5-mL eppendorf tube and the peptide is added at the desired concentration.
2. After peptide treatment, add 150 μL of NaPB and pour the bacterial suspension on a poly(L-lysine)-coated cover glass, previously placed in a well of a 24-wells polystyrene plate.
3. Incubate the plate for 90 min at 30°C to allow the adhesion of bacteria.
4. Wash each cover glass with 800 μL of NaPB to remove non-adherent bacteria. Repeat washing four times (*see* **Note 21**).

3.2.1.2. Fixation

1. Withdraw NaPB and add 800 μL of 2.5% glutaraldehyde solution to each well.
2. Incubate the plate for 2 h at 4°C.
3. Remove the glutaraldehyde solution.
4. Rinse cover glasses with 1 mL KPB (kept at 4°C).
5. Leave the plate at 4°C for 10 min.
6. Remove buffer.
7. Repeat Steps 4–6 three times

3.2.1.3. Dehydration

1. Add 1 mL of 30% ethanol to each well, incubate at rt for 2 min, and remove the alcohol. Repeat twice.
2. Add 1 mL of 50% ethanol to each well, incubate at rt for 2 min, and then remove the alcohol. Repeat twice.
3. Add 1 mL of 70% ethanol to each well, incubate at rt for 2 min, and then remove the alcohol. Repeat twice.

4. Add 1 mL of 80% ethanol to each well, incubate at rt for 5 min, and then remove the alcohol. Repeat twice.
5. Add 1 mL of 96% ethanol to each well, incubate at rt for 5 min, and then remove the alcohol. Repeat twice.
6. Add 1 mL of pure (100%) ethanol to each well, incubate at rt for 5 min, and then remove the alcohol.
7. Repeat twice and leave cover glasses in 2 mL of 100% ethanol (*see* **Note 22**).

3.2.1.4. Critical Point Drying

1. Put a small envelope made of optical paper (*see* **Note 23**) in a petri dish filled with pure ethanol. If more than one sample has to be dried at the same time, label the envelope outside with a pencil (*see* **Note 24**).
2. Insert each single cover glass coated with bacteria into the paper envelope.
3. Close the envelope with a stapler.
4. Transfer samples into the chamber of the CPD device (*see* **Note 25**).
5. Process for CPD following manufacturer's instructions.

3.2.1.5. Mounting

1. Remove samples from CPD device.
2. Open the paper envelope with the help of tweezers and remove each cover glass.
3. By using a toothpick, apply a very thin layer of conductive glue on the stubs.
4. Mount each cover glass on the stub (*see* **Note 26**).
5. Wait for the glue to dry.

3.2.1.6. Gold Coating and Sample Observation

1. Insert mounted samples on a sputter coater device.
2. Select sputter parameters to obtain a conductive film of about 15 nm.
3. View sample under SEM. An example is given in **Fig. 16.2**.

3.2.2. Transmission Electron Microscopy

With regard to TEM, the sample has to be of such a low density to permit electrons (that have relatively low penetrating power) to travel through it. The two most frequently employed procedures to prepare biological samples for TEM observation are resin embedding and ultrathin sectioning and negative staining.

For resin embedding and ultrathin sectioning, after fixation and dehydration, the sample is embedded in a plastic resin that subsequently polymerizes into a solid hard plastic block. The block is cut into ultrathin sections (50–100 nm thick) that are placed on a metal grid and then stained with heavy metal salts.

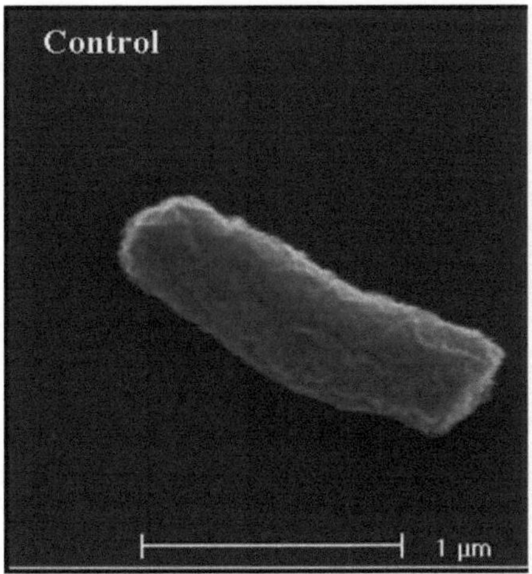

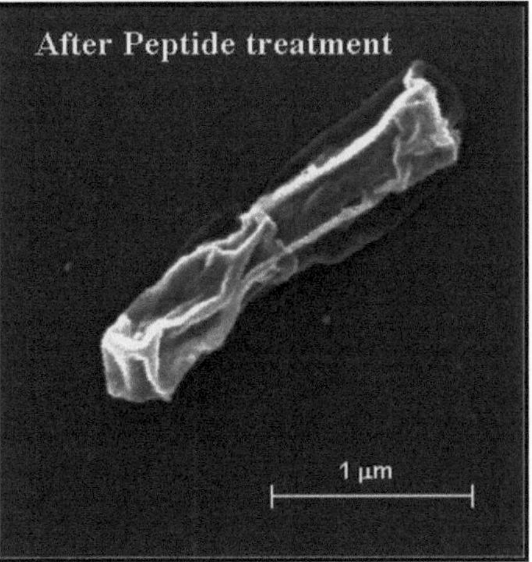

Fig. 16.2. SEM observation of *E. coli* cells before and after 15 min treatment with temporin L, at a lethal concentration (50 μM). Each figure has been magnified × 15,000 or × 30,000. (Reproduced with permission, from Mangoni et al. (7). © the Biochemical Society.).

These compounds can react with various molecular components of the cell and, because of the good capacity to scatter electrons of their metal atoms, significantly improve the image contrast in the microscope. By following this procedure, in-depth ultrastructural analysis of the sample can be performed.

In the case of negative staining, this faster and less expensive method is suitable for imaging suspensions of small and particulate samples (e.g., bacteria, viruses, cell components and purified macromolecules). The sample is spread onto a grid coated with a plastic-support film and mixed with a "negative stain" solution (phosphotungstic acid/neutralized potassium phosphotungstate, ammonium molybdate, or uranyl acetate). Importantly, an advantage of this method is that, beside providing contrast between specimen and background, the layer of negative stain preserves the macromolecular structure of the sample and protects it from the collapse due to the dehydration process as well as from irradiation damage.

3.2.2.1. Preparation of Resin-Embedded Samples (Cell Processing, Fixation, Dehydration, and Resin Embedding)

1. Exponentially growing bacteria in LB medium (*see* **Section 3.1.1**) are harvested by centrifugation, washed twice in NaPB, and resuspended in the same buffer. Approximately 2×10^7 cells are transferred into a 1.5-mL eppendorf tube where the peptide is then added.
2. After peptide treatment, centrifuge bacteria at $300 \times g$ for 20 min.
3. Wash twice with NaPB, then remove NaPB, and replace it with 2.5% glutaraldehyde in CB. Resuspend the cells in the fixative solution and keep the sample at 4°C for 1 h.

4. Spin the bacterial suspension at 13,500×*g* in 0.3 mL polypropylene microcentrifuge tubes (with a microcentrifuge) for 15 min at 4°C. Discard supernatant by inverting the tube and remove the resulting pellets by cutting off the bottom end of the tubes with a clean razor blade and place them into small glass beakers (*see* **Note 27**).
5. Wash the pellets in CB (5 times × 10 min).
6. Post-fix with 1% (v/v) osmium tetroxide in CB for 1 h at 4°C.
7. Rinse twice with CB for 5 min.
8. Dehydrate samples through ascending ethanol solution series: 30, 50, 70, and 96% ethanol, for 5 min each at 4°C, then pure (100%) ethanol for 15 min at 4°C, and finally pure ethanol for 15 min at rt.
9. Replace ethanol with propylene oxide (3 times × 20 min).
10. Replace propylene oxide with Durcupan® ACM resin/propylene oxide mixture, 1:1 (v/v). To facilitate resin infiltration into the samples, place the beakers on a slow turning wheel device: the slow rotatory movement will provide the necessary mixing during incubation. Leave for 1 h at rt, under a fume hood. Then, replace with fresh resin/propylene oxide mixture, 1:1 (v/v), and leave overnight.
11. Remove 1:1 (v/v) resin/propylene oxide mixture and replace with 3:1 (v/v) resin/propylene oxide mixture (2 × 1 h at rt on the slow turning wheel, under a fume hood).
12. Remove 3:1 (v/v) resin/propylene oxide mixture and replace it with pure resin (2 × 1 h). Put the beakers on the slow turning wheel, in an oven, at 37°C.
13. By carefully using a toothpick or a suitable needle, transfer the pellets to fresh resin in molds and leave them in the oven at 37°C for 24 h, then at 70°C for 72 h, for polymerization.

3.2.2.2. Semi-thin and Ultrathin Section Staining

1. Once the blocks have polymerized, they are removed from the molds and may be appropriately trimmed and used for semi-thin sectioning with an ultramicrotome: semi-thin sections (about 1 μm thick) are cut using a glass knife and placed on a drop of water over a glass slide. The glass slide is then put on a hotplate at about 80°C to ensure adhesion of the sections to the glass surface upon water evaporation, and these are finally stained with 1% (w/v) toluidine blue in 1% (w/v) borax solution for 1 min at 60°C. After rinsing with distilled water and drying, the semi-thin sections can be observed with a light microscope to select a suitable area of the sample for ultrathin sectioning.

2. After further trimming of the blocks, ultrathin sections (about 50–100 nm thick) are cut and picked up onto electron microscopy copper grids (usually, several sections may be collected on each single grid).
3. For the classic "double staining" method, ultrathin sections are "stained" first with 5% (w/v) uranyl acetate in 70% (v/v) ethanol: place a droplet of the staining solution on a clean piece of parafilm (protected from direct light exposure); put the grid on the droplet, with the sections side down, and leave for 15–25 min at rt. Rinse thoroughly the grid with filtered distilled water.
4. After uranyl acetate, the ultrathin sections are treated with Reynold's lead citrate. In this case, a droplet of the staining solution is placed in a petri dish containing a few pellets of sodium hydroxide, which has the ability to absorb atmospheric carbon dioxide, thus protecting the lead acetate solution and avoiding lead carbonate precipitation and section contamination. Again, the grid is placed, with the sections side down, on the droplet of staining solution and left for 15–25 min at rt, before rinsing thoroughly with filtered, carbonate-free distilled water.
5. After staining and washing, the grids are allowed to dry in a desiccator before TEM observation.

3.2.2.3. Negative Staining (Drop Method)

1. Exponentially growing bacteria in LB medium (*see* **Section 3.1.1**) are harvested by centrifugation, washed twice in 10 mM NaPB, and resuspended in the same buffer. Approximately 2×10^7 cells (in 100 μL) are transferred into a 1.5-mL eppendorf tube and treated with the peptide.
2. After peptide treatment, centrifuge cells at 300×*g* for 20 min.
3. Resuspend the pellet in 300 μL of NaPB (*see* **Note 28**).
4. Place a small droplet of the bacterial suspension onto a formvar-coated grid (*see* **Note 29**).
5. After a few seconds (usually 10–30 s should be enough to allow the absorption of bacterial cells to the formvar film), "wash" the grid with about 10 droplets of the negative staining PTA solution (*see* **Note 30**).
6. Leave the last drop of PTA solution on the grid for 10 s, then remove the excess of liquid by touching the edge of the grid with a piece of filter paper.
7. Allow the grid to dry in a desiccator, until the following day, before microscopic observation with TEM (*see* **Note 31**). An example is given in **Fig. 16.3**.

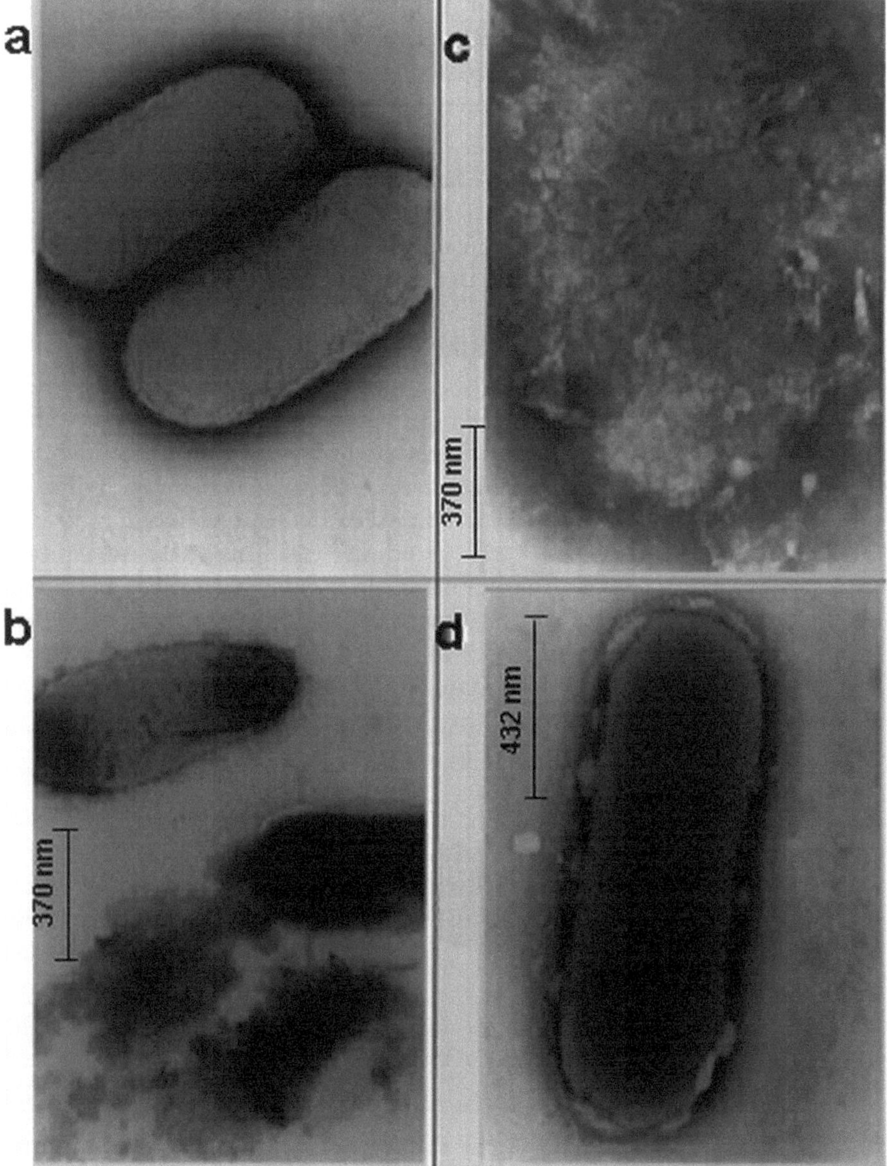

Fig. 16.3. Negative staining of *E. coli* cells for TEM observation. **(a)**, Intact bacteria; **(b)**, bacteria in the presence of the synthetic antimicrobial peptide K5L7 at the minimum inhibitory concentration (MIC); **(c)**, bacteria in the presence of K5L7 (enlargement); and **(d)**, bacteria in the presence of K5L7 at a concentration of less than MIC. Contributed by Yechiel Shai, Weizmann Institute, Rehovot, Israel.

4. Notes

1. Use heat to solubilize these salts.
2. Be careful when NaOH is dissolved in water, because this is an exergonic reaction, which generates heat. Do not store this solution in glass bottles.

3. Wrap staining solutions with aluminum foils to protect them from light.
4. Mowiol-mounting medium can be stored at rt not more than 1 month. Before usage, it is recommended to warm it at rt to avoid bubble formation.
5. Fluorescently labeled cells and tissues exhibit a characteristic photobleaching curve in response to excitation. Much of the photobleaching can be attributed to the generation of free radicals. The use of free radical scavengers has been shown to decrease the rate of photobleaching. Common scavengers include *n*-propyl gallate, *p*-phenylenediamine, and DABCO, whereas live systems have been reported to reduce photobleaching in the presence of vitamin C. If possible, liquid embedding media should be used; samples mounted in liquid media seem not to shrink in volume over time, when sealed properly, and can be stored at –20°C.
6. Handle glutaraldehyde carefully. It is an irritant for skin and mucosa (ocular and first respiratory tracts). Avoid prolonged breathing and contact with eyes. Glutaraldehyde should be used under a fume hood.
7. This is supplied in the form of a suspension of very fine conducting carbon particles in a volatile carrier. Good for forming conducting bonds between specimen and stub.
8. Sodium cacodylate is highly toxic. Always wear gloves and work under a fume hood.
9. Osmium tetroxide is easily reduced by impurities and light exposure. Use only acid-cleaned glassware (thoroughly rinsed in distilled water), store solutions in the dark (at 4°C for prolonged periods) but discard if a reddish color develops. When using osmium tetroxide, always wear gloves and work under a fume hood to avoid exposure to its dangerous vapors.
10. Propylene oxide is very reactive, extremely volatile, and carcinogenic, so caution should be observed when using this solvent. Always wear suitable gloves and work under a fume hood.
11. Be careful in handling epoxy resins, since they may cause severe irritation upon prolonged or repeated contact. Wear suitable gloves and work under a fume hood.
12. Uranyl acetate is slightly radioactive and a very toxic substance, with cumulative effects. Follow appropriate precautions to limit exposure to it, wear gloves, and possibly work under a fume hood.
13. Wear gloves when using lead citrate and avoid excessive exposure to this toxic compound. Lead citrate solutions readily absorb carbon dioxide, with formation of a

precipitate of lead carbonate that may contaminate sections. Care should be taken not to breathe (thus emitting CO_2) on lead citrate during staining. Reynold's lead citrate can be kept at 4°C for several months if tightly stoppered. Discard it if precipitate appears.

14. PTA is corrosive and causes burns. Negative stains are all heavy metals and should be handled as toxic and hazardous materials (the usage of suitable protective clothes, gloves, eye/interface protection as well as a fume hood is recommended). Note that negative staining does not kill pathogens. Thus, the infectious agent should be inactivated prior to negative staining and the grid should be sterilized by gas or ultraviolet radiation. The negative stain solution can be stored for 1–3 weeks at 4°C.
15. To completely remove the alkaline solution, it is important that each cover glass remains separated by the others. This is to avoid the formation of a white mark on the glass after sterilization by flame.
16. Be careful, do not flame cover glass until alcohol is not completely evaporated.
17. To avoid attachment of the cover glass to the plastic, place it on the surface of a 96-wells plate. It is important that both sides of the glass are in contact with air.
18. During washing, make sure to pipette the buffer solution on the wall of the well and not directly on the top of the cover glass.
19. Since air drying of hydrated biological specimens can result in collapse and shrinkage, due to the surface tension between water and vapor, CPD can avoid this problem by placing the specimen in an environment where the fluid, within the specimen, can pass from the liquid to gas phase with zero surface tension. The best transition fluid is represented by carbon dioxide, because its critical point is reached under conditions that do not damage the biological sample (31°C and 73.8 bar). Once carbon dioxide is fully converted to gas, the specimen is dry. In practice, it is necessary to fix the specimen and then to transfer it into an "intermediate fluid" (typically ethanol, as water is not miscible with CO_2), according to the scheme below:

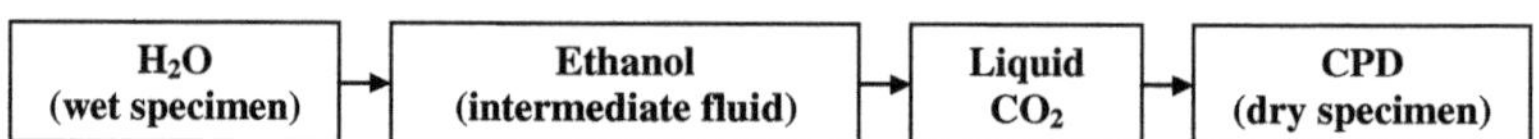

20. Chemical dehydration by hexamethyldisilazane (HMDS) can be carried out as a good alternative to CPD. Compared with CPD, HMDS drying takes a shorter time, less effort,

no equipments, and lower costs (10). The procedure is as follows:
(a) after dehydration phase with ethanol, transfer dry cover glasses into the wells of a new 24-wells polystyrene plate; (b) add 1 mL of HMDS (Sigma) in each well. HMDS is toxic and harmful by inhalation; carry out all these steps under a fume hood; (c) incubate the plate at rt for 30 min; (d) remove HMDS with a Pasteur pipette; (e) allow cover glasses to dry until the next day.

21. In this phase of bacteria fixation, cells are still alive.
22. At this point, it is crucial to avoid that cover glasses run dry.
23. Make the envelope by closing (with a stapler) three sides of a folded sheet of optical paper around each cover glass.
24. It is very important not to forget to properly position the cover glass into the envelope: the poly(L-lysine)/bacteria side should be directed toward the labeled side of the envelope.
25. In all these phases, samples must be kept wet.
26. When you put the cover glass on the stub, make sure that the emptied side of the cover glass (without bacteria) is in contact with the glue, otherwise you will loose your sample!
27. After primary aldehyde fixation, bacteria cell pellets should be cohesive enough to be handled during subsequent processing steps, as if they were tissue pieces. If this is difficult to achieve, or if too small pellets are obtained because of the limited number of cells which are present in the sample, the suspensions of glutaraldehyde-fixed cells can be pelleted and embedded in agar. Different methods can be used for this purpose, but the common net result is the formation of agarized cell pellets, sufficiently hard and large for further handling.
28. A good negative staining preparation should ideally produce a single layer of individual, well-separated bacterial cells on the coated grid. If the original suspension is too concentrated, several grids may be prepared from serial dilutions of the sample.
29. Formvar, otherwise known as polyvinyl formal, appears to be the most widely used support film for TEM grids worldwide. During this step, you can keep the grid a bit raised from the bench by using tweezers.
30. Negative stains do not interact with biological samples; they rather adsorb to the specimen particles and, upon drying, enclose them forming a structureless, smooth glassy matrix that surrounds external/accessible internal surfaces

of each particle and penetrate, to a certain extent, into their irregularities.

31. Typically, stain clouds or "shadows" are obtained around small portions of the sample, producing a strong contrast with respect to the background. This is because of the different electron density (i.e., electron scattering power) between the organic matter and the surrounding metal atoms. As a consequence, the sample will appear as a negative picture (light areas on a dark background) when observed under TEM, and its morphology and fine structural details will be clearly distinguishable (*see* **Fig. 16.3**).

Acknowledgments

We thank Prof. Donatella Barra for careful reading the manuscript. Work reported here has been supported in part by grants from the Italian Ministry of Education, University and Research and from the Università di Roma "La Sapienza".

References

1. Boman, H. G. (1995) Peptide antibiotics and their role in innate immunity. *Annu. Rev. Immunol.* **13**, 61–92.
2. Hancock, R. E. and Sahl, H. G. (2006) Antimicrobial and host-defense peptides as new anti-infective therapeutic strategies. *Nat. Biotechnol.* **24**, 1551–1557.
3. Zasloff, M. (2002) Antimicrobial peptides of multicellular organisms. *Nature* **415**, 389–395.
4. Jenssen, H., Hamill, P., and Hancock, R. E. (2006) Peptide antimicrobial agents. *Clin. Microbiol. Rev.* **19**, 491–511.
5. Nicolas, P. and Mor, A. (1995) Peptides as weapons against microorganisms in the chemical defense system of vertebrates. *Annu. Rev. Microbiol.* **49**, 277–304.
6. Mookherjee, N. and Hancock, R. E. (2007) Cationic host defence peptides: innate immune regulatory peptides as a novel approach for treating infections. *Cell. Mol. Life Sci.* **64**, 922–933.
7. Mangoni, M. L., Papo, N., Barra, D., Simmaco, M., Bozzi, A., Di Giulio, A., and Rinaldi, A. C. (2004) Effects of the antimicrobial peptide temporin L on cell morphology, membrane permeability and viability of *Escherichia coli. Biochem. J.* **380**, 859–865.
8. Rodriguez, G. G., Phipps, D., Ishiguro, K., and Ridgway, H. F. (1992) Use of a fluorescent redox probe for direct visualization of actively respiring bacteria. *Appl. Environ. Microbiol.* **58**, 1801–1808.
9. Severin, E., Stellmach, J., and Nachtigall, H. M. (1985) Fluorimetric assay of redox activity in cells. *Anal. Chim. Acta* **170**, 341–346.
10. Nation, J. L. (1983) A new method using hexamethyldisilazane for preparation of soft insect tissues for scanning electron microscopy. *Stain Technol.* **58**, 347–351.

Chapter 17

Molecular Simulations of Antimicrobial Peptides

Allison Langham and Yiannis N. Kaznessis

Abstract

Recent advances in molecular dynamics (MD) simulation methods and in available computational resources have allowed for more reliable simulations of biological phenomena. From all-atom MD simulations, we are now able to visualize in detail the interactions between antimicrobial peptides (AMPs) and a variety of membrane mimics. This helps us to understand the molecular mechanisms of antimicrobial activity and toxicity. This chapter describes how to set up and conduct molecular dynamics simulations of AMPs and membrane mimics. Details are given for the construction of systems of interest for studying AMPs, which can include simulations of peptides in water, micelles, or lipid bilayers. Explanations of the parameters needed for running a simulation are provided as well.

Key words: Molecular dynamics, CHARMM, NAMD, micelles, lipid bilayers.

1. Introduction

1.1. Why Computer Simulations?

Molecular dynamics (MD) simulations can provide some of the molecular level details necessary to understand the mechanism of action of antimicrobial peptides (AMPs). For many AMPs, it is apparent that the cell membrane is the target. The general interaction between positively charged AMPs and negatively charged bacterial membranes is well understood, as it is the need for hydrophobic content to allow for interactions between the peptide and the membrane core. However, these basic facts are not enough to allow for design of AMPs of therapeutic value. There is no simple, linear correlation between peptide charge or hydrophobicity and antimicrobial activity. Consequently, intuitive design rules for novel AMPs are hard to deduce. The wide variety of sequences and structures of naturally occurring AMPs suggests

A. Giuliani, A.C. Rinaldi (eds.), *Antimicrobial Peptides*, Methods in Molecular Biology 618,
DOI 10.1007/978-1-60761-594-1_17, © Springer Science+Business Media, LLC 2010

that there is most certainly not one mechanism of action by which all peptides act.

In this chapter, we discuss the use of MD simulations to study AMPs. The goals in using MD simulations are to both understand the mechanism of action by which these peptides kill bacteria and develop predictive models that can be used to design highly potent, yet non-toxic peptides.

We should note that computer simulations have as steep a learning curve as any sophisticated experimental method. The logistics of running computer simulations can also be challenging, from setting up the system, to choosing numerical integration schemes and statistical ensembles, to managing and analyzing large volumes of data. Nonetheless, as with any sophisticated experimental method, the pleasure of discovery awaits on the other side of a tortuous effort.

1.2. Theory

For a thorough review of the underlying theory of MD simulations, the reader is directed to Allen and Tildesley's "Computer Simulation of Liquids" (1). MD simulations are based on classical mechanics or, more specifically, on Newton's second equation of motion. If the force acting on each atom in a system is known, the acceleration can be determined. Using this acceleration, the equations of motion can be integrated, resulting in a trajectory of the positions, velocities, and accelerations of each particle in the system. The force F_i on an atom of mass m_i, which is at position r_i, is determined from the potential energy of the system, V, as shown in equation [1]. The potential can be used to directly determine the position of a particle as a function of time, as in equation [2].

$$F = -\nabla V \qquad [1]$$

$$\frac{\mathrm{d}V}{\mathrm{d}r_i} = m_i \frac{\mathrm{d}^2 r_i}{\mathrm{d}t^2} \qquad [2]$$

The equations of motion are solved for each atom in the system. The forces acting on the atoms in the new positions are calculated and the simulation continues for as many time steps as necessary.

Empirical energy functions are used to approximate the force fields for large biological systems composed of many atoms. Some common force fields are AMBER (2), OPLS/AMBER (3), CHARMM (4), and GROMOS (5). We focus here on the CHARMM (Chemistry at HARvard Macromolecular Mechanics) force field, which has been developed over the last 25 years, and is continuously revised to better match new experimental data. There are also several commercially available MD simulation engines for use empowered with the above force fields. We recommend the use of the CHARMM MD simulation engine or NAMD (NAnoscale Molecular Dynamics) (6). A number of CHARMM scripts for the design and analysis of a variety of simulation systems are available on the CHARMM forum (www.charmm.org). However, NAMD is designed to take advantage of advances in parallel computing. This permits longer simulations of larger systems than are possible with CHARMM.

We should note that more recently a graphics user interface has been developed for CHARMM that significantly simplifies the set up of simulation systems and parameters. The software is available at http://www.charmm-gui.org/.

1.2.1. The Potential Function

The potential function is a sum of interaction energies. The value of the potential is determined by summing the bonded terms with the non-bonded potential terms, V_{bonded} and $V_{\text{non-bonded}}$, respectively:

$$V(\underline{R}) = V_{\text{bonded}} + V_{\text{non-bonded}} \qquad [3]$$

1.2.1.1. Non-bonded Terms

The non-bonded energy terms account for interactions between non-bonded atoms and atoms that are more than three covalent bonds away from each other in a molecule. These are modeled by the van der Waals energy and electrostatic energy, the first and second summation terms in equation [4], respectively.

$$V_{\text{non-bonded}} = \sum_{\text{non-bonded pairs}} 4\varepsilon\left[\left(\frac{\sigma_{ik}}{r_{ik}}\right)^{12} - \left(\frac{\sigma_{ik}}{r_{ik}}\right)^{6}\right] + \sum_{\text{non-bonded pairs}} \frac{q_i q_j}{Dr_{ij}} \qquad [4]$$

The calculation of the non-bonded terms in the potential function is the most time-consuming part of the MD simulation. In principle, the interactions between every pair of atoms should be calculated explicitly, meaning that for an N-atom system, N^2 calculations would be required. To decrease this number, methods were developed to ignore the interactions between two atoms separated by a distance greater than a specified cutoff distance. For the van der Waals interactions, the potential is "shifted" off over a period r_{on} to r_{off}: Atoms farther from each other than the distance r_{off} are assumed not to interact.

For the electrostatic interactions, cutoff schemes have been shown to be inaccurate when computing electrostatic interactions (7). Instead, the Ewald summation is used, which separates the potential into a slowly decaying long-range component and also a quickly varying short-range component. There are several parameters that need to be optimized when solving the Ewald summation. For large systems, it is best to fix the real sum cut off, R_{cutoff}, and choose a value for the width of the Gaussians, α, that is large enough to accelerate convergence of the real sum. The value of α determines the relative rate of convergence of the real and reciprocal sums. The accuracy of the sum is independent of α. In practice, when the Ewald sum is used, the Gaussian source distributions centered on point ions are sampled onto a finite grid and the resulting Fourier integral is approximated by a Fourier series, which is evaluated numerically using Fast Fourier Transforms. Complex exponentials in the reciprocal sum are evaluated by β-spline interpolation. Therefore, the grid point spacing and the β-spline order must be optimized. The grid point spacing dictates the number of grids in each direction in the simulation cell onto which the Gaussian charges are interpolated. The β-spline is used to interpolate charges onto the grids. Higher values give more accurate sums but at a greater computational cost. In order for Ewald's method to be appropriately applied to a simulation, the system must have no net charge.

1.2.1.2. Bonded Terms

The bonded terms include the bonds, angles, and bond rotations in the molecule (**Fig. 17.1**) and are computed using equation [5]:

$$V_{\text{bonded}} = \sum_{\text{pairs}} K_{\text{b}}(b - b_0)^2 + \sum_{\text{angles}} K_\theta(\theta - \theta_0)^2 + \sum_{\text{dihedral pairs}} K_\phi(1 - \cos(n\phi)) \qquad [5]$$

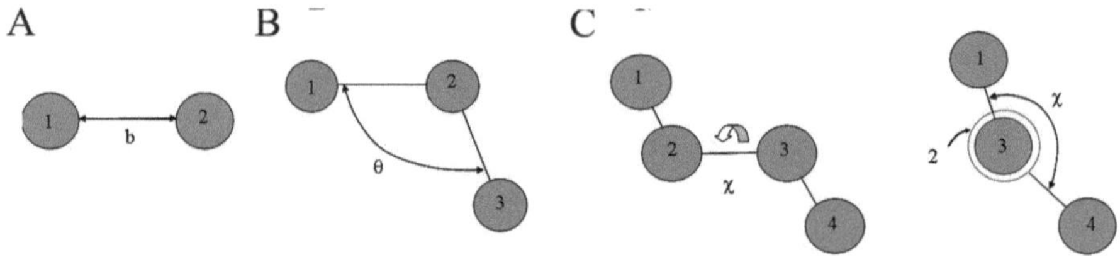

Fig. 17.1. Bonded interaction motions: (**a**) bond stretch is the movement between atoms directly bound to each other; (**b**) angle bend is a function of the relative position of atoms separated by one additional atom (1–3 pairs); (**c**) rotate-along-bond (dihedrals) is a function of the relative position of atoms separated by two additional atoms (1–4 pairs).

These terms are calculated using approximations for each type of motion. The bond-stretch term is a harmonic potential approximated as a function of displacement from the ideal bond length, b_0 for 1,2-pairs of atoms. The force constant, K_{b}, determines the strength of the bond and is determined by the atoms involved in the bond. The angle-bend term is associated with the movement a bond angle, θ, from its ideal value θ_0 and is also represented by a harmonic potential with a force constant, K_θ, which depends on the atoms involved. The rotate-along-bond energy is associated with the torsion angle potential function, which takes into account the steric barriers between 1,4-pairs of atoms (atoms separated by three covalent bonds). This potential expressed as a cosine function.

1.2.1.3. Additional Potentials

There are two additional potential terms in the CHARMM force field. The improper dihedrals term (equation [6]) accounts for out of plane bending, where k_ω is the force constant and $\omega - \omega_o$ is the out of plane angle. The other component is the Urey–Bradley potential (equation [7]), which accounts for angle bending using 1,3 non-bonded interactions. In this term, k_U is the respective force constant and $u - u_o$ is the distance between the 1,3 atoms in the harmonic potential:

$$E_{\text{impropers}} = \sum K_\omega(\omega - \omega_0)^2 \quad [6]$$

$$E_{\text{Urey-Bradley}} = \sum K_u(u - u_0)^2 \quad [7]$$

The CHARMM parameter file contains the parameters for each atom type and bond type. It provides the equilibrium bond lengths and van der Waals radii.

1.2.2. Integration Algorithms

Because there are so many atoms in a MD simulation system and because the energy function is semi-empirical, there is no analytical solution to the equations they formulate. There are many possible integration algorithms. Most are based on Taylor series expansions of the particle position and its derivatives. In CHARMM, the "leapfrog" Verlet algorithm is used. The equations solved in this algorithm are

$$r(t + \delta t) = r(t) + v\left(t + \frac{1}{2}\delta t\right)\delta t \quad [8]$$

$$v\left(t + \frac{1}{2}\delta t\right) = v\left(t + \frac{1}{2}\delta t\right) + a(t)\delta t \quad [9]$$

The velocities are calculated at time $t + 1/2\ \delta t$. These velocities are then used to calculate positions r at the time $t + \delta t$, so that the velocities "leap" over the positions, then the positions "leap" over the velocities (1).

In NAMD, the velocity Verlet algorithm is used. In this method, which minimizes round-off error, the positions, velocities, and accelerations are stored at time t(1). The equations used are

$$r(t + \delta t) = r(t) + \delta t v(t) + \frac{1}{2}\delta t^2 a(t) \quad [10]$$

$$v\left(t+\frac{1}{2}\delta t\right)=v(t)+\frac{1}{2}\delta t[a(t)+a(t+\delta t)] \quad [11]$$

The time step, δt, used with these algorithms is of importance. A smaller time step means slower dynamics. On the other hand, a large time step can cause instability due to large jumps in the conformation negatively impacting the energy of the system. The time step must be much smaller than the fastest characteristic motion within the system. For biological systems, this motion is usually the vibration of hydrogen bonds. A time step of 2 fs (2×10^{-15} s) has been used widely for biological molecules when the SHAKE algorithm is employed to constrain the hydrogen bonds.

1.3. Temperature and Pressure Control

One can show that simply solving Newton's equations of motion for a system of *N* particles in volume *V* conserves the total energy of the system. Indeed any classical mechanical set of equation of motion samples the microcanonical, or *NVE* ensemble, where *N*, *V*, and *E* are constant.

It is more sensible to simulate biological systems under constant pressure and temperature constraints. The equations of motion need to then be modified appropriately. Temperature is often controlled by using Nose–Hoover–Langevin dynamics. In Langevin dynamics, a stochastic force term is included to introduce the effects of random interactions such as friction between molecules and an occasional high-velocity collision to mimic perturbations that would occur in an experimental system. For constant pressure and temperature simulations in which Langevin dynamics are used to control the temperature, the pressure can be controlled in NAMD with a modified Nose–Hoover method. This method is a combination of the methods described in (8) and (9). The revised equations of motion for this method are given below (equations [12], [13], [14], [15], [16], [17], and [18]), where *W* is the mass of the piston, τ is the oscillation period, *R* is the noise on the atoms, and R_e is the noise on the piston (10):

$$r' = p/m + e'r \quad [12]$$

$$p' = F - e'p - gp + R \quad [13]$$

$$V' = 3Ve' \quad [14]$$

$$e'' = 3\,V/W(P - P_0) - g_e e' + R_e/W \quad [15]$$

$$W = 3\,N\tau^2 kT \quad [16]$$

$$\langle R^2\rangle = 2mgkT/h \quad [17]$$

$$\langle R_e^2\rangle = 2Wg_e kT/h \quad [18]$$

1.4. Atomistic Versus Coarse-Grained Simulations

The previous discussion assumes the implementation of fully atomistic simulations. This is the preferred method for the study of antimicrobial peptides, as the ability of the force fields to reproduce structural and conformational changes diminishes with coarse-grained force fields. However, coarse-grained simulations offer a computationally cheaper approach. In these simulations, the atoms are not represented individually, but are grouped into "pseudo-atoms." This reduction of the system can allow for the simulation of larger systems or for longer timescales.

The level to which the system is reduced can be from the simplest united-atom model, in which the two hydrogens of a methylene group are represented together with the carbon, to groupings of pseudo-atom "beads" that represent two to four methylene groups. As with all-atom simulations, the parameterization for the interactions between beads is empirical. At higher levels of coarse graining, the accuracy of the system behavior is less reliable. For example, simulation of pore formation by the frog AMP magainin has been carried out with a coarse-grained method (11). Simulations of this type might be useful in some applications, but are not the focus of this chapter.

1.5. Implicit Solvation

Implicit solvation (or continuum solvation) is another method for simplifying a fully atomistic simulation. Instead of using fully atomistic "explicit" water molecules, the solvent is represented as a continuous medium, wherein a mean field approach can be used to approximate the behavior of individual solvent molecules. Two common methods use the Poisson–Boltzmann equation or the Generalized Born model. For a more thorough review of the different implicit solvation models, the reader is referred to (12). The Poisson–Boltzmann equation models the electrostatic environment of a solute as a polarizable continuum. In order to further reduce computational expense, more approximations are usually made. For instance, the Generalized Born model is an approximation to the linearized Poisson–Boltzmann equation (13). It is based on modeling the solute as a sphere with different internal dielectric constant than the solvent. Any type of implicit solvation can be used to reduce the number of atoms in the system, but results in the loss of data for the movement of the individual ions and solvent molecules in the system.

1.6. Common Simulation Systems

It has been proposed that AMPs act to kill bacteria through a series of steps: attraction, aggregation, penetration, and lysis (14). Simulations of this process would provide the ultimate level of detail in understanding the mechanism of action of AMPs; however, the timescales on which these processes occur exceed what is currently possible with all-atom MD simulations. Formation of pores has been observed in coarse-grained MD simulations (15, 16), but such simulations do not provide the same resolution as atomistic approaches, and may not accurately reflect system properties and behavior. To accommodate the current limitations on computational power, we can explore various steps of this process individually (**Fig. 17.2**)

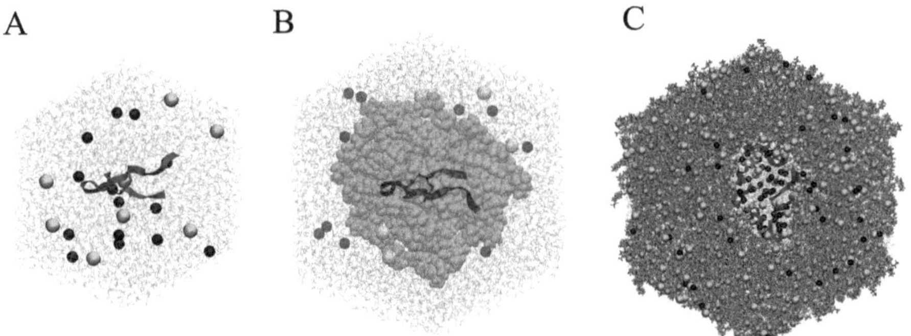

Fig. 17.2. The three types of systems typically investigated. Peptide in water (with ions) is shown in (**a**). This type of simulation can be completed in days. View (**b**) shows a typical peptide–micelle simulation. A typical micelle simulation can be completed in a span of several weeks. A lipid bilayer is shown in (**c**). With current computational resources, a lipid bilayer simulation takes several months to complete.

1.6.1. Simulations in Water

One of the simplest systems used in MD of AMPs consists of peptides in water. The simulation box is composed of a single peptide, water, counterions (if necessary to achieve charge neutrality), and additional salt if desired. This can be done simply to study the conformational motions of the peptide, or to measure physicochemical properties that might be related to the activity or toxicity of the peptide. When a set of data is available for a large number of related peptides, predictive quantitative structure–activity relationships (QSARs) can be developed. This is an approach widely used in computer-aided drug design (as described in 17 and demonstrated in 18–23). QSARs, obtained through multiple linear regression, correlate properties calculated from the peptide–water simulations to toxicity of the peptide. The hypothesis is that the properties of monomeric peptides can be related to their hemolytic activity and cytotoxicity. Simulations in water require significantly less computational resources than AMPs in environments mimicking lipid membrane, but can nevertheless provide meaningful results.

1.6.2. Simulations in Micelles

Simulations in water can provide a method for rapid screening of new sequences based on properties of single peptides, but in order to obtain information as to how the peptides interact with membrane mimics, and to elucidate the structural or sequence basis for the varying levels of toxicity in several peptide analogues, their interactions with membrane-mimicking micelles can be observed. There are several reasons to use micelles as membrane mimics. Like the lipid bilayers of real membranes, micelles possess a well-defined hydrophobic core and a flexible, hydrophilic interface and are commonly used in place of monolayers or bilayers in experimental methods such as NMR spectroscopy (24–27). Simulations in micelles are much less time and resource consuming than simulations in lipid bilayers, due to their smaller size and faster relaxation times, which have been shown through both direct experiments and simulations to be on the order of 500–1,000 ps for micelles (28–32), in contrast with lipid bilayers, which require tens of nanoseconds (33, 34).

Molecular dynamics studies of micelles have been performed for many types of micelles, including anionic micelles such as those formed by sodium dodecyl sulfate (SDS) (29, 35–38); zwitterionic micelles (e.g., dodecylphosphocholine, DPC) (30, 31, 39–41); and a few mixed composition micelles (42, 43). DPC micelles are considered to be acceptable models of eukaryotic cell membranes, which are generally rich in zwitterionic phospholipids. SDS has been used as a mimic for the negatively charged molecules found in bacterial membranes, because SDS micelles possess an anionic exterior and a hydrophobic interior. The differences of peptide behavior between the SDS micelle and the DPC micelle provide a basis for the study of the activity and toxicity of selected AMPs. Indeed, it is quite possible that AMPs work through different mechanisms to kill bacterial or mammalian cells. In the case of the unstructured indolicidin, for example, the peptide takes on different conformations when interacting with the two types of micelles (44).

1.6.3. Simulations in Lipid Bilayers

The long timescales on which aggregation and penetration of the bilayer by AMPs occur have prohibited the use of fully atomistic MD simulations to study these phenomena. The computational power needed to observe such phenomena using fully atomistic simulations is currently not available. It is possible, however, to study systems for which an initial orientation has been suggested from experimental data. For example, the pore structure of protegrin-1 in a mixed phosphatidylethanolamine (POPE): phosphatidylglycerol (POPG) bilayer was determined from NMR experiments. This was used as a starting configuration for MD simulations of the pore system, allowing a molecular level view of the interactions between the pore and the bilayer system, while eliminating the problems associated with initialization of the system (45).

2. Materials

1. PDB (Protein Data Bank, http://www.rcsb.org) file for AMP of interest with atomic coordinates.
2. Coordinate files for micelles or lipids, as needed.
3. Visualization software, such as VMD or Rasmol.
4. CHARMM topology and parameter sets (available for free download at http://www.cems.umn.edu/research/kaznessis/antimicrobial.htm.)
5. Simulation engine (NAMD or CHARMM recommended).
6. Access to supercomputers or UNIX machines.

3. Methods

3.1. System Construction

1. The initial configuration of the peptide should be based on an experimentally determined structure (e.g., through X-ray crystallography or NMR), since simulations are not at the point to be able to properly fold a peptide in a reasonable amount of time. The initial peptide coordinate file can be obtained from the Protein Data Bank for many AMPs. The type of simulation will dictate the next steps of system construction.
2. For simulations carried out in water, typically a large box of water is first constructed. This is often done in CHARMM by copying and translating a box of water. Two pre-made boxes of TIP3P water molecules are available, one containing 125 molecules and the other 216 molecules. The peptide is then placed in the box of water, and the molecules that overlap with the peptide are removed. Counterions and any additional ions should be added at this time. They can be added by replacing water molecules.
3. For simulations in micelles, coordinates exist for two common types of micelles, DPC and SDS. DPC coordinates can be obtained from the web site of Professor Tuck Wong (http://www.chem.missouri.edu/wong/). SDS coordinates have been published by Alex MacKerell (36). The peptide is can be placed in the center of the micelle with the two centers of mass overlapping. This peptide–micelle complex is then solvated as described in Step 2.
4. A detailed method for building lipid bilayer simulations, with all necessary CHARMM scripts,

is available on Benoit Roux's web site (http://thallium.bsd.uchicago.edu/RouxLab/membrane.html). In this method, the bilayer is built around the peptides, with bad contacts between peptides and lipids or among lipid molecules reduced through a series of rotations and translations. Libraries of lipid conformations are available for a few common lipids, such as dipalmitoylphosphatidylcholine (DPPC). A brief summary of Roux's method is outlined below:

4.1. Use sys1.inp to find the cross-sectional area of the peptide or oligomer. This information is necessary to determine the size of the lipid bilayer that will be needed.

4.2. Use sys2.inp to place dummy atoms into preliminary top and bottom leaflets. These atoms are replaced by the head group atoms of the lipids in subsequent steps. The height of the bilayer must be specified based on the length of lipids being used.

4.3. Sys3.inp minimizes the dummy atoms.

4.4. Sys4.inp requires the library of lipid conformers. In this step, each dummy atom is replaced with a lipid molecule. Each lipid is then systematically translated and rotated to remove bad contacts (such as overlapping atoms).

4.5. Use sys5.inp to minimize the system, which at this time is composed of peptide(s) and lipids.

4.6. Use sys6.inp and sys7.inp to make a sheet and then a box of water that is of the appropriate size for the system. Sys8.inp is then used to position the water box on each side of the bilayer.

4.7. Use sys9.inp to further minimize the system.

4.8. Additional minimization scripts (sys10.inp to sys18.inp) can be used.

5. Create a.psf file for the fully constructed system.

3.2. Running a NAMD Simulation

1. Convert the prepared coordinate file to a NAMD.pdb file. This can be done using the program crd2pdb, which is available on the NAMD web site (http://www.ks.uiuc.edu/Research/namd/).

2. Convert the primary structure file (.psf) to a NAMD.psf file using charmm2namd, which is also available on the NAMD web site.

3. Create the simulation script. The following parameters need to be set (*see* **Table 17.1**):

   3.1. Dimensions for the simulation box.

Table 17.1
Description of each term to be specified in the NAMD script. Typical values for each term are given as well

Term	Description	Typical value
Outputenergies	Number of time steps between energy outputs	1,000
xstFreq	Number of time steps between to the extended system coordinate file	1,000
dcdFreq	Number of time steps between writing to the trajectory file	1,000
wrapAll	Use periodic boundary conditions on all atoms	On
wrapNearest	Use nearest image to wrap the coordinates	On
rigidbonds	Fixes hydrogen bond lengths to the nominal value	All
timestep	Time step in femtoseconds	2
cutoff	Distance at which non-bonded and bonded interactions are cutoff	11
nonBondedFreq	Number of time steps between evaluation of the non-bonded interactions	1
fullElectFrequency	Number of time steps between evaluation of electrostatic interactions	1
pairlistdist	Distance between pairs for inclusion on the pairlist	14
stepsPerCycle	Number of time steps between atom reassignments	20
switching	Use switching for the non-bonded van der Waals interactions	On
switchDist	Distance at which the switching function should start to switch	8
exclude	Which pairs of bonded atoms to exclude	Scaled 1–4
1-4scaling	Constant factor to modify the electrostatic interactions for the pairs that are modified by exclude	1
langevin	Use Langevin dynamics	On
langevinDamping	Damping coefficient for Langevin dynamics	10
langevinTemp	Temperature for Langevin calculations	$temp
langevinHydrogen	Do not apply Langevin dynamics to hydrogens	No
langevinPiston	Use Langevin piston pressure control	On
langevinPistonTarget	Target pressure for Langevin piston pressure control	1.01325
langevinPistonPeriod	Barostat oscillation time in femtoseconds	200
langevinPistonDecay	Damping timescale in femtoseconds	10
langevinPistonTemp	Noise temperature	$temp

3.2. Pressure and temperature at which to run the simulation.

3.3. The number of time steps between each output (OutputEnergies, xstFreq, and dcdFreq). They need to be small enough to allow for capture of the details of interest in the simulation but large enough so that the trajectory file does not become too large for storage.

3.4. Specify cell basis vectors for periodic boundary conditions. For hexagonal cells, the vectors are $[(a, 0, 0), (a/2, a\sqrt{3}/2, 0), (0, 0, h)]$, where a is the width of the box (defined as the distance from the center of the simulation to the center of the adjacent image box) and h is the height. For rhombic dodecahedron cells, the vectors are $[(a, 0, 0), (0, a, 0), (a/2, a/2, a\sqrt{2}/2)]$, where a is the box length (defined as the distance from the center of the simulation to the center of the adjacent image box).

3.5. Turn on wrapAll and wrapNearest if using periodic boundary conditions. If wrapAll is specified, when a molecule leaves the simulation box, its coordinates are translated to the other side of the cell when they are output. If wrapNearest is specified, then the coordinates are wrapped to the nearest image to the origin, not the diagonal unit cell centered on the origin.

3.6. Turn on rigidbonds so that a 2 fs time step can be used.

3.7. Specify the cutoff distance.

3.8. Specify nonBondedFreq and fullElectFrequency (the number of steps between calculations of the non-bonded interactions).

3.9. Specify the distance for the pair list (pairlistdist).

3.10. Specify how often the pair list should be updated (Stepspercycle).

3.11. Turn on switching for the van der Waals interactions.

3.12. Specify the distance at which the switching function should be activated.

3.13. Specify which non-bonded interactions to exclude (typically exclude 1–2 and 1–3 bonded atoms from the non-bonded interactions). Also specify if the 1–4 interactions should be scaled.

3.14. Set up the temperature and pressure controllers. Specify the Langevin damping coefficient, target pressure, piston oscillation period, and piston damping coefficient.

3.15. Specify the particle mesh Ewald summation parameters. The function needs to be specified "on" and grid sizes must be provided. The grid should be chosen so that that there is approximately one point per Angstrom.

4. Minimize the system. The default minimization algorithm in NAMD combines conjugate gradient and line search methods. The amount of minimization necessary is dependent on the size of the system and how well the system was built.
5. Heat the system to the desired temperature. This should be done gradually. The Langevin piston is turned on and given a low temperature set point (usually less than 100 K). Langevin dynamics are run for a few thousand steps and then the temperature is gradually increased. Increasing by 30 K in each cycle to the final temperature set point should allow for the system to remain stable during the heating process.
6. Run the simulation. The simulation should be carried out until all properties of interest have stabilized.

4. Notes

1. There are a number of issues that can arise in the construction of the system. In particular, bad contacts can be built into the system in such a way that they cannot be overcome. In particular, one must be careful about the arrangement of lipid or detergent molecules near aromatic residues. The carbon chains can be placed so that they are inserted through the phenyl ring and will remain there for the duration of the simulation. Similar care must be taken for the position of the carbon chains near constrained loops, such as in β-hairpin peptides.
2. The initial configuration of the system is extremely important. There are still issues in the initialization of peptide–lipid systems because, in general, convergence of peptide–lipid systems is difficult to achieve due to the slow timescales on which the peptides are able to re-orient themselves when binding to the bilayer. Because of this, the proper orientation of the peptides at the start of the simulation is crucial. The need for prior knowledge of the binding conformation, and the restrictions on the accessible timescales, has hindered the predictive ability of lipid

bilayer simulations. Specific information about the location of the peptide in a lipid bilayer is rarely available and so one has to assume a starting orientation and position for the peptide. This choice has a significant influence on the behavior of the peptide and on the conformation of the system at the end of the simulation. This problem can be overcome, in part, by using micelles. By starting with the peptide inside the micelle, the peptide "sees" the same surroundings regardless of the initial orientation (**Fig. 17.3**)

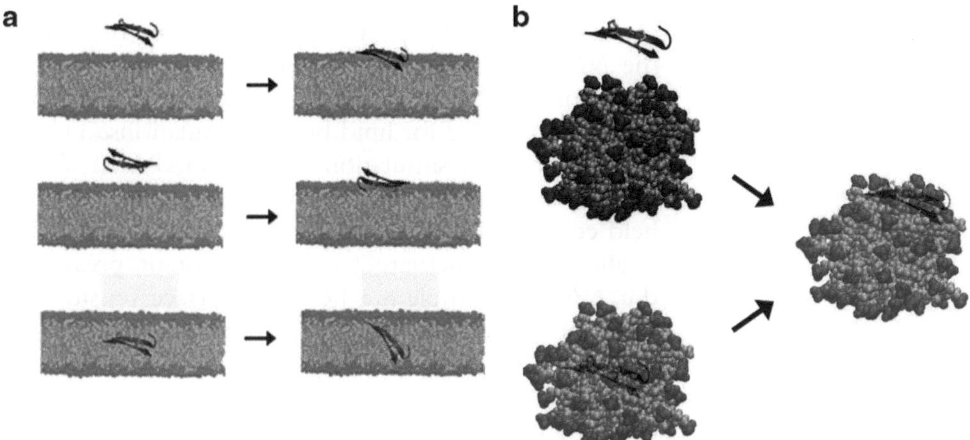

Fig. 17.3. Initialization issues. (**a**) For lipid bilayer simulations, the initial configuration has a strong bias on the final conformation observed. All three starting configurations should reach the same final conformation, the lower right, which is supported by experimental data. (**b**) For micelles, the final conformation can be reached with the peptide starting inside or outside the micelle, although again, it is unlikely to observe a peptide rotating while embedded into the micelle, so for simulations with the peptide starting outside of the micelle, the proper "face" must be positioned nearest the micelle surface. Using the conformation with the center of mass of the peptide overlapping the micelle center of mass reduces the bias.

3. Setup for systems can be done without using CHARMM. Scripts have been written for building systems using VMD (Visual Molecular Dynamics) and are available at the NAMD/VMD web site (http://www.ks.uiuc.edu/Research/vmd/).
4. The orientation of the simulation box vectors is different in CHARMM than in NAMD. For example, the hexagonal simulation box, recommended for lipid bilayer simulations, is rotated 15 degrees counterclockwise in CHARMM as compared to NAMD. Care must be taken in the start of the simulation of a system that has been built in CHARMM but is being run in NAMD.
5. The setup of the system should be checked visually at each step using VMD or Rasmol to ensure the proper orientation of molecules relative to each other.

6. When constructing a lipid bilayer simulation for a pore, care must be taken in the initial stages to prevent the lipids from entering the interior of the pore. MMFP (miscellaneous mean field potential; see mmfp.doc in the CHARMM documentation) constraints are used in sys2.inp in the construction of any lipid bilayer to maintain the proper bilayer thickness. Cylindrical constraints can be used to prevent the lipids from entering the pore.
7. For simulations of peptides in water or interacting with micelles, the constant pressure–temperature ensemble is recommended. For lipid bilayer simulations, there is some discussion over the proper ensemble (46–48). Although the *NPT* (constant pressure–constant temperature) ensemble is used in most simulations, problems with the area per lipid can arise for lipid bilayer simulations. This is particularly true for simulations using CHARMM, for which NP_zAT ensemble – where the area in the *x–y* plane is held constant but the *z*-dimension, normal to the bilayer, is allowed to fluctuate to retain constant pressure – or the $NP\gamma T$ ensemble – where the surface tension is held constant – is recommended. In the simulations that have been presented in the literature, the simulation cell in *NPT* simulations using CHARMM or NAMD regularly shrinks, reducing the area per lipid. In 2004, Jensen and coworkers simulated a lipid bilayer containing 36 DPPC lipids per leaflet in the *NPT* ensemble and observed that the *x–y* dimensions of the cell decreased for the entire 15-ns simulation (49). Images of this system show an unrealistic expansion of the lipid in the *z*-direction, such that the two leaflets have separated. Feller and Pastor explained the unnatural behavior of lipid bilayers in simulations as being due to the small sizes of the bilayers that are studied (46). Long wavelength undulations of the bilayer are absent in small simulation cells, causing an increase in the surface tension above what would normally be expected from a macroscopic bilayer. The discussion on this topic is still open, but the options to run with constant area or a constant ratio of the *x*- and *y*-dimensions are available in NAMD.
8. For lipid bilayer simulations, the hydration of the lipids and the concentration of ions can have strong effects on the results. When experimental data are available, it should be matched as closely as possible.
9. Analysis can be done using CHARMM. A number of scripts for analysis of simulations are available in the script archive on the CHARMM Forum (www.charmm.org).

Analysis will vary depending on the type of simulation carried out.

10. All simulations can provide information about the structure of the peptide. The secondary structure can be measured for each time point of the simulation by calculation of the dihedral angles. It is possible to compare the root mean square deviation of the peptide through the simulation to the initial structure or some other chosen structure. The dipole moment of the peptide is easily calculated, as well as the surface area. The area exposed to the solvent can also be easily determined. The shape of the peptide can be measured in terms of eccentricity or the maximum or average dimensions. The moment of inertia and radius of gyration are also easily calculated. The interaction energies between the residues within the peptide or the peptide and water can also be calculated and broken down further into the van der Waals interaction energy and the electrostatic interaction energy. Additionally, though hydrogen bonds are not explicitly modeled in CHARMM, the energy due to hydrogen bond interactions that have been defined by atom type, distance, and angle can be calculated.
11. Simulations in micelles allow for calculation of peptide properties as well as interaction energies (van der Waals and electrostatic) between peptide and self, peptide and micelle, or peptide and water. Hydrogen bond energy can also be calculated. Micelle shape can be analyzed over the course of the simulation, by monitoring the moments of inertia.
12. Radial distribution functions with micelle or bilayer core or head groups can provide information about the strength of interaction for each residue and the desired system component.
13. For bilayer simulations, properties such as thickness, tilt of lipids, order parameters, trans-gauche conformations, hydrogen bonding between head groups, and interactions with ions are often of importance.

References

1. Allen, M. P. and Tildesley, D. J. (1987) *Computer Simulations of Liquids.* Oxford: Clarendon Press.
2. Pearlman, D. A., Case, D. A., Caldwell, J. W., Ross, W. S., Cheatham, T. E., III, DeBolt, S., Ferguson, D., Seibel, G., and Kollman, P. (1995) Amber, a computer program for applying molecular mechanics, normal mode analysis, molecular dynamics and free energy calculations to elucidate the structures and energies of molecules. *Comp. Phys. Commun.* **91**, 1–41.
3. Jorgensen, W. L. and Tirado-Rives, J. (1988) The OPLS [optimized potentials for lipid simulations] potential functions for proteins, energy minimizations for crystals of cyclic peptides and crambin. *J. Am. Chem. Soc.* **110**, 1657–1666.
4. Brooks, B. R., Bruccoleri, R. E., Olafson, B. D., States, D. J., Swaminathan, S., and

Karplus, M. (1983) CHARMM: a program for macromolecular energy, minimization, and dynamics simulations. *J. Comp. Chem.* **4**, 187–217.

5. Berendsen, H. J. C., van der Spoel, D. L., and van Drunen, R. (1995) GROMACS: a message-passing parallel molecular dynamics implementation. *Comp. Phys. Comm.* **91**, 43–56.
6. Phillips, J. C., Braun, R., Wang, W., Gumbart, J., Tajkhorshid, E., Villa, E., Chipot, C., Skeel, R. D., Kale, L., and Schulten, K. (2005) Scalable molecular dynamics with NAMD. *J. Comput. Chem.* **26**, 1781–1802.
7. Patra, M., Karttunen, M., Hyvönen, M. T., Falck, E., Lindqvist, P., and Vattulainen, I. (2003) Molecular dynamics simulations of lipid bilayers: major artifacts due to truncating electrostatic interactions. *Biophys J.* **84**, 3636–3645.
8. Martyna, G. J., Tobias, D. J., and Klein, M. L. (1994) Constant pressure molecular dynamics algorithms. *J. Chem. Phys.* **101**, 4177–4189.
9. Feller, S. E., Zhang, Y., Pastor, R. W., and Brooks, B. R. (1995) Constant pressure molecular dynamics simulation: the Langevin piston method. *J. Chem. Phys.* **103**, 4613–4621.
10. Bhandarkar, M., Brunner, R., Chipot, C., Dalke, A., Dixit, S., Grayson, P., Gullingsrud, J., Gursoy, A., Hardy, D., Hénin, J., Humphrey, W., Hurwitz, D., Krawetz, N., Kumar, S., Nelson, M., Phillips, J., Shinozaki, A., Zheng, G., and Zhu, F. (2005) *NAMD User's Guide Version 2.6b1*. Urbana: Theoretical Biophysics Group, University of Illinois and Beckman Institute.
11. Leontiadou, H., Mark, A. E., and Marrink, S. J. (2006) Antimicrobial peptides in action. *J. Am. Chem. Soc.* **128**, 12156–12161.
12. Roux, B. and Simonson, T. (1999) Implicit solvent models. *Biophys. Chem.* **78**, 1–20.
13. Dominy, B. N. (1999) Parameterization and application of an implicit solvent model for macromolecules. *Mol. Simulat.* **24**, 259–274.
14. Brogden, K. A. (2005) Antimicrobial peptides: pore formers or metabolic inhibitors in bacteria?. *Nat. Rev. Microbiol.* **3**, 238–250.
15. Kandasamy, S. K. and Larson, R. G. (2007) Binding modes of protegrin-1, a beta-strand antimicrobial peptide, in lipid bilayers. *Mol. Simulat.* **33**, 799–807.
16. Leontiadou, H., Mark, A. E., and Marrink, S. J. (2007) Ion transport across transmembrane pores. *Biophys. J.* **92**, 4209–4215.
17. Parvu, L. (2003) QSAR – a piece of drug design. *J. Cell. Mol. Med.* 7, 333–335.
18. Jenssen, H., Fjell, C. D., Cherkasov, A., and Hancock, R. E. (2008) QSAR modeling and computer-aided design of antimicrobial peptides. *J. Pept. Sci.* **14**, 110–114.
19. Kaznessis, Y. N., Snow, M. E., and Blankley, C. J. (2001) Prediction of blood-brain partitioning using Monte Carlo simulations of molecules in water. *J. Comput. Aid. Mol. Des.* **15**, 697–708.
20. Lejon, T., Stiberg, T., Strøm, M. B., and Svendsen, J. S. (2004) Prediction of antibiotic activity and synthesis of new pentadecapeptides based on lactoferricins. *J. Pept. Sci.* **10**, 329–335.
21. Lejon, T., Strom, M. B., and Svendsen, J. S. (2001) Antibiotic activity of pentadecapeptides modelled from amino acid descriptors. *J. Pept. Sci.* 7, 74–81.
22. Lejon, T., Svendsen, J. S., and Haug, B. E. (2002) Simple parameterization of non-proteinogenic amino acids for QSAR of antibacterial peptides. *J. Pept. Sci.* **8**, 302–306.
23. Marrero-Ponce, Y., Torrens, F., Alvarado, Y. J., and Rotondo, R. (2006) Bond-based global and local (bond, group and bond-type) quadratic indices and their applications to computer-aided molecular design. 1. QSPR studies of diverse sets of organic chemicals. *J. Comput. Aid. Mol. Des.* **20**, 685–701.
24. Beswick, V., Guerois, R., Cordier-Ochsenbein, F., Coïc, Y. M., Tam, H. D., Tostain, J., Noël, J. P., Sanson, A., and Neumann, J. M. (1999) Dodecylphosphocholine micelles as a membrane-like environment: new results from NMR relaxation and paramagnetic relaxation enhancement analysis. *Eur. Biophys. J.* **28**, 48–58.
25. Kallick, D. A., Tessmer, M. R., Watts, C. R., and Li, C. Y. (1995) The use of dodecylphosphocholine micelles in solution NMR. *J. Magn. Reson. B* **109**, 60–65.
26. Kloosterman, D. A., Goodwin, J. T., Burton, P. S., Conradi, R. A., Stockman, B. J., Scahill, T. A., and Blinn, J. R. (2000) An NMR study of conformations of substituted dipeptides in dodecylphosphocholine micelles: implications for drug transport. *Biopolymers* **53**, 396–410.
27. Lauterwein, J., Bösch, C., Brown, L. R., and Wüthrich, K. (1979) Physicochemical studies of the protein-lipid interactions in melittin-containing micelles. *Biochim. Biophys. Acta* **556**, 244–264.
28. Fernandez, C. and Wuthrich, K. (2003) NMR solution structure determination of membrane proteins reconstituted in detergent micelles. *FEBS Lett.* **555**, 144–150.

29. Rakitin, A. R. and Pack, G. R. (2004) Molecular dynamics simulations of ionic interactions with dodecyl sulfate micelles. *J. Phys. Chem. B.* **108**, 2712–2716.
30. Wymore, T., Gao, X. F., and Wong, T. C. (1999) Molecular dynamics simulation of the structure and dynamics of a dodecylphosphocholine micelle in aqueous solution. *J. Mol. Struct.* **485**, 195–210.
31. Wong, T. C. and Kamath, S. (2002) Molecular dynamics simulations of adrenocorticotropin (1–24) peptide in a solvated dodecylphosphocholine (DPC) micelle and in a dimyistoylphosphatidylcholine (DMPC) bilayer. *J. Biomol. Struct. Dyn.* **20**, 39–57.
32. Tieleman, D. P., van der Spoel, D., and Berendsen, H. J. C. (2000) Molecular dynamics simulations of dodecylphosphocholine micelles at three different aggregate sizes: micellar structure and lipid chain relaxation. *J. Phys. Chem. B* **104**, 6380–6388.
33. Shepherd, C. M., Schaus, K. A., Vogel, H. J., and Juffer, A. H. (2001) Molecular dynamics study of peptide-bilayer adsorption. *Biophys. J.* **80**, 579–596.
34. Kandasamy, S. K. and Larson, R. G. (2004) Binding and insertion of alpha-helical antimicrobial peptides in POPC bilayers studied by molecular dynamics simulations. *Chem. Phys. Lipids* **132**, 113–132.
35. Bruce, C. D., Berkowitz, M. L., Perera, L., and Forbes, M. D. E. (2002) Molecular dynamics simulation of sodium dodecyl sulfate micelle in water: micellar structural characteristics and counterion distribution. *J. Phys. Chem. B.* **106**, 3788–3793.
36. MacKerell, A. D. (1995) Molecular dynamics simulation analysis of a sodium dodecyl sulfate micelle in aqueous solution: decreased fluidity of the micelle hydrocarbon interior. *J. Phys. Chem.* **99**, 327–341.
37. Wymore, T. and Wong, T. C. (1999) Molecular dynamics study of substance P peptides partitioned in a sodium dodecylsulfate micelle. *Biophys. J.* **76**, 1213–1227.
38. Yang, S. T., Yub Shin, S. Y., Kim, Y. C., Kim, Y., Hahm, K. S., and Kim, J. I. (2002) Conformation-dependent antibiotic activity of tritrpticin, a cathelicidin-derived antimicrobial peptide. *Biochem. Biophys. Res. Commun.* **296**, 1044–1050.
39. Bond, P. J., Cuthbertson, J., and Sansom, M. S. (2005) Simulation studies of the interactions between membrane proteins and detergents. *Biochem. Soc. Trans.* **33**, 910–912.
40. Bond, P. J., Cuthbertson, J. M., Deol, S. S., and Sansom, M. S. (2004) MD simulations of spontaneous membrane protein/detergent micelle formation. *J. Am. Chem. Soc.* **126**, 15948–15949.
41. Bond, P. J. and Sansom, M. S. (2003) Membrane protein dynamics versus environment: simulations of OmpA in a micelle and in a bilayer. *J. Mol. Biol.* **329**, 1035–1053.
42. Marrink, S. J. and Mark, A. E. (2002) Molecular dynamics simulations of mixed micelles modeling human bile. *Biochemistry* **41**, 5375–5382.
43. Wendoloski, J. J., Kimatian, S. J., Schutt, C. E., and Salemme, F. R. (1989) Molecular dynamics simulation of a phospholipid micelle. *Science* **243**, 636–638.
44. Khandelia, H. and Kaznessis, Y. (2007) Cation-pi interactions stabilize the structure of the antimicrobial peptide indolicidin near membranes: molecular dynamics simulations. *J. Phys. Chem. B* **111**, 242–250.
45. Langham, A. A., Ahmad, A. S., and Kaznessis, Y. N. (2008) On the nature of antimicrobial activity: a model for protegrin-1 pores. *J. Am. Chem. Soc.* **130**, 4338–4346.
46. Feller, S. E. and Pastor, R. W. (1996) On simulating lipid bilayers with an applied surface tension: periodic boundary conditions and undulations. *Biophys. J.* **71**, 1350–1355.
47. Jahnig, F. (1996) What is the surface tension of a lipid bilayer membrane? *Biophys. J.* **71**, 1348–1349.
48. Roux, B. (1996) Commentary: surface tension of biomembranes. *Biophys. J.* **71**, 1346–1347.
49. Jensen, M. O., Mouritsen, O. G., and Peters, G. H. (2004) Simulations of a membrane-anchored peptide: structure, dynamics, and influence on bilayer properties. *Biophys. J.* **86**, 3556–3575.

Chapter 18

Computer-Based Analysis, Visualization, and Interpretation of Antimicrobial Peptide Activities

Ralf Mikut

Abstract

This chapter describes a computer-based method for analyzing the quantitative structure–activity relationships (QSAR) of antimicrobial peptides. Quantitative or qualitative activity measurements and known peptide sequences are used as input for the analysis. The analysis steps consist of the preprocessing which specifically deals with dilution series from an assay with luminescent bacteria, transformation of quantitative activity values into activity classes, a feature extraction method using molecular descriptors for amino acids, feature selection methods, visualization strategies, the classifier model design for discrimination of active and inactive peptides, and the in silico design of promising new peptide candidates.

Key words: Antimicrobial peptides, molecular descriptors, QSAR, data analysis.

1. Introduction

The application of quantitative structure–activity relationship (QSAR) models (1) is a promising means of analyzing the relationship between a peptide sequence and its specific activities, such as antimicrobial action against single or groups of microbes, including Gram-positive or Gram-negative bacteria (2–4).

For the data-based design of QSAR models, structural information about the peptide and its activity-related information are required. The structural information is mostly based on different kinds of features derived from atomic or molecular descriptors of the peptide sequence. Activity-related information can include the following:

A. Giuliani, A.C. Rinaldi (eds.), *Antimicrobial Peptides*, Methods in Molecular Biology 618,
DOI 10.1007/978-1-60761-594-1_18, © Springer Science+Business Media, LLC 2010

1. A qualitative ranking based on a live/dead staining or zone of inhibition study that results in an "active–inactive" classification for each peptide.
2. A single quantitative measurement for each peptide, such as the minimal inhibitory concentration (*MIC*) of a substance, the half maximal inhibition (*IC50*) or an inhibition of 75% (*IC75*), and inhibitory concentrations relative to a control peptide (e.g., *RelIC50, RelIC75*). The peptide activity increases with lower values because the desired effect (e.g., a 75% inhibition of bacterial activity) occurs at a lower peptide concentration.
3. Separate activity measurements for different peptide concentrations in a dilution series. A typical example is luminescence measurements that indicate the remaining activity of luminescent bacteria exposed to different peptide concentrations, resulting from synthesis using SPOT technology (5). This assay is described in detail by Hilpert in this book (6). Especially in the case of dilution series, careful preprocessing and data validation are key to obtaining reliable results.

A QSAR model can be optimized solely either for prediction power or for a compromise between prediction power and model simplicity. Both approaches require a large number of candidate features based on molecular descriptors. A compromise between prediction power and simplicity is based on drastic reduction of candidate features, using mathematical evaluation measures for feature selection (7).

This chapter describes the complete analysis process, starting with preprocessing of the dilution series, followed by feature extraction using molecular descriptors, then feature selection, model design, and finishing with the in silico design of promising and novel peptide candidates. All steps can be computed using the Peptide Extension Package of the MATLAB Toolbox Gait-CAD (8) (*see* **Section 3**).

2. Materials

In the design of a QSAR model for antimicrobial peptides, sequences of the investigated peptides (as one-letter code) and the activity-related information of each peptide are necessary. The latter information can be given in different forms (*see* **Section 1**) as follows:

1. A qualitative activity class as a text description (e.g., peptides are fully active, similar to positive control, worse than control, inactive) or by a numerical class code (e.g., "–1"

for lower and "1" for higher activity relative to a positive control).

Illustrative example:

Sequence	Activity class
TQEHSNRWN	worse than control
KWRWRVKKR	fully active

2. A quantitative value characterizing activity in the form of a critical peptide concentration.
Illustrative example:

Sequence	IC75
TQEHSNRWN	0.4653
KWRWRVKKR	0.0491

3. A series of activity-related measurements, $A(c)$, depending on the absolute or relative peptide concentration, c. These measurements should include a positive and negative control with known activity for each plate. A low $A(c)$ value means a low bacterial activity and a high activity of the peptide.
Illustrative example:

Sequence	**Plate**	***A*(1)**	***A*(0.5)**	***A*(0.25)**	***A*(0.125)**	***A*(0.0625)**	***A*(0)**
TQEHSNRWN	1	10	2,000	10,000	9,850	10,000	9,900
KWRWRVKKR	1	25	46	63	142	1,696	10,000
Positive control	1	10	10	100	1,050	9,000	9,900
Negative control	1	8,000	8,500	8,800	9,500	9,800	10,000

3. Methods

The following steps can be calculated using the Peptide Extension Package of the MATLAB Toolbox Gait-CAD (8). MATLAB® is a registered trademark of The MathWorks, Inc., Gait-CAD and the Peptide Extension Package are open source packages, licensed under the conditions of the GNU General Public License (GNU-GPL) of The Free Software Foundation. The download is available under http://sourceforge.net/projects/gait-cad. Both packages require MATLAB version 5.3 or later (*see* http://www.mathworks.com/).

After starting MATLAB and Gait-CAD, a text file containing tabulator-separated values for peptide sequence and activity-relative information (*see* **Section 2**) can be imported by executing *Peptides – Standard procedure (published in Humana Press Book) – Step 0 – Load data* (*see* screenshot in **Fig. 18.1**). Sample files

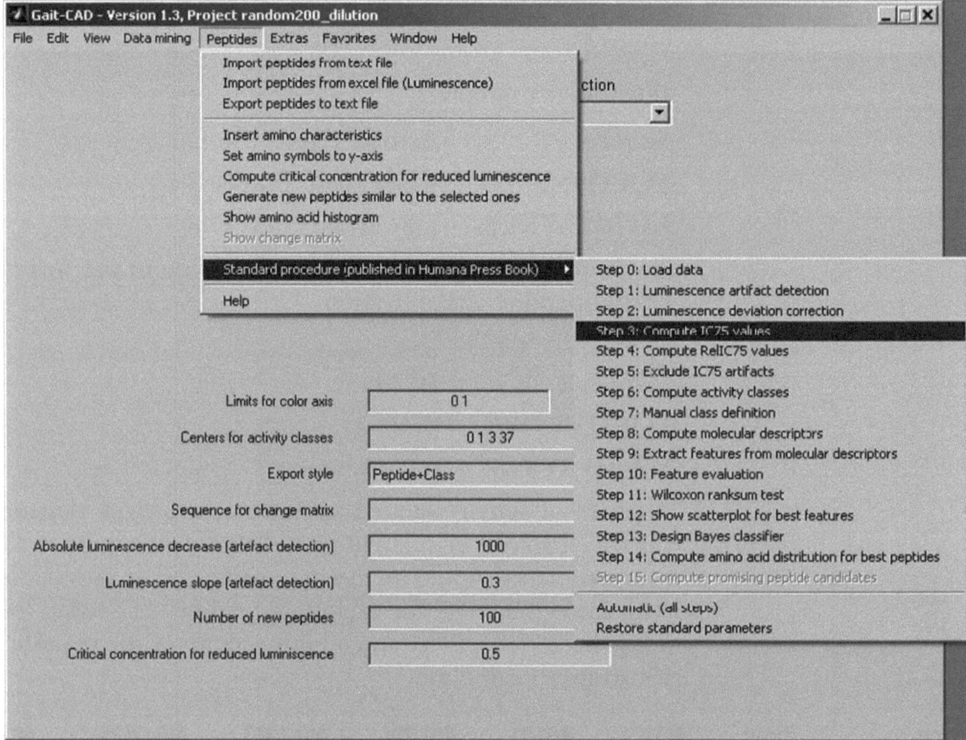

Fig. 18.1. Screenshot of the Peptide Extension Package from the MATLAB toolbox Gait-CAD.

for import are contained in the toolbox. After importing, Steps 1–15 can be performed by executing the related menu items. The results are shown as figures and protocol text files in the project directory. Parameters can be modified in the graphical user interface of the program: a reset to the proposed standard parameters is achieved by *Peptides – Standard Procedure (published in Humana Press Book) – Restore standard parameters.*

The starting point for the following steps to applying a QSAR model depends on the type of activity information, as discussed in **Section 2** (*see* **Table 18.1**). For example, the first three steps are only necessary if the activity is in the form of a dilution series.

1. It is assumed that bacterial activity is not positively correlated with peptide concentration. All data showing serious violations are handled as artifacts and are excluded from subsequent data analysis. We define a violation as a minimum increase of 30% of the activity value at lower concentrations and more than 1% of the maximal luminescence value in the study.

   Illustrative example: The maximal activity value recorded for a luminescent bacterium in a study is 10,000. For an activity value $A(0.5) = 2{,}000$ and concentration $c = 0.5$, a peptide is considered an artifact if the bacterial

Table 18.1
Necessary steps depending on different types of activity information

	Necessary steps in Section 3				
	1–3	4–5	6	7	8–15
Dilution series with control peptides	×	×	×		×
Dilution series without control peptides	×			×	×
RelIC75 activity values			×		×
Other activity values				×	×
Activity classes					×

activity at the next higher concentration (here: 1) is $A(1) > 2{,}600$ (30% increase, more than $0.01 \times 10{,}000 = 100$, *see* **Fig. 18.2**). For an activity value of $A(0.5) = 100$, concentration 0.5, the artifact limit is $A(1) > 200$ (more than 30% increase, more than 100).

2. In this next step, all smaller deviations from the assumptions of Step 1 are corrected. In order to avoid an increasing bacterial activity value with increasing peptide concentrations, each activity value is set to the maximum of all activity values for the same concentration, *c*, and of all activity values at higher concentrations.

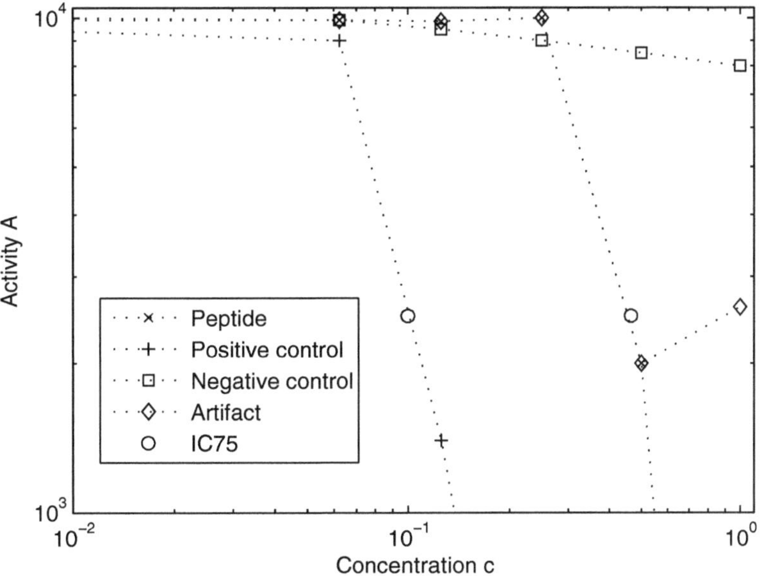

Fig. 18.2. Dilution series and associated bacterial activity with illustrative examples for one test peptide, one positive and one negative control, an artifact, and estimated *IC75* values for the test peptide and the positive control.

Illustrative example: $A(0) = 9{,}900$, $A(0.0625) = 10{,}000$, $A(0.125) = 9{,}850$, $A(0.25) = 10{,}000$, $A(0.5) = 2{,}000$, $A(1) = 10$ (*see* example in **Fig. 18.2**) is corrected to $A(0) = 10{,}000$, $A(0.125) = 10{,}000$.

3. The next aim is to calculate a single quantitative value that characterizes peptide activity from concentration-dependent bacterial activity values. For the computation of the *IC75*, we look for the two concentrations surrounding the *IC75* (*see* **Note 1**) with values just below and above the 75% reduction. Between these two concentrations, the *IC75* value is estimated using a logarithmic description of the activity values:

$$IC75 = ((\log_{10}(A(C_1)) - \log_{10}(A_{IC75})) \times c_2 + (\log_{10}(A_{IC75}) - \log_{10}(A(c_2))) \times c_1))/(\log_{10}(A(c_1)) - \log_{10}(A(c_2)))$$

Illustrative example: The activity value A_{IC75} is calculated using 25% of the maximal concentration of the dilution series (here: $A(0) = 10{,}000$): $A_{IC75} = 2{,}500$. The value is located between $c_1 = 0.25$ and $c_2 = 0.5$. The result is $IC75 = (\log_{10}(10{,}000) - \log_{10}(2{,}500)) \times 0.5 + (\log_{10}(2{,}500) - \log_{10}(2{,}000)) \times 0.25)/(\log_{10}(10{,}000) - \log_{10}(2{,}000)) = 0.4653$ (*see* **Fig. 18.2**).

4. The *IC75* value of each peptide is then compared with the *IC75* of the positive control on the same plate (*see* **Note 2**). The relative value, *RelIC75*, is computed by dividing the *IC75* of each peptide by the *IC75* on the same plate: *RelIC75* = *IC75*/*IC75* (positive control). A peptide with *RelIC75* = 1 has the same activity as the positive control; smaller values indicate a stronger activity.

   Illustrative example: On Plate 1, the *IC75* value for the positive control is 0.1. The *IC75* of a test peptide on the same plate is 0.4653. Consequently, the test peptide's *RelIC75* value is 4.653, which indicates lower activity than the positive control.

5. All plates must be handled as artifact and discarded from the analysis when they exhibit extreme deviations from the *IC75* of the positive control or value $A(1)$ of the negative control. An extreme *IC75* deviation of a positive control is defined by less than 50% and more than 200% of the median value of all positive controls. An extreme deviation for $A(1)$ of a negative control is defined by a value smaller than 25% of the median $A(1)$ of all negative controls. The median value is the value separating the lower and higher half of a group of values.

Illustrative example: The median of all *IC75* values is 0.08 for the positive control, and the median of the activity values for the negative control $A(1) = 8{,}000$. All plates with an *IC75* value for the positive control larger than 0.16 and lower than 0.04 or an activity value for the negative control $A(1) < 2{,}000$ are labeled as artifact.

6. We recommend four different classes using the *RelIC75* and the luminescence value $A(1)$ of test peptides. These classes are "fully active" ($RelIC75 \leq 0.5$), "similar to positive control" ($0.5 < RelIC75 \leq 2$), "worse than control" ($2 < RelIC75 \leq 20$), and "inactive" ($RelIC75 > 20$ or $A(1) > 0.9 \times A(1)$ of the negative control on the same plate). In the case of a dilution series, the additional condition $A(1)$ of a peptide larger than $0.9 \times A(1)$ of the negative control ensures that peptides without a relevant luminescence reduction are classified as inactive. This condition helps to handle plates with anomalous values, such as an unusually high *IC75* value for the positive control.

   Illustrative example: A peptide with a *RelIC75* value of 4.653 is classified as "worse than control"; if the negative control on a plate has an activity of $A(1) = 8{,}000$, all peptides with $A(1) > 7{,}200$ on the same plate are classified as inactive.

7. In the event that different forms of a quantitative value result from 2. In **Sections 1** and **2**, the class definition differs slightly. Here, a meaningful class definition must be established together with the design of each experiment. While class names can be the same, the values have to be defined relative to the corresponding values of a known positive control (similar to Step 6).

   Illustrative example: An alternative class definition uses *MIC* values instead of *RelIC75* values. A known positive control has a *MIC* value of 0.4; in the experimental design, *MIC* values lower than 0.2 were defined as fully active; therefore, a peptide with a *MIC* value of 0.1 is assigned to the class "fully active."

8. The following steps extract structural information from the peptide sequences. Besides the sequence itself, molecular descriptors can be used for assigning corresponding descriptor values to each amino acid. Typical molecular descriptors include the isoelectric point, the hydrophobicity following the Hopp–Woods scale or the Goldman–Engelmann–Steitz scale, molecular mass (all values according to (9)), or the Octanol-Interface scale (10) to describe the behavior by the peptide interaction with membranes (*see* **Table 18.2** and **Note 3**). New peptide sequences are

Table 18.2
Molecular descriptors for amino acids

Name	Symbol	Isoelectric point	Hopp–woods	Goldman–Engelman–Steitz	Molecular mass	Octanol-Interface scale
Alanine	A	6.01	−0.50	−1.60	89.00	0.33
Arginine	R	10.76	3.00	12.30	174.00	1
Asparagine	N	5.41	0.20	4.80	132.00	2.41
Aspartic acid	D	2.77	3.00	9.20	133.00	0.43
Cysteine	C	5.07	−1.00	−2.00	121.00	0.22
Glutamic acid	E	3.22	3.00	8.20	147.00	1.61
Glutamine	Q	5.65	0.20	4.10	146.00	0.19
Glycine	G	5.97	0.00	−1.00	75.00	1.14
Histidine	H	7.59	−0.50	3.00	155.00	−0.06
Isoleucine	I	6.02	−1.80	−3.10	131.00	−0.81
Leucine	L	5.98	−1.80	−2.80	131.00	−0.69
Lysine	K	9.74	3.00	8.80	146.00	1.81
Methionine	M	5.74	−1.30	−3.40	149.00	−0.44
Phenylalanine	F	5.48	−2.50	−3.70	165.00	−0.58
Proline	P	6.48	0.00	0.20	115.00	−0.31
Serine	S	5.68	0.30	−0.60	105.00	0.33
Threonine	T	5.87	−0.40	−1.20	119.00	0.11
Tryptophan	W	5.00	−3.40	−1.90	204.00	−0.24
Tyrosine	Y	5.66	−2.30	0.70	181.00	0.23
Valine	V	5.97	−1.50	−2.60	117.00	−0.53

generated by replacing the amino acids in a sequence with their corresponding molecular descriptors.

Illustrative example: For a sequence KWRWRVKKR, the sequence of isoelectric point is (9.74, 5.00, 10.76, 5.00, 10.76, 5.97, 9.74, 9.74, 10.76).

9. In order to apply data analysis tools, features must be extracted from the sequences of amino acids or molecular descriptors. For the amino acid sequence, relative frequencies of amino acid n-mers in a sequence are calculated: we use always $n = 1$ (frequency of amino acids) and $n = 2$ (frequency of amino acid pairs). Mean values, minimal and maximal values, and standard deviations are computed for molecular descriptors (*see* **Note 4**).

   Illustrative example: Using sequence KWRWRVKKR, the frequencies of amino acids are 3/9 for K and R, 2/9 for W, and 1/9 for V. The frequencies of 2-mers are 2/8 for WR and 1/8 for KW, RW, RV, VK, KK, and KR.

For the sequence of isoelectric points (9.74, 5.00, 10.76, 5.00, 10.76, 5.97, 9.74, 9.74, 10.76), the mean value is 8.61, the standard deviation is 2.37, the minimal value is 5.00, and the maximal value is 10.76.

10. Feature evaluation and selection can be performed using any statistical measure to analyze relevant differences between the different activity classes. For a fast check, a univariate analysis of variance for each feature and a multivariate analysis of variance for two and three features simultaneously (ANOVA, respectively, MANOVA) are sufficient. These measures evaluate the mean values of selected features for all peptides belonging to the different classes, the covariance matrices of these classes, and the total variance of the complete data set (*see* **Note 5**). Features with large distances between the mean values of different classes and small covariances within the classes are preferred. These methods are described in statistical textbooks (for reference *see* (11)), and they are implemented in many statistical software packages.

    Illustrative example: In the study (12), using 1,609 peptides, the standard deviation of the Hopp–Woods descriptor (ANOVA value 0.23) is detected as the best single feature, while the standard deviation of the Hopp–Woods descriptor and mean value of the isoelectric point are one promising pair of features for multivariate analysis (MANOVA value 0.26).

11. The best features of Step 10 are not always characterized by statistically relevant differences between activity classes. The relevance of generated best features should be analyzed using a non-parametric Wilcoxon rank sum test for class pairs, especially using the feature distributions of fully active peptides against peptides of all other classes. This test estimates the probability (p value) that the population of fully active peptides has an identical distribution to the populations of the three other classes describing less active peptides. The test results are more confident compared to a t test because most features do not fulfill the assumption of a normal distribution.

    Illustrative example: Using data from study (7), the relevance of different features (e.g., standard deviation of Hopp–Woods descriptor) is shown with a p value of <0.05.

12. The best combination of relevant features is now visualized using a two-dimensional scatterplot. This visualization is used to analyze the position of fully active peptides in

comparison to the other classes. This analysis must answer the following questions:

- Position and form of the region with data points representing the class of fully active peptides: spread out over the complete figure = no relevant information, one compact region = indicates one unified mode of action, multiple separated regions = indicates different modes of action
- Parameters of regions (high, middle, or low feature values)
- Existence of outliers (possible measurement errors or hint for different modes of action; must be tested with additional experiments)

Illustrative example: In a study with 1,609 peptides from (12), in **Fig. 18.3**, the class of fully active peptides is fairly compact, located around 2.2–2.8 of the standard deviation of Hopp–Woods (higher than normal) and 7–8 for the mean value of the isoelectric point (higher than normal). There are some outliers beyond these limits, especially with higher values for the isoelectric point (8.2–8.8) and the standard deviation of Hopp–Woods (3.2–3.3).

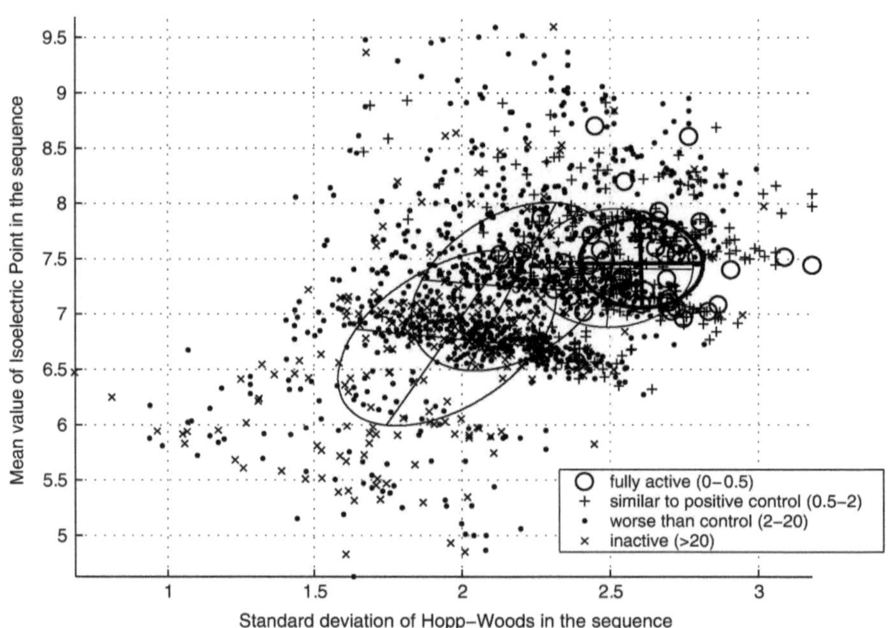

Fig. 18.3. Scatterplot showing the best features standard deviation for the sequence of Hopp–Woods, mean value of the sequence of isoelectric point; position of the mean value and the covariance matrix of the class "fully active" for the Bayes classifier (*bolded ellipsoid* with *bolded cross*).

13. Based on the two or three selected features, a Bayes classifier is designed to discriminate the classes. This technique is used to estimate the probability that a certain class is associated with given feature values and vice versa. For our scenario, an estimation of normal distributions containing parameters for mean values and the covariance matrix for each class is sufficient if the class regions in the figure according to Step 12 are approximately compact and normally distributed. The design of Bayes classifiers is described in statistical textbooks, such as (11), and is implemented in many statistical software packages. The classifier predicts group membership for each peptide.

    Illustrative example: Using the study in (12), mean values and covariance matrices for all classes are shown as ellipsoids in **Fig. 18.3**. The ellipsoid for the class "fully active" is bolded.

14. The generation of promising new peptides is achieved in two separate steps (#14 and #15 following). First, the amino acid distribution of the best peptides (e.g., the class "fully active") is calculated. The peptide frequencies are added up, resulting in regions between 0 and 1 for each amino acid. To compute a new peptide candidate, a vector of random, equally distributed numbers with values between 0 and 1 is generated. This vector is translated into corresponding amino acids using the cumulated frequencies. Here, a tenfold of the desired number of promising peptide candidates after the final step is generated, e.g., 1,000 for 100 after Step 15.

    Illustrative example: In a hypothetical study, the class "fully active" has amino acid frequencies of 0.4 for R and W, 0.2 for K, and 0 for all other amino acids. The cumulated frequencies are 0.4 for R, 0.8 for W, 1 for K. To translate the random sequence, all values below 0.4 are set to R, all values between 0.4 and 0.8 to W and the rest to K. For a sequence of random numbers (0.95, 0.23, 0.61, 0.49, 0.89, 0.76, 0.45, 0.02, 0.82), the generated amino acid sequence is KRWWKWWRK.

15. In the last step of model peptide generation, using the sequences of the newly generated peptides from Step 14, the selected features according to Steps 8–11 are computed. Now, the Bayes classifier model from Step 13 can be applied to all generated peptides. With this model, the probability that a newly generated peptide belongs to the class of fully active peptides can be computed. The most promising peptides are the peptides with the highest probabilities for the class "fully active." Finally, these peptides (e.g., the best 100/1,000 generated peptides in Step

14) are synthesized and their activity values are measured. These measurements are now used as input for Step 1, to begin a new iteration of peptide design.

4. Notes

1. Different alternatives for the chosen activity indicator or the methods for the computation are possible here; however, the more commonly used *IC50* value is more sensitive against disturbances and oscillations of the luminescence values. The same problem occurs in model estimation of the four parameter model of Hill (13).
2. All luminescence experiments are very sensitive to differences in bacteria, peptide concentration, and purity, as well as the living conditions of the bacteria during the experiment (e.g., temperature, humidity). The proposed normalization strategy assumes similar bacteria concentrations and living conditions during one experiment for one plate. Consequently, all activity measures are based on relative measures according to the positive control. One drawback of this normalization method is a high sensitivity of *RelIC75* to concentration differences in the positive control on different plates.
3. The molecular descriptors shown in **Table 18.2** are only a small subset of possible scales. As examples, other hydrophobicity scales (*see* values from (9)), membrane-oriented scales (10), or scales based on a principal component analysis of other molecular descriptors (2, 15) can be used. In addition, atomic descriptors (3) or features to describe three-dimensional characteristics (16) can be included.
4. Mean, minimum, maximum, and standard deviation values for the whole sequence are only the simplest features used to extract sequence information. One further option is frequencies of fuzzy terms (7). In addition, the descriptors can be analyzed for each position in a sequence or as segments, such as the first or last two amino acids (12).
5. These evaluation measures are only reliable if all classes are covered by a sufficient number of peptides. Under normal conditions, each class should include more than 50 peptides. A formal statistical evaluation of the results is difficult due to the large candidate feature set after feature extraction, resulting from Step 8. A strong evaluation would require the use of Bonferroni-like corrections for multiple tests. Since the features used in the model are strongly correlated, this leads

to extremely conservative results due to an overestimation of the degrees of freedom. The best way to compensate for this is the extraction of candidate features, the model-based design of promising peptides, and its verification by a newly synthesized library.

The evaluation of features using univariate or multivariate analysis of variance is only one possible option. Another alternative is a regression model, which regresses activity with the candidate features.

References

1. Perkins, R., Fang, H., Tong, W., and Welsh, W. (2003) Quantitative structure-activity relationship methods: perspectives on drug discovery and toxicology. *Environ. Toxicol. Chem.* **22**, 1666–1679.
2. Hellberg, S., Sjostrom, M., Skagerberg, B., and Wold, S. (1987) Peptide quantitative structure-activity relationships, a multivariate approach. *J. Med. Chem.* **30**, 1126–1235.
3. Cherkasov, A., Hilpert, K., Jenssen, H., Fjell, C. D., Waldbrook, M., Mullaly, S. C., Volkmer, R., and Hancock, R. E. (2008) Use of artificial intelligence in the design of small peptide antibiotics effective against a broad spectrum of highly antibiotic-resistant superbugs. *ACS Chem. Biol.*, **4**, 65–74.
4. Frecer, V. (2006) QSAR analysis of antimicrobial and haemolytic effects of cyclic cationic antimicrobial peptides derived from protegrin-1. *Bioorgan. Med. Chem.* **14**, 6065–6674.
5. Hilpert, K., Winkler, D. F. H., and Hancock, R. E. W. (2007) Peptide arrays on cellulose support: SPOT synthesis – a time and cost efficient method for synthesis of large numbers of peptides in a parallel and addressable fashion. *Nat. Protoc.* **2**, 1333–1349.
6. Hilpert, K. (2009) High-throughput screening for antimicrobial peptides using the SPOT Technique. In *Part II: Analysis, Properties, and Mechanisms of Antimicrobial Peptides* (from this book).
7. Mikut, R. and Hilpert, K. (2009) Interpretable features for the activity prediction of short antimicrobial peptides using fuzzy logic. *Int. J. Pept. Res. Ther.* **15 (2)**, 129–137.
8. Mikut, R., Burmeister, O., Braun, S., and Reischl, M. (2008a). The open source MATLAB toolbox Gait-CAD and its application to bioelectric signal processing. In *Proceedings of DGBMT-Workshop Biosignal Analysis, Potsdam*, 109–111.
9. Hildebrand, P. (2006). *Zur Strukturvorhersage der Membranproteine.* PhD thesis, Humboldt-Universität zu Berlin, Mathematisch-Naturwissenschaftliche Fakultät.
10. White, S. and Wimley, W. (1999) Membrane protein folding and stability: physical principles. *Annu. Rev. Biophys. Biomol. Struct.* **28**, 319–365.
11. Tatsuoka, M. M. (1988) *Multivariate Analysis.* New York: Mac Millan.
12. Mikut, R., Reischl, M., Ulrich, A., and Hilpert, K. (2008b). Data-based activity analysis and interpretation of small antibacterial peptides. In *Proceedings of the 18th Workshop Computational Intelligence, Universitätsverlag Karlsruhe*, 189–203.
13. DeLean, A. (1978) Simultaneous analysis of families of sigmoidal curves: application to bioassay, radioligand assay, and physiological dose-response curves. *Am. J. Physiol. Gastr. L. Physiol.* **235**, 97–102.
14. Sandberg, M., Eriksson, L., Jonsson, J., Sjostrom, M., and Wold, S. (1998) New chemical descriptors relevant for the design of biologically active peptides: a multivariate characterization of 87 amino acids. *J. Med. Chem.* **41**, 2481–2491.
15. Mekenyan, O., Nikolova, N., and Schmieder, P. (2003) Dynamic 3D QSAR techniques: applications in toxicology. *J. Mol. Struct. THEOCHEM* **622**, 147–165.

Section III

Antimicrobial Peptides as Therapeutic Agents

Chapter 19

Potential Therapeutic Application of Host Defense Peptides

Lijuan Zhang and Timothy J. Falla

Abstract

Host defense peptides (HDPs) are relatively small, mostly cationic, amphipathic, and of variable length, sequence, and structure. The majority of these peptides exhibit broad-spectrum antimicrobial activity and often activity against viruses and some cancer cell lines. In addition, HDPs also provide a range of immunomodulatory activities related to innate immunity defense, inflammation, and wound healing. The development of these multi-faceted molecules and their bioactivities into clinically important therapeutics is being pursued using a number of different approaches. Here we review the role of HDPs in nature and application of this role to the development of novel therapeutics.

Key words: Host defense peptides, HDPs, antimicrobial peptides, anti-infectives, antifungals, anti-sepsis, topical therapeutics, immunomodulators.

1. Introduction

Host defense peptides (HDPs), traditionally referred to as antimicrobial peptides, are ribosomally synthesized, relatively small (<10 kDa), usually cationic, amphipathic, and of variable length, sequence, and structure (1). The classification of HDPs has been based on circular dichroism studies many of which have been supported by nuclear magnetic resonance (NMR) data and include five major classes, namely α-helical, β-sheet (with multiple disulfide bridges), looped (characterized by a single disulfide bridge), extended, and cyclic (2). The mechanism of action of HDPs has primarily been investigated by observing their effect upon model membranes and whole bacterial cells. Subsequently HDPs can be divided into membrane active and non-membrane active, a division based upon both the physicochemical parameters of the

A. Giuliani, A.C. Rinaldi (eds.), *Antimicrobial Peptides*, Methods in Molecular Biology 618,
DOI 10.1007/978-1-60761-594-1_19, © Springer Science+Business Media, LLC 2010

peptide and the properties of the membrane system itself (3, 4). Indolicidin is a 13-residue HDP rich in tryptophan and proline isolated from bovine neutrophils (5). It has a relatively broad spectrum of antimicrobial activity and is the basis of an investigational drug (Omiganan^{TM}) currently in clinical development. The NMR structure of indolicidin when complexed with dodecylphosphocholine (DPC) reveals a central proline–tryptophan hydrophobic core bracketed by positively charged regions near the peptide termini (6). The tryptophan side chains, with one exception, fold flat against the peptide backbone, thus giving the molecule a wedge shape (6). The mode of action of this peptide can be best described as a stepwise interaction with the membrane and influenced greatly by specific electrostatic interactions, lipid fluidity, and peptide concentration (7). In contrast to a detergent or "carpet-like" effect it induces a continuous shrinking and thinning of the planar bilayers. At low peptide concentrations, indolicidin forms an amorphous layer on the fluid domains when such domains contain anionic lipids. At high peptide concentrations, indolicidin appears to initiate a lowering of the gel-phase domains independent of the presence of anionic lipid (8). Dye leakage experiments also suggest that indolicidin induces an increase in fluidity of the lipid bilayer leading to dye release (7). Such remodeling of the membrane is consistent with the actions of amphiphilic α-helical membranolytic peptide, melittin, and arginine-rich cell-penetrating peptide TAT (9). For non-membrane-active HDPs multiple targets have been proposed, ranging from inhibition of intracellular macromolecular synthesis to inhibition of cell division to changes in gene regulation (10). It should be noted that such activities have also been observed for membrane-active peptides when at subminimal inhibitory concentrations (10).

Most HDPs are broad-spectrum antimicrobials with activity against bacteria, viruses, fungi, and parasites and, in a number of cases, their activity extends to tumor cells and spermatozoa. They are found in a wide variety of species ranging from bacteria through insects to humans and they can be expressed constitutively or induced in response to infection, trauma, inflammatory mediators, and cytokines. The increased frequency in the number of isolated drug-resistant pathogenic microbes in hospitals worldwide has prompted both research institutions and the pharmaceutical industry to design and develop novel drugs targeted against resistant pathogens. Interest has focused on HDPs primarily due to their broad-spectrum activity, low potential to induce resistance, and diverse range of activities. This chapter will focus on the role of HDPs in nature and their therapeutic potential. Peptides that have been actively pursed by companies in preclinical and clinical studies will also be mentioned.

2. Role of HDPs in Nature and Areas for Potential Therapy

2.1. Mammalian HDPs

Mammals produce two major classes of HDPs which are defensins and cathelicidins. Defensins consist of three subclasses, α-, β-, and θ-defensins; details of the structure and distribution have been discussed else where (11). Several lines of research suggest that there has been rapid evolution of defensin genes in mammals through duplication and diversification, which supports an evolutionary response of the immune system to the ever-changing microbial ecology of the host habitat (12). Cathelicidins are characterized by a conserved N-terminal region that shares homology in the proregion with cathelin, whereas the C-terminal antimicrobial domain varies considerably and has no homology among different members of cathelicidins (13). Humans and mice have only one cathelicidin gene, whereas a number of mammals have been found to have multiple cathelicidin genes such as pigs producing PR-39 and protegrins and cows and ruminants indolicidins and bactenecins (14). The human cathelicidin, hCAP18, is present in the secondary granules of neutrophils. The C-terminal antimicrobial peptide, LL-37, is liberated by proteinase cleavage coincident with degranulation and secretion (13). Humans also produce several other HDPs including hepcidins which are small, cysteine-rich peptides that also function as hormonal regulators of iron homeostasis (15) and dermcidin that is expressed in eccrine sweat glands and transported via sweat to the epidermal surface (16). Histatins, also found in man, are a family of histidine-rich cationic peptides that are secreted into the saliva from parotid and submandibular salivary duct cells (17).

In addition, peptide derived from host proteins also participates in innate defense including bactericidal permeability – increasing protein (BPI), lysozyme, lactoferrin, hemoglobin, ribonuclease, complement, neuropeptides, and peptide hormones. Some of these have entered the drug development process and are discussed in more detail below.

There is increasing attention on the role of chemokines in innate defense. Chemokines are a family of structurally related peptides with conserved amino-terminal motifs. They regulate leukocyte trafficking during both health and disease. Some chemokines have been shown to exert direct antibacterial activity. Conversely, several granule-bound antibacterial proteins of granulocytes and epithelium-expressed antibacterial polypeptides possess chemotactic activity and stimulate cells of the adaptive immune system. Human macrophage inflammatory protein 3α (MIP-3α), also known as CCL20, is a 70-amino-acid chemokine which exclusively binds to chemokine receptor 6. This polypeptide has direct antimicrobial, antifungal, and antiviral activities.

The solution structure of a synthetic peptide consisting of the last 20 residues of CCL20 exhibits a highly amphipathic α-helical structure, reminiscent of a large number of antimicrobial peptides, and antimicrobial activities against *Escherichia coli* (*E. coli*) and *Staphylococcus aureus* (*S. aureus*) suggesting a significant role in its broad anti-infective activity (18). CXCL14 is another chemokine with broad-spectrum antimicrobial activity for cutaneous Gram-positive bacteria and *Candida albicans* (*C. albicans*) as well as the Gram-negative enterobacterium *E. coli* (18). Increased salt concentrations and skin-typical pH conditions do not decrease its antimicrobial function. CXCL14 is highly abundant in the epidermis and dermis of healthy human skin but is downregulated under conditions of inflammation and disease supporting its defense role as a part of skin immunity (19). Also present in the epidermis, catestatin is a functional HDP derived from the neuroendocrine protein chromogranin A. In human skin, chromogranin A is proteolytically processed into the antimicrobial fragment catestatin and its level is upregulated upon injury (20). The peptide is antimicrobial with activity against relevant skin microbes and its mechanism of action involves membrane disruption. In vivo expression of catestatin in murine skin increases in response to injury and infection, demonstrating a direct link between the neuroendocrine and cutaneous immune systems (20).

The inflamed epithelium of human pharynx produces CXC chemokines MIG/CXCL9, IFN-inducible protein-10/CXCL10, and IFN-inducible T-cell alpha-chemoattractant/CXCL11, all showing antibacterial activity against *Streptococcus pyogenes* (*S. pyogenes*) (21, 22). Inhibition of MIG/CXCL9 expression reduces the antibacterial activity at the surface of inflamed pharyngeal cells (21). Interestingly, this chemokine is also highly expressed in the epithelium of testis, epididymis, and prostate with activity against the urogenital pathogen *Neisseria gonorrhoeae* (*N. gonorrhoeae*) (21). The male reproductive tract expresses granulocyte chemotactic protein 2 (GCP-2)/CXCL6 in the epithelium of the testis, epididymis, and seminal vesicles (23). However, no GCP-2/CXCL6 chemokine has been detected in blood plasma of healthy donors, indicating that expression is localized (24). Both MIG/CXCL9 and GCP-2/CXCL6 have been suggested to play important roles in both host defense of the male urogenital tract and during fertilization (25).

2.2. Skin and Potential Topical Therapeutics

In human skin HDPs are produced in many cell types including keratinocytes, sweat glands, neutrophils, monocytes, NK cells, and mast cells, and the level of certain HDPs, such as LL-37, in skin is directly associated with susceptibility to cutaneous infections (26). In several human skin diseases there is an inverse correlation between severity of the disease and the level of HDP production. Atopic dermatitis (AD) is a common chronic inflammatory skin disease that has increased in prevalence over the last half century. A growing body of evidence suggests that in such cases there are a variety of defects in the innate immune response resulting in diminished recruitment of innate immune cells (PMNs, pDC, and NK cells) to the skin, epithelial barrier disruption, and TLR2 defects are just some of the credible explanations for AD patients' susceptibility to pathogens such as *S. aureus*, herpes simplex virus, and vaccinia virus (27). AD patients also have a reduced amount of dermcidin in their sweat which correlates with an impaired innate defense of human skin in vivo (16). In addition to AD, eczema herpeticum and eczema vaccinatum are all relevant to deficient expression of LL-37 in skin biopsies partially explaining the pathophysiology of these diseases (28, 29). In contrast overexpression of HDPs can lead to increased protection against skin infections as seen in patients with psoriasis and rosacea, inflammatory skin diseases which rarely result in superinfection. Transgenic mice expressing porcine cathelicidin peptide PR-39 showed increased resistance to group A streptococcal skin infection (30). Restoration of innate peptides to normal levels via external application would provide a compelling therapeutic application. The clinical application of HDPs is being actively pursued by a number of companies in the areas of acne and acne rosacea.

Acne is a disorder of the pilosebaceous unit which is made up of a hair follicle, sebaceous gland, and a hair. The human pilosebaceous units produce peptides of the β-defensin family and LL-37 and in combination with other antimicrobial agents, including lipids, provide active defense against infection. In patients with acne vulgaris, increased levels of HDPs are often found in inflamed or infected skin areas indicating a role for these peptides in fighting infection (31). Sebocytes also produce the functional LL-37 peptide (32). Similar to keratinocytes the expression of LL-37 in sebocytes can be induced by vitamin D or extracts of the bacterium *Propionibacterium acnes* (*P. acnes*). Acne patients have more *P. acnes* in their follicles than people without acne, and the bacterium has become resistant to a wide range of antibiotics. As acne lesions are associated with moderate-to-severe inflammation a treatment involving both antimicrobial and anti-inflammatory would be of significant benefit. HDPs are well known for binding and neutralizing bacterial debris including the lipoteichoic acid of the Gram-positive cell wall responsible for inflammation, resulting in downregulation of pro-inflammatory cytokines. Well demonstrated are granulysin-derived peptides that not only are antimicrobial against *P. acnes* but also suppress *P. acnes*-stimulated cytokine release (33). Peptides including MBI

594AN (www.migenix.com), an indolicidin variant, and HB64 (www.helixbiomedix.com), a de novo-designed α-helical peptide, have demonstrated benefit as part of acne treatment in humans.

S. aureus is the most common pathogen in cutaneous and systemic infections (34). The virulence factors produced by *S. aureus* can disrupt the epithelial barrier, inhibit opsonization by antibody and complement, interfere with neutrophil chemotaxis, cytolysis of neutrophils, and inactivation of HDPs (35). Staphylococcal wound infection is of increasing concern in all parts of clinical wound management. Today's standard of care is still systemic antibiotic therapy. The use of host defense peptides as an alternative for the prevention and treatment of wound infections is an expanding field. Ovispirin-1 is the N-terminal 18 amino acids of the ovine cathelicidin SMAP-29 (36). The variant peptide, proline-novispirin G10, demonstrated broad-spectrum antimicrobial activity with moderate hemolytic and cytotoxic activities. In the infected wound chamber model it demonstrated a 4 log (10) reduction in bacterial counts, indicating the potential of P-novispirin G10 as an alternative in future antimicrobial wound care (36).

Chronic wounds such as venous calf and diabetic foot ulcers are frequently contaminated and colonized by bacteria. Topical antimicrobial therapy of infected diabetic foot ulcers provides a targeted high-concentration therapeutic without the adverse effects of systemic anti-infective agents. Pexiganan acetate, an investigational drug, is a synthetic analog of magainin 2. The drug is thought to act by disturbing the permeability of the cell membrane or cell wall. In the early 1990s two Phase III multicenter, randomized, double-blind trials were conducted in diabetic patients with infected foot ulcers and both achieved clinical cure or improvement in approximately 90% of patients after topical Pexiganan 1% and oral Ofloxacin 800 mg/day (37). Similar clinical efficacy (within the 95% confidence interval) was obtained more recently in two consecutive, double-blind, controlled trials for topical Pexiganan and oral Ofloxacin in clinical improvement rates (38). Interestingly bacterial resistance to Ofloxacin emerged in some patients who received Ofloxacin, but no significant resistance to Pexiganan emerged among patients receiving Pexiganan (38). Topical Pexiganan might be an effective alternative to oral antibiotic therapy in treating diabetic patients with a mildly infected foot ulcer and potentially reduces the risk of selecting antimicrobial-resistant bacteria (38).

Central venous catheter (CVC) users suffer from infection that is due to colonization by bacteria and fungi around the catheter insertion site that leads to bloodstream infection after the seeding of microbes from the colonized catheter into blood. The Centers for Disease Control and Prevention estimate that CVC-related bloodstream infection occurs in 250,000–400,000 patients each year in the United States alone, resulting in about 70,000 deaths. The development of the indolicidin variant, omiganan pentahydrochloride, for such an application has been advanced initially by Migenix Inc.

2.3. Mucosal Defense and Potential Application

HDPs are expressed in all human epithelia including the oral cavity, lung, respiratory tract, gastrointestinal tract, and vagina (39). The microbes found in the oral cavity comprise more than 700 species arranged as both single-species and multispecies biofilms (40). Both α-defensin and β-defensin, LL-37, histatins, and other antimicrobial proteins play crucial roles in the control of oral microbes including periodontitis-related bacteria, *Candida*, and papilloma virus. In addition the commensal bacteria are potent inducers of HDPs and pro-inflammatory cytokines in oral epithelial cells (41, 42). The protective role of HDPs in humans in oral health is well illustrated by Morbus Kostmann, a severe form of congenital neutropenia with HDP deficiency which is associated with the occurrence of infections and periodontal disease (43). Recently a peptide mimetic meta-phenylene ethynylene (mPE), based on the structure of frog peptide magainin, was developed by Polymedix Inc. and was shown to prevent the formation of biofilms by oral pathogenic bacteria at nanomolar concentrations (44). Peptides and peptide mimetics could provide benefit as new generation of oral hygiene products with therapeutic potential for a healthy oral environment. To this end, a chewing gum containing a synthetic decapeptide KSL-W displayed promising in vitro/in vivo release profiles and provides a novel and effective way to deliver peptide to the various compartments of the oral cavity (45).

Oral candidiasis is an infection caused by the genus *Candida* on the mucous membranes of the mouth. Patients with oral candidiasis show reduced salivary defensins (46). Immunosuppressive drugs also suppress inducible β-defensin expression resulting in subsequent *Candida* colonization among head and neck chemotherapy patients (47). Histatins are a family of histidine-rich peptides that are secreted into saliva from the parotid and submandibular salivary ducts (17). These peptides exhibit potent fungicidal activity against *Candida* species, *Saccharomyces cerevisiae*, and *Cryptococcus neoformans* (48). Among at least 50 histatin peptides, derived from posttranslational proteolytic processing, histatin 5 (Hst 5), which contains 24 amino acids, has the highest level of activity against *C. albicans* (49, 50). HIV-positive patients have significantly lower levels of histatin-5, suggesting that the reduced expression of this HDP may contribute to the enhanced predisposition of this population to oral candidiasis (51). A shorter analog, P-113 (amino acid residues 4–16 of Hst 5), is as active as the full-length Hst 5 (52). This peptide has been developed in its D-amino acid form (P-113D) by Demegen Pharmaceuticals and licensed to Pacgen Biopharmaceuticals Corporation (Vancouver, BC, Canada) for treatment of HIV-associated oral candidiasis. The company announced positive results in 2008 from a Phase IIb dose-ranging trial and demonstrated that PAC-113 (formally P-113D) is effective in the treatment of oral candidiasis and compares favorably to the efficacy by the current standard of care (for details see below).

Defensins and cathelicidins are the main families of HDPs that are present in airway secretions and that are expressed by the airway epithelium. In addition to microbes, cytokines, growth factors, and active vitamin D have been implicated as major regulators of HDPs (53). Historically impaired innate defense has been the center of debate in the cystic fibrosis (CF) community related to colonization with *Pseudomonas aeruginosa* (*P. aeruginosa*) and other bacteria in the chronically infected and inflammatory lungs of the CF patient (54). The protective role of HDPs has been demonstrated recently using transgenic animals. Transgenic mice that ectopically expressed the HDP PG-1 not only had better survival rates post *Actinobacillus suis* challenge but also had significantly lower bacterial loads in their lungs with reduced numbers of pulmonary pathological lesions (55). The antimicrobial function of PG-1 was confirmed in vitro by using fibroblast cells isolated from the transgenic mice but not the wild-type mice. Moreover, differential blood cell counts in bronchoalveolar lavage fluid indicated a greater number of neutrophils in PG-1 transgenic mice than in their WT littermates after bacterial challenge. Therefore ectopic expression of PG-1 in mice confers enhanced resistance to bacterial infection (55). Topical delivery of three α-helical peptides demonstrated a significant reduction of viable *P. aeruginosa* in a rat chronic lung infection model treated daily for 3 days with peptide versus vehicle alone (56).

P. aeruginosa is not only the most prevalent cause of life-threatening infections in the lungs of CF patients but the third leading cause of severe hospital-acquired infections. Strains of *Acinetobacter baumannii*, *P. aeruginosa*, and *klebsiella pneumoniae* that exhibit resistance to almost all available antibiotics, except polymyxins, have emerged as common causes of hospital-acquired infections in critically ill patients. The increasing problem of antimicrobial resistance has forced the medical community to evaluate alternative preventive and therapeutic strategies for nosocomial pneumonia, an infection that is associated with considerable morbidity and mortality. Polymyxin, which is a peptide antibiotic produced by *Bacillus* spp., has become a drug of last resort to fight multidrug-resistant Gram-negative pathogens and offers hope to critical care units (57). In addition to bacterial lung infection pulmonary aspergillosis is also common among immunosuppressed patients. A synthetic lipopeptide, palmitoyl-lys-ala-DAla-lys, that is composed of only four L and D amino acids linked to fatty acids not only shows a potent, broad spectrum of in vitro antimicrobial activity against human-pathogenic yeast, fungi, and bacteria but exerts specific antifungal activity toward pathogenic *A. fumigatus* fungi in immunosuppressed C57BL/6 wild-type mice and increases the survival rate with low toxicity effects and no damage to the treated lung tissues (58).

The gastrointestinal tract of mammals is heavily colonized with a complex and dynamic microbial community. Intestinal epithelial cells, particularly Paneth cells, are the major producers of antimicrobial arsenal including defensins, cathelicidins, lysozyme, phospholipases, and proteases that exhibit a broad range of actions. A disturbed antimicrobial defense has been suggested to be a critical factor in the pathogenesis of Crohn's disease, an inflammatory disease of the intestinal tract. Patients that suffer from Crohn's disease not only have impaired alpha-defensins in Paneth cells, but impaired induction of mucosal β-defensins likely due to a low copy number of the beta-defensin gene cluster among the patients (59). In both ileal and colonic Crohn's disease, the lack of HDPs seems responsible for a broadly diminished antibacterial killing by the mucosa (60). It should be noted that pathogenic microorganisms have developed multiple modalities to combat host immune response. *S. flexneri* actively dampens host innate immune responses by suppressing transcription of specific HDP genes, and both *Vibrio cholerae* and enterotoxigenic *E. coli* (ETEC) can suppress LL-37 and β-defensin-1 expression in the intestinal epithelial cells (61). Cholera toxin and labile toxin, the major virulence proteins of *V. cholerae* and ETEC, respectively, are predominantly responsible for these effects (61).

Topical microbicides is a female-controlled strategy for preventing the acquisition and transmission of STIs/HIV infections by killing the virus or inhibiting its uptake and fusion with cells. A vast majority of mucosal HIV exposures do not result in productive infection, implying that innate mucosal immune defenses are highly protective, and failure of these mucosal defenses may have contributed to over 3 million new HIV infections in 2006 (62). The simultaneous presence of other genital infections, including classical sexually transmitted infections (STIs), can enhance HIV susceptibility (63). Co-infections that act as important HIV cofactors include human cytomegalovirus (CMV), Herpes simplex virus type 2, *Neisseria gonorrhoeae*, and many others (63). Bacterial vaginosis is a common condition associated with increased risk of STIs/HIV. Data suggest that vaginal lavage fluid from women with bacterial vaginosis is deficient in HDPs and antimicrobial activity compared to fluid from healthy women or women with vulvovaginal candidiasis, and this deficiency could predispose individuals to STIs/HIV (64). A recent report suggests, although need confirming, that increased levels of human β-defensins mRNA in sexually HIV-1 exposed but uninfected individuals may contribute to the natural resistance mechanisms against HIV infection (65). This is not surprising given that a complex synergism among HDPs constitutes the majority of the intrinsic anti-HIV-1 activity in cervicovaginal secretions (66). In addition to defensins, LL-37 inhibits HIV-1 replication in PBMC, including primary CD4(+) T cells, and the inhibition was independent of FPRL-1 signaling and readily reproduced using various HIV-1 isolates (67).

As mentioned all mammals express α- and β-defensins but certain non-human primates also express θ-defensins. Humans do not produce θ-defensins due to a point mutation resulting in stop codons in the otherwise open-reading frame of the θ-defensin precursor (68). Retrocyclins are the counterpart of the recreated θ-defensins based on genetic blueprint. One of the analogs, RC-101, prevents HIV-1 infection in an organ-like construct of human cervicovaginal tissue and retained full activity in the presence of vaginal fluid (69). The peptide remained within the cervicovaginal tissues throughout the 9-day incubation period without altering tissue viability, inducing damage, or inducing the release of inflammatory cytokines (69), supporting the potential development of RC-101 as a topical microbicide to prevent HIV-1 infection and transmission. Due to the high cost of preclinical studies in primates none of the HDPs mentioned has been advanced to efficacy study. But identification and enhancement of "natural" innate immune defenses still remain attractive avenues for development of safe, non-toxic microbicides. Some safety data have been obtained using small animals. Nisin is a HDP produced by *Lactibacilli* spp. In rabbit repeated intravaginal application of high dose of nisin gel for 14 days on cervicovaginal epithelium suggested no treatment-related abnormalities either in the vaginal cell morphology or structural abnormalities in the mucosal epithelium (70). Also nisin gel did not induce irritation and/or inflammation in the vaginal epithelium (70).

2.4. Anti-tumor Activity and Potential Therapy

Human β-defensin-1 (hBD-1) is a candidate tumor suppressor, and cancer-specific loss of hBD-1 is found in 90% of renal clear cell carcinomas and in 82% of prostate cancers indicating that hBD-1 may have anti-tumor activities and its loss may contribute to prostate tumorigenesis (71). hBD-1 promoter mutations are found in renal and prostate cancer clinical samples and such mutations may be responsible for cancer-specific loss of hBD-1 expression (72). In vitro hBD-1 causes rapid induction of cytolysis and caspase-mediated apoptosis in prostate cancer cells (73). PAX2, an oncogene and a transcriptional regulator normally expressed during early development, has been implicated in carcinomas of the kidney, prostate, breast, and ovary (74). Recent study reveals that PAX2 suppresses hBD1 expression in prostate cancer and knock-down of PAX2 expression results in the re-expression of hBD1 and subsequently prostate cancer cell death (75).

In addition to immune surveillance of HDPs in cancer defense, they also showed therapeutic activity against tumor in both in vitro and in vivo models. Early studies demonstrated that magainin and its D-amino acid enantiomer were effective in treatment of ovarian cancer in mice when injected intraperitoneally (76). In another study it is demonstrated that a membrane disrupting lytic peptide conjugate destroys hormone-dependent and hormone-independent breast cancer cells in xenografts of human breast cancer cells as well as prostate cancer in nude mice (77, 78). In fact, peptides of various origins exhibit a broad spectrum of cytotoxic activity against cancer cells. Brevinin-2R is a defensin type of HDP isolated from the skin of the frog *Rana ridibunda*. It exhibits preferential cytotoxicity toward malignant cells as compared to normal primary cells (79). Brevinin-2R-triggered cancer cell death involves decrease in mitochondrial membrane potential and total cellular ATP levels and increase in reactive oxygen species production (79). Cecropin A and B exert selective cytotoxic and antiproliferative efficacy in bladder cancer cells while causing no harm to normal cells (80). Gomesin is a HDP isolated from hemocytes of the spider *Acanthoscurria gomesiana*. This peptide shows direct cytotoxicity to melanoma B16F10-Nex2 cells and several human tumor cell lineages (81). Topical treatment of subcutaneous murine melanoma with gomesin incorporated in a cream base significantly delayed tumor growth (81). This represents a novel and potential use for gomesin as a topical agent against unsuccessfully treated intradermal and epithelial skin cancers (81). Gene therapy with a plasmid that expresses a secreted human α-defensin-1 (HNP1) in A549 tumor cells inhibits the growth of human lung adenocarcinoma xenograft in nude mice (82). The HNP1-treated mice show both decreased microvessel density and increased lymphocyte infiltration in tumor tissue, and the antiangiogenesis effect of HNP1 may also contribute to its tumor inhibitory activity in vivo (82). Some companies are advancing lytic peptide technology for potential tumor therapy (see below).

2.5. Immunomodulators Against Infection and Sepsis

Immunomodulation represents a novel approach to antimicrobial therapy that depends on bolstering host immunity, rather than direct antimicrobial activity. HDPs of mammalian origin including defensins, cathelicidins, and dermcidin are all able to stimulate expression of cytokines, chemokines that bridge innate and adaptive immunity (83, 84). LL-37 is particularly interesting because it possesses both pro-inflammatory and anti-inflammatory roles upon environmental stimuli (85, 86). Sepsis, a life-threatening complication of infections and the most common cause of death in intensive care units, is characterized by a hyperactive and out-of-balance network of endogenous pro-inflammatory cytokines. None of the current therapies are entirely effective. Recent data suggest that in contrast to healthy individuals and critically ill non-septic patients, ex vivo inducibility of human hBD-2 in peripheral blood cells from septic patients is reduced and such impaired inducibility may contribute to the complex immunological dysfunction in patients with severe sepsis (87). HDPs have long been known to have endotoxin-neutralizing properties. Recent data suggest that they can modulate the inflammatory cascade by targeting multiple inflammatory mediators. LL-37 has modest direct antimicrobial activity under physiological conditions, but in a rat intra-abdominal sepsis model it significantly downregulates pro-inflammatory cytokines and improves the outcome of survival rate (88). In a mouse model of allergic contact dermatitis, the disease can be either enhanced in the absence of cathelicidin in the *Cnlp*$^{-/-}$ mouse or inhibited by the exogenous introduction of recombinant cathelicidin during the sensitization phase (89). Such anti-inflammatory role in host immunity in vivo is due to inhibition of TLR4 signaling through an alteration of cell membrane function (89). In a *P. aeruginosa* sepsis murine model in the context of a competent immune system, a de novo-derived peptide, WLBU2, was prophylactic and therapeutic against *P. aeruginosa* bacteremia in mice (90). Inimex Pharmaceuticals (Vancouver, BC, Canada) is actively pursuing innate defense regulators (IDRs) as novel immunomodulatory therapeutics. IDR-1, built based on the same theme, has no direct antimicrobial activity, but is effective in restricting many types of infection, while limiting pro-inflammatory responses (86). The ability of these peptides to selectively suppress harmful pro-inflammatory responses, while maintaining beneficial infection-fighting components of host innate defenses, makes them a potential therapeutic for anti-sepsis therapies that merit further investigation (86). hLF1-11 is another selective immune response amplifier that acts by stimulating the innate immune system. It comprises the first 11 N-terminal amino acids of human lactoferrin. Oral administration of lactoferrin protects animals against infectious diseases and changes can be observed in the number of cells in the leukocyte subsets in the peripheral blood and spleens of mice after 1 day of oral administration (91). hLF1-11 exhibits in vivo effects against a wide range of bacteria and fungi. It is currently under development by AM-Pharma (Bunnik, The Netherlands).

2.6. Novel Technology and Potential Applications

Cost and stability remain the biggest hurdle in peptide drug development. New technology allows the easy and versatile synthesis of a broad variety of facially amphiphilic polymers and these non-peptide mimics of HDPs utilize the same mechanism chosen by nature to control bacterial growth but are cost-effective. Oxanorbornene-derived polymers can be 50 times more selective for Gram-positive over Gram-negative bacteria, while other polymers surprisingly showed the opposite preference (92). Peptoid oligomers are synthesized with a peptoid spacer chain to allow an adhesive peptide moiety for easy and robust immobilization onto substrata. Confocal microscopy images showed that the membranes of adherent *E. coli* cells can be damaged after 2-h exposure to the modified substrata, suggesting that the peptoid oligomers retain antimicrobial properties even when immobilized on substrata (93). Surface modification techniques that create surfaces capable of killing adherent bacteria are promising solutions to infections associated with medical devices. Melimine is produced by combining portions of the insect HDP mellitin and protamine. The new hybrid is not only potent against *Stapylococcus* and *Pseudomonas* but stable to heat sterilization with 250 times more improved hemolytic activity than parent peptide melittin (94). Melimine significantly reduces bacterial adhesion to contact lenses when adsorbed or covalently attached, suggesting that it is an excellent candidate for development as an antimicrobial coating for devices (94). Ceragenins (CSAs), also called cationic steroid antibiotics, are synthetically produced small molecule compounds comprised of a sterol backbone with amino acids and other chemical groups attached to them to mimic the cationic, facially amphiphilic structures of most HDPs (95). These compounds have a net positive charge and are able to rapidly disrupt the target membranes leading to rapid cell death (96). The CSAs are active against clinical isolates of vancomycin intermediate *S. aureus* (VISA), heterogeneous vancomycin intermediate *S. aureus* (hVISA), as well as vancomycin-resistant *S. aureus* (VRSA) (97, 98). The Ceragenin™ technology is currently advanced by Ceragenix Pharmaceutical Inc. to formulate Cerashield antimicrobial coatings for medical devices in their preclinical studies.

3. Companies Advancing HDP Therapeutics

The current status of developmental programs by companies is outlined in this section. Information was based primarily on publications, public disclosures, press releases, and company's web site information.

3.1. Access Pharmaceuticals

Access Pharmaceuticals (www.accesspharma.com) acquired Pexiganan in 2009 from Macrochem Corporation who initially licensed the rights to Pexiganan from Genaera in 2007. As mentioned above Pexiganan is a novel topical broad-spectrum antibiotic being developed for the treatment of mild-to-moderate diabetic foot ulcer infections. Pexiganan has completed two Phase III clinical trials.

3.2. Agennix

Agennix (www.agennix.com) is a private biotechnology company developing Talactoferrin, a recombinant human lactoferrin produced in *Aspergillus niger* and structurally identical to the native lactoferrin in all material respects, differing only in its glycosylation. It is an orally active protein that mediates its activity through the gut and the GALT (gut-associated lymphoid tissue). It acts through a mechanism of dendritic cell recruitment, maturation, and activation. Following oral administration, Talactoferrin is transported into the Peyer's patches of the small intestine, where it recruits circulating immature dendritic cells (DCs), bearing tumor antigens, to the GALT and induces their maturation. DC maturation in the presence of tumor antigens and lymphoid effector cells induces a strong systemic innate and adaptive immune response mediated by anti-cancer natural killer (NK) cells, CD8+ lymphocytes, and NK-T cells. This results in the activation of tumor-draining lymph nodes, cellular infiltration of distant tumors, and tumor cell death. Mounting the initial immune response in the GALT, away from the primary tumor, and using a physiologically important pathway minimizes the effect of the cancer's local immunosuppressive defenses. The company is developing an oral formulation of Talactoferrin for cancer indications and a topical gel formulation for the treatment of diabetic foot ulcers. The company is also investigating the activity of the oral formulation of Talactoferrin in patients with severe sepsis in an NIH-funded clinical trial.

3.3. AM-Pharma

AM-Pharma Holding B.V. (www.am-pharma.com) also focuses on the development of human lactoferrin-based technology. The lead compound hLF1-11 comprises the first 11 N-terminal amino acids of the endogenous human lactoferrin. The compound exerts both antibacterial and antifungal activities against a broad range of microorganisms. However, it appears that hLF1-11 also acts by selectively stimulating the innate immune system. Overall safety has been established in preclinical and clinical studies, and hLF1-11 is currently being tested for its ability to prevent bacterial infections in hematopoietic stem cell transplant (HSCT) patients. In the first part of a Phase IIa study, a non-randomized, open-label safety study, where a 0.5 mg dose of hLF1-11 was given to HSCT patients for 10 consecutive days, no serious drug-related safety issues were encountered. This study will be followed by a safety study in HSCT recipients using a multiple daily dose regime.

3.4. BiolineRx

BioLineRx Ltd (www.biolinerx.com) licensed BL-2060, a novel antimicrobial peptidomimetic, in 2006 from Dr. Amram Mor at the Department of Biotechnology and Food Engineering, Technion, Israel. A naturally occurring antimicrobial peptide represents less than an ideal drug candidate due to poor bioavailability issues, potential immunogenicity, potential toxicity, and high production costs. Mor developed the use of peptidomimetics based on oligo-acyl-lysines (OAKs) to overcome the limitations inherent in the physical characteristics of such peptides (99). OAKs represent a new class of antibacterial and antimalarial peptidomimetic oligomers, composed of alternating aminoacyl chains and cationic amino acids (100). Like potent HDPs, OAKs can display rapid, nonhemolytic, broad-spectrum microbicidal properties, and they do not induce the emergence of resistance. The company is currently advancing BL-2060 for the treatment of resistant bacterial infections.

3.5. Cadence

Cadence Pharmaceuticals, Inc. (www.cadencepharm.com) (formerly Strata Pharmaceuticals, Inc.) is a biopharmaceutical company focused on in-licensing, developing, and commercializing proprietary product candidates principally for use in the hospital setting. The company in-licensed the exclusive rights to Omigard™ for North American and Europe in 2004 from Migenix Pharmaceuticals (Vancouver, BC, Canada) for the prevention and treatment of device-related, surgical wound-related, and burn-related infections. Omigard™, omiganan pentahydrochloride 1% aqueous gel, is an analog of indolicidin that has been demonstrated to be rapidly bactericidal and fungicidal with prolonged duration of activity against microorganisms commonly found on the skin surface (101). Importantly, meaningful resistance to Omigard™ has not been induced in the laboratory after extensive study nor has Omigard™ demonstrated the potential to induce cross-resistance to other antimicrobial therapeutics. Omigard™ has previously completed a Phase III trial. While this study did not show statistical significance for the primary end point (prevention of catheter-related bloodstream infections), statistically significant outcomes were demonstrated for the reduction of local catheter site infection (LCSI) and catheter colonization. In August 2005, Cadence initiated CLIRS, or the Central Line Infection Reduction Study. CLIRS is a confirmatory Phase III clinical trial with a primary end point of the reduction of LCSI. The company expects the results of this study to be released in 2009.

3.6. Ceragenix Pharmaceuticals

Ceragenix (www.ceragenix.com) is a biopharmaceutical and medical device company developing non-peptide functional mimics of endogenous antimicrobial peptides such as the cathelicidin LL-37 and the α- and β-defensins. Ceragenix has licensed the exclusive worldwide rights to a new class of patented small molecule compounds from Dr. Paul Savage at the Department of Chemistry and Biochemistry, Brigham Young University. These compounds are aminosterols that bear a positive charge and are termed "Cationic Steroid Antibiotics" or CSAs or Ceragenins™. Extensive preclinical testing has shown that the lead compound, CSA-13, is highly effective at clinically relevant concentrations against a broad spectrum of bacterial pathogens, including multidrug-resistant organisms such as *Pseudomonas*, MRSA, and VRSA. The compounds have activity not just against bacteria, but are also active against certain fungi (Candida), viruses (orthopox family), and certain cancers due to the fact that all these cells share in common the presence of negatively charged phospholipids on their cell membrane surfaces. The CSAs are electrostatically attracted to these membranes and induce apoptosis by rapid depolarization of the cell membranes. Unlike most antibiotics, which are bacteristatic, the CSAs are bactericidal. CSA-13 is the active component in the company's CeraShield™ technology that is designed to prevent bacterial colonization and biofilm development on medical devices. Preclinical testing suggests that the company's CeraShield™-coated silicone urinary catheter was able to provide complete protection against *E. coli* colonization for the duration of a 21-day study.

3.7. Cutanea

Cutanea Life Science (www.cutanealife.com) licensed Omiganan from Migenix for the treatment of dermatological conditions. The company announced Phase II clinical data in October 2007 for the treatment of rosacea. The trial compared Omiganan 2.5 and 1% topical gel to vehicle in subjects with papulopustular rosacea. After 9 weeks of treatment, once-daily Omiganan 2.5% gel showed superior lesion count reductions and treatment success compared to 1% Omiganan and vehicle. Omiganan provided greater improvements compared to vehicle among patients with a more severe condition at baseline. Lesion counts continued to drop at all evaluations over the duration of the study, indicating that further improvements may be expected with a duration of treatment exceeding 9 weeks. Based on the results from this study, Cutanea has planned to take a once-daily dose of Omiganan 2.5% into Phase III.

3.8. Exponential Biotherapies

Exponential Biotherapies (EBI) (www.expobio.com) is developing a family of synthetic peptides. The peptide biomolecules have exhibited various properties and capacities to modulate the cascade of the immune response in isolated cells and in various animal models of disease. The compounds thus represent a new class of therapeutic agent and can be described as having modulating or controlling functions over the immune response – a capability that makes them clinically relevant for many important disease states characterized by aberrant immune activation. Severe hemorrhagic shock (HS) followed by resuscitation induces a massive inflammatory response, which may culminate in systemic inflammatory response syndrome, multiple organ dysfunction syndrome, and possibly death. The company demonstrated that synthetic oligopeptides related to the primary structure of human chorionic gonadotropin (HCG), in particular LQGV, have therapeutic potential due to attenuating life-threatening inflammation and organ damage associated with (trauma) HS and resuscitation in animal model with high-dose LPS-induced inflammation (102). In addition the synthetic peptides show a remarkable spectrum of biological effects such as modulation of angiogenesis, inhibition of septic shock syndrome, prevention of diabetes, and reduction of ischemia-reperfusion damage (103). The company completed a Phase I trail where human volunteers were administered increasing single or multiple doses of the company's compound EA-230. All the doses were tolerated without significant adverse events. The Phase I trials were expanded into an additional study design in which healthy volunteers were exposed to a fixed dose infusion of LPS. Control subjects received placebo while all the test subjects were infused with EA-230. Results indicated that a variety of reactions caused by LPS were countered by the drug. Importantly, these reactions were not only physically observable, such as control of fever, but also biochemically detectable, such as changes in several chemical substances in the body associated with inflammation in sepsis. The first study involving human patients is scheduled for 2009. EBI received approval from German and Dutch regulatory authorities to initiate an exploratory Phase II study to assess the effect of EA-230 on renal function and on safety and tolerability in patients at increased risk of developing renal failure following major cardiac on-pump surgery.

3.9. Helix BioMedix

Helix BioMedix (www.helixbiomedix.com) is developing novel, broad-spectrum, lipohexapeptides or "small molecule peptides" that specifically combine the attributes of small molecule natural products with the advantages of HDPs for the treatment of skin and wound infections and the prevention of *S. aureus* infections including those caused by MRSA. As with traditional HDPs, the company's lead lipohexapeptides (HB1345 and HB1275) are rapidly cidal, fail to engender resistance in vitro, are readily synthesized, and do not exhibit cross-resistance with other antibiotics (104). In addition, these molecules also have the advantage of being more stable, safer, and more cost-effective to manufacture than traditional HDPs. Primarily due to acylation (addition of a lipid), these molecules are significantly more active in complex biological environments such as serum or wound fluid. As a result, lipohexapeptides exhibit potent activity in animal infection models. In pre-clinical testing the lipopeptides exhibited broad-spectrum antimicrobial activity against *S. aureus*, *S. pyogenes*, and *P. aeruginosa* and also pathogenic fungi such as *Candida* and *Trichophyton* species (104, 105). This activity was maintained against antibiotic-resistant organisms such as MRSA and vancomycin-resistant *Enterococci*. In a *S. aureus*-abraded skin infection model, the lipohexapeptides significantly reduced the number of bacteria following 3 days of once-daily dosing, and in many cases, the peptide eradicated the pathogen (104). In a guinea pig dermatophytosis model, the company's lead peptide candidates significantly reduced pathogen count and delivered clinical benefits comparable to Terbinafine, a drug approved by the FDA for onchomycosis (105). In both animal models, toxicity was not significantly different from vehicle alone.

3.10. Inimex Pharmaceutics

Inimex Pharmaceuticals (www.inimexpharma.com) focuses on the development and commercialization of Innate Defense Regulators (IDRs) that selectively trigger the body's innate defenses without causing inflammation. Inimex' lead IDR product IMX942 targets a broad spectrum of life-threatening hospital infections, many of which are caused by antibiotic-resistant bacteria such as MRSA or vancomycin-resistant *Enterococcus* (VRE). IMX942 is active against both fully sensitive and drug-resistant bacteria and its ability to stimulate host innate immune defenses is not susceptible to bacterial antibiotic resistance mechanisms. Genomics studies confirm that Inimex's peptides caused increases in gene products associated with infection-clearing aspects of innate immunity without causing upregulation of pro-inflammatory cytokines such as TNF-α. Inimex's peptides are able to clear lung infections with *S. pneumoniae*, the most common cause of bacterial pneumonia. In this pneumonia model, Inimex compounds demonstrate activity comparable to azithromycin (Zithromax®, Pfizer), a leading prescriptive for pneumonia. The first clinical indication for IMX942 will be the prevention of infections associated with cancer chemotherapy-induced immune suppression.

3.11. Lytix Biopharma AS

Lytix Biopharma AS (www.lytixbiopharma.com) develops peptide-based antimicrobials and cancer therapeutics. The company's lead compound, LTX-109, is an antimicrobial membrane lytic peptide currently under late-stage preclinical development. LTX-109 shows a dose-dependent bactericidal effect in a mouse infection model and demonstrates efficacy against antibiotic-resistant species such as MRSA, VRE, and multiresistant *Pseudomonas* isolates. It demonstrates no in vitro target-specific cross-resistance with other classes of antibiotics. Some of the company's peptides possess anti-tumor activity due to the fact that similar to bacterial cells many cancer cell types overexpress certain negatively charged molecules compared to normal healthy cells. In in vitro testing the company's peptides target these negatively charged cell surface molecules causing necrosis of the tumor and the induction of anti-tumor immune responses.

3.12. Migenix

Migenix (www.migenix.com) developed bovine indolicidin-based technologies for topical treatment of bacterial infections. Both Cadence's Omigard^{TM} for catheter infections and Cutenea's CLS-001 for treatment of acne rosacea were initially developed by and licensed from Migenix (see above).

3.13. Novozymes A/S

Nonozymes A/S (www.novozymes.com) is developing plectasin-based technology for treatment of infections. Plectasin is the first defensin-like HDP isolated from a fungus and has potent in vitro and in vivo activity against Gram-positive bacteria (106). The company's lead compound NZ2114 was a variant identified in a high-throughput mutation and screening process with improved antimicrobial potency against *Staphylococci* and *Streptococci* (107). NZ2114 showed significantly higher CSF penetration than the parent peptide in an experimental meningitis model. Peak concentrations of NZ2114 in purulent CSF was observed approximately 3 h after an intravenous bolus infusion of either 20 or 40 mg/kg and exceeded the MIC (>10-fold) during a 6-h study period. Treatment with NZ2114 caused a significantly higher reduction in CSF bacterial concentrations as compared to therapy with ceftriaxone (107). Also, significantly more rabbits had sterile CSF at 5 and 10 h when treated with NZ2114 as compared to therapy with ceftriaxone (107). Novozymes entered in December 2008 a global licensing and collaboration agreement with Sanofi-Aventis for the development and marketing of NZ2114 for the treatment of severe infections such as pneumonia, septicemia, and complicated skin and soft tissue infections.

3.14. Pacgen Biopharmaceuticals

Pacgen Biopharmaceuticals Corporation (www.pacgenbiopharm.com) focuses on the development of peptide/protein-based therapeutics for the treatment of infectious and inflammatory diseases. The company's lead compound, PAC-113, is an antifungal peptide in Phase II clinical trials for the treatment of oral candidiasis. The peptide is based on histatin technology initially developed by Demgen Pharmaceuticals (Pittsburgh, PA). Pacgen recently reported positive results from its Phase IIb dose-ranging clinical trial of PAC-113.

3.15. PolyMedix

PolyMedix (www.polymedix.com) is currently advancing a de novo drug design approach to design small molecule compounds that mimic or regulate the activity of protein targets. PMX-30063 was designed to mimic the amphiphilic structure of defensin, but in the form of a synthetic, non-peptide, small molecule structure. It is bactericidal and kills microbes by direct biophysical disruption of the bacterial cell membrane. It is active against both Gram-positive and Gram-negative bacteria, in particular against 146 different strains of *Staphylococcus* including clinical isolates of multiple vancomycin and methicillin-resistant strains. An efficacious systemic dose of 2 mg/kg BID in the mouse *S. aureus* thigh model PMX-30063 has received approval in 2008 from Health Canada for a Phase I clinical trial to assess safety for treatment of systemic infections. In preclinical toxicology testing the NOAEL (no observable adverse effect level) was found to be 24 mg/kg/day for PMX-30063 which provides potential for a meaningful therapeutic index in the clinic.

3.16. Polyphor Ltd

Polyphor Ltd (www.polyphor.com) developed peptide mimetics of β-sheet HDPs, such as protegrin-1, creating, for example, POL-7080, which has selective activity toward Gram-negative bacteria and, in particular, *Pseudomonas* spp. (108). The compound is currently in preclinical development for the treatment of *Pseudomonas* infections.

References

1. Reddy, K. V., Yedery, R. D., and Aranha, C. (2004) Antimicrobial peptides: premises and promises. *Int. J. Antimicrob. Agents* **24**, 536–547.
2. Hancock, R. E. (1997) Peptide antibiotics. *Lancet* **349**, 418–422.
3. Toke, O. (2005) Antimicrobial peptides: new candidates in the fight against bacterial infections. *Biopolymers* **80**, 717–735.
4. Toke, O., Cegelski, L., and Schaefer, J. (2006) Peptide antibiotics in action: investigation of polypeptide chains in insoluble environments by rotational-echo double resonance. *Biochim. Biophys. Acta* **1758**, 1314–1329.
5. Selsted, M. E., Novotny, M. J., Morris, W. L., Tang, Y. Q., Smith, W., and Cullor, J. S. (1992) Indolicidin, a novel bactericidal tridecapeptide amide from neutrophils. *J. Biol. Chem.* **267**, 4292–4295.
6. Rozek, A., Friedrich, C. L., and Hancock, R. E. (2000) Structure of the bovine antimicrobial peptide indolicidin bound to dodecylphosphocholine and sodium dodecyl sulfate micelles. *Biochemistry* **39**, 15765–15774.
7. Askou, H. J., Jakobsen, R. N., and Fojan, P. (2008) An atomic force microscopy study of the interactions between indolicidin and supported planar bilayers. *J. Nanosci. Nanotechnol.* **8**, 4360–4369.
8. Shaw, J. E., Alattia, J. R., Verity, J. E., Prive, G. G., and Yip, C. M. (2006) Mechanisms of antimicrobial peptide action: studies of indolicidin assembly at model membrane interfaces by *in situ* atomic force microscopy. *J. Struct. Biol.* **154**, 42–58.
9. Shaw, J. E., Epand, R. F., Hsu, J. C., Mo, G. C., Epand, R. M., and Yip, C. M. (2008) Cationic peptide-induced remodelling of model membranes: direct visualization by *in situ* atomic force microscopy. *J. Struct. Biol.* **162**, 121–138.
10. Otvos, L., Jr. (2005) Antibacterial peptides and proteins with multiple cellular targets. *J. Pept. Sci.* **11**, 697–706.
11. Selsted, M. E. and Ouellette, A. J. (2005) Mammalian defensins in the antimicrobial immune response. *Nat. Immunol.* **6**, 551–557.
12. Radhakrishnan, Y., Hamil, K. G., Yenugu, S., Young, S. L., French, F. S., and Hall, S. H. (2005) Identification, characterization, and evolution of a primate beta-defensin gene cluster. *Genes Immun.* **6**, 203–210.
13. Zanetti, M., Gennaro, R., and Romeo, D. (1995) Cathelicidins: a novel protein family with a common proregion and a variable

C-terminal antimicrobial domain. *FEBS Lett.* **374**, 1–5.
14. Nizet, V. and Gallo, R. L. (2003) Cathelicidins and innate defense against invasive bacterial infection. *Scand. J. Infect. Dis.* **35**, 670–676.
15. Hilton, K. B. and Lambert, L. A. (2008) Molecular evolution and characterization of hepcidin gene products in vertebrates. *Gene* **415**, 40–48.
16. Rieg, S., Steffen, H., Seeber, S., et al. (2005) Deficiency of dermcidin-derived antimicrobial peptides in sweat of patients with atopic dermatitis correlates with an impaired innate defense of human skin in vivo. *J. Immunol.* **174**, 8003–8010.
17. Oppenheim, F. G., Xu, T., McMillian, F. M., et al. (1988) Histatins, a novel family of histidine-rich proteins in human parotid secretion. Isolation, characterization, primary structure, and fungistatic effects on *Candida albicans*. *J. Biol. Chem.* **263**, 7472–7477.
18. Chan, D. I., Hunter, H. N., Tack, B. F., and Vogel, H. J. (2008) Human macrophage inflammatory protein 3alpha: protein and peptide nuclear magnetic resonance solution structures, dimerization, dynamics, and anti-infective properties. *Antimicrob. Agents Chemother.* **52**, 883–894.
19. Maerki, C., Meuter, S., Liebi, M., et al. (2009) Potent and broad-spectrum antimicrobial activity of CXCL14 suggests an immediate role in skin infections. *J. Immunol.* **182**, 507–514.
20. Radek, K. A., Lopez-Garcia, B., Hupe, M., et al. (2008) The neuroendocrine peptide catestatin is a cutaneous antimicrobial and induced in the skin after injury. *J. Invest. Dermatol.* **128**, 1525–1534.
21. Egesten, A., Eliasson, M., Johansson, H. M., et al. (2007) The CXC chemokine MIG/CXCL9 is important in innate immunity against *Streptococcus pyogenes*. *J. Infect. Dis.* **195**, 684–693.
22. Eliasson, M., Frick, I. M., Collin, M., Sorensen, O. E., Bjorck, L., and Egesten, A. (2007) M1 protein of *Streptococcus pyogenes* increases production of the antibacterial CXC chemokine MIG/CXCL9 in pharyngeal epithelial cells. *Microb. Pathog.* **43**, 224–233.
23. Linge, H. M., Collin, M., Nordenfelt, P., Morgelin, M., Malmsten, M., and Egesten, A. (2008) The human CXC chemokine granulocyte chemotactic protein 2 (GCP-2)/CXCL6 possesses membrane-disrupting properties and is antibacterial. *Antimicrob. Agents Chemother.* **52**, 2599–2607.
24. Collin, M., Linge, H. M., Bjartell, A., Giwercman, A., Malm, J., and Egesten, A. (2008) Constitutive expression of the antibacterial CXC chemokine GCP-2/CXCL6 by epithelial cells of the male reproductive tract. *J. Reprod. Immunol.* **79**, 37–43.
25. Linge, H. M., Collin, M., Giwercman, A., Malm, J., Bjartell, A., and Egesten, A. (2008) The antibacterial chemokine MIG/CXCL9 is constitutively expressed in epithelial cells of the male urogenital tract and is present in seminal plasma. *J. Interferon Cytokine Res.* **28**, 191–196.
26. Di Nardo, A., Yamasaki, K., Dorschner, R. A., Lai, Y., and Gallo, R. L. (2008) Mast cell cathelicidin antimicrobial peptide prevents invasive group A *Streptococcus* infection of the skin. *J. Immunol.* **180**, 7565–7573.
27. De Benedetto, A., Agnihothri, R., McGirt, L. Y., Bankova, L. G., and Beck, L. A. (2009) Atopic dermatitis: a disease caused by innate immune defects? *J. Invest. Dermatol.* **129**, 14–30.
28. Howell, M. D., Wollenberg, A., Gallo, R. L., et al. (2006) Cathelicidin deficiency predisposes to eczema herpeticum. *J. Allergy Clin. Immunol.* **117**, 836–841.
29. Howell, M. D., Jones, J. F., Kisich, K. O., Streib, J. E., Gallo, R. L., and Leung, D. Y. (2004) Selective killing of vaccinia virus by LL-37: implications for eczema vaccinatum. *J. Immunol.* **172**, 1763–1767.
30. Lee, P. H., Ohtake, T., Zaiou, M., et al. (2005) Expression of an additional cathelicidin antimicrobial peptide protects against bacterial skin infection. *Proc. Natl. Acad. Sci. USA* **102**, 3750–3755.
31. Schittek, B., Paulmann, M., Senyurek, I., and Steffen, H. (2008) The role of antimicrobial peptides in human skin and in skin infectious diseases. *Infect. Disord. Drug Targets* **8**, 135–143.
32. Lee, D. Y., Yamasaki, K., Rudsil, J., et al. (2008) Sebocytes express functional cathelicidin antimicrobial peptides and can act to kill *Propionibacterium acnes*. *J. Invest. Dermatol.* **128**, 1863–1866.
33. McInturff, J. E., Wang, S. J., Machleidt, T., et al. (2005) Granulysin-derived peptides demonstrate antimicrobial and anti-inflammatory effects against *Propionibacterium acnes*. *J. Invest. Dermatol.* **125**, 256–263.
34. Temple, M. E. and Nahata, M. C. (2000) Pharmacotherapy of lower limb diabetic ulcers. *J. Am. Geriatr. Soc.* **48**, 822–888.

35. Iwatsuki, K., Yamasaki, O., Morizane, S., and Oono, T. (2006) Staphylococcal cutaneous infections: invasion, evasion and aggression. *J. Dermatol. Sci.* **42**, 203–214.
36. Jacobsen, F., Mohammadi-Tabrisi, A., Hirsch, T., et al. (2007) Antimicrobial activity of the recombinant designer host defence peptide P-novispirin G10 in infected full-thickness wounds of porcine skin. *J. Antimicrob. Chemother.* **59**, 493–498.
37. Lamb, H. M. and Wiseman, L. R. (1998) Pexiganan acetate. *Drugs* **56**, 1047–1052, discussion 1053–1054.
38. Lipsky, B. A., Holroyd, K. J., and Zasloff, M. (2008) Topical versus systemic antimicrobial therapy for treating mildly infected diabetic foot ulcers: a randomized, controlled, double-blinded, multicenter trial of pexiganan cream. *Clin. Infect. Dis.* **47**, 1537–1545.
39. Chung, W. O., Dommisch, H., Yin, L., and Dale, B. A. (2007) Expression of defensins in gingiva and their role in periodontal health and disease. *Curr. Pharm. Des.* **13**, 3073–3083.
40. Aas, J. A., Paster, B. J., Stokes, L. N., Olsen, I., and Dewhirst, F. E. (2005) Defining the normal bacterial flora of the oral cavity. *J. Clin. Microbiol.* **43**, 5721–5732.
41. Dale, B. A. and Fredericks, L. P. (2005) Antimicrobial peptides in the oral environment: expression and function in health and disease. *Curr. Issues Mol. Biol.* **7**, 119–133.
42. Vankeerberghen, A., Nuytten, H., Dierickx, K., Quirynen, M., Cassiman, J. J., and Cuppens, H. (2005) Differential induction of human beta-defensin expression by periodontal commensals and pathogens in periodontal pocket epithelial cells. *J. Periodontol.* **76**, 1293–1303.
43. Putsep, K., Carlsson, G., Boman, H. G., and Andersson, M. (2002) Deficiency of antibacterial peptides in patients with morbus Kostmann: an observation study. *Lancet* **360**, 1144–1149.
44. Beckloff, N., Laube, D., Castro, T., et al. (2007) Activity of an antimicrobial peptide mimetic against planktonic and biofilm cultures of oral pathogens. *Antimicrob. Agents Chemother.* **51**, 4125–4132.
45. Faraj, J. A., Dorati, R., Schoubben, A., et al. (2007) Development of a peptide-containing chewing gum as a sustained release antiplaque antimicrobial delivery system. *AAPS PharmSciTech* **8**, 26.
46. Tanida, T., Okamoto, T., Okamoto, A., et al. (2003) Decreased excretion of antimicrobial proteins and peptides in saliva of patients with oral candidiasis. *J. Oral Pathol. Med.* **32**, 586–594.
47. Meyer, J. E., Harder, J., Gorogh, T., et al. (2004) Human beta-defensin-2 in oral cancer with opportunistic *Candida* infection. *Anticancer Res.* **24**, 1025–1030.
48. Helmerhorst, E. J., Reijnders, I. M., van't Hof, W., Simoons-Smit, I., Veerman, E. C., and Amerongen, A. V. (1999) Amphotericin B- and fluconazole-resistant *Candida* spp., *Aspergillus fumigatus*, and other newly emerging pathogenic fungi are susceptible to basic antifungal peptides. *Antimicrob. Agents Chemother.* **43**, 702–704.
49. Yin, A., Margolis, H. C., Grogan, J., Yao, Y., Troxler, R. F., and Oppenheim, F. G. (2003) Physical parameters of hydroxyapatite adsorption and effect on candidacidal activity of histatins. *Arch. Oral Biol.* **48**, 361–368.
50. Castagnola, M., Inzitari, R., Rossetti, D. V., et al. (2004) A cascade of 24 histatins (histatin 3 fragments) in human saliva. Suggestions for a pre-secretory sequential cleavage pathway. *J. Biol. Chem.* **279**, 41436–41443.
51. Torres, S. R., Garzino-Demo, A., Meiller, T. F., Meeks, V., and Jabra-Rizk, M. A. (2009) Salivary histatin-5 and oral fungal colonisation in HIV+ individuals. *Mycoses* **52**, 11–15.
52. Rothstein, D. M., Spacciapoli, P., Tran, L. T., et al. (2001) Anticandida activity is retained in P-113, a 12-amino-acid fragment of histatin 5. *Antimicrob. Agents Chemother.* **45**, 1367–1373.
53. Hiemstra, P. S. (2007) The role of epithelial beta-defensins and cathelicidins in host defense of the lung. *Exp. Lung Res.* **33**, 537–542.
54. Bals, R., Wang, X., Wu, Z., et al. (1998) Human beta-defensin 2 is a salt-sensitive peptide antibiotic expressed in human lung. *J. Clin. Invest.* **102**, 874–880.
55. Cheung, Q. C., Turner, P. V., Song, C., et al. (2008) Enhanced resistance to bacterial infection in protegrin-1 transgenic mice. *Antimicrob. Agents Chemother.* **52**, 1812–1819.
56. Zhang, L., Parente, J., Harris, S. M., Woods, D. E., Hancock, R. E., and Falla, T. J. (2005) Antimicrobial peptide therapeutics for cystic fibrosis. *Antimicrob. Agents Chemother.* **49**, 2921–2927.
57. Falagas, M. E., Kasiakou, S. K., Tsiodras, S., and Michalopoulos, A. (2006) The use of intravenous and aerosolized polymyxins for the treatment of infections in critically ill patients: a review of the recent literature. *Clin. Med. Res.* **4**, 138–146.

58. Vallon-Eberhard, A., Makovitzki, A., Beauvais, A., Latge, J. P., Jung, S., and Shai, Y. (2008) Efficient clearance of *Aspergillus fumigatus* in murine lungs by an ultrashort antimicrobial lipopeptide, palmitoyl-lys-ala-DAla-lys. *Antimicrob. Agents Chemother.* **52**, 3118–3126.
59. Wang, G., Stange, E. F., and Wehkamp, J. (2007) Host-microbe interaction: mechanisms of defensin deficiency in Crohn's disease. *Expert Rev. Anti-infect. Ther.* **5**, 1049–1057.
60. Wehkamp, J., Koslowski, M., Wang, G., and Stange, E. F. (2008) Barrier dysfunction due to distinct defensin deficiencies in small intestinal and colonic Crohn's disease. *Mucosal Immunol.* **1**(Suppl 1), S67–S74.
61. Chakraborty, K., Ghosh, S., Kole, H., et al. (2008) Bacterial exotoxins downregulate cathelicidin (hCAP18/LL37) and human beta-defensin 1 (HBD-1) expression in the intestinal epithelial cells. *Cell Microbiol.* **10**, 2520–2537.
62. Iqbal, S. M. and Kaul, R. (2008) Mucosal innate immunity as a determinant of HIV susceptibility. *Am. J. Reprod. Immunol.* **59**, 44–54.
63. Kaul, R., Pettengell, C., Sheth, P. M., et al. (2008) The genital tract immune milieu: an important determinant of HIV susceptibility and secondary transmission. *J. Reprod. Immunol.* 77, 32–40.
64. Valore, E. V., Wiley, D. J., and Ganz, T. (2006) Reversible deficiency of antimicrobial polypeptides in bacterial vaginosis. *Infect. Immun.* **74**, 5693–5702.
65. Zapata, W., Rodriguez, B., Weber, J., et al. (2008) Increased levels of human beta-defensins mRNA in sexually HIV-1 exposed but uninfected individuals. *Curr. HIV Res.* **6**, 531–538.
66. Cole, A. M. and Cole, A. L. (2008) Antimicrobial polypeptides are key anti-HIV-1 effector molecules of cervicovaginal host defense. *Am. J. Reprod. Immunol.* **59**, 27–34.
67. Bergman, P., Walter-Jallow, L., Broliden, K., Agerberth, B., and Soderlund, J. (2007) The antimicrobial peptide LL-37 inhibits HIV-1 replication. *Curr. HIV Res.* **5**, 410–415.
68. Cole, A. M., Hong, T., Boo, L. M., et al. (2002) Retrocyclin: a primate peptide that protects cells from infection by T- and M-tropic strains of HIV-1. *Proc. Natl. Acad. Sci. USA* **99**, 1813–1818.
69. Cole, A. L., Herasimtschuk, A., Gupta, P., Waring, A. J., Lehrer, R. I., and Cole, A. M. (2007) The retrocyclin analogue RC-101 prevents human immunodeficiency virus type 1 infection of a model human cervicovaginal tissue construct. *Immunology* **121**, 140–145.
70. Aranha, C. C., Gupta, S. M., and Reddy, K. V. (2008) Assessment of cervicovaginal cytokine levels following exposure to microbicide Nisin gel in rabbits. *Cytokine* **43**, 63–70.
71. Donald, C. D., Sun, C. Q., Lim, S. D., et al. (2003) Cancer-specific loss of beta-defensin 1 in renal and prostatic carcinomas. *Lab. Invest.* **83**, 501–505.
72. Sun, C. Q., Arnold, R., Fernandez-Golarz, C., et al. (2006) Human beta-defensin-1, a potential chromosome 8p tumor suppressor: control of transcription and induction of apoptosis in renal cell carcinoma. *Cancer Res.* **66**, 8542–8549.
73. Bullard, R. S., Gibson, W., Bose, S. K., et al. (2008) Functional analysis of the host defense peptide Human Beta Defensin-1: new insight into its potential role in cancer. *Mol. Immunol.* **45**, 839–848.
74. Gibson, W., Green, A., Bullard, R. S., Eaddy, A. C., and Donald, C. D. (2007) Inhibition of PAX2 expression results in alternate cell death pathways in prostate cancer cells differing in p53 status. *Cancer Lett.* **248**, 251–261.
75. Bose, S. K., Gibson, W., Bullard, R. S., and Donald, C. D. (2008) PAX2 oncogene negatively regulates the expression of the host defense peptide human beta defensin-1 in prostate cancer. *Mol. Immunol.* **46**, 1140–1148.
76. Baker, M. A., Maloy, W. L., Zasloff, M., and Jacob, L. S. (1993) Anticancer efficacy of Magainin 2 and analogue peptides. *Cancer Res.* **53**, 3052–3057.
77. Hansel, W., Leuschner, C., and Enright, F. (2007) Conjugates of lytic peptides and LHRH or betaCG target and cause necrosis of prostate cancers and metastases. *Mol. Cell Endocrinol.* **269**, 26–33.
78. Leuschner, C., Enright, F. M., Gawronska, B., and Hansel, W. (2003) Membrane disrupting lytic peptide conjugates destroy hormone dependent and independent breast cancer cells in vitro and in vivo. *Breast Cancer Res. Treat.* **78**, 17–27.
79. Ghavami, S., Asoodeh, A., Klonisch, T., et al. (2008) Brevinin-2R(1) semi-selectively kills cancer cells by a distinct mechanism, which involves the lysosomal-mitochondrial death pathway. *J. Cell. Mol. Med.* **12**, 1005–1022.
80. Suttmann, H., Retz, M., Paulsen, F., et al. (2008) Antimicrobial peptides of the Cecropin-family show potent antitumor

activity against bladder cancer cells. *BMC Urol.* **8**, 5.
81. Rodrigues, E. G., Dobroff, A. S., Cavarsan, C. F., et al. (2008) Effective topical treatment of subcutaneous murine B16F10-Nex2 melanoma by the antimicrobial peptide gomesin. *Neoplasia* **10**, 61–68.
82. Xu, N., Wang, Y. S., Pan, W. B., et al. (2008) Human alpha-defensin-1 inhibits growth of human lung adenocarcinoma xenograft in nude mice. *Mol. Cancer Ther.* 7, 1588–1597.
83. Yang, D., Chertov, O., and Oppenheim, J. J. (2001) Participation of mammalian defensins and cathelicidins in anti-microbial immunity: receptors and activities of human defensins and cathelicidin (LL-37). *J. Leukoc. Biol.* **69**, 691–697.
84. Niyonsaba, F., Suzuki, A., Ushio, H., Nagaoka, I., Ogawa, H., and Okumura, K. (2008) The human antimicrobial peptide dermcidin activates normal human keratinocytes. *Br. J. Dermatol.* **160**, 243–249.
85. Mookherjee, N., Brown, K. L., Bowdish, D. M., et al. (2006) Modulation of the TLR-mediated inflammatory response by the endogenous human host defense peptide LL-37. *J. Immunol.* **176**, 2455–2464.
86. Mookherjee, N., Rehaume, L. M., and Hancock, R. E. (2007) Cathelicidins and functional analogues as antisepsis molecules. *Expert Opin. Ther. Targets* **11**, 993–1004.
87. Book, M., Chen, Q., Lehmann, L. E., et al. (2007) Inducibility of the endogenous antibiotic peptide beta-defensin 2 is impaired in patients with severe sepsis. *Crit. Care* **11**, R19.
88. Torossian, A., Gurschi, E., Bals, R., Vassiliou, T., Wulf, H. F., and Bauhofer, A. (2007) Effects of the antimicrobial peptide LL-37 and hyperthermic preconditioning in septic rats. *Anesthesiology* **107**, 437–441.
89. Di Nardo, A., Braff, M. H., Taylor, K. R., et al. (2007) Cathelicidin antimicrobial peptides block dendritic cell TLR4 activation and allergic contact sensitization. *J. Immunol.* **178**, 1829–1834.
90. Deslouches, B., Gonzalez, I. A., De Almeida, D., et al. (2007) De novo-derived cationic antimicrobial peptide activity in a murine model of *Pseudomonas aeruginosa* bacteraemia. *J. Antimicrob. Chemother.* **60**, 669–672.
91. Wakabayashi, H., Takakura, N., Yamauchi, K., and Tamura, Y. (2006) Modulation of immunity-related gene expression in small intestines of mice by oral administration of lactoferrin. *Clin. Vaccine Immunol.* **13**, 239–245.
92. Lienkamp, K., Madkour, A. E., Musante, A., Nelson, C. F., Nusslein, K., and Tew, G. N. (2008) Antimicrobial polymers prepared by ROMP with unprecedented selectivity: a molecular construction kit approach. *J. Am. Chem. Soc.* **130**, 9836–9843.
93. Statz, A. R., Park, J. P., Chongsiriwatana, N. P., Barron, A. E., and Messersmith, P. B. (2008) Surface-immobilised antimicrobial peptoids. *Biofouling* **24**, 439–448.
94. Willcox, M. D., Hume, E. B., Aliwarga, Y., Kumar, N., and Cole, N. (2008) A novel cationic-peptide coating for the prevention of microbial colonization on contact lenses. *J. Appl. Microbiol.* **105**, 1817–1825.
95. Lai, X. Z., Feng, Y., Pollard, J., et al. (2008) Ceragenins: cholic acid-based mimics of antimicrobial peptides. *Acc. Chem. Res.* **41**, 1233–1240.
96. Epand, R. F., Savage, P. B., and Epand, R. M. (2007) Bacterial lipid composition and the antimicrobial efficacy of cationic steroid compounds (Ceragenins). *Biochim. Biophys. Acta.* **1768**, 2500–2509.
97. Chin, J. N., Rybak, M. J., Cheung, C. M., and Savage, P. B. (2007) Antimicrobial activities of ceragenins against clinical isolates of resistant *Staphylococcus aureus*. *Antimicrob. Agents Chemother.* **51**, 1268–1273.
98. Van Bambeke, F., Mingeot-Leclercq, M. P., Struelens, M. J., and Tulkens, P. M. (2008) The bacterial envelope as a target for novel anti-MRSA antibiotics. *Trends Pharmacol. Sci.* **29**, 124–134.
99. Rotem, S. and Mor, A. (2009) Antimicrobial peptide mimics for improved therapeutic properties. *Biochim. Biophys. Acta* **1788**, 1582–1592.
100. Sarig, H., Rotem, S., Ziserman, L., Danino, D., and Mor, A. (2008) Impact of self-assembly properties on antibacterial activity of short acyl-lysine oligomers. *Antimicrob. Agents Chemother.* **52**, 4308–4314.
101. Fritsche, T. R., Rhomberg, P. R., Sader, H. S., and Jones, R. N. (2008) Antimicrobial activity of omiganan pentahydrochloride against contemporary fungal pathogens responsible for catheter-associated infections. *Antimicrob. Agents Chemother.* **52**, 1187–1189.
102. van den Berg, H. R., Khan, N. A., van der Zee, M., et al. (2009) Synthetic oligopeptides related to the [beta]-subunit of human chorionic gonadotropin attenuate inflammation and liver damage after (trauma) hemorrhagic shock and resuscitation. *Shock* **31**, 285–291.

103. Benner, R. and Khan, N. A. (2005) Dissection of systems, cell populations and molecules. *Scand. J. Immunol.* **62**(Suppl 1), 62–66.
104. Falla, T., Harris, S. M., and Zhang, L. (2007) Novel antiinfective for the treatment and prevention of wound infection, ASM Biodefense and emerging diseases, Washington, DC, February 27th–March 2nd.
105. Zhang, L., Harris, S. M., and Falla, T. J. (2007) Lipohexapeptides as topical therapeutics for fungal infections, 107th American Society of Microbiology General Meeting, Toronto Canada, May 21–25.
106. Mygind, P. H., Fischer, R. L., Schnorr, K. M., et al. (2005) Plectasin is a peptide antibiotic with therapeutic potential from a saprophytic fungus. *Nature* **437**, 975–980.
107. Ostergaard, C., Sandvang, D., Frimodt-Moller, N., and Kristensen, H. H. (2009) High CSF penetration and potent CSF bactericidal activity of NZ2114 – a novel Plectasin variant – during experimental pneumococcal meningitis. *Antimicrob. Agents Chemother.* **53**, 1581–1585.
108. Robinson, J. A., Shankaramma, S. C., Jetter, P., et al. (2005) Properties and structure-activity studies of cyclic beta-hairpin peptidomimetics based on the cationic antimicrobial peptide protegrin I. *Bioorg. Med. Chem.* **13**, 2055–2064.

Chapter 20

Therapeutic Potential of HDPs as Immunomodulatory Agents

Håvard Jenssen and Robert E.W. Hancock

Abstract

One of the most significant advances in medical history is the discovery and development of antibiotics, which in the middle of last century was flourishing and appeared to be the ultimate solution to the treatment of life-threatening human bacterial diseases. However, lately there has been a huge decline in the rate of discovery of new antimicrobial intervention strategies in parallel with an increasing incidence of multidrug-resistant pathogens; if these circumstances do not change we will continue to approach the end of the antibiotic era. Facing this dark future, scientists are considering new strategies for intervention tailored around the appropriate (selective) stimulation of the host's immune system, and particularly rapid acting innate immunity, as an alternative to direct targeting of microbial pathogens. One recent player in such an immunomodulatory strategy is the naturally occurring host defence peptides (HDP) and their synthetic innate defence regulator (IDR) analogues. In this chapter, we will discuss the potential therapeutic use of HDPs and IDRs as immunomodulatory agents.

Key words: Cationic host defence peptides, anti-infective therapy, antimicrobial peptides, innate defence regulators, immune stimulation, immunomodulator, anti-inflammatory.

1. Introduction

The initiation of the therapeutic use of penicillin during the Second World War followed by the discovery and development of other antibiotics targeting bacterial, and later also viral and fungal pathogens, has undoubtedly had a tremendous impact on human life. However, recently there has been a tremendous decline in the discovery and development of new drugs, e.g. only three new chemical classes of antibacterials have entered the market in the past 40 years (i.e., lipopeptides, oxazolidinones and streptogramins, all targeting

A. Giuliani, A.C. Rinaldi (eds.), *Antimicrobial Peptides*, Methods in Molecular Biology 618,
DOI 10.1007/978-1-60761-594-1_20, © Springer Science+Business Media, LLC 2010

Gram-positive infections), while the frequency of isolation of drug-resistant pathogens is rapidly increasing. Globally it is estimated that infectious diseases still account for one-third of all human deaths, and big players like respiratory infections, HIV, malaria, and tuberculosis are all found among the top eight killers (www.who.int/whosis/whostat/2008/en/index.html). A novel strategy for the development of new interventions is to target the host's own immediate defence system, innate immunity. However, the development of drugs targeting the innate immune system has experienced some difficulties, and sceptics have argued against the potential of this strategy, largely due to the potential downside offered by inflammation (especially if too vigorous or too prolonged) and the complexity and consequent lack of mechanistic biological understanding of this defence machinery. Despite this there are several exciting potential targets for development of new anti-infectives and a number of targets are currently being targeted by different strategies, e.g. antibodies, biologicals including proteins and peptides, and small-molecule agonists and antagonists; the success of these ventures has been reviewed elsewhere (1–6). Another promising drug class is so-called host defence peptides (HDPs; also termed cationic antimicrobial peptides – AMPs when they have direct antimicrobial activity) and synthetic derivative thereof. The therapeutic use and potential of these HDPs as potential anti-infective immune-modulators are discussed in detail in this chapter.

2. Targeting the Host Innate Immune System

To prevent and cure infections by targeting the innate immune system is an innovative but ambitious idea. This intrinsically complex and conserved signalling system has developed throughout evolution to respond and defeat all types of infections. It is highly effective, given our constant exposure to pathogens but relatively infrequent symptomatic infections; at the same time it is delicately tuned such that it seldom results in prolonged inflammation. The strategy of designing anti-infectives for selective modulation of the innate immune defence system has been made possible by the discovery of specific pathogen recognition receptors, such as Toll-like receptors, and intracellular sensors of microbial components such as the Nod-like receptors and retinoic-acid inducible gene I-like receptors (7). Exposure to pathogenic signalling molecules results in a rapid response that normally returns to homeostasis within hours to days; if unsuccessful in clearing the infection a smooth transition to adaptive (acquired immunity) occurs. However, under some circumstances this control of inflammation can break down and/or uncontrolled so-called chronic inflammation can ensue; if the inflammatory response is too vigorous or maintained for a prolonged period of time it may lead to pathological consequences and many human disease syndromes demonstrate the symptoms of uncontrolled inflammation including excessive bleeding and chronic pain. Hence the optimal anti-infective therapy based on innate immunity should employ tailored immunomodulators that offer the potential to tip the balance back in favour of the host, by either (selectively) boosting or inhibiting selected elements of the immune response at the same time as exploiting and enhancing the efficiency of the powerful and multifaceted effector mechanisms that have evolved specifically for the purpose of pathogen clearance. It is also important to remember that there are many aspects of the pathogen-sensing, response and control mechanisms that we do not understand, giving hope that in the future even more novel targets for anti-infective therapies will be revealed to assist us in facing the increasing drug resistance challenge. New knowledge and sophisticated technological advances will help accelerate the development of novel anti-infective immunomodulators, e.g. systems biology approaches, genomic libraries for siRNA gene silencing, the high-throughput development of gene knockout mice, studies of single-nucleotide polymorphisms associated with disease and the profiling of transcriptional, micro-RNA and protein interaction networks will all promote new insights into innate immune response mechanisms (8, 9). In this regard, a recently launched powerful tool is the publicly-available bioinformatics resource, Innate DB (www.innatedb.com) (10) and IIDB (http://db.systemsbiology.net/IIDB) (11), which are innate immunity-specific databases that include data analysis resources to facilitate the functional analysis of innate immunity responses, as well as the IIPGA program (www.innateimmunity.net), which is a collaborative effort to analyse polymorphisms in human innate immunity genes.

3. Host Defence Peptides

The constant battle between the host and the pathogens through evolution has resulted in a finely tuned defence system involving innate (germ line encoded) and adaptive (antigen specific due to gene rearrangements and acquired due to pathogenic challenge) immunity. Host defence peptides have been demonstrated to play a pivotal role in the orchestration of innate immune and inflammatory responses of mammals, amphibians, and insects (12–14). Recently, increasing evidence is supporting the involvement of these antimicrobial peptides in bridging these two immune defence strategies.

Being key players in the fundamental interplay between host and microbe makes these HDPs/AMPs signature molecules of host defence exhibiting substantial diversity even within a single host and occurring in moderate to high concentrations. Virtually all species of life produce them, with more than 1,000 natural occurring peptides having been described to date, and the majority of these are described in databases for eukaryotic host defence peptides: e.g. the site at the University of Trieste (http://www.bbcm.units.it/~tossi/pag1.htm) and the AMPer site (http://www.cnbi2.com/cgi-bin/amp.pl) (15).

These peptides generally fall into four major structural categories based on their amphiphilic conformations that often occur only after membrane interaction; namely β-structures with two to four β-strands and amphipathic α-helices, and less commonly loop and extended structures (**Fig. 20.1**). These peptides are typically short (12–50 amino acids), carrying a net positively charged (+ 2 to 9) due to excess basic arginine and/or lysine residues, and contain up to 50% hydrophobic amino acids (permitting membrane interaction and enabling membrane damage and/or cell penetration). Although their potency varies substantially, depending in part on antagonism by physiological concentrations of mono- and divalent cations and polyanions they often have broad spectra of activity that can encompass antibacterial, antifungal, antiviral and anti-parasitic activities (16). Equally important is

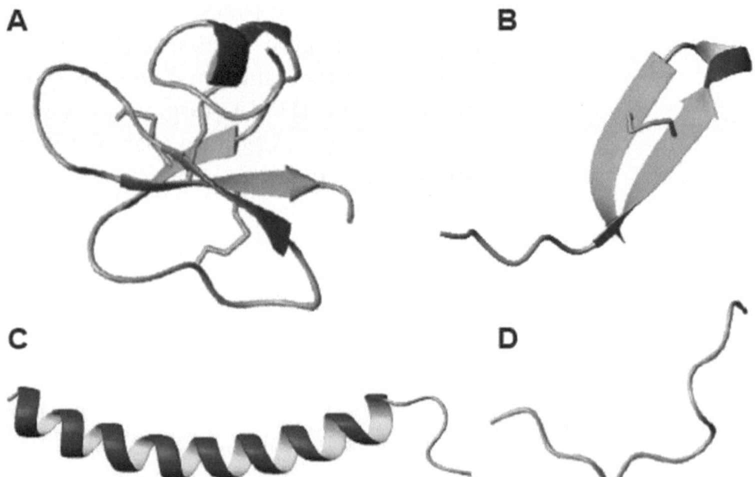

Fig. 20.1. Structural classes of antimicrobial peptides. **(A)** Mixed structure of human β-defensin-2 (HBD-2) (PDB code 1FQQ) (82), **(B)** looped thanatin (PDB code 8TFV) (83), **(C)** α-helical human cathelicidin LL-37 (PDB code 2k6o) (84), **(D)** extended indolicidin (PDB code 1G89) (85). The figures have been prepared with use of the graphic program MolMol 2 K.1 (86).

their ability to recruit and activate or modulate effectors of the innate immune system (14, 17).

4. Host Defence Peptides Mode of Action

Initially cationic peptides were investigated due to their direct antimicrobial activity. Though the peptides usually demonstrate lower potency than conventional antibiotics, one strength lies in their ability to kill multidrug-resistant bacteria. They are able to act extremely rapidly and can engage and inhibit multiple bacterial targets (18). Initially the peptides were believed to act only by perforating bacterial membranes through, e.g. “aggregate” (19), “toroidal pore” (20–23), “barrel-stave” (24) or “carpet” mechanistic models (25). Later studies also demonstrated the ability of many peptides to translocate across the cytoplasmic membrane of bacteria (26) targeting DNA/RNA synthesis (27–30), protein synthesis (28, 29, 31, 32), enzymatic activities (33, 34) and cell wall regeneration (35) among others.

In addition to this direct antimicrobial activity, there is strong evidence that a broad range of cationic peptides can stimulate the host immune system as a mode of promoting pathogen clearance, thus giving rise to the new term host defence peptides (HDPs) (**Fig. 20.2**). The terms AMPs and HDPs are to some extent interchangeable, though the term HDPs is used here for natural peptides with known immunomodulatory properties, while AMPs are known for their direct antimicrobial effects; however, a single peptide may have both features. For example, human peptides are likely directly antimicrobial when found in high concentrations (e.g. in neutrophil granules or in the crypts of the intestine) and exhibit prominent immunomodulatory properties when found at lower concentrations, e.g. after release and dissemination from these sites, e.g. at mucosal surfaces. There are several examples indicating that HDPs are key players in innate immunity. For example, inflammatory bowel disease, also known as Crohn’s disease, is generally characterized by chronic inflammation of the intestine, commonly in the distal ileum and/or colon (36). Recent studies have demonstrated that patients suffering from Crohn’s disease have a deficiency in the epithelial cell secretion of β-defensins (i.e. human β-defensins 2 and 3) (37, 38) as well as expression of human cathelicidin LL-37 (39). This results in an imbalance between luminal bacteria and the HDP concentrations, allowing intestinal bacteria from the microbial flora to trigger inflammation with associated adherence and invasion into the mucosa (40). HDP expression in the skin is also known to protect against invasive bacterial infections. Transgenic mice

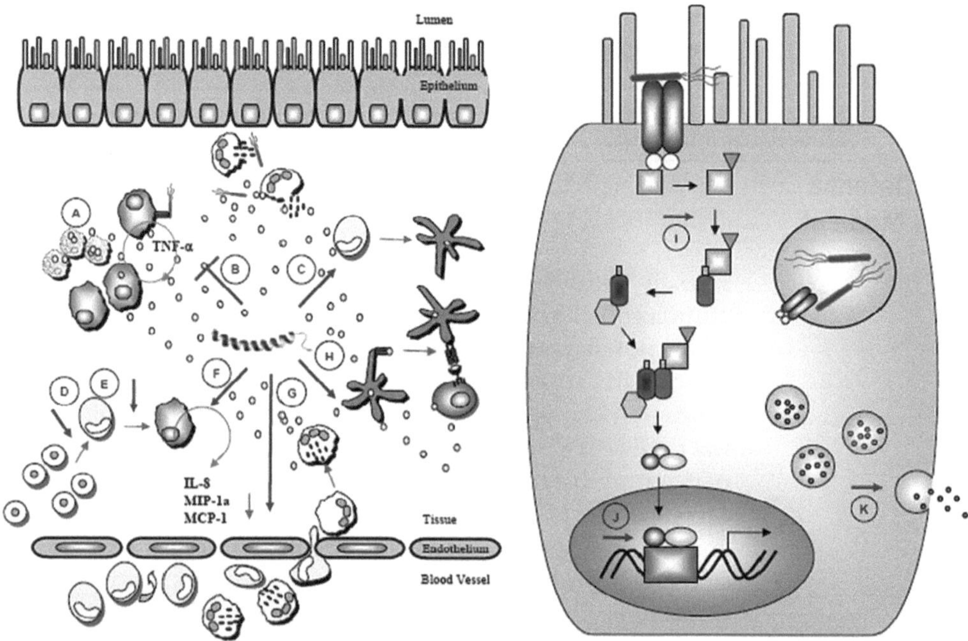

Fig. 20.2. Immunomodulatory properties of host defense peptides. The figure illustrates possible targets at the systemic level (*left*) and in a single cell (*right*). Several HDPs', e.g. LL-37, CRAMP, Indolicidin and SMAP-29 can alter toll like receptor-induced responses, e.g. suppress the expression of pro-inflammatory molecules (e.g. tumor necrosis factor α) and inhibit apoptosis of neutrophils (A & B) (47, 51, 87–91). Dendritic cell differentiation may be targeted (C) (92) as well as several steps in the maturation and differentiation of monocytes to macrophages (D & E). LL-37 activates macrophages (F), leading to release of effector molecules, primarily chemokines. These chemokines, as well as HDP's alone (e.g. CRAMP, HNP-1, HNP-2, LL-37), promotes chemotaxis and migration of leukocytes to the afflicted area (G) (93–97). LL-37 promotes the expression of co-stimulatory molecules on dendritic cell and promotes expression of T_h1 cytokine interleukin-12 (H). Several immune cells are know to be targeted by HDPs. Cell signalling may be affected at several levels in the signalling cascade (I) (73, 98), interfering with gene transcription (J), altering the cellular levels of different effector molecules and cellular degranulation (K) (47, 88, 89, 92, 99–103). The figure is reprinted and modified from Hamill et al., 2008 (1) with permission from authors and Current Opinion in Biotechnology, Copyright © 2008, Elsevier.

mutated in the gene for the mouse peptide CRAMP, a homolog of human LL-37 (41, 42), have been observed to develop significantly more severe skin infections than control mice, though the isolated leukocytes from these CRAMP-deficient mice were functionally competent and similar to wild-type leukocytes (43). Similarly, patients suffering from morbus Kostmann disease and associated severe gum inflammation have an associated deficiency in neutrophil levels of LL-37, as well as reduced concentrations of α-defensins (i.e. human neutrophil peptides-1, -2 and -3) (44). Similarly the ability of peptides to mediate the attraction of various immune cell types (either directly as chemokines or indirectly by stimulating epithelial and monocytic cells to secrete conventional chemokines) to demonstrate anti-inflammatory/anti-endotoxic activity, stimulate blood vessel growth (angiogenesis), promote wound healing, and promote and polarize adaptive immune responses have all been demonstrated in mouse models and human or mouse tissue or cell systems (43, 45–52). Indeed it can be argued that the systemic effects found with any cationic peptide may be primarily due to the often-undescribed immunomodulatory properties of the peptide, given that the direct antimicrobial activity of peptides, but usually not their immunomodulatory activities, is strongly antagonized by physiological salt concentrations, i.e. 100 mM monovalent- and 2 mM divalent cations (as well as polyanions like glycosaminoglycans such as heparin) (46). In our view the systemic action of peptides that are deliberately added for therapeutic purposes may involve either direct (i.e. antibiotic) or indirect (i.e. immunomodulatory) antimicrobial activity. For example, the most clinically advanced peptide developed by Migenix, primarily as an antimicrobial, has recently been demonstrated to have activity in phase III clinical trails in preventing catheter-associated tunnel infections, as well as activity against the inflammatory disease Rosacea, for which there is no microbial association (**Table 20.1**).

5. Commercial Production of Host Defence Peptides

The clinical use of HDPs has been limited by issues like cost of goods. Several commercial participants have tried to circumvent this issue by moving from traditional solid-phase synthesis to solution-phase methods, and further to a hybrid methodology, e.g. a convergent strategy of synthesizing peptides, i.e., selecting peptide segments and assembling them together typically using a solution phase methodology. The best example of this is Roche (http://www.roche.com/index.htm) and Trimeris' (http://www.trimeris.com) joint success in large-scale production of the HIV fusion inhibitor, enfuvirtide (Fuzeon, T-20), by solid- and solution phase hybrid synthesis (53, 54). Enfuvirtide is a 36-amino-acid peptide that binds to a region of the HIV envelope glycoprotein gp41, thus blocking viral entry into CD^{4+} T cells. The drug is licensed for use in patients in whom other HIV medications are losing effectiveness, in part due to the limited annual production capacity (3.7 metric tons in 2005 providing treatment for ~47,000 patients). Aside from traditional peptide synthesis, other biotechnological approaches may be used for commercial production of HDPs, e.g. recombinant systems built on bacterial-, fungal- or mammalian-expression platforms. Novozymes (Bagsvœrd, Denmark, http://www.novozymes.com/en) have demonstrated great success using a fungal-based system to recombinantly

Table 20.1
Immunomodulatory and antimicrobial peptides in clinical trials or developmental stages

Drug	Sequence	Description/status/results	Company & reference
Plectasin NZ2114	$GFGC_1NGPWDEDDMQC_2HNHC_3KSIKGYKGGYC_1AKGGFVC_2KC_3Y$-COOH (Plectasin)	*Preclinical*: A variant of plectasin which has demonstrated potent Gram-positive effect in systemic pneumococcal and streptococcal infections	Novozymes AS/Sanofi-Aventis (Bagsvœrd, Denmark), www.novozymes.com
PAC-113	AKRHHGYKRKFH-$CONH_2$	*Phase IIb* : A 12-mer segment present in histatin 3 and 5, which reduces oral candidiasis (0.15% mouth rinse formulation)	Pacgen (Vancouver, BC, Canada), www.pacgenbiopharm.com
XOMA 629/XMP.629	KLFR-(D-naph-A)-QAK-(D-naph-A)	*Phase IIa*: A 9-mer peptide derived from a human HDP (bactericidal/permeability-increasing protein). Developed as a topical gel for treatment of common skin disease, impetigo, with demonstrated potent antibacterial activity against several difficult-to-control bacterial strains including methicillin-resistant *Staphylococcus aureus* Note: The XOMA 629 development program was recently suspended (Press release 10 November 2008)	XOMA (Berkeley, CA, USA), www.xoma.com (104)
Omiganan®/ CP-226/ MBI-226/ CLS001	ILRWPWWPWRRK-$CONH_2$	*Phase III* : Analogue of bovine AMP (indolicidin). Omiganan has been demonstrated in Phase II trials to have anti-inflammatory properties vs. Rosacea and acne *Phase IIIb*: Direct antibacterial properties for prevention of catheter related infections.	Migenix/Cutanea (Vancouver, BC, Canada), www.migenix.com, www.cutanealife.com (105)

IMX942	KSRIVPAIPVSLL-$CONH_2$ (IDR-1)	*Preclinical*: 5-mer peptide modelled on IDR-1. IDR-1 was demonstrated to work through selective stimulation of innate immunity, up-regulating protective immunity while suppressing pro-inflammatory cytokine production in response to bacterial TLR agonists	Inimex (Burnaby, BC, Canada), www.inimexpharma.com
hLF1-11	GRRRRSVQWCA-COOH	*Phase IIa*: N-terminal fragment of human HDP (lactoferrin) has indicated immunomodulatory activity. Trials address allogeneic bone marrow stem cell transplantation–associated infections and other drug resistant hospital acquired infections	AM-Pharma (Bunnvik, The Netherlands), www.ampharma.com
CZEN-002	(Ac-CKPV-$CONH_2$)$_2$	*Phase IIb*: A synthetic 8-mer peptide derived from α-melanocyte-stimulating hormone. Inhibits vulvovaginal candidiasis, commonly known as vaginal yeast infection	Zengen/Zensano (Woodland Hills, CA, USA), www.biospace.com
PTX002	SIQDLNVSMKLFRKQAKWKIIV KLNDGRELSLD-COOH	*Discovery phase*: 33-mer peptide with two distinct β-sheet folds and a 12-mer peptide derivative thereof. Has demonstrated broad-spectrum antimicrobial and antiendotoxin neutralizing activity in both in vitro and in vivo experiments	PepTx (St. Paul, MN, USA), www.peptx.com (106, 107)
PTX005/SC4	KLFKRHLKWKII-$CONH_2$		

Note: Amino-acid sequences are given in one-letter code. Cysteines forming disulfide bonds are numbered with subscripts to indicate their pairings. N- and C-terminal modifications are indicated before the hyphen in the front or after the hyphen at the end of the sequence.

produce plectasin in large scale with therapeutic purity under good manufacturing practice (GMP) conditions (55). Plectasin is a fungal defensin-like 40-amino-acid peptide, with broad-spectrum antibacterial activity (56) and no affect on cell viability or interleukin-8 production (57). A derivative of plectasin (Plectasin NZ2114) is currently in preclinical development as a very potent direct Gram-positive antimicrobial in a collaboration between Novozymes and Sanofi-Aventis (Bridgewater, NJ, USA) (**Table 20.1**).

Aside from the cost of goods, questions have been raised regarding poor pharmacokinetic properties of these peptides, in addition to their lack of oral availability (58); however, formulation and/or chemical modification, e.g. N-methylation, of peptides may rationally improve key pharmacokinetic characteristics. Multiple N-methylation has been demonstrated to significantly improve the oral bioavailability, metabolic stability and intestinal permeability of peptides (58, 59). Advances in formulation and delivery systems, e.g. implantable scaffolds, hydrogels and micro- or nano- particle systems, will also help expedite the progression of immunomodulators into clinical use (60).

Despite these solutions, the biggest challenge has probably been peptide in vivo stability. Development of a pegylated form of interferon-α for chronic hepatitis C treatment, that displays greatly enhanced stability, illustrates how technological advances in peptide derivatization can improve the clinical usefulness of biologic-based therapies (61). Another approach to increase metabolic stability is through chemically altering the peptide backbone structure, changing the traditional peptide to a so-called peptidomimetic, which can be defined as "a compound that, as the ligand of a receptor, can imitate or block the biological effect of a peptide at the receptor level" (62), or "chemical structures designed to convert the information contained in peptides into small non-peptide structures" (63). Peptidomimetics can be designed to mimic peptides and participate in protein–protein and protein–nucleic acid interactions. They often have an intrinsic ability to form stable helical conformations (64) which is important since helical motifs constitute the largest class of protein secondary structure and play a major role in mediating protein–protein interactions. At the same time these small synthetic molecules that mimic natural peptides are not susceptible to proteolytic degradation, an advantage they hold over their natural counterparts. Conversely, at the same time it can be argued that potent immunomodulatory substances should ideally be biodegradable to limit overstimulation of the immune system leading to inflammation or hyperstimulation. Thus it might be hypothesized that when moving from directly antimicrobial AMPs to host stimulatory HDPs these issues may be of lesser importance. Proteolytic degradation may in fact lead to shorter peptide fragments with superior immunomodulatory activity, thus eliciting responses not documented through in vitro screening. Similar phenomena have been described earlier when Domagk in the early 1930s discovered that the red textile dye Prontosil effectively inhibited streptococcal and staphylococcal infections in vivo, while being inactive against the bacteria in vitro (65). It was later concluded that a metabolic degradation product of Prontosil, a sulfanilamide, possessed the antimicrobial activity (66), illustrating one of the problems of screening for active anti-infectives in a host-free system.

6. HDPs in Clinical Development

Host defence peptides are an interesting class of agents with under-explored potential as anti-infective therapeutics. Early work explored the peptides for topical applications, exploiting their direct antimicrobial properties. Biotech companies like Magainin Pharmaceuticals (now Genaera; http://www.genaera.com), Micrologix (now Migenix; http://www.migenix.com) and IntraBiotics (http://www.intrabiotics.com) started designing therapeutic peptides that only differ from their natural homologues by a few amino acids. However, despite promising results from the early clinical trials several failed at later stages, e.g. the Genaera developed Pexiganan (MSI-78) a synthetic 22-amino-acid variant of the amphibian peptide magainin-2 (67), and the IntraBiotics Pharmaceuticals developed pig protegrin analogue Iseganan (IB-367) (68–70).

Despite these disappointing early failures of cationic peptide drugs, several other new ventures have been launched, largely derived from academic labs worldwide. The first antimicrobial peptide to show statistically significant clinical effects was a bovine indolicidin homolog MX-226 (also known as Omiganan or CPI-226; originally developed by Migenix Inc, Vancouver, BC, Canada) (**Table 20.1**). In Phase III clinical trails, the topical application of MX-226 pentahydrochloride in a 1% gel (Omigard) led to a significant 21% reduction of colonization of central venous catheters and a 50% reduction in tunnel infections (www.migenix.com/prod_226.html) (71, 72). Intriguingly Omiganan (as CLS001) has also demonstrated efficacy as an anti-inflammatory peptide, in suppressing the effects of acute acne and Rosacea in Phase II clinical trials, confirming the multifaceted actions of these HDP derivatives. Oral candidiasis in immunocompromised patients is been tackled in a Phase IIb study by Pacgen (Vancouver, BC, Canada), using a short 12-mer peptide segment (PAC-113) originally isolated from human histatin 3 and 5. Much in the line of this Xoma (Berkley, CA, USA; http://www.xoma.com) is also developing a synthetic peptide from bactericidal/permeability-increasing protein, a human host defence protein. The peptide XOMA 629 has demonstrated great direct antibacterial activity and is currently undergoing clinical Phase IIa trials for topical eradication of common skin infection pathogens in addition to methicillin-resistant *Staphylococcus aureus*.

Synthetic peptides designed based on knowledge about the immunomodulatory properties of LL-37 and other HDPs are currently being explored for their therapeutic potential (12) due to their unique ability to promote protective innate immunity while suppressing potentially harmful inflammatory responses (**Fig. 20.2**). A good example on the success of these studies is the described innate defence regulator-1 (IDR-1), an anti-infective peptide that selectively modulates the innate immune response (73). Optimization of immunomodulatory properties resulted in depletion of the peptides direct antibacterial properties, underlining the complex nature of the natural HDPs. However, despite lacking direct antibacterial activity, IDR-1 confers protection against multiple bacterial pathogens (i.e. methicillin-sensitive/-resistant *S. aureus*, vancomycin-sensitive/-resistant *Enterococcus* and *Salmonella enterica*) in in vivo mice models. Studies concluded that this protection resulted from activation of monocyte-macrophage cells, stimulation of monocyte produced chemokine expression and dampening of pro-inflammatory cytokine responses (**Fig. 20.2**). Further optimization studies on IDR-1 has led to a 5-amino-acid peptide derivative (IMX942) that currently have been launched into preclinical development by Inimex Pharmaceuticals (Burnaby, BC, Canada) (**Table 20.1**). Based on the results there are plans for continuation evaluating efficacy in targeting pneumonia, surgical site infections and chemotherapy induced neutropenia.

The human milk protein-derived peptide hLF1-11 was initially being developed by AM Pharma (http://www.ampharma.com) as a direct antimicrobial peptide. However, after the peptide demonstrated immunomodulatory properties the focus changed, and hLF1-11 is currently in clinical trials for evaluation of efficacy in protection and prevention of infection during allogeneic stem cell transplantation (**Table 20.1**).

Mimicking the structural features of endogenous host defence peptides by other chemical structures has been looked upon as an attractive way to circumvent several of the drawbacks linked to peptides and peptide chemistry. One type of HDP mimics is called Ceragenins. These are synthetically produced small-molecule compounds comprised of a sterol backbone with amino acids and/or other chemical groups attached (74). Several Ceragenins have been developed by Ceragenix

Pharmaceutical Inc. (Denver CO, USA) and have been demonstrated to possess broad-spectrum activities against multidrug-resistant clinically isolated bacteria (75, 76). However, there has been unsatisfactory safety reports for cationic steroid antibiotics (77), thus restricting the use of Ceragenins to topical application or after immobilizing at the surface of medical devices for prevention of biofilms (78). Consequently Ceragenix Pharmaceutical have used their broad-spectrum antimicrobial compounds and combined it with the Ceragenin™ technology to formulate CeraShield™, antimicrobial coatings for medical devices. Preclinical studies with CeraShield™ are currently ongoing for a variety of medical devices, from tubes and intravenous connectors to different catheters and orthopaedic and surgical implants (http://www.ceragenix.com/).

Another class of mimetic has been developed by PolyMedix Inc. (Radnor, PA, USA). PolyMedix has tailored small chemical molecules that mimic several of the well HDPs. On example from their portfolio is a peptide mimetic whose design was based on the structure of the well-described HDP magainin. This mimetic mPE (meta-phenylene ethynylene) has demonstrated activity against a variety of bacterial and *Candida* species down to nano-molar concentrations (79). Another of their compounds, a small non-peptide mimetic (PMX-30063), mimics the amphiphilic structure of defensines. PMX-30063 is claimed to be bactericidal against both Gram-positive and Gram-negative bacteria by disruption of the bacterial cell membrane, and the company has also other small molecule mimetic that possess similar activities (80). PMX-30063 has demonstrated efficacy in a mice model of systemic *S. aureus* infection at 2 mg/kg, and a Phase I clinical safety study with PMX-30063 has just been completed, demonstrating a safe and well-tolerated single doses with no observable adverse effects up to 24 mg/kg per day (Press release).

7. Conclusion

We are entering the “post-antibiotic era” in which many previously successful drug regimes are becoming ineffective, consequently the time has come to vigorously pursue alternative potential treatment approaches toward infectious diseases. Development of HDPs as immunomodulatory therapeutic alternatives is questionably subject to unique difficulties. The innate immune system is intrinsically complex and there is great genetic variation between individuals; additionally there is fundamental interspecies differences (81) cluttering the correlation between animal models and human trials. However, the complexity of the innate immune system through its evolution has also resulted in huge redundancy, thus almost duplicating critical response cascades, hence creating a safety net for drug developers while also undermining the effectiveness of single targeting drug candidates.

The success of peptides as drugs, and especially HDPs as anti-infectives, has not carried significant fruits yet. However, this concept is in its infancy and with continued research and adaptation of new technology platforms especially introducing bioinformatics approaches, it may enable this strategy to solve some of the forthcoming challenges we will be facing from emerging infections diseases. Any disease is per definition caused by a pathogen's ability to overcome or circumvent the host immune response, thus we argue that development of novel anti-infectives directly targeting the host immune system has a reasonable chance of succeeding, while a reduced chance of resistance development.

Acknowledgments

This manuscript is dedicated to the memory of Aaron W.J. Wyatt who tragically passed away on 24 December 2008. Aaron was not only a superb colleague but also a good friend.

The author's research in this area was supported by the Foundation for the National Institutes of Health, Gates Foundation and Canadian Institutes for Health Research through two separate Grand Challenges in Global Health Initiatives and by Genome British Columbia for the Pathogenomics of Innate Immunity Research Program. REWH is the recipient of a Canada Research Chair.

References

1. Hamill, P., Brown, K., Jenssen, H., and Hancock, R. E. W. (2008) Novel anti-infectives: is host defence the answer?. *Curr. Opin. Biotechnol.* **19**, 628–636.
2. Lai, Y. and Gallo, R. L. (2008) Toll-like receptors in skin infections and inflammatory diseases. *Infect. Disord. Drug Targets* **8**, 144–155.
3. O'Neill, L. A. (2006) Targeting signal transduction as a strategy to treat inflammatory diseases. *Nat. Rev. Drug Discov.* **5**, 549–563.
4. Kanzler, H., Barrat, F. J., Hessel, E. M., and Coffman, R. L. (2007) Therapeutic targeting of innate immunity with Toll-like receptor agonists and antagonists. *Nat. Med.* **13**, 552–559.
5. Romagne, F. (2007) Current and future drugs targeting one class of innate immunity receptors: the Toll-like receptors. *Drug Discov. Today* **12**, 80–87.
6. Wales, J., Andreakos, E., Feldmann, M., and Foxwell, B. (2007) Targeting intracellular mediators of pattern-recognition receptor signalling to adjuvant vaccination. *Biochem. Soc. Trans.* **35**, 1501–1503.
7. Creagh, E. M. and O'Neill, L. A. (2006) TLRs, NLRs and RLRs: a trinity of pathogen sensors that co-operate in innate immunity. *Trends Immunol.* **27**, 352–357.
8. Alper, S., Laws, R., Lackford, B., Boyd, W. A., Dunlap, P., Freedman, J. H., and Schwartz, D. A. (2008) Identification of innate immunity genes and pathways using a

comparative genomics approach. *Proc. Natl. Acad. Sci. USA* **105**, 7016–7021.
9. Tegner, J., Nilsson, R., Bajic, V. B., Bjorkegren, J., and Ravasi, T. (2006) Systems biology of innate immunity. *Cell. Immunol.* **244**, 105–109.
10. Lynn, D. J., Winsor, G. L., Chan, C., Richard, N., Laird, M. R., Barsky, A., Gardy, J. L., Roche, F. M., Chan, T. H., Shah, N., Lo, R., Naseer, M., Que, J., Yau, M., Acab, M., Tulpan, D., Whiteside, M. D., Chikatamarla, A., Mah, B., Munzner, T., Hokamp, K., Hancock, R. E. W., and Brinkman, F. S. (2008) InnateDB: facilitating systems-level analyses of the mammalian innate immune response. *Mol. Syst. Biol.* **4**, 218.
11. Korb, M., Rust, A. G., Thorsson, V., Battail, C., Li, B., Hwang, D., Kennedy, K. A., Roach, J. C., Rosenberger, C. M., Gilchrist, M., Zak, D., Johnson, C., Marzolf, B., Aderem, A., Shmulevich, I., and Bolouri, H. (2008) The innate immune database (IIDB). *BMC Immunol.* **9**, 7.
12. Hancock, R. E. W. and Sahl, H. G. (2006) Antimicrobial and host-defense peptides as new anti-infective therapeutic strategies. *Nat. Biotechnol.* **24**, 1551–1157.
13. Oppenheim, J. J. and Yang, D. (2005) Alarmins: chemotactic activators of immune responses. *Curr. Opin. Immunol.* **17**, 359–365.
14. Zasloff, M. (2002) Antimicrobial peptides of multicellular organisms. *Nature* **415**, 389–395.
15. Fjell, C. D., Hancock, R. E. W., and Cherkasov, A. (2007) AMPer: a database and an automated discovery tool for antimicrobial peptides. *Bioinformatics* **23**, 1148–1155.
16. Jenssen, H., Hamill, P., and Hancock, R. E. W. (2006) Peptide antimicrobial agents. *Clin. Microbiol. Rev.* **19**, 491–511.
17. Hancock, R. E. W. (2001) Cationic peptides: effectors in innate immunity and novel antimicrobials. *Lancet Infect. Dis.* **1**, 156–164.
18. Brogden, K. A. (2005) Antimicrobial peptides: pore formers or metabolic inhibitors in bacteria?. *Nat. Rev. Microbiol.* **3**, 238–250.
19. Wu, M., Maier, E., Benz, R., and Hancock, R. E. W. (1999) Mechanism of interaction of different classes of cationic antimicrobial peptides with planar bilayers and with the cytoplasmic membrane of *Escherichia coli*. *Biochemistry* **38**, 7235–7242.
20. Hallock, K. J., Lee, D. K., and Ramamoorthy, A. (2003) MSI-78, an analogue of the magainin antimicrobial peptides, disrupts lipid bilayer structure via positive curvature strain. *Biophys. J.* **84**, 3052–3060.
21. Henzler Wildman, K. A., Lee, D. K., and Ramamoorthy, A. (2003) Mechanism of lipid bilayer disruption by the human antimicrobial peptide, LL-37. *Biochemistry* **42**, 6545–6558.
22. Matsuzaki, K., Murase, O., Fujii, N., and Miyajima, K. (1996) An antimicrobial peptide, magainin 2, induced rapid flip-flop of phospholipids coupled with pore formation and peptide translocation. *Biochemistry* **35**, 11361–11368.
23. Yang, L., Harroun, T. A., Weiss, T. M., Ding, L., and Huang, H. W. (2001) Barrel-Stave model or Toroidal model? A case study on melittin pores. *Biophys. J.* **81**, 1475–1485.
24. Ehrenstein, G. and Lecar, H. (1977) Electrically gated ionic channels in lipid bilayers. *Q. Rev. Biophys.* **10**, 1–34.
25. Pouny, Y., Rapaport, D., Mor, A., Nicolas, P., and Shai, Y. (1992) Interaction of antimicrobial dermaseptin and its fluorescently labeled analogues with phospholipid membranes. *Biochemistry* **31**, 12416–12423.
26. Zhang, L., Rozek, A., and Hancock, R. E. W. (2001) Interaction of cationic antimicrobial peptides with model membranes. *J. Biol. Chem.* **276**, 35714–35722.
27. Park, C. B., Kim, H. S., and Kim, S. C. (1998) Mechanism of action of the antimicrobial peptide buforin II: buforin II kills microorganisms by penetrating the cell membrane and inhibiting cellular functions. *Biochem. Biophys. Res. Commun.* **244**, 253–257.
28. Patrzykat, A., Friedrich, C. L., Zhang, L., Mendoza, V., and Hancock, R. E. W. (2002) Sublethal concentrations of pleurocidin-derived antimicrobial peptides inhibit macromolecular synthesis in escherichia coli. *Antimicrob. Agents Chemother.* **46**, 605–614.
29. Subbalakshmi, C. and Sitaram, N. (1998) Mechanism of antimicrobial action of indolicidin. *FEMS Microbiol. Lett.* **160**, 91–96.
30. Lehrer, R. I., Barton, A., Daher, K. A., Harwig, S. S., Ganz, T., and Selsted, M. E. (1989) Interaction of human defensins with *Escherichia coli*. Mechanism of bactericidal activity. *J. Clin. Invest.* **84**, 553–561.
31. Boman, H. G., Agerberth, B., and Boman, A. (1993) Mechanisms of action on *Escherichia coli* of cecropin P1 and PR-39, two antibacterial peptides from pig intestine. *Infect. Immun.* **61**, 2978–2984.
32. Friedrich, C. L., Rozek, A., Patrzykat, A., and Hancock, R. E. W. (2001) Structure and mechanism of action of an indolicidin peptide derivative with improved activity against

gram-positive bacteria. *J. Biol. Chem.* **276**, 24015–24022.

33. Kragol, G., Lovas, S., Varadi, G., Condie, B. A., Hoffmann, R., and Otvos, L., Jr. (2001) The antibacterial peptide pyrrhocoricin inhibits the ATPase actions of DnaK and prevents chaperone-assisted protein folding. *Biochemistry* **40**, 3016–3026.
34. Otvos, L., Jr., O, I., Rogers, M. E., Consolvo, P. J., Condie, B. A., Lovas, S., Bulet, P., and Blaszczyk-Thurin, M. (2000) Interaction between heat shock proteins and antimicrobial peptides. *Biochemistry* **39**, 14150–14159.
35. Hechard, Y. and Sahl, H. G. (2002) Mode of action of modified and unmodified bacteriocins from Gram-positive bacteria. *Biochimie* **84**, 545–557.
36. Podolsky, D. K. (2002) Inflammatory bowel disease. *N. Engl. J. Med.* **347**, 417–429.
37. Wehkamp, J., Fellermann, K., Herrlinger, K. R., Baxmann, S., Schmidt, K., Schwind, B., Duchrow, M., Wohlschlager, C., Feller, A. C., and Stange, E. F. (2002) Human beta-defensin 2 but not beta-defensin 1 is expressed preferentially in colonic mucosa of inflammatory bowel disease. *Eur. J. Gastroenterol. Hepatol.* **14**, 745–752.
38. Wehkamp, J., Harder, J., Weichenthal, M., Mueller, O., Herrlinger, K. R., Fellermann, K., Schroeder, J. M., and Stange, E. F. (2003) Inducible and constitutive beta-defensins are differentially expressed in Crohn's disease and ulcerative colitis. *Inflamm. Bowel. Dis.* **9**, 215–223.
39. Schauber, J., Rieger, D., Weiler, F., Wehkamp, J., Eck, M., Fellermann, K., Scheppach, W., Gallo, R. L., and Stange, E. F. (2006) Heterogeneous expression of human cathelicidin hCAP18/LL-37 in inflammatory bowel diseases. *Eur. J. Gastroenterol. Hepatol.* **18**, 615–621.
40. Wehkamp, J., Schmid, M., and Stange, E. F. (2007) Defensins and other antimicrobial peptides in inflammatory bowel disease. *Curr. Opin. Gastroenterol.* **23**, 370–378.
41. Gudmundsson, G. H., Agerberth, B., Odeberg, J., Bergman, T., Olsson, B., and Salcedo, R. (1996) The human gene FALL39 and processing of the cathelin precursor to the antibacterial peptide LL-37 in granulocytes. *Eur. J. Biochem.* **238**, 325–332.
42. Gallo, R. L., Kim, K. J., Bernfield, M., Kozak, C. A., Zanetti, M., Merluzzi, L., and Gennaro, R. (1997) Identification of CRAMP, a cathelin-related antimicrobial peptide expressed in the embryonic and adult mouse. *J. Biol. Chem.* **272**, 13088–13093.
43. Nizet, V., Ohtake, T., Lauth, X., Trowbridge, J., Rudisill, J., Dorschner, R. A., Pestonjamasp, V., Piraino, J., Huttner, K., and Gallo, R. L. (2001) Innate antimicrobial peptide protects the skin from invasive bacterial infection. *Nature* **414**, 454–457.
44. Putsep, K., Carlsson, G., Boman, H. G., and Andersson, M. (2002) Deficiency of antibacterial peptides in patients with morbus Kostmann: an observation study. *Lancet* **360**, 1144–1149.
45. Cherkasov, A., Hilpert, K., Jenssen, H., Fjell, C. D., Waldbrook, M., Mullaly, S. C., Volkmer, R., and Hancock, R. E. W. (2008) Use of artificial intelligence in the design of small peptide antibiotics effective against a broad spectrum of highly antibiotic-resistant superbugs. *ACS Chem. Biol.* **1**, 65–74.
46. Bowdish, D. M., Davidson, D. J., Lau, Y. E., Lee, K., Scott, M. G., and Hancock, R. E. W. (2005) Impact of LL-37 on anti-infective immunity. *J. Leukoc. Biol.* **77**, 451–459.
47. Scott, M. G., Davidson, D. J., Gold, M. R., Bowdish, D., and Hancock, R. E. W. (2002) The human antimicrobial peptide LL-37 is a multifunctional modulator of innate immune responses. *J. Immunol.* **169**, 3883–3891.
48. Fukumoto, K., Nagaoka, I., Yamataka, A., Kobayashi, H., Yanai, T., Kato, Y., and Miyano, T. (2005) Effect of antibacterial cathelicidin peptide CAP18/LL-37 on sepsis in neonatal rats. *Pediatr. Surg. Int.* **21**, 20–24.
49. McGwire, B. S., Olson, C. L., Tack, B. F., and Engman, D. M. (2003) Killing of African trypanosomes by antimicrobial peptides. *J. Infect. Dis.* **188**, 146–152.
50. Joly, S., Maze, C., McCray, P. B., Jr., and Guthmiller, J. M. (2004) Human beta-defensins 2 and 3 demonstrate strain-selective activity against oral microorganisms. *J. Clin. Microbiol.* **42**, 1024–1029.
51. Giacometti, A., Cirioni, O., Ghiselli, R., Mocchegiani, F., D'Amato, G., Circo, R., Orlando, F., Skerlavaj, B., Silvestri, C., Saba, V., Zanetti, M., and Scalise, G. (2004) Cathelicidin peptide sheep myeloid antimicrobial peptide-29 prevents endotoxin-induced mortality in rat models of septic shock. *Am. J. Respir. Crit. Care Med.* **169**, 187–194.
52. Brogden, K. A., Nordholm, G., and Ackermann, M. (2007) Antimicrobial activity of cathelicidins BMAP28, SMAP28, SMAP29, and PMAP23 against *Pasteurella multocida* is more broad-spectrum than host species specific. *Vet. Microbiol.* **119**, 76–81.

53. Andersson, L., Blomberg, L., Flegel, M., Lepsa, L., Nilsson, B., and Verlander, M. (2000) Large-scale synthesis of peptides. *Biopolymers* **55**, 227–250.
54. Schneider, S. E., Bray, B. L., Mader, C. J., Friedrich, P. E., Anderson, M. W., Taylor, T. S., Boshernitzan, N., Niemi, T. E., Fulcher, B. C., Whight, S. R., White, J. M., Greene, R. J., Stoltenberg, L. E., and Lichty, M. (2005) Development of HIV fusion inhibitors. *J. Pept. Sci.* **11**, 744–753.
55. Mygind, P. H., Fischer, R. L., Schnorr, K. M., Hansen, M. T., Sonksen, C. P., Ludvigsen, S., Raventos, D., Buskov, S., Christensen, B., De Maria, L., Taboureau, O., Yaver, D., Elvig-Jorgensen, S. G., Sorensen, M. V., Christensen, B. E., Kjaerulff, S., Frimodt-Moller, N., Lehrer, R. I., Zasloff, M., and Kristensen, H. H. (2005) Plectasin is a peptide antibiotic with therapeutic potential from a saprophytic fungus. *Nature* **437**, 975–980.
56. Gottlieb, C. T., Thomsen, L. E., Ingmer, H., Mygind, P. H., Kristensen, H. H., and Gram, L. (2008) Antimicrobial peptides effectively kill a broad spectrum of *Listeria monocytogenes* and *Staphylococcus aureus* strains independently of origin, sub-type, or virulence factor expression. *BMC Microbiol.* **8**, 205.
57. Hara, S., Mukae, H., Sakamoto, N., Ishimoto, H., Amenomori, M., Fujita, H., Ishimatsu, Y., Yanagihara, K., and Kohno, S. (2008) Plectasin has antibacterial activity and no affect on cell viability or IL-8 production. *Biochem. Biophys. Res. Commun.* **374**, 709–713.
58. Chatterjee, J., Gilon, C., Hoffman, A., and Kessler, H. (2008) N-methylation of peptides: a new perspective in medicinal chemistry. *Acc. Chem. Res.* **41**, 1331–1342.
59. Biron, E., Chatterjee, J., Ovadia, O., Langenegger, D., Brueggen, J., Hoyer, D., Schmid, H. A., Jelinek, R., Gilon, C., Hoffman, A., and Kessler, H. (2008) Improving oral bioavailability of peptides by multiple N-methylation: somatostatin analogues. *Angew Chem. Int. Ed. Engl.* **47**, 2595–2599.
60. Kobsa, S. and Saltzman, W. M. (2008) Bioengineering approaches to controlled protein delivery. *Pediatr. Res.* **63**, 513–519.
61. Barnard, D. L. (2001) Pegasys (Hoffmann-La Roche). *Curr. Opin. Investig. Drugs* **2**, 1530–1538.
62. Giannis, A. and Kolter, T. (1993) Pepidomimetics for receptor ligands – discovery, development, and medical perspectives. *Angew. Chem. Int. Ed.* **32**, 24.
63. Wiley, R. A. and Rich, D. H. (1993) Peptidomimetics derived from natural products. *Med. Res. Rev.* **13**, 327–384.
64. Sanborn, T. J., Wu, C. W., Zuckermann, R. N., and Barron, A. E. (2002) Extreme stability of helices formed by water-soluble poly-N-substituted glycines (polypeptoids) with alpha-chiral side chains. *Biopolymers* **63**, 12–20.
65. Domagk, G. (1935) A report on the chemotherapy of bacterial infections. *Deut. Med. Woch.* **Ixi:250**.
66. Trefouel, J., Nitti, F., and Bovet, D. (1935) Activity of p-aminophenylsulfamide in the experimental streptococcal infections of the mouse and rabbit. *CR Seances Soc. Biol.* **120**, 756.
67. Lamb, H. M. and Wiseman, L. R. (1998) Pexiganan acetate. *Drugs* **56**, 1047–1052, discussion 1053–1054.
68. Trotti, A., Garden, A., Warde, P., Symonds, P., Langer, C., Redman, R., Pajak, T. F., Fleming, T. R., Henke, M., Bourhis, J., Rosenthal, D. I., Junor, E., Cmelak, A., Sheehan, F., Pulliam, J., Devitt-Risse, P., Fuchs, H., Chambers, M., O'Sullivan, B., and Ang, K. K. (2004) A multinational, randomized phase III trial of iseganan HCl oral solution for reducing the severity of oral mucositis in patients receiving radiotherapy for head-and-neck malignancy. *Int. J. Radiat. Oncol. Biol. Phys.* **58**, 674–681.
69. van Saene, H., van Saene, J., Silvestri, L., de la Cal, M., Sarginson, R., and Zandstra, D. (2007) Iseganan failure due to the wrong pharmaceutical technology. *Chest* **132**, 1412.
70. Kollef, M., Pittet, D., Sanchez Garcia, M., Chastre, J., Fagon, J. Y., Bonten, M., Hyzy, R., Fleming, T. R., Fuchs, H., Bellm, L., Mercat, A., Manez, R., Martinez, A., Eggimann, P., Daguerre, M., and Luyt, C. E. (2006) A randomized double-blind trial of iseganan in prevention of ventilator-associated pneumonia. *Am. J. Respir. Crit. Care Med.* **173**, 91–97.
71. Fritsche, T. R., Rhomberg, P. R., Sader, H. S., and Jones, R. N. (2008) Antimicrobial activity of omiganan pentahydrochloride tested against contemporary bacterial pathogens commonly responsible for catheter-associated infections. *J. Antimicrob. Chemother.* **61**, 1092–1098.
72. Fritsche, T. R., Rhomberg, P. R., Sader, H. S., and Jones, R. N. (2008) Antimicrobial activity of omiganan pentahydrochloride against contemporary fungal pathogens responsible for catheter-associated infections. *Antimicrob. Agents Chemother.* **52**, 1187–1189.

73. Scott, M. G., Dullaghan, E., Mookherjee, N., Glavas, N., Waldbrook, M., Thompson, A., Wang, A., Lee, K., Doria, S., Hamill, P., Yu, J. J., Li, Y., Donini, O., Guarna, M. M., Finlay, B. B., North, J. R., and Hancock, R. E. W. (2007) An anti-infective peptide that selectively modulates the innate immune response. *Nat. Biotechnol.* **25**, 465–472.
74. Lai, X. Z., Feng, Y., Pollard, J., Chin, J. N., Rybak, M. J., Bucki, R., Epand, R. F., Epand, R. M., and Savage, P. B. (2008) Ceragenins: cholic acid-based mimics of antimicrobial peptides. *Acc. Chem. Res.* **41**, 1233–1240.
75. Chin, J. N., Rybak, M. J., Cheung, C. M., and Savage, P. B. (2007) Antimicrobial activities of ceragenins against clinical isolates of resistant *Staphylococcus aureus. Antimicrob. Agents Chemother.* **51**, 1268–1273.
76. Van Bambeke, F., Mingeot-Leclercq, M. P., Struelens, M. J., and Tulkens, P. M. (2008) The bacterial envelope as a target for novel anti-MRSA antibiotics. *Trends Pharmacol. Sci.* **29**, 124–134.
77. Savage, P. B., Li, C., Taotafa, U., Ding, B., and Guan, Q. (2002) Antibacterial properties of cationic steroid antibiotics. *FEMS Microbiol. Lett.* **217**, 1–7.
78. Savage, P. B., Pollard, J., Feng, Y., Reddy, L. K., and Genberg, C. (2008): Use of a Ceragenin-Based Coating to Prevent Bacterial Colonization of Urinary Catheters. In *Interscience Conference on Antimicrobial Agents & Chemotherapy* pp. Poster K–1479.
79. Beckloff, N., Laube, D., Castro, T., Furgang, D., Park, S., Perlin, D., Clements, D., Tang, H., Scott, R. W., Tew, G. N., and Diamond, G. (2007) Activity of an antimicrobial peptide mimetic against planktonic and biofilm cultures of oral pathogens. *Antimicrob. Agents Chemother.* **51**, 4125–4132.
80. Scott, R. W., DeGrado, W. F., and Tew, G. N. (2008) De novo designed synthetic mimics of antimicrobial peptides. *Curr. Opin. Biotechnol.* **19**, 620–627.
81. Viola, A. and Luster, A. D. (2008) Chemokines and their receptors: drug targets in immunity and inflammation. *Annu. Rev. Pharmacol. Toxicol* **48**, 171–197.
82. Sawai, M. V., Jia, H. P., Liu, L., Aseyev, V., Wiencek, J. M., McCray, P. B., Jr., Ganz, T., Kearney, W. R., and Tack, B. F. (2001) The NMR structure of human beta-defensin-2 reveals a novel alpha-helical segment. *Biochemistry* **40**, 3810–3816.
83. Mandard, N., Sodano, P., Labbe, H., Bonmatin, J. M., Bulet, P., Hetru, C., Ptak, M., and Vovelle, F. (1998) Solution structure of thanatin, a potent bactericidal and fungicidal insect peptide, determined from proton two-dimensional nuclear magnetic resonance data. *Eur. J. Biochem.* **256**, 404–410.
84. Wang, G. (2008) Structures of human host defense cathelicidin LL-37 and its smallest antimicrobial peptide kr-12 in lipid micelles. *J. Biol. Chem.* **283**, 32637–32643.
85. Rozek, A., Friedrich, C. L., and Hancock, R. E. W. (2000) Structure of the bovine antimicrobial peptide indolicidin bound to dodecylphosphocholine and sodium dodecyl sulfate micelles. *Biochemistry* **39**, 15765–15774.
86. Koradi, R., Billeter, M., and Wuthrich, K. (1996) MOLMOL: a program for display and analysis of macromolecular structures. *J. Mol. Graph.* **14**, 51–55, 29–32.
87. Bowdish, D. M., Davidson, D. J., Scott, M. G., and Hancock, R. E. W. (2005) Immunomodulatory activities of small host defense peptides. *Antimicrob. Agents Chemother.* **49**, 1727–, 1732.
88. Mookherjee, N., Brown, K. L., Bowdish, D. M., Doria, S., Falsafi, R., Hokamp, K., Roche, F. M., Mu, R., Doho, G. H., Pistolic, J., Powers, J. P., Bryan, J., Brinkman, F. S., and Hancock, R. E. W. (2006) Modulation of the TLR-mediated inflammatory response by the endogenous human host defense peptide LL-37. *J. Immunol.* **176**, 2455–2464.
89. Mookherjee, N., Wilson, H. L., Doria, S., Popowych, Y., Falsafi, R., Yu, J. J., Li, Y., Veatch, S., Roche, F. M., Brown, K. L., Brinkman, F. S., Hokamp, K., Potter, A., Babiuk, L. A., Griebel, P. J., and Hancock, R. E. W. (2006) Bovine and human cathelicidin cationic host defense peptides similarly suppress transcriptional responses to bacterial lipopolysaccharide. *J. Leukoc. Biol.* **80**, 1563–1574.
90. Bowdish, D. M. and Hancock, R. E. W. (2005) Anti-endotoxin properties of cationic host defence peptides and proteins. *J. Endotoxin Res.* **11**, 230–236.
91. Ohgami, K., Ilieva, I. B., Shiratori, K., Isogai, E., Yoshida, K., Kotake, S., Nishida, T., Mizuki, N., and Ohno, S. (2003) Effect of human cationic antimicrobial protein 18 Peptide on endotoxin-induced uveitis in rats. *Invest Ophthalmol. Vis. Sci.* **44**, 4412–4418.
92. Davidson, D. J., Currie, A. J., Reid, G. S., Bowdish, D. M., MacDonald, K. L., Ma, R. C., Hancock, R. E. W., and Speert, D. P. (2004) The cationic antimicrobial peptide LL-37 modulates dendritic cell differentiation and dendritic cell-induced T cell polarization. *J. Immunol.* **172**, 1146–1156.

93. Tjabringa, G. S., Ninaber, D. K., Drijfhout, J. W., Rabe, K. F., and Hiemstra, P. S. (2006) Human cathelicidin LL-37 is a chemoattractant for eosinophils and neutrophils that acts via formyl-peptide receptors. *Int. Arch. Allergy Immunol.* **140**, 103–112.
94. Chertov, O., Michiel, D. F., Xu, L., Wang, J. M., Tani, K., Murphy, W. J., Longo, D. L., Taub, D. D., and Oppenheim, J. J. (1996) Identification of defensin-1, defensin-2, and CAP37/azurocidin as T-cell chemoattractant proteins released from interleukin-8-stimulated neutrophils. *J. Biol. Chem.* **271**, 2935–2940.
95. Kurosaka, K., Chen, Q., Yarovinsky, F., Oppenheim, J. J., and Yang, D. (2005) Mouse cathelin-related antimicrobial peptide chemoattracts leukocytes using formyl peptide receptor-like 1/mouse formyl peptide receptor-like 2 as the receptor and acts as an immune adjuvant. *J. Immunol.* **174**, 6257–6265.
96. Territo, M. C., Ganz, T., Selsted, M. E., and Lehrer, R. (1989) Monocyte-chemotactic activity of defensins from human neutrophils. *J. Clin. Invest.* **84**, 2017–2020.
97. Djanani, A., Mosheimer, B., Kaneider, N. C., Ross, C. R., Ricevuti, G., Patsch, J. R., and Wiedermann, C. J. (2006) Heparan sulfate proteoglycan-dependent neutrophil chemotaxis toward PR-39 cathelicidin. *J. Inflam. (Lond)* **3**, 14.
98. Yu, J., Mookherjee, N., Wee, K., Bowdish, D. M., Pistolic, J., Li, Y., Rehaume, L., and Hancock, R. E. W. (2007) Host defense peptide LL-37, in synergy with inflammatory mediator IL-1beta, augments immune responses by multiple pathways. *J. Immunol.* **179**, 7684–7691.
99. Bowdish, D. M., Davidson, D. J., Speert, D. P., and Hancock, R. E. W. (2004) The human cationic peptide LL-37 induces activation of the extracellular signal-regulated kinase and p38 kinase pathways in primary human monocytes. *J. Immunol.* **172**, 3758–3765.
100. Lau, Y. E., Rozek, A., Scott, M. G., Goosney, D. L., Davidson, D. J., and Hancock, R. E. W. (2005) Interaction and cellular localization of the human host defense peptide LL-37 with lung epithelial cells. *Infect. Immun.* **73**, 583–591.
101. Niyonsaba, F., Someya, A., Hirata, M., Ogawa, H., and Nagaoka, I. (2001) Evaluation of the effects of peptide antibiotics human beta-defensins-1/-2 and LL-37 on histamine release and prostaglandin D(2) production from mast cells. *Eur J. Immunol.* **31**, 1066–1075.
102. Li, J., Post, M., Volk, R., Gao, Y., Li, M., Metais, C., Sato, K., Tsai, J., Aird, W., Rosenberg, R. D., Hampton, T. G., Sellke, F., Carmeliet, P., and Simons, M. (2000) PR39, a peptide regulator of angiogenesis. *Nat. Med.* **6**, 49–55.
103. Gallo, R. L., Ono, M., Povsic, T., Page, C., Eriksson, E., Klagsbrun, M., and Bernfield, M. (1994) Syndecans, cell surface heparan sulfate proteoglycans, are induced by a proline-rich antimicrobial peptide from wounds. *Proc. Natl. Acad. Sci. USA* **91**, 11035–11039.
104. Schroeder, J. M. and Harder, J. (2006) Antimicrobial peptides in skin disease. *Drug Discov. Today* **3**, 8.
105. Melo, M. N., Dugourd, D., and Castanho, M. A. (2006) Omiganan pentahydrochloride in the front line of clinical applications of antimicrobial peptides. *Recent Patents Anti-Infect Drug Disc.* **1**, 201–207.
106. Ilyina, E., Roongta, V., and Mayo, K. H. (1997) NMR structure of a de novo designed, peptide 33mer with two distinct, compact beta-sheet folds. *Biochemistry* **36**, 5245–5250.
107. Mayo, K. H., Haseman, J., Young, H. C., and Mayo, J. W. (2000) Structure-function relationships in novel peptide dodecamers with broad-spectrum bactericidal and endotoxin-neutralizing activities. *Biochem. J.* **349 Pt 3**, 717–728.

Chapter 21

Assay Systems for Measurement of Anti-inflammatory Activity

Evelina Rubinchik and Christopher Pasetka

Abstract

It is widely accepted that cationic antimicrobial peptides possess potent microbicidal properties. Recent studies show that in addition to their antimicrobial action, these peptides can exhibit anti-inflammatory activity. The purpose of this chapter is to describe in vivo ear inflammation models that can be used for evaluating the anti-inflammatory activity of antimicrobial peptides. The models are based on different mechanisms of inflammation development and include irritant dermatitis (a model induced by a single application of 12-*o*-tetradecanoylphorbol acetate [TPA]) and allergic dermatitis, or delayed type hypersensitivity reaction (a model induced by repetitive application of oxazolone).

Key words: Antimicrobial peptides, irritant dermatitis model, allergic dermatitis model, TPA, oxazolone.

1. Introduction

It is widely accepted that cationic antimicrobial peptides possess potent microbicidal properties. Recent studies show that in addition to their antimicrobial action, these peptides can exhibit anti-inflammatory activity. For example, in vitro, antimicrobial peptides were shown to regulate genes associated with anti-inflammatory or pro-inflammatory mediators (1, 2) and reduce cytokine production (3–5). In vivo, the anti-inflammatory activity was also demonstrated in murine models of ear inflammation induced by TPA (6, 7) and oxazolone (7).

The purpose of this chapter is to describe two simple and reliable in vivo inflammation models that can be used for

A. Giuliani, A.C. Rinaldi (eds.), *Antimicrobial Peptides*, Methods in Molecular Biology 618,
DOI 10.1007/978-1-60761-594-1_21, © Springer Science+Business Media, LLC 2010

evaluating the anti-inflammatory activity of antimicrobial peptides. The first model is TPA-induced acute ear edema which is a model of irritant dermatitis (8). The mechanism by which TPA causes inflammation is related to increased vascular permeability mediated by the release of prostaglandins, leukotrienes, histamine, and serotonin (9). The main advantage of the TPA model is that it is fast, relatively easy to perform, and well characterized. One of the disadvantages is that inflammation develops very rapidly, thus only prophylactic treatment can be evaluated. Another disadvantage is that the model is known to generate false-positive results as agents that are not recognized as anti-inflammatory compounds in humans have shown activity in it (10). Therefore, the use of secondary models should be considered. One of the more popular models used is ear inflammation induced by oxazolone, which is an example of an allergic dermatitis model (9). The mechanism of action is a delayed hypersensitivity reaction that requires mice to be first sensitized by application of oxazolone and then challenged several days later by a second application of oxazolone. This second application results in the development of an inflammatory response.

2. Materials

2.1. Irritant Dermatitis Model

1. 95% ethanol, USP (Sigma). Store at room temperature.
2. 12-*o*-tetradecanoylphorbol acetate (TPA), (Sigma). Store at −20°C.
3. Dimethyl sulfoxide (DMSO), (Sigma). Store at room temperature.
4. Dexamethasone (Sigma). Store at 2–8°C.
5. Acetone (Sigma). Store at room temperature.
6. Stock TPA solution: 0.5% (w/v) TPA solution in DMSO (prepare in a glass vial). Distribute into 100–200 μL aliquots and store at −20°C for up to 6 months.
7. Final TPA solution (*see* **Note 1**): 0.005% TPA solution in acetone prepared by diluting 1 volume of the stock 0.5% TPA solution with 99 volumes of acetone (*see* **Note 2**). Keep at room temperature and use within 2 h of preparation.
8. Peptide solutions: dissolve the peptide to the desired concentrations (typically in the 0.1–5.0% (w/v) range) in 95% ethanol (*see* **Note 3**). Keep protected from light at room temperature and use on the day of preparation (unless stability is characterized and supports longer storage).

9. Positive control (dexamethasone) solution: 0.1% (w/v) dexamethasone solution in 95% ethanol.

2.2. Allergic Dermatitis Model

1. 4-Ethoxymethylene-2-phenyloxazone (oxazolone) (Sigma). Store at 2–8°C.
2. Acetone (Sigma). Store at room temperature.
3. Corn oil (Sigma). Store at room temperature.
4. 95% Ethanol, USP grade (Sigma). Store at room temperature.
5. Oxazolone induction solution: 3% (w/v) oxazolone solution prepared in a 5:1 (v/v) mixture of acetone:corn oil carrier. Use within 2 h of preparation (*see* **Note 2**).
6. Oxazolone challenge solution (*see* **Note 4**): 1% (w/v) oxazolone solution in acetone. Use within 2 h of preparation (*see* **Note 2**).
7. Peptide solutions: refer to Step 8 of **Section 2.1**.
8. Positive control (dexamethasone) solution: refer to Step 9 of **Section 2.1.**

2.3. Animals

CD-1 female mice that are 5–6 weeks of age (21–27 g) are used in these studies. All animals have free access to a standard certified pelleted diet and water.

2.4. Supplies

1. 8-mm dermal biopsy punch (Miltex Instrument Company, Inc.).
2. 2 or 4 mL glass vials (made of low extractable borosilicate glass) with screw-up caps (Wheaton, VWR).
3. 1.7 mL polypropylene microcentrifuge tubes, pre-weighed (VWR or Fisher Scientific).

3. Methods

3.1. Irritant Dermatitis Model

1. One day before the start of the experiment weigh the animals and evaluate their health status. Any animals in poor health or out of the specified weight range should not be assigned to experimental groups and must be replaced by spare animals obtained from the same batch and maintained under the same environmental conditions.
2. Randomize the animals into groups of 5–10 mice making sure that each group has the same mean body weight and house each group in a separate cage(s) labeled with the group identification.

3. To induce ear inflammation, apply a 20 μL volume of 0.005% (w/v) TPA solution as a single topical application on the dorsal area of the right ear of each animal using a 20 μL pipette. Spread the volume evenly over the whole surface of the ear.
4. Apply 20 μL of the peptide or dexamethasone (positive control) solutions topically to the dorsal area of the right ear immediately after TPA application (*see* **Notes 5** and 6). Spread the volume evenly over the whole surface of the ear. Treat the animals in the vehicle control group with 20 μL of 95% ethanol.
5. Leave the left ears untreated (no TPA or treatment applications) to serve as background controls.
6. At 6.0 h ± 10 min after administration of TPA, euthanize mice by CO_2 asphyxiation. Remove both the left and right ears and obtain an 8-mm biopsy from each ear using a dermal biopsy punch. Ensure that the biopsies are removed from the exact same area of the ear in all animals.
7. Place the biopsy into a pre-weighed microcentrifuge tube (*see* **Note** 7) labeled with the animal identification number and ear site (left or right). Determine biopsy weights and analyze results as described in **Section 3.3.** below.

3.2. Allergic Dermatitis Model

1. Evaluate and randomize the animals as described in Steps 1 and 2 of **Section 3.1.**
2. Clip the fur from each animal's abdomen the day before the experiment is to start, taking care not to abrade the skin.
3. On the day of the experiment (day 0), apply a 50 μL volume of 3% oxazolone induction solution to all mice as a single topical application on the shaved abdomen using an appropriately sized pipette. Spread the volume evenly over the application area.
4. On day 5, apply a 20 μL volume of 1% oxazolone challenge solution (*see* **Note 4**) as a single topical application on the dorsal area of the right ear using a 20 μL pipette. Spread the solution evenly over the whole surface of the ear.
5. One hour after the oxazolone challenge, apply 20 μL of the peptide solution, vehicle control or positive control (dexamethasone) solution topically to the dorsal area of the right ear (*see* **Notes 5** and **6**). Spread the volume evenly over the whole surface of the ear. The animals in the vehicle control group will receive 20 μL of the vehicle.
6. At 24 h ± 10 min after application of oxazolone, euthanize the animals by CO_2 asphyxiation. Remove both left and right ears and obtain 8-mm biopsies from each ear making

sure that the biopsies are collected from the exact same area of the ear in all animals.

7. Place the biopsy into a pre-weighed microcentrifuge tube (*see* **Note 7**) labeled with the animal number and biopsy location (left or right ear). Determine the biopsy weights and analyze the results as described in **Section 3.3**.

3.3. Data Analysis

1. Calculate the difference in biopsy weights between the right (treated) and left (untreated) ears and determine the percent of ear weight increase according to the formula:

$$\frac{[\text{Right ear weight (mg)} - \text{Left ear weight (mg)}]}{\text{Left ear weight (mg)}} \times 100\%$$

2. Present the data as mean ± standard deviation and perform a statistical analysis to compare biopsy data from the treated group with the vehicle control group.

3.4. Study Design and Experimental Strategy

1. It is a good idea to perform a model development study before the initiation of drug testing to ensure that the optimal concentrations of TPA and oxazolone are used (*see* **Notes 1** and **4**).
2. Use 5–6 animals per group in the exploratory studies (e.g., dose range-finding experiments) and 10 animals per group in the definitive studies as, even with the best techniques, the results can be variable.
3. Ensure that the drug effect is determined by comparing to the response in the vehicle-treated group and not to the untreated group as some formulation excipients may also exhibit inhibitory activity, especially in the TPA model (*see* **Note 3**). It is also a good idea to determine the vehicle effect in a preliminary experiment to ensure that it has a minimal or no effect on inflammation.
4. We also recommend testing activity of the peptide alone to ensure that it does not cause irritation on its own when used at high concentrations (see example in Step 5 of **Section 3.4** below).
5. An example of the evaluation of the anti-inflammatory activity of an antimicrobial peptide in the TPA and oxazolone models is shown in **Fig. 21.1**. The peptide 11J68CN is a novel, synthetic, antimicrobial peptide (indolicidin analog) comprised of 13 amino acid residues with the following primary sequence: W (Dab) W W (Dab) P (Dab) W (Dab) W P (Dab) W-amide, where W=tryptophan, Dab=2,4-diamino butyric acid, and P=proline. The peptide has demonstrated in vitro activity against a wide variety of Gram-positive and Gram-negative bacteria and fungi

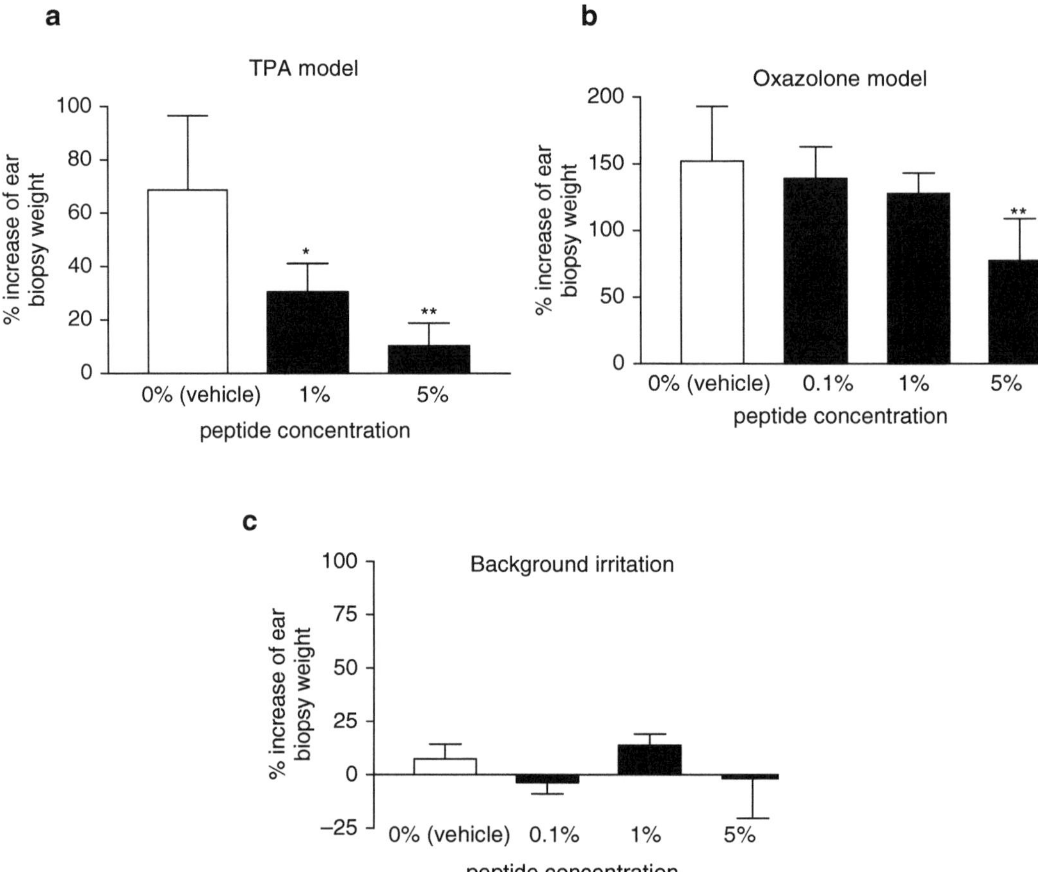

Fig. 21.1. Evaluation of anti-inflammatory activity of antimicrobial peptide 11J68CN. The peptide reduces ear edema induced by TPA (**a**) and oxazolone (**b**) and does not cause irritation when used alone (**c**). * – $P < 0.05$, ** – $P < 0.001$.

(11). In vivo, the peptide inhibited, in a dose-dependent manner, TPA-induced inflammation (up to 85% inhibition, **Fig. 21.1a**) and oxazolone-induced inflammation (up to 49% inhibition, **Fig. 21.1b**). In addition, the effect of peptide alone was evaluated with the results indicating that the peptide was not an irritant (application of vehicle or peptide solutions resulted in insignificant (±15%) changes in the ear biopsy weights, **Fig. 21.1c**).

4. Notes

1. Before starting peptide testing, we recommend confirming in a preliminary experiment that the 0.005% TPA concentration results in an optimal inflammatory response

as differences between mouse breeds and treatment application techniques may affect the outcome of edema development. The goal of such a preliminary experiment is to determine the TPA concentration that results in significant but manageable ear inflammation. A very high TPA dose may result in irreversible damage and ear necrosis (12). An example of such an experiment is shown in **Fig. 21.2**. In this study, a 20 μL volume of the vehicle solution (acetone) or 0.0005, 0.005, and 0.05% TPA solution was applied to separate groups of mice. Three mice from each group were sampled at 1, 3, 6, and 24 h after TPA administration. The results (**Fig. 21.2**) indicate that the 0.0005% TPA concentration was too low to induce inflammation, while the 0.05% TPA solution resulted in inflammation that was too significant (i.e., persisted for more than 24 h). In contrast, the 0.005% TPA solution appeared to be an optimal concentration as it produced considerable but controllable inflammation.

2. Acetone is a highly volatile vehicle that, if left in an open container, may evaporate to a significant degree thus increasing the concentration of the agent dissolved in it. For preparation of TPA and oxazolone solutions in acetone, use a glass vial with a tightly closing cap and a volume that is equal to or just slightly greater than the final solution volume. The vial must be kept tightly closed between applications in order to prevent evaporation of the acetone.
3. Ethanol is a typical vehicle of choice for treatment formulations as it helps with drug solubility and also enhances skin penetration. Moreover, ethanol alone has no significant effect on TPA- and oxazolone-induced ear inflammation. If other vehicles are to be used, their effect on inflammatory response must be determined in a preliminary experiment since high activities may be observed with some

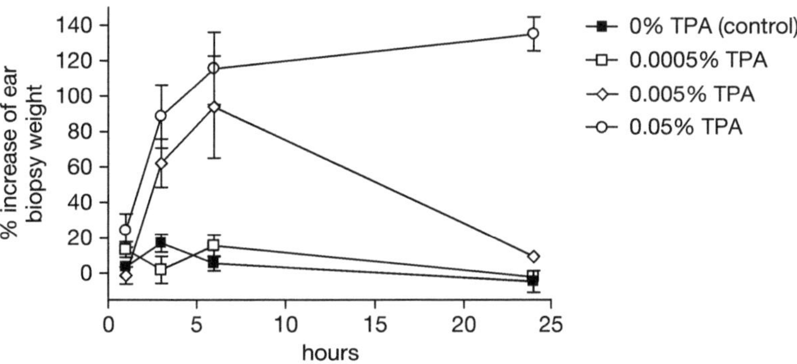

Fig. 21.2. Ear edema development induced by various concentrations of TPA.

formulations. Such vehicles must be avoided and alternative formulations should be used to ensure that a true drug effect is determined. The TPA model is especially sensitive with significant non-specific inhibitory responses occasionally observed.

4. Similarly to the TPA model, it is important to determine the optimal oxazolone concentration in a preliminary experiment. The goal of such a preliminary experiment is to determine the oxazolone concentration that results in significant ear inflammation but is not accompanied by the direct irritation that can be observed at high oxazolone concentrations. An example of such an experiment is shown in **Fig. 21.3**. In this study, the animals were induced with a 3% oxazolone solution and challenged with either 0.25, 0.5, 1, or 2% oxazolone solutions. A separate group of mice was induced with 0% oxazolone (vehicle only) and challenged with 0.25–2% oxazolone solutions to monitor for direct irritation. The effect was evaluated at 24 h post challenge. The results (**Fig. 21.3**) indicate that 1% oxazolone induced a significant inflammatory effect without being a direct irritant while irritation was observed with 2% oxazolone.

5. Peptide treatments can be applied at times that differ from those indicated in the methods above. In addition, peptides can be formulated in vehicles different from those recommended above. The following information should be taken into consideration when making a decision on the timing of treatment application and the type of formulation to be

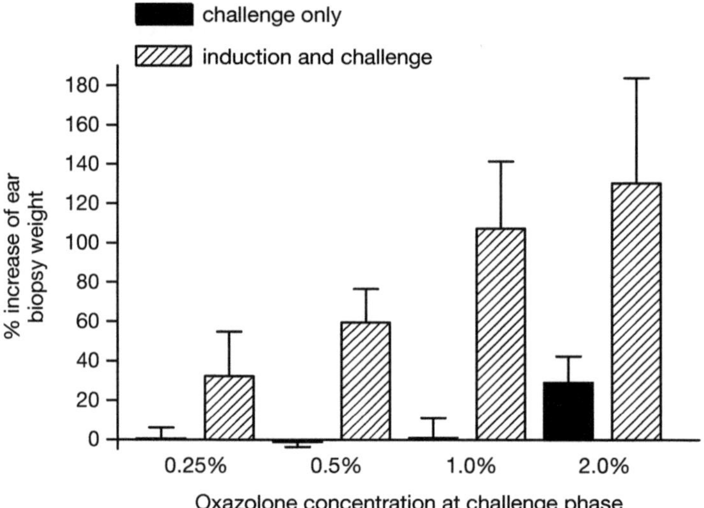

Fig. 21.3. Ear edema development induced by various concentrations of oxazolone at challenge stage.

used. The inflammatory reaction in the TPA model develops very rapidly, thus the treatments should be administered either immediately after (most typical approach) or no later than 15–30 min post induction. Such a rapid inflammatory response also requires rapid penetration of the applied treatment. In order to ensure significant skin penetration, the treatments are typically formulated in vehicles that enhance skin penetration such as acetone or ethanol. We recommend avoiding acetone since it is a more volatile vehicle than ethanol. Since inflammation takes longer to develop in the oxazolone model, a delay in the treatment application is possible. We usually apply treatment at 1 h post oxazolone application but we also employed treatments at 3 and/or 5 h post challenge. Regarding formulation, both pure solvents (usually 95% ethanol) and complex topical formulations can be employed.

6. Treatment with 0.1% (w/v) dexamethasone serves as a positive control and is expected to result in significant inhibition of TPA- and oxazolone-induced ear edema. In our hands, we typically observed a statistically significant inhibition of ≥90% for the TPA model and 60–90% for the oxazolone model. If dexamethasone does not show a significant inhibitory effect, this usually indicates that the inflammation is too severe to be abrogated and the model conditions should be adjusted.
7. The weight of ear biopsies may be affected by the leakage of the extracellular exudate from the inflamed tissues. Our preliminary experiments indicate that while control (untreated) biopsy weights remain stable over 2 h, the biopsies from inflamed tissues lose up to 9% of their weight. Therefore, we recommend placing biopsies in pre-weighed eppendorf tubes immediately after collection and weighing them together with any exudate to ensure accurate measurements.

References

1. Scott, M. G., Rosenberger, C. M., Gold, M. R., Finlay, B. B., and Hancock, R. E. (2000) An alpha-helical cationic peptide selectively modulates macrophage responses to lipopolysaccharide and directly alters macrophage gene expression. *J. Immunol.* **165**, 3358–3365.
2. Scott, M. G., Davidson, D. J., Gold, M. R., Bowdish, D., and Hancock, R. E. (2002) The human antimicrobial peptide LL-37 is a multifunctional modulator of innate immune responses. *J. Immunol.* **169**, 3883–3891.
3. McInturff, J. E., Wang, S. J., Machleidt, T., Lin, T. R., Oren, A., Hertz, C. J., Krutzik, S. R., Hart, S., Zeh, K., Anderson, D. H., Gallo, R. L., Modlin, R. L., and Kim, J. (2005) Granulysin-derived peptides demonstrate antimicrobial and anti-inflammatory effects against *Propionibacterium acnes*. *J. Invest. Dermatol.* **125**, 256–263.
4. Masera, R. G., Bateman, A., Muscettola, M., Solomon, S., and Angeli, A. (1996) Corticostatins/defensins inhibit in vitro NK activity and cytokine production by human

peripheral blood mononuclear cells. *Regul. Pept.* **62**, 13–21.
5. Guarna, M. M., Coulson, R., and Rubinchik, E. (2006) Anti-inflammatory activity of cationic peptides: application to the treatment of acne vulgaris. *FEMS Microbiol. Lett.* **257**, 1–6.
6. Scott, M. G., Lee, K., Maclean, S., Gong, J., Finlay, B. B., and Hancock, R. E. W. (2003) Immunomodulatory peptides that protect in murine models of infection. *Abstr 43rd Intersci. Conf. Antimicrob. Agents Chemother.* **43**, 62.
7. McNicol, P. J., Rubinchik, E., Pasetka, C., Barger, R., Lopez, J., and Friedland, H. D. (2001) MBI 594AN is active against *Acne vulgaris* organisms, does not induce resistance and has anti-inflammatory activity in vivo. *Abstr 41st Intersci. Conf. Antimicrob. Agents Chemother.* **41**, 205.
8. Rao, T. S., Currie, J. L., Shaffer, A. F., and Isakson, P. C. (1993) Comparative evaluation of arachidonic acid (AA)- and tetradecanoylphorbol acetate (TPA)-induced dermal inflammation *Inflammation* **17**, 723–741.
9. Tramposch, K. M. (1999) Skin inflammation. In *In vivo models of inflammation*. Morgan, D. W., Marshall, L. A. (Eds.), pp. 179–204. Basel, Switzerland: Birkhauser Verlag.
10. Carlson R. P., O'Neill-Davis, L., Chang, J., and Lewis, A. J. (1985) Modulation of mouse ear edema by cyclooxygenase and lipoxygenase inhibitors and other pharmacologic agents *Agents Actions* **17**, 197–204.
11. Krieger, T. J., and McNicol, P. J. (2004) Antimicrobial cationic peptides and formulations thereof. Patent number US 6,835, 536 B2.
12. Young, J. M., Wagner, B. M., and Spires, D. A. (1983) Tachyphylaxis in 12-o-tetradecanoylphorbol acetate- and arachidonic acid-induced ear edema *J. Invest. Dermatol.* **80**, 48–52.

Chapter 22

Ex Vivo Skin Infection Model

Evelina Rubinchik and Christopher Pasetka

Abstract

Over the past decade, the emergence of bacterial strains resistant to conventional antibiotics has necessitated the discovery and development of new antimicrobial therapies. This chapter describes a skin infection model that is based on the use of excised skin derived from the domestic pig. The model conditions mimic the environment of human skin and efficiently support the growth of clinically relevant bacterial and fungal species, thus making it useful for evaluating the activity of antimicrobial peptides and other antibiotics as well as their respective formulations.

Key words: Antimicrobial peptides, skin infection model, Gram-positive bacteria, Gram-negative bacteria, fungi.

1. Introduction

Several different in vivo skin infection models have been described in the literature (1–3); however, none have entered into widespread use. It is well known that the skin of rodents, i.e., species typically employed in such studies, differs significantly from human skin and offers a poor support for the growth of clinically relevant human pathogens. Treatment applications are also challenging and require occlusion to prevent animals from licking and grooming the infected site.

The purpose of this chapter is to describe a simple and reliable ex vivo skin infection model that significantly reduces the use of experimental animals while allowing for the simple and reproducible evaluation of a compound's antimicrobial activity. The model is based on the use of excised skin derived from the domestic pig. The skin pieces are disinfected and inoculated with

A. Giuliani, A.C. Rinaldi (eds.), *Antimicrobial Peptides*, Methods in Molecular Biology 618,
DOI 10.1007/978-1-60761-594-1_22, © Springer Science+Business Media, LLC 2010

human pathogens. The treatment is then applied and the number of colony forming units (CFU) per skin piece is determined after incubation. This model provides several advantages. First, the morphological and functional characteristics of pig skin are comparable to those of human skin (4). Second, large quantities of pig skin can be easily obtained and kept frozen for long periods of time, thus allowing for flexibility in experiment initiation. Third, the skin can be inoculated with various human pathogens under conditions (temperature, humidity) that mimic human skin colonization and promote organism survival and growth. Finally, the model allows for accurate quantitative determination of the antimicrobial or antifungal activity of both antimicrobial peptides and their topical formulations.

2. Materials

2.1. Pig Skin

Pig skin obtained from female or male domestic pigs (*see* **Note 1**).

2.2. Microorganisms

Gram-positive and Gram-negative bacteria and fungi obtained from ATCC (methicillin-sensitive *Staphylococcus aureus* ATCC 14154; *Escherichia coli* ATCC 25922; *Enterobacter cloacae* ATCC 13047; *Serratia marcescens* ATCC 13880, *Candida albicans* ATCC 14053) or from clinical laboratories (methicillin-resistant *S. aureus, S. epidermidis,* and *Propionibacterium acnes*). The organisms listed above were successfully used in this model in our laboratory but other organisms and/or strains can also be employed. Store vials with bacterial and fungal cultures at –70°C. Do not thaw.

2.3. Microbiology Supplies

Microbiology supplies can be purchased from any general scientific supply company such as Sigma, Fisher, or VWR.

1. 5% sheep blood agar plates; trypticase soy agar (TSA) plates for culture of *S. aureus, S. epidermidis, S. marcescens,* and *C. albicans*; McConkey agar plates for culture of *E. coli* and *E. cloacae;* Wilkins–Chalgren agar plates for culture of *P. acnes.* Store plates inverted at 4°C (*see* **Notes 2 and 3**).
2. Sterile Petri plates (15 × 100 mm, polystyrene).
3. Inoculating loops (sterile, single-use plastic loops or reusable metal loop).
4. Bacterial cell spreaders (25.4 mm spreader head, metal).
5. Inoculating turntable.
6. 0.5 McFarland turbidity standard (16 × 125 mm glass tube). Store at 2–30°C, protected from light.

7. Sterile cotton swabs (152.4 mm, wood).
8. Glass culture tubes (16 × 125 mm, borosilicate glass).
9. Plastic tubes (12 × 75 mm, 5 mL, snap-cap, polystyrene).
10. Anaerobic jar and Gas Pak® kit with catalyst (BBL Gas-Pak system).

2.4. Reagents

1. 70% ethanol, denatured (VWR). Store at room temperature.
2. NaCl. Store at room temperature.
3. Isotonic saline solution: 0.9% NaCl solution in Milli-Q or deionized water. Sterilize by filtration through 0.2 μm filter. Store at room temperature for up to 1 month.
4. Polyanetholesulfonic acid (PSA), (Sigma). Store at room temperature.
5. PSA solution: 1% (w/v) PSA solution in isotonic saline. Sterilize by filtration through 0.2 μm filter. Prepare the solution fresh on the day of the experiment.
6. Peptide formulations: dissolve the peptides to the desired concentration (typically in the 0.1–5.0% (w/v) range) in 0.9% saline. Adjust pH to 4–8 to avoid extreme values. Keep protected from light at room temperature and use on the day of preparation (unless stability is characterized and supports longer storage). In addition to saline, more complex formulations such as creams, gels, or ointments can be employed, allowing for a more accurate simulation of the treatment conditions in the clinical setting.

3. Methods

3.1. Preparation of Pig Skin Samples

1. Immediately after animal sacrifice, clip the pig skin free of hair and swab thoroughly with 70% ethanol (*see* **Note 4**).
2. Surgically remove the required number of skin sections (approximately 20 × 20 cm each). Store the sections at –20°C in re-sealable plastic bags, pre-swabbed with 70% ethanol. Expel the air from the bags to reduce skin dehydration.
3. On the day of the experiment, remove any remaining fat layers from the frozen skin sections with a scalpel and cut the skin into 2 × 2 cm pieces using a sterile metal ruler as a measuring guide.
4. Sterilize the thawed skin pieces by immersing them in 70% alcohol (*see* **Note 4**).

5. Pat dry the skin pieces with sterile (autoclaved) paper towels and distribute into experimental groups. Place the pieces from the same treatment group into sterile Petri plates (up to 5 pieces/plate) labeled with group identification numbers (*see* **Note 5**).

3.2. Preparation of Bacterial Inoculum

1. For plate inoculation, use either disposable plastic loops or a reusable metal inoculating loop. If a metal loop is used, sterilize it by holding it in a gas or ethanol burner flame until it is red hot. Cool the metal loop by touching it to a sterile portion on the surface of an agar plate until it stops sizzling.
2. Remove the tube with the frozen bacterial or fungal culture from the freezer, open the cap, and scrape the frozen surface with the inoculation loop. Do not allow the contents of the vial to thaw.
3. Streak bacteria or fungi across one third of a 5% sheep blood agar plate. Using a new plastic loop or sterilized metal loop, pass the loop through the first streak and streak across a new third of the plate. Repeat one more time to inoculate the remaining third of the plate.
4. Incubate the plates (inoculated surface facing down) with aerobic organisms (e.g., *S. aureus*) for 24–48 h at 35±2°C or until colonies become visible. Incubate the plates with anaerobic organisms (e.g., *P. acnes*) for 48–72 h at 35±2°C in an anaerobic environment i.e., in a securely closed anaerobic jar with a Gas Pak® CO_2 generator. Ensure that a palladium catalyst has been attached to the underside of the anaerobic jar lid prior to sealing to remove residual oxygen.
5. At the end of the incubation period, examine the plate to ensure that no contaminating colonies are present. Collect several isolated colonies using a sterile inoculation loop, suspend them in 5 mL of sterile saline in a sterile glass culture tube, and vortex well.
6. Adjust the concentration of the suspension to 10^8 colony forming units (CFU) per mL by comparing the turbidity to that of a 0.5 McFarland standard (*see* **Note 6**). Ensure that both the McFarland standard and bacterial or fungal suspensions are well suspended prior to visual comparison.
7. Dilute the bacterial or fungal suspension with sterile saline to obtain a final inoculum of 10^3–10^7 CFU/mL (*see* **Note 7**) and use this inoculum immediately after preparation to infect the skin pieces as described in **Section 3.3**.
8. Verify the concentration of the final inoculum by preparing one-log (10-fold) serial dilutions of the inoculum. The pro-

cedure described below is based on a 10^7 CFU/mL final inoculum size but it can be easily adjusted for other inoculum sizes. The dilutions of the final inoculum are prepared by mixing 100 μL of the 10^7 CFU/mL bacterial or fungal suspension into 900 μL of sterile saline (10^{-1} dilution). Following vortexing, a 100 μL aliquot of the 10^{-1} dilution is transferred into 900 μL of sterile saline to yield a 10^{-2} dilution. The process is repeated until dilutions ranging from 10^{-1} to 10^{-5} have been prepared. Ensure that each suspension is mixed well by vortexing immediately prior to the removal of an aliquot for dilution.

9. Pipet 100 μL of each dilution onto a dry, pre-labeled agar plate. Sterilize a metal spreader by dipping the triangular portion into a container with 70% ethanol and passing it through a burner flame to ignite the ethanol and allowing the flame to burn out. Cool the spreader by touching it to the surface of an agar plate (without bacterial cells). With the inoculated plate on the turntable, spin the turntable and spread the 100 μL bacterial aliquot over the entire agar surface.
10. After application, allow plates to dry and then incubate as described in Step 4 of **Section 3.2**. During incubation, each bacterial cell will grow into a separate CFU.
11. Determine the final inoculum size from the number of CFUs observed on the diluted plates. Specifically, select the plates that show 10 to 200 well-separated colonies per plate and count all colonies on the plate. This is the number of live organisms that is present in a 100 μL aliquot. By multiplying this number by 10, the number of CFUs per 1 mL can be determined. This CFU/mL value is then multiplied by the corresponding dilution factor to determine the final inoculum concentration. For example, if 12 colonies were observed on the plate corresponding to the 10^{-5}-fold dilution, then the number of bacteria in 1 mL of the final inoculum (undiluted) is $12 \times 10 \times 10^5 = 1.2 \times 10^7$ CFU/mL.

3.3. Skin Inoculation and Treatment

1. Perform all procedures under aseptic conditions. Following distribution of the skin pieces into each treatment group (described in Step 5 of **Section 3.1**), apply a 50 μL aliquot of the bacterial inoculum (10^3–10^7 CFU/mL, *see* **Note 7**) to the dry epidermal side of the skin and spread uniformly over the entire area using a sterile pipet tip.
2. Replace the plate lid and incubate the plate with inoculated skin pieces at 35±2°C in a humidified chamber (i.e., a plastic container humidified by placing paper towels soaked in water

on the bottom) until the time of treatment initiation. For *P. acnes*, perform the incubation under anaerobic conditions.

3. Depending on the goal of the experiment, the treatment can be initiated either immediately or with a 1–8 h delay postinoculation. We recommend delaying the treatment initiation (e.g., for 2 h) to allow the bacteria to establish an infection.
4. Apply 25–100 μL of the peptide formulation onto the infected side of each skin piece and spread uniformly over the entire area using a sterile pipet tip (*see* **Notes 8 and 9)**
5. Treat the samples from the vehicle group with an identical volume (25–100 μL) of the vehicle formulation. Leave the skin pieces from the control group without treatment.
6. Incubate the Petri plates containing the skin pieces in a humidified chamber for up to 24 h at 35±2°C. For *P. acnes*, perform the incubation under anaerobic conditions.

3.4. Determination of Bacterial Counts

1. Following the desired incubation period (1–24 h), remove the skin samples from the humidified chamber.
2. Label sterile tubes with sample identification numbers and add 1 mL of sampling solution (0.9% NaCl supplemented with 1% PSA). Other concentrations of PSA can be selected as described in **Section 3.5**.
3. To collect bacteria or fungi from a skin piece, dip a sterile cotton swab into 1 mL of sampling solution and swab the pig skin surface 5–10 times, making sure that the whole area is covered by swabbing. Repeat the swabbing procedure three times in total.
4. After the final swab, vortex the tube holding the cotton tip in the solution. Squeeze the solution out of the cotton tip into the sampling tube and discard the swab.
5. Prepare one-log (10-fold) serial dilutions of the sampling solution in 0.9% saline as described in Step 8 of **Section 3.2**.
6. Plate 100 μL of each dilution onto a pre-labeled agar plate as described in Steps 9 and 10 of **Section 3.2** (*see* **Note 10**).
7. Incubate the plates at 35±2°C for 24 h (or until colonies are clearly visible). Incubate plates inoculated with *P. acnes* for 48 to 72 h at 35±2°C in an anaerobic environment. Quantify the number of CFUs as described in Step 11 of **Section 3.2**. Assign the low level of detection value (10 CFU/site) to a sample in the event that no viable colonies are detected on the plates.
8. Calculate the CFUs and $\log_{10}$ CFU/skin piece and determine $\log_{10}$ CFU/site ± standard deviation (SD) for each group.

3.5. Inhibition of Residual Antimicrobial Activity

When performing swabbing of the skin to collect bacteria, the antimicrobial peptide may also be transferred into the sampling solution. This could result in false-positive activity because of a peptide carry-over effect (not associated with activity on the skin surface). Therefore, any residual antimicrobial activity must be abrogated by the addition of an appropriate inhibitor to the sampling solution (5). The selection of inhibitor for antimicrobial peptides is based on the fact that they are cationic and will therefore interact with negatively charged molecules (e.g., PSA) resulting in precipitation, rendering the peptide inactive.

1. Apply the antimicrobial peptide (at the highest concentration to be used in the study) to 12 pig skin pieces as described in Step 4 of **Section 3.3** and incubate for the desired amount of time.
2. At the end of incubation, swab skin pieces into 1 mL of sterile saline containing different concentrations of the selected inhibitor (e.g., 0.01, 0.1, and 1% (w/v) PSA). Use three skin pieces per group. Swab the remaining three pieces into saline without inhibitor.
3. Prepare a bacterial or fungal lawn by spreading 100 μL of a 10^7–10^8 CFU/mL inoculum over the agar plate (employ the organism that you intend to use in your experiments). Allow plates to dry.
4. Apply a 10–20 μL aliquot of the swabbed sampling solutions (peptide + inhibitor) onto the bacterial lawn at the pre-labeled sites of the plate. In addition, apply the peptide solution without inhibitor and the inhibitor solution without the peptide.
5. Incubate the plate for 18–24 h at 35±2°C and examine for inhibition of bacterial growth at the site of application. A clear (no bacterial growth) or less dense area will be observed at the site of peptide application in the event of a carry-over. Select the concentration of the inhibitor that results in complete inhibition of the carry-over.
6. Ensure that the inhibitor itself has no effect on bacterial growth. A less dense or a completely clear area observed at the site of inhibitor application would suggest a negative effect on bacterial growth. The use of such inhibitors should be rejected.

3.6. Study Design and Experimental Strategy

1. Use 5–6 skin pieces per group in exploratory studies (e.g., dose range-finding experiments) and 10–12 pieces per group in confirmatory studies that are powered for statistical significance.
2. Ensure that the effect of the vehicle alone on bacterial growth is evaluated, as some formulation excipients may also exhibit antimicrobial activity.

3. An example of a study design that evaluated the dose–response of an antimicrobial peptide is shown in **Table 22.1**. Omiganan (formerly MBI 594AN or MBI 226), an antimicrobial peptide currently in clinical development, was used in this study. Skin obtained from antibiotic-free pigs was prepared as described above, inoculated with *P. acnes* at 2.5×10^6 CFU/skin, and treated 2 h later with 50 μL of 0.1–5% omiganan solutions or vehicle alone (hydro-alcoholic formulation). The control group remained untreated. Following a 24-h incubation at 35±2°C in a humidified, anaerobic chamber, the bacterial counts were determined using the swabbing solution supplemented with the peptide inhibitor (1% PSA). Our preliminary experiments indicated that this concentration was sufficient to inhibit all residual peptide activity without causing any negative effect on bacterial growth. This model design resulted in characterization of the dose–response activity of omiganan and also showed that the formulation alone has a small inhibitory effect on bacterial counts (due to the presence of ethanol). The potent antimicrobial activity that omiganan demonstrated in this *P. acnes* infection model correlated well with the therapeutic activity of omiganan observed in the phase II clinical trial against acne vulgaris (6).
4. An example of the time–response study design is shown in **Table 22.2**. In this experiment the cationic peptide omiganan was prepared in a 1% gel formulation and applied (100 μL/site) to pig skin pieces infected with *S. epidermidis* (50 μL of 2×10^6 CFU/mL). Preliminary

Table 22.1
Dose–response activity of antimicrobial peptide omiganan against *P. acnes* in pig skin infection model

Treatment	Mean $\log_{10}$ CFU/skin ± SD at 24 h posttreatment	Log_{10} CFU/skin reduction vs. control
No treatment (control)	6.26 ± 0.06*	
Vehicle solution	5.16 ± 0.37	–1.10
0.1% omiganan solution	4.12 ± 0.18*	–2.14
0.25% omiganan solution	4.06 ± 0.07*	–2.20
0.75% omiganan solution	3.19 ± 0.56*	–3.07
1.25% omiganan solution	3.35 ± 0.58*	–2.91
2.5% omiganan solution	2.51 ± 0.34*	–3.75
5.0% omiganan solution	2.40 ± 0.77*	–3.86

* $p < 0.001$ vs. vehicle solution.

Table 22.2
Time–response activity of antimicrobial peptide omiganan against *S. epidermidis* in pig skin infection model

Treatment	Inoculum (log_{10} CFU/site) 0 h	Log_{10} CFU/site mean ± SD			
		1 h	4 h	8 h	24 h
Vehicle gel	4.44	4.37±0.57	4.50±0.37	5.21±0.09	8.33±0.27
Omiganan 1.0% gel		1.64±0.69*	1.71±0.66*	1.56±0.77*	3.14±1.24*

*$p < 0.001$ vs. vehicle gel.

experiments showed that the vehicle gel had no effect on bacterial growth (data not shown). Pig skin pieces were swabbed at 1, 4, 8, and 24 h posttreatment into saline solution containing 1% PSA. This model design allowed for the demonstration that the antibacterial effect of 1% omiganan was very rapid, with significant killing observed at 1 h posttreatment (2.7 log_{10} CFU/site reduction). At 24 h, a 5.2 log_{10} reduction was observed as compared to the control sites treated with vehicle alone. When 1% omiganan gel was evaluated in human volunteers, a ≥3 log_{10} reduction in skin microbial colonization was observed at different time points (7), again indicating a good correlation of the model results with clinical outcomes.

4. Notes

1. Pig skin can be obtained from laboratory pigs or from slaughter houses. The animals should be free from any antibiotic treatment and should not have consumed any antibiotic-supplemented diet for at least 7 days prior to skin sampling. Our preliminary experiments demonstrated that the presence of residual antibiotics in pig skin may result in partial or significant inhibition of bacterial growth. If antibiotic-free animals are impossible to obtain or if information on antibiotic consumption is not available, then the skin should be pretested for its ability to support bacterial growth.
2. Ensure that the plate surface does not have excessive condensation. If the agar surface is even slightly wet, the streaked bacteria may float away and the colonies may not

be well separated. The plates can be dried by leaving the lids off in a laminar flow hood.

3. Sheep blood agar plates support the growth of all microorganisms and should be used for initial bacterial culture (from stock) to ensure the absence of contamination with other organisms. When growing organisms swabbed from the skin pieces, use specific agar plates (such as TSA, McConkey, or Wilkins–Chalgren agar) that preferably support the growth of the selected organism and thus reducing the possibility of contamination (and sources of error) if the skin was not thoroughly sterilized.
4. Use of disinfectants other than alcohol (for example, iodine, chlorhexidine) is not recommended as they may leave residual amounts on the skin surface, resulting in the inhibition of bacterial growth.
5. If skin collected from several pigs is to be used in one study, skin pieces derived from different animals must be evenly distributed between groups to remove any potential bias related to differences in the skin sources.
6. In addition to the McFarland standard, a spectrophotometer can be used for the adjustment of the bacterial culture in accordance with the manufacturer's instructions (for example, Spectronic spectrophotometer, model 20D+, Milton Roy).
7. Depending on the goals of the experiment, the final inoculum may vary from 10^3 CFU/mL (mild infection) to 10^7 CFU/mL (severe infection). Many organisms will grow rapidly resulting in significant skin infection at 24 h postinoculation, regardless of the initial inoculum size. The exceptions are some very slow growing organisms (e.g., *P. acnes*) that will show no increase or only a small increase over a 24 h period. An example of typical bacterial growth following infection with low and high inocula is shown in **Table 22.3**.

Table 22.3
Bacterial growth following skin infection with low and high inocula of *S. aureus* ATCC 19636

Applied at time 0 (log_{10} CFU/site)	*n*	Recovered at 24 h (log_{10} CFU/site)	
		Mean ± SD	Range
2.70	12	7.83±0.37	7.10–8.47
6.70	12	8.24±0.32	7.77–8.89

8. For screening purposes, cationic peptides can be formulated in a simple vehicle that is known to have no effect on bacterial growth (e.g., sterile saline). If more complex formulations (such as gels, creams, or ointments) are used, their effect on bacterial growth should be evaluated. See an example of the effect of the ethanol-containing vehicle on bacterial counts (**Table 22.1**).
9. Smaller volumes (25–50 μL) are applied if liquid formulations are used. Larger volumes (100 μL) are used when more viscous formulations are employed (gels, creams, or ointments). The general rule is to select a volume that can be evenly spread over the whole skin area as one thin layer. Excessive treatment volumes that result in dripping outside or pooling on the skin must be avoided.
10. A spiral plater (Spiral Biotech) can be used instead of manual plating resulting in significant savings with respect to time and supplies (plates, tubes).

References

1. Ray, T.L. (1985) Animal models of experimental Candida infection of the skin. In *Models in Dermatology*. Maibach, H.I., Lowe, N.J. (Eds.), Vol. 1, pp. 41–50. Basel: S. Karger AG.
2. Jordan, W.E. (1970) The experimental induction of superficial cutaneous infections in guinea pigs *J. Invest. Derm.* **55**, 149–152.
3. Sedlock, D.M. (1985) Animal skin degerming models. In *Models in Dermatology*. Maibach, H.I., Lowe, N.J. (Eds.), Vol. 2, pp. 131–144. Basel: S. Karger AG.
4. Meyer, W., Schwarz, R., and Neurand, K. (1978) The skin of domestic mammals as a model for the human skin, with special reference to the domestic pig *Curr. Probl. Dermatol.* 7, 39–52.
5. Boon, R. J., Beale, A. S., and Sutherland R. (1985) Efficacy of topical mupirocin against an experimental *Staphylococcus aureus* surgical wound infection. *J. Antimicrob. Chemother.* **16**, 519–526.
6. Friedland, H.D., Sharp, D.D., and Robinson, J.R. (2003) Double-blind, randomized, vehicle-controlled study to assess the safety and efficacy of MBI 594AN in the treatment of acne vulgaris. *Abstr. 61st Ann. Meeting Am. Acad. Derm.* **61**, 22.
7. Isaacson R. E. (2003) MBI-226 Micrologix/Fujisawa *Curr. Opin. Invest. Drugs* **4**, 999–1003.

Chapter 23

Measuring Antimicrobial Peptide Activity on Epithelial Surfaces in Cell Culture

Gill Diamond, Sunghan Yim, Isaura Rigo, and Laura McMahon

Abstract

To more accurately assess the activity and role of epithelial cell-derived antimicrobial peptides in their native settings, it is essential to perform assays at the surfaces under relevant conditions. In order to carry this out, we utilize three-dimensional cultures of airway and gingival epithelium, which are grown at an air–liquid interface. Under these conditions, the cultures can be subjected to challenge with a variety of factors known to cause an increase in antimicrobial peptide gene expression. The functional relevance of this induction can then be assessed by quantifying antibacterial activity either directly on the surface of the cells or using the fluid secreted onto the apical surface of the cultures. The relative contribution of the peptides can also be measured by pre-incubation of the secreted fluid with specific inhibitory antibodies. Thus, a relatively inexpensive in vitro model can be used to evaluate the role of antimicrobial peptides in mucosal epithelium.

Key words: Airway, oral epithelium, vitamin D, cathelicidin.

1. Introduction

In general, most antimicrobial peptides have been identified by protein purification methods from tissues, followed by in vitro assays to identify the active fraction (see (1–3), for example). Alternatively, DNA sequences coding for homologous peptides are identified through computational or experimental methods (e.g., (4, 5)). Subsequently, the purified natural peptides or synthetic peptides based on the DNA sequences are characterized by in vitro antimicrobial assays against target organisms, which provide important information regarding the potential for the

A. Giuliani, A.C. Rinaldi (eds.), *Antimicrobial Peptides*, Methods in Molecular Biology 618,
DOI 10.1007/978-1-60761-594-1_23, © Springer Science+Business Media, LLC 2010

peptide's use as a therapeutic or give clues to its role in innate defense. Furthermore, the expression of many host defense peptide genes is induced by a variety of factors both in cell culture (e.g., (6, 7)) and in vivo (8, 9), supporting their role in innate host defense of mucosal epithelia such as the airway, the oral cavity, and the intestinal tract. However, to accurately assess the relationship between the induction of a peptide gene (as measured by mRNA levels) and an increase in antibacterial activity in a tissue is difficult.

Recently, novel methods for cell culture which provide three-dimensional model systems that recapitulate the mucosal epithelium for different tissues have been described (10, 11). The systems rely on culturing epithelial cells in an air–liquid interface (ALI) in serum-free medium, leading to well-differentiated cultures, often covered with a cell-derived secretion. This provides an environment that allows for quantification of antimicrobial activity in mucosal secretions under conditions that are more representative of the original tissue and allows the investigator to modulate antimicrobial peptide gene expression with exogenous factors. Assay conditions can then be developed to determine the role of endogenous antimicrobial peptides on mucosal surfaces.

2. Materials

2.1. Airway Epithelial Cell Culture

1. Bronchial epithelial growth media basal medium without growth factors, cytokines, or supplements (BEBM) (Lonza, Walkersville, MD).
2. Bronchial epithelial growth media (BEGM) SingleQuots Kit supplement and growth factors added directly to medium. For 500 mL medium, add bovine pituitary extract (BPE) 2 mL, insulin 0.5 mL, hydrocortisol 0.5 mL, gentamicin sulfate and amphotericin-B 0.5 mL, retinoic acid 0.5 mL, transferrin 0.5 mL, triiodothyronine 0.5 mL, epinephrine 0.5 mL, and hEGF 0.5 mL (Lonza, Walkersville, MD). Antibiotics may be added, but must be removed prior to antimicrobial assays (*see* **Note 1**).
3. Costar Transwell permeable supports 12-mm insert, 12-well plate.
4. Collagen from human placenta, type VI. A total of 10 mg is dissolved in 20 mL of dH_2O and 40 μL of concentrated acetic acid is added. The collagen is then incubated at 37°C for 15–30 min for the collagen to fully dissolve. The stock solution is diluted 1:10 with dH_2O to coat the Transwell inserts.

5. Normal human bronchial epithelial (NHBE) (Lonza, Walkersville, MD). Other primary cultures may be used (adapted from 11).

2.2. Gingival Epithelial Cell Culture

1. Dulbecco's Modified Eagle's Medium with glucose and L-glutamine, supplemented with 10% bovine serum and penicillin–streptomycin.
2. Collagen, type I (rat tail) at 1.1 mg/mL in water with 43 μL concentrated acetic acid (33%) per 5 mL of solution is incubated at 37°C for 30 min to dissolve the collagen.
3. 10X DMEM (without sodium bicarbonate) powder is prepared in water at 13.48%, filter sterilized, and aliquots are stored at –20°C.
4. 10X reconstitution buffer: 22 mg/mL sodium bicarbonate, 20 mM HEPES, 0.62 N NaOH. Aliquots are stored at –20°C.
5. Keratinocyte serum-free medium (KSFM) supplemented with L-glutamine. Calcium chloride is added to 0.03 M. Bovine pituitary extract and epithelial growth factor are supplied with the medium and are added as per the manufacturer's instructions.
6. Costar Transwell permeable supports: 24-mm insert, polyester membrane, 6-well polystyrene plate.
7. OKF6/TERT oral keratinocyte cells are obtained with material transfer agreement from the laboratory of Dr. James Rhinewald, Harvard University (adapted from 12).

2.3. Antimicrobial Assays

1. 10X phosphate-buffered saline solution.
2. 1,25-Dihydroxyvitamin D_3 10 μg is made to 10^{-5} M concentration by dissolving it in 100% ethanol. Vitamin D is added to the BEGM medium to make a final concentration of 10^{-8} M to induce the expression of LL-37. Ethanol is used as a control.
3. LB broth Miller used to grow *Pseudomonas aeruginosa* in liquid culture and agar plates.
4. Bordet–Gengou agar used to grow *Bordetella bronchiseptica* on agar plates.
5. Stainer–Scholte medium, used to grow *B. bronchiseptica* in liquid culture.
6. AAGM (30 g/L of trypticase soy broth or 40 g/L of trypticase soy agar, 6 g/L yeast plus 0.75% dextrose [filter sterilized], and 0.4% sodium bicarbonate [filter sterilized] added after autoclaving) used to grow *Aggregatibacter actinomycetemcomitans* in liquid culture and agar plates.

3. Methods

Beta-defensins and cathelicidins are antimicrobial peptides expressed in mucosal epithelial cells (reviewed in (13, 14)). Their expression is induced in response to a variety of agents including bacterial lipopolysaccharide (LPS), interleukin (IL)-1β, and the active form of vitamin D, 1,25$(OH)_2$ D_3 (reviewed in (15, 16)).

To assess the activity of these peptides in airway epithelial cells, primary cultures of bronchial epithelial cells are grown in an air–liquid interface and are allowed to mature and differentiate for 20 days before any experiments are performed. The bronchial epithelial cells are then basolaterally treated with an inducing agent, such as IL-1β (100 ng/mL) or vitamin D at a concentration of 10^{-8} M. The airway surface fluid (ASF) is then collected by washing the cells with 50 μL of filter-sterilized 1X PBS. As a control for the poorly water soluble 1,25$(OH)_2$ D_3, control cells are treated with an equal volume of ethanol. The effect of the inducing agents on the bactericidal activity of ASF is studied using airway pathogens such as *B. bronchiseptica* or *P. aeruginosa*. To determine whether specific peptides are responsible for all or part of the activity, the ASF can be pretreated with inhibitory antibodies prior to incubation with bacteria. A relative increase in the number of colonies relative to control (pretreated with nonspecific IgG) suggests a role for that peptide in killing.

To quantify the activity of peptides on the surface of gingival epithelial cells, a three-dimensional culture system is used whereby epithelial cells are cultured on Transwells coated with a feeder layer of cells in collagen (10). Once the top layer is confluent, surface medium is removed and the epithelial cells are cultured similar to airway cells in an air–liquid interface. Activity against bacteria is measured using challenge with the periodontal pathogen *A. actinomycetemcomitans*.

3.1. Airway Epithelial Cell Culture

1. Normal human bronchial epithelial (NHBE) cells are initially grown in monolayer culture by standard tissue culture techniques in BEGM until they reach confluence.
2. To prepare the air–liquid interface system, 12-mm Transwell inserts (*see* **Note 2**) are coated with 200 μL of diluted type VI collagen. Coated inserts are dried in a laminar flow hood overnight. After the collagen dries the plates are exposed to 30 min of UV light in the hood.
3. Confluent cultures of NHBE cells are washed with 1X HBSS and trypsinized. The trypsin is neutralized with 10% serum-based medium. The cells are then centrifuged at low speed for 5 min and resuspended in BEGM. Cell suspensions are

counted and seeded in the 12-mm Transwell plates with approximately 250,000 cells per well. Once the cells reach confluence in the 12-mm Transwell inserts the medium is removed from the apical surface of the cells. Cells are kept in a 37°C, humidified 5% CO_2 incubator. The cells in the Transwell inserts are allowed 20 days to fully mature and differentiate before any experiments are done. The medium in the basolateral chamber is changed every 2–3 days. The apical surface is washed with either PBS or 10 mM phosphate buffer when the medium is changed.

4. Two days before the experiment is carried out, the apical surface is washed twice with 1X PBS and the basolateral medium is replaced with antibiotic-free BEGM.
5. The cells are treated basolaterally with an inducing agent such as $1{,}25(OH)_2$ D_3 at a final concentration of 10^{-8} M or with IL-1β at 100 ng/mL for up to 48 h. The ASF is collected in 1.5 mL tubes after induction by washing the apical surface of the cells with 50 μL of filter-sterilized 1X PBS. ASF is stored in –80°C.

3.2. Airway Surface Fluid Antimicrobial Assay

1. *P. aeruginosa* is streaked from a frozen stock to a LB plate and grown in a 37°C incubator overnight. *B. bronchiseptica* is streaked from a frozen stock to a Bordet–Gengou (BG) agar plate grown in a 37°C incubator overnight.
2. A single colony of bacteria is inoculated and grown in liquid culture in a 37°C shaking incubator overnight. Bacteria from the overnight liquid culture is diluted 1:10 in fresh medium in a new tube and incubated for 3–6 h in a 37°C shaking incubator to obtain bacteria in the mid-log phase of bacterial growth. These bacteria are then diluted to 5×10^4 CFU/mL in 10 mM phosphate buffer.
3. 500 CFU of bacteria (10 μL) are mixed with 90 μL of the ASF in a 1.5 mL tube. The mixture is incubated in a 37°C water bath while shaking at 75 rpm. The contents of the tube are seeded on LB plates (for *P. aeruginosa*) or BG agar (for *B. bronchiseptica*) overnight.
4. Bacterial colonies are counted to determine activity of the ASF against the bacteria. An example of the results is shown in **Fig. 23.1**.

3.3. Inhibition of Activity by Specific Antibodies

1. Bacteria are prepared as above.
2. ASF (90 μL) is incubated with 1 μL anti-LL-37 or control serum for 1 h at 37°C.
3. Bacteria (in 10 μL) are added to the ASF mixture and incubated as above. An example of the results is shown in **Fig. 23.2**.

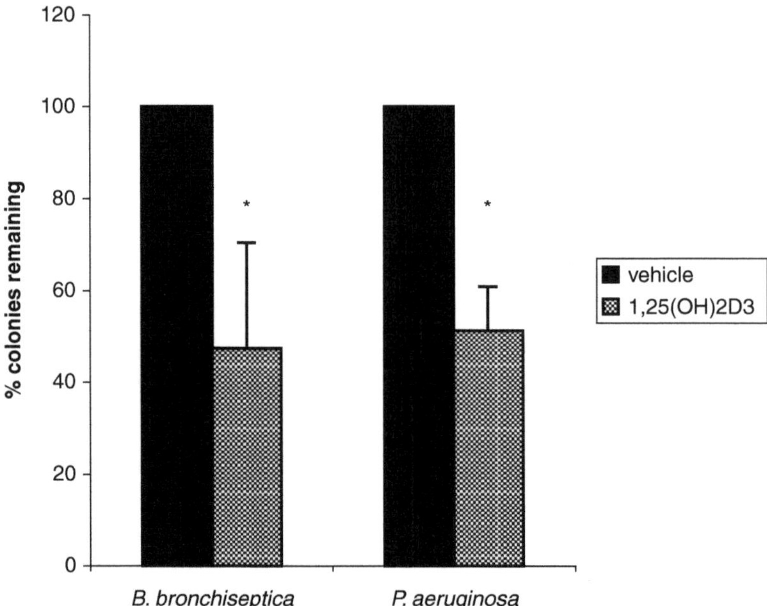

Fig. 23.1. Induction of antibiotic activity of airway surface fluid by 1,25(OH)$_2$ D$_3$. NHBE cells were grown on Transwell CM inserts and exposed to 10^{-8} M 1,25(OH)$_2$ D$_3$ basolaterally for 0, 24, and 48 h. Antimicrobial activity of ASL against *B. bronchiseptica* and *P. aeruginosa* was determined by counting the number of colonies on 5% BG plates supplemented with 5% sheep blood and LB plates, respectively, after incubation with the ASL was collected from the NHBE (Reproduced from (7) with permission from Elsevier Science.).

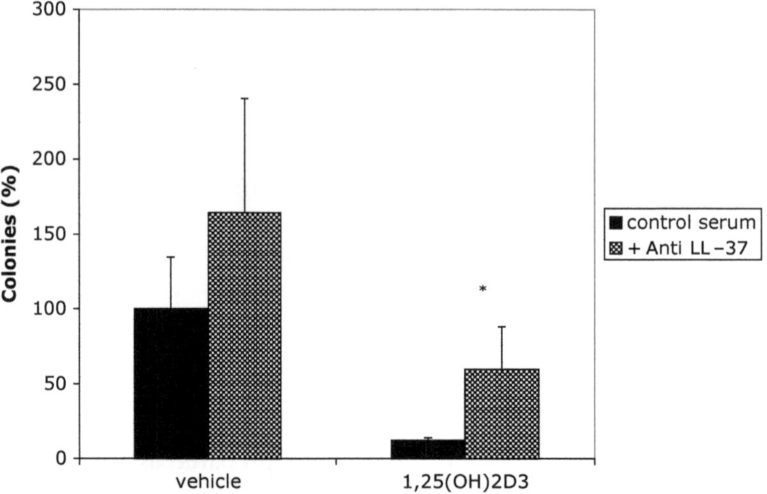

Fig. 23.2. Inhibition of antimicrobial activity with LL-37 antibody. ASF was pretreated with 1 μL anti-LL-37 antibody (gift of R. Gallo, UCSD) or control serum for 1 h prior to addition of *P. aeruginosa* in antimicrobial assays as described above. Results are reported as means ± SEM (*$p < 0.05$) of at least three independent experiments. (Reproduced from (7) with permission from Elsevier Science.).

3.4. Direct Activity on Airway Epithelial Cell Surfaces

1. The bronchial epithelial cells in airway–liquid interface are stimulated basolaterally for 24 h.
2. *P. aeruginosa* is grown in liquid culture overnight (*see* **Note 3**) and then diluted 1:10 in new LB broth and incubated for 3–6 h to obtain bacteria in mid-log phase.
3. The mid-log phase bacteria are diluted to 500 CFU in 0.5 μL and deposited on the apical surface of the bronchial epithelial cells. After 5 h the apical surface of the cells is washed with 100 μL of PBS three times and plated to determine the number of viable colonies.

3.5. Gingival Epithelial Cell Culture

1. Murine fibroblast cells NIH3T3 are grown as a monolayer in a 75 cm^2 tissue culture flask at 37°C, 5% CO_2 in DMEM/bovine serum to 90% confluence using standard tissue culture methods. Normal human gingival cells (OKF-6/TERT) are grown as a monolayer in 25 cm^2 tissue culture flasks in KSFM to confluence at 37°C, 5% CO_2 using standard tissue culture techniques.
2. To prepare the air–liquid interface, 24-mm Transwell inserts in a 6-well plate are first coated with 0.8 mL/Transwell of an acellular collagen mixture (*see* **Note 4**). For 6 mL (one plate) combine in the following order: 0.6 mL 10X DMEM, 0.6 mL 10X reconstitution buffer, 0.54 mL FBS, 60 μL 200 mM L-glutamine, and 4.2 mL collagen solution (to 0.77 mg/mL). Mix by *gently* swirling and inverting the tube. Try to avoid bubbles. Add 0.8 mL to each Transwell insert. Allow the collagen to solidify at room temperature for 30–60 min.
3. The cellular collagen mixture is then added to the Transwells. For each Transwell, 2.3 mL of the cellular (fibroblast) collagen mixture, containing 5–7 × 10^5 cells, is required. For one plate combine in the following order:

   1.40 mL 10X DMEM

   1.40 mL 10X reconstitution buffer

   1.56 mL FBS

   140 μL L-glutamine (200 mM)

   9.5 mL collagen solution (for a final concentration of 0.70 mg/mL)

   (a) Add 1.0 mL of 3T3 cells at 3–4 × 10^6/mL, for a final volume of 15 mL. The 3T3 cells are harvested using trypsin EDTA and standard tissue culture methods. After centrifugation, cells are re-suspended in 1 mL DMEM/bovine serum, counted and diluted to 3–4 × 10^6/mL with medium. Mix the cellular collagen solution gently by inverting the tube.

(b) Add 2.3 mL per Transwell on top of the acellular collagen layer and leave in the hood for 45–60 min without moving the plate. Then transfer to 37°C, 5% CO_2 for 45–60 min, until the collagen gel solidifies.

(c) Add DMEM/bovine serum medium: 1.5 mL each inside and outside of the Transwell insert. Continue incubation at 37°C, 5% CO_2.

4. At 2–3 days slide a sterile spatula around the edge of the gels, to move them away from the sides of the Transwell inserts. After 5–7 days, the collagen gels should be completely contracted away from the sides of the inserts. Change the medium every 2–3 days.

5. Once the collagen gels have contracted, the epithelial (OKF6/TERT) cells are seeded on the top of the cellular (fibroblast) collagen gels.
   (a) OKF6/TERT cells are harvested using trypsin EDTA and standard tissue culture techniques, counted, and resuspended to 15×10^6 cells/mL in KSFM.
   (b) Remove the medium from inside the inserts. Remove about half the medium from outside the inserts. Leave enough so that the bottom of the insert rests completely on the medium, but not so much that the medium rises into the insert.
   (c) Add 60 mL of the OKF6/TERT suspension on top of the middle of each collagen gel. Confirm OKF6/TERT cell growth by also seeding the same amount of cells into 1 well of a 12-well plate.
   (d) Incubate at 37°C, 5% CO2 for 2 h to allow the OKF6/TERT cells to attach. Then add 2 mL of KSFM to the inserts and 1.2 mL DMEM/bovine serum to the wells and continue incubation at 37°C, 5% CO_2.

6. Replace the medium in the inserts with fresh KSFM on the next day and continue at 37°C, 5% CO_2, replacing both media every 2 days.

7. After approximately 3–4 days (when the control well has grown to confluence) the inserts can be moved to the air–liquid interface plate containing differentiation medium (BEGM).
   (a) Add 1.6 mL per well BEGM with antibiotics to a new 6-well tissue culture plate.
   (b) Remove the medium from inside the Transwells and move each to the new plate. The bottom of the insert should be completely covered with medium, but the medium should not rise into the insert, allowing the cells to be fed from below and the apical surface of the epithe-

lium to be exposed to air. Incubate the cells at 37°C, 5% CO_2 for 15 days. Feed every 2–3 days by removing as much medium as possible and add 0.65 mL per well.

8. Equilibrate the cells in antibiotic-free medium.
   (a) Add 1.6 mL BEGM without antibiotics to each well of a new 6-well tissue culture plate.
   (b) Remove any medium from inside the Transwells and move each to the new plate.
   (c) Incubate the cells at 37°C, 5% CO_2 for 6 days. Every 2–3 days remove as much medium as possible and add 0.65 mL per well BEGM without antibiotics.

3.6. Direct Activity on Gingival Epithelial Cell Surfaces

1. Grow *A. actinomycetemcomitans* from frozen culture by streaking on an AAGM plate (*see* **Note 3**) and incubating at 37°C, 10% CO_2 for 3–4 days.
2. Use a single colony to inoculate 30 mL of AAGM broth in a 75 cm^2 tissue culture flask. Incubate at 37°C, 10% CO_2 for 40–48 h. The bacteria will grow as a biofilm on the bottom of the flask.
3. The next day, move the Transwell inserts containing the differentiated OKF6/TERT cells equilibrated in antibiotic-free medium to a new 6-well plate containing BEGM without antibiotics plus either 1,25-dihydroxyvitamin D_3 at 10^{-8} M or an equal volume of ethanol. Make at least duplicate wells for each treatment.
4. The following day, prepare a single-cell suspension of *A. actinomycetemcomitans.*
   (a) Remove the medium from the flask and wash three times with PBS.
   (b) Add 1 mL PBS, scrape the bacteria into the liquid, and transfer to a microtube.
   (c) Vortex on the highest speed for 1 min and then allow to settle for 10 min.
   (d) Transfer the top 750 μL, which is the single-cell suspension, to a new microtube.
   (e) Make a 1:10 dilution and find its absorbance at 595 nm.
   (f) An OD = 1 corresponds to 2.9×10^6 cfu/mL.
5. Dilute the single-cell suspension to 1×10^3 cfu/μL in AAGM broth.
6. At 18 h after 1,25-dihydroxyvitamin D_3 treatment of the Transwell insert cells, add 500 cfu (0.5 μL) to the center of the apical surface of each.
7. Incubate at 37°C, 10% CO_2 for 1–3 h.

8. Move each Transwell insert to an unused well. Wash the apical epithelial surface with 50 μL of PBS.
9. Remove the PBS wash and plate it on an AAGM agar plate. Repeat for each Transwell insert.
10. Incubate the plates at 37°C, 10% CO_2 for 3–4 days.
11. Count the colonies present on each plate.

4. Notes

1. Do not forget to change to antibiotic-free medium at least 2 days prior to the assay.
2. Some airway epithelial cells (such as NHBE) can be grown and differentiated in ALI on larger (e.g., 24 mm) inserts, while others require the smaller, 12-mm filters. More ASF can be obtained from the larger filters, so it is best to first determine which size is optimal. The number of cells seeded on the larger filters is increased accordingly.
3. It helps to use bacteria that are resistant to at least one antibiotic. They can be streaked on antibiotic-containing plates, grown in medium with antibiotic, but the antibiotics should be washed out during processing prior to assays. Final plating on antibiotic-containing plates will then only identify the bacteria that were intentionally added to the assay.
4. For the gingival epithelial cell Transwell preparation:
   (a) Bovine collagen type I can also be used. If bought as a prepared solution (Organogenesis is one source), it is usually stored at 4°C. Test to see if it solidifies at room temperature. If it does, use pre-chilled pipettes and keep the solutions on ice. Do a trial run to see if this will cause the fibroblasts to lyse. If the collagen solution does not solidify at room temperature, warm at 37°C for 15 min before use and keep all reagents at room temperature during preparation of the collagen layers.
   (b) Adjust the amount of FBS in both the acellular and cellular collagen mixtures to maintain final collagen concentrations of 0.77 and 0.70 mg/mL, respectively.
   (c) Observe the collagen as it is mixed into each solution to see that it mixes in completely.
   (d) A confluent 75 cm^2 tissue culture flask of 3T3 cells (fibroblasts) contains more than enough cells for two Transwell plates.

(e) After 2–3 days, free up the edges of the collagen gels using either a sterile pipette tip or a sterile metal spatula.

(f) For the epithelial layer (TERT cells), two to three confluent 25 cm^2 flasks are needed for *one* Transwell plate.

Acknowledgments

The authors would like to thank Drs. Leslie Fulcher and Scott Randell (University of North Carolina, Chapel Hill) and Helena Kashleva and Anna Dongari-Bagtzoglou (University of Connecticut) for assistance in growing the ALI cultures of airway and gingival cells, respectively. GD is supported by grants from the Cystic Fibrosis Foundation and the NIH (DE18781).

References

1. Cole, A. M., Weis, P., and Diamond, G. (1997) Isolation and characterization of pleurocidin, an antimicrobial peptide in the skin secretions of winter flounder. *J. Biol. Chem.* **272**, 12008–12013.
2. Diamond, G., Zasloff, M., Eck, H., Brasseur, M., Maloy, W. L., and Bevins, C. L. (1991) Tracheal antimicrobial peptide, a novel cysteine-rich peptide from mammalian tracheal mucosa: peptide isolation and cloning of a cDNA. *Proc. Natl. Acad. Sci. USA* **88**, 3952–3956.
3. Harder, J., Bartels, J., Christophers, E., and Schroder, J. -M. (1997) A peptide antibiotic from human skin. *Nature* **387**, 861.
4. Bals, R., Goldman, M. J., and Wilson, J. M. (1998) Mouse β-defensin 1 is a salt-sensitive antimicrobial peptide present in epithelia of the lung and urogenital tract. *Infect. Immun.* **66**, 1225–1232.
5. Garcia, J. R., Krause, A., Schulz, S., Rodrigues-Jimenez, F. J., Kluver, E., Adermann, K., Forssmann, U., Frimpong-Boateng, A., Bals, R., and Forssmann, W. G. (2001) Human beta-defensin 4: a novel inducible peptide with a salt-sensitive spectrum of antimicrobial activity. *FASEB J.* **15**, 1819–1821.
6. Diamond, G., Russell, J. P., and Bevins, C. L. (1996) Inducible expression of an antibiotic peptide gene in lipopolysaccharide-challenged tracheal epithelial cells. *Proc. Natl. Acad. Sci. USA* **93**, 5156–5160.
7. Yim, S., Dhawan, P., Ragunath, C., Christakos, S., and Diamond, G. (2007) Induction Of Cathelicidin In Normal And CF Bronchial Epithelial Cells By 1,25-Dihydroxyvitamin D3. *J. Cystic Fibrosis* **6**, 403–410.
8. Caverly, J. M., Diamond, G., Gallup, J. M., Brogden, K. A., Dixon, R. A., and Ackermann, M. R. (2003) Coordinated expression of tracheal antimicrobial peptide and inflammatory-response elements in the lungs of neonatal calves with acute bacterial pneumonia. *Infect. Immun.* **71**, 2950–2955.
9. Stolzenberg, E. D., Anderson, G. M., Ackermann, M. R., Whitlock, R. H., and Zasloff, M. (1997) Epithelial antibiotic induced in states of disease. *Proc. Natl. Acad. Sci. USA* **94**, 8686–8690.
10. Dongari-Bagtzoglou, A. and Kashleva, H. (2006) Development of a highly reproducible three-dimensional organotypic model of the oral mucosa. *Nat. Prot.* **1**, 2012–2018.
11. Fulcher, M. L., Gabriel, S., Burns, K. A., Yankaskas, J. R., and Randell, S. H. (2005) Well-differentiated human airway epithelial cell cultures. *Methods Mol. Med.* **107**, 183–206.
12. Dongari-Bagtzoglou, A. and Kashleva, H. (2006) Development of a novel three-dimensional in vitro model of oral Candida infection. *Microb. Pathog.* **40**, 271–278.
13. Brogden, K. A. (2005) Antimicrobial peptides: pore formers or metabolic inhibitors in bacteria?. *Nat. Rev. Microbiol.* **3**, 238–250.

14. Hancock, R. E. W. and Diamond, G. (2000) The role of cationic antimicrobial peptides in innate host defences. *Trends Microbiol.* **8**, 402–410.
15. Diamond, G., Beckloff, N., and Ryan, L. K. (2008) Host defense peptides in the oral cavity and the lung: similarities and differences. *J. Dent. Res.* **87**, 915–927.
16. Diamond, G., Laube, D., and Klein-Patel, M. E. (2004) In *Mammalian Host Defence Peptides*. R. E. W. Hancock and D. A. Devine (Eds.), Vol. 6, pp. 111–138. Cambridge, UK: Cambridge University Press.

Chapter 24

Antimicrobial and Antibiofilm Activity of Quorum Sensing Peptides and Peptide Analogues Against Oral Biofilm Bacteria

Karen LoVetri and Srinivasa Madhyastha

Abstract

Widespread antibiotic resistance is a major incentive for the investigation of novel ways to treat or prevent infections. Much effort has been put into the discovery of peptides in nature accompanied by manipulation of natural peptides to improve activity and decrease toxicity. The ever increasing knowledge about bacteria and the discovery of quorum sensing have presented itself as another mechanism to disrupt the infection process. We have shown that the natural quorum sensing (QS) peptide, competence-stimulating peptide (CSP), used by the caries causing bacteria *Streptococcus mutans* when used in higher than normally present concentrations can actually contribute to cell death in *S. mutans*. Using an analogue of this quorum sensing peptide (KBI-3221), we have shown it to be beneficial at decreasing biofilm of various *Streptococcus* species. This chapter looks at a number of assay methods to test the inhibitory effects of quorum sensing peptides and their analogues on the growth and biofilm formation of oral bacteria.

Key words: Streptococci, quorum sensing (QS) peptides, competence stimulating peptide (CSP), analogue, antibiofilm.

1. Introduction

Bacteria use a cell-to-cell communication system called "quorum sensing" (QS) to obtain information on bacterial density in their environment for regulating the expression of genes involved in virulence and biofilm formation in response to fluctuations in cell density (1). Gram-positive bacteria such as Streptococci use short peptides called "competence stimulating peptide" (CSP), a quorum sensing (QS) molecule, which requires a two-component (histidine kinase receptor and response regulator) transduction system for detection and function (2). Although Streptococci

A. Giuliani, A.C. Rinaldi (eds.), *Antimicrobial Peptides*, Methods in Molecular Biology 618,
DOI 10.1007/978-1-60761-594-1_24, © Springer Science+Business Media, LLC 2010

constitute 60–90% of the early colonizers of dental plaque, *Streptococcus mutans* is the principal etiological agent of dental cavities (3). As the quorum sensing in *S. mutans* is known to regulate a number of physiological activities including biofilm development (2), inhibition of the quorum sensing pathway would be the best approach for attenuating *S. mutans* virulence. We have designed and synthesized a peptide analogue (KBI-3221) based on the *S. mutans* CSP molecule to specifically target the quorum sensing system and inhibit the process of biofilm development (4).

The procedures outlined in this chapter include 96-well and 12-well microtiter plate assays to determine the inhibitory effects of quorum sensing peptide and its analogue on planktonic growth as well as biofilm formation of oral bacteria and also on biofilm-embedded oral bacteria. In addition, the hydroxyapatite disk assay is used to mimic the tooth surface as an adhesive surface for the typical oral bacteria, and colony counts are used to assess the effectiveness of the peptide and/or analogue in question. Furthermore, confocal scanning laser microscopy is used to visualize the effect of the CSP peptide on biofilm formation in *S. mutans*.

2. Materials

2.1. Cell Culture and Peptide Synthesis

1. Glycerol (15%) stock of oral bacteria. Stored –80°C. Streptococcal strains used: *S. mutans* (UA159), *S. oralis* (NCTC11472), and *S. sobrinus* (HNG909S).
2. Todd Hewitt Yeast Extract (THYE) Agar: Todd Hewitt Broth (Oxoid, Nepean, ON, Canada), 3% yeast extract, and 15% agar.
3. Sterile 100 × 15 mm petri plates.
4. Todd Hewitt Yeast Extract Broth + mucin: Todd Hewitt Broth, 3% yeast extract, 0.01% hog gastric mucin, pH 7.0.
5. Sterile 12 mL culture tubes (Simport, Thermo Fisher Scientific, Nepean, ON, Canada).
6. Anaerobic incubation conditions: GENbag Anaer bags and sachets (bioMerieux, Montreal, Canada) and/or BD GasPak EZ Anaerobe (Thermo Fisher Scientific, Nepean, ON, Canada).
7. Spectrophotometer Genesys 10uv (Thermo Fisher Scientific Nepean, ON, Canada).
8. CSP (SGSLSTFFRLFNRSFTQALGK) and its analogue (*see* **Note 1**), KBI-3221 (SGSLSTFFRLFNASFTQALGK) are synthesized by Mimotopes (Roseville, MN, USA) using the

following parameters. The qualitative analysis of the peptides is done by electrospray mass spectrometry (ESMS) and purified to 90% by reverse phase-high performance liquid chromatography (RP-HPLC) on a 150 × 4.6 mm Monitor C_{18} column with a gradient of acetonitrile (10–66.6%) in 0.1% trifluoroacetic acid for 25 min. The peptides are readily solubilized in sterile double-distilled reverse osmosis water and stored in aliquots at –20°C.

2.2. 96-Well Biofilm Assay

1. Semi-defined minimal (SDM) medium is prepared in five parts as shown in **Table 24.1**. All stock solutions are prepared in sterile double-distilled reverse osmosis water unless otherwise indicated. To prepare SDM medium with supplements, add 2.0 mL of Casamino acids, 1.0 mL of glucose, 1.0 mL of amino acid mixture, 1.0 mL of vitamins to every 10 mL of SDM (*see* **Note 2**). Adjust the pH of the mixture to 7.0 with 1.0 N HCl, followed by filter sterilization.
2. 96-well (flat bottom) polystyrene microtiter plates (costar 3598, Corning Inc., Corning, NY).

Table 24.1
Composition of the semi-defined minimal (SDM) medium and additives

Composition	Stock concentration g/L	Composition	Stock concentration mg/100 mL
SDM[a]		Amino acids[b] (10× stock)	
Potassium phosphate dibasic (K_2HPO_4)	10.1	L-Glutamic acid	588.52
Potassium phosphate (KH_2PO_4)	2.0	L-Arginine HCl	210.66
Ammonium sulfate (NH_4SO_4)	1.3	L-Cysteine HCl	204.9
Sodium chloride (NaCl)	2.0	L-Tryptophan	2.04
Magnesium sulfate ($MgSO_4 \cdot 7\ H_2O$)	0.2		
0.2% Casamino acids[a]		Vitamins[b] (10× stock)	
Casamino acids	10	Nicotinic acid	4.92
		Pyridoxine HCl	20.56
		Pantothenic acid	2.38
200 mM Glucose[c]		Riboflavin	0.376
Dextrose ($C_6H_{12}O_6$)	27	Thiamine HCl	0.1
		D-Biotin	0.12

[a] Mix, autoclave, and store at room temperature (*see* **Note 3**).
[b] Mix, filter sterilize, and store at 4°C (*see* **Note 4**).
[c] Mix, filter sterilize, and store at room temperature (*see* **Note 4**).

3. Microtiter plate reader (Multiskan Ascent, Labsystems, Helsinki, Finland).
4. Stain solution: 0.4% (w/v) crystal violet. Heat on low and stir until it completely dissolves. Let it cool and store at 4°C.
5. Solubilizing solution: 33% (v/v) acetic acid. Store at room temperature.

2.3. 12-Well Biofilm Assay

1. 0.25% THYE broth + 0.01% hog gastric mucin medium and adjust to pH 7.0 using 1 N HCl, autoclave, store at room temperature.
2. 12-well polystyrene microtiter plate (Costar 3513, Corning Inc., Corning, NY).
3. Phosphate-buffered saline (PBS): Prepare 10 × stock with 1.37 M NaCl, 27 mM KCl, 100 mM Na_2HPO_4, 18 mM KH_2PO_4 (adjust to pH 7.4 with 1 N HCl if necessary) autoclave, store at room temperature. Dilute 100 mL with 900 mL sterile double-distilled reverse osmosis water for use.
4. Sonicator FS 20 (Fisher Scientific, Nepean, ON, Canada).

2.4. Hydroxyapatite Disk (HAD) Assay

1. HAD (Clarkson Chromatography Products Inc., South Williamsport, PA, 0.38″ dia. × 0.06–0.08″ thick).
2. Physiological saline: 0.9% (w/v) sodium chloride. Autoclave and store at room temperature.
3. Sterile 12 mL culture tubes.
4. Sonicator FS20.
5. Vortex mixer (Fisher Scientific, Nepean, ON, Canada).

2.5. Confocal Scanning Laser Microscopy (CSLM)

1. Fluorescent stain solution: 2% (w/v) Wheat germ agglutinin–Alexa Fluor 488® conjugate (W11261 Invitrogen, Burlington, ON, Canada) in PBS (*see* **Note 5**).
2. Olympus IX-70 CSLM with an argon laser (Carson Scientific, Markham, ON, Canada).
3. Fluoview software and Image Pro-Plus software (Carson Scientific, Markham, ON, Canada).

3. Methods

It is important to have rapid, sensitive, and reliable methods to test the effects of quorum sensing peptides and their analogues on the growth and biofilm formation of oral bacteria. The 96-well assay provides a simple and rapid method for the initial screening of peptides for antimicrobial as well as antibiofilm activity. The results of the 96-well assay can be further confirmed by the

12-well assay wherein you count the surviving cells in terms of colony forming units in order to confirm the peptide activity. The HAD assay is particularly relevant to testing a peptide activity against oral bacteria whereby the tooth surface is mimicked by using the hydroxyapatite disk thus allowing for the oral bacteria to behave in a more natural manner in terms of adhesion. Additionally, it is possible to visualize the bacterial biofilm formation on HAD in the presence and absence of peptide with the confocal scanning laser microscopy which is a further proof of biofilm inhibition.

3.1. Bacterial Culture for Assays

1. All *Streptococcus* spp. cultures are started from glycerol stock by streaking onto THYE agar plates.
2. The plates are incubated at 37°C for 48 h under anaerobic conditions in the GENbag system. After using the plate, it can be stored for a further 48 h at 4°C (*see* **Note 6**). Overnight cultures are prepared by aliquoting 3 mL of THYE & mucin broth into a 12 mL sterile tube, inoculating heavily from the plate into the broth.
3. Static incubation is carried out at 37°C for 16–18 h under anaerobic conditions.
4. The optical density of culture is measured at 600 nm and the cultures having an optical density between 1.0 and 1.3 are chosen (*see* **Note 7**).

3.2. 96-Well Biofilm Assay

1. The 96-well biofilm assay is based on a previously described method of Cvitkovitch et al. (4). Biofilm is developed in 96-well (flat bottom) polystyrene microtiter plates.
2. Growth of static biofilm is initiated by inoculation of overnight culture at 3.3% into the semi-defined minimal (SDM) medium with supplements. Biofilm is grown in the presence of CSP or KBI-3221 and in water for control (20 μL of peptide or water in a total volume of 200 μL/well).
3. Plates are statically incubated at 37°C for 16 h under anaerobic conditions.
4. Growth of planktonic cells is determined by measuring absorbance at 600 nm using a microtiter plate reader.
5. Biofilm is measured by discarding medium with planktonic cells (*see* **Note 8**) and rinsing with 200 μL of water (once). Air dry the plate overnight (*see* **Note 9**).
6. The bound cells are stained for 15 min by adding 200 μL of stain solution (*see* **Note 8**).
7. Stain solution is removed (*see* **Note 8**) and wells are rinsed with 200 μL water (three times) and let the plate air dry for 15 min.

8. Add 200 μL of solubilizing solution and let the plate sit for 15 min and measure absorbance at 630 nm using a microtiter plate reader.
9. For each experiment, background staining is corrected by subtracting the staining solution bound to controls. All comparative analyses are conducted by incubating strains in the same microtiter plate to minimize variability **Fig. 24.1**.

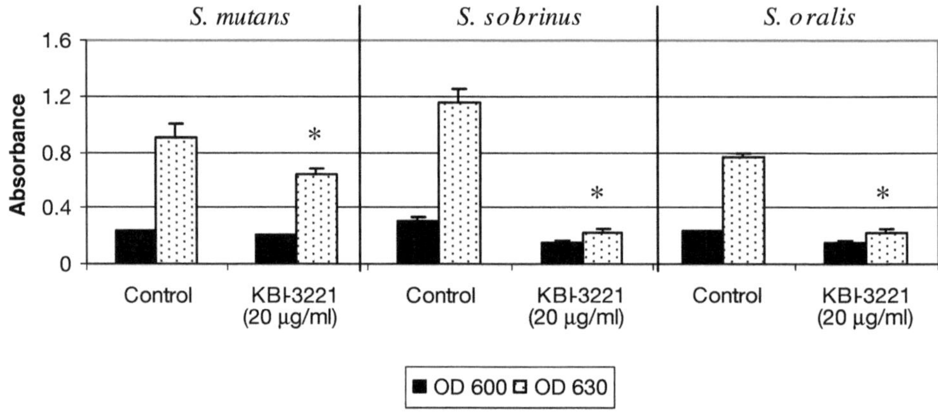

Fig. 24.1. Antibiofilm activity of KBI-3221 against oral streptococci (as determined by the 96-well biofilm assay. Planktonic growth and stained biofilm were measured at 600 and 630 nm, respectively. Values are expressed as mean ± SD. Asterisks (*) indicate a significant difference ($P < 0.05$) in biofilm inhibition between treated and untreated).

3.3. 12-Well Biofilm Assay

1. A slightly modified 12-well biofilm assay is based on a previously described method of Li et al (2).
2. The static biofilms are developed in a 12-well polystyrene microtiter plate.
3. Growth of biofilm is initiated by inoculation of overnight culture at 1.0% into 0.25 dilute THYE + 0.01% hog gastric mucin medium.
4. Wells contain either CSP or KBI-3221 or water as control (200 μL of peptide or water in a total volume of 2 mL/well).
5. Plates are statically incubated at 37°C under anaerobic conditions for 20 h.
6. Planktonic cells are removed (*see* **Note 10**) and wells are rinsed once with 2 mL PBS.
7. Biofilm cells are suspended in 2 mL of PBS and then plate is sonicated (*see* **Note 11**) for 15 s.
8. The cells are serially diluted (*see* **Note 12**) in PBS and plated on THYE agar. Plates are incubated at 37°C under anaerobic conditions and viable cell counts are taken after 48 h (**Fig. 24.2**).

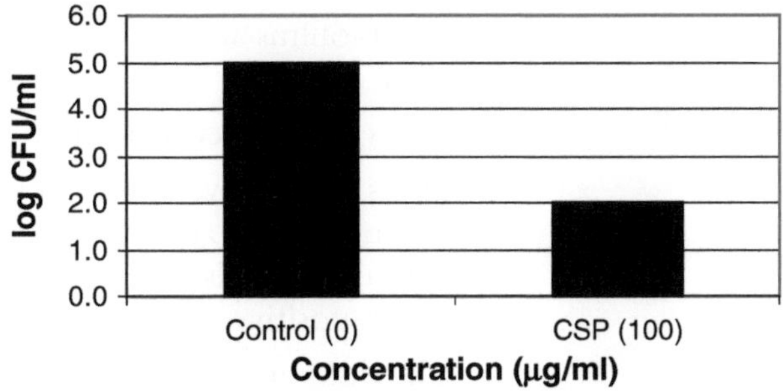

Fig. 24.2. Antibiofilm activity of CSP against biofilm-embedded *S. mutans* (as determined by the 12-well biofilm assay. Biofilm-embedded bacteria was measured by CFU counts).

3.4. Hydroxyapatite Disk (HAD) Assay

1. The HAD assay is based on a previously described method (5). The experiment is carried out using 12 mL sterile culture tubes.
2. Growth of biofilm is initiated by inoculation of overnight culture at 1.0% into 0.25 dilute THYE + mucin medium.
3. The 12 mL tubes contain 4.5 mL inoculum and either 500 μL of CSP or KBI-3221 or control (water).
4. The sterile HAD was added to the tubes and was incubated for 24 h at 37°C under stationary anaerobic conditions.
5. The disks are washed (three times) in 5 mL physiological saline (*see* **Note 13**).
6. Washed disks are added using sterile tweezers to 5 mL fresh physiological saline.
7. The disks are sonicated for 30 s, followed by 1 min of vortexing.
8. Plate appropriate dilutions (*see* **Note 12**) onto THYE agar.
9. Plates are incubated at 37°C under anaerobic conditions and viable cell counts are taken after 48 h (**Fig. 24.3**).

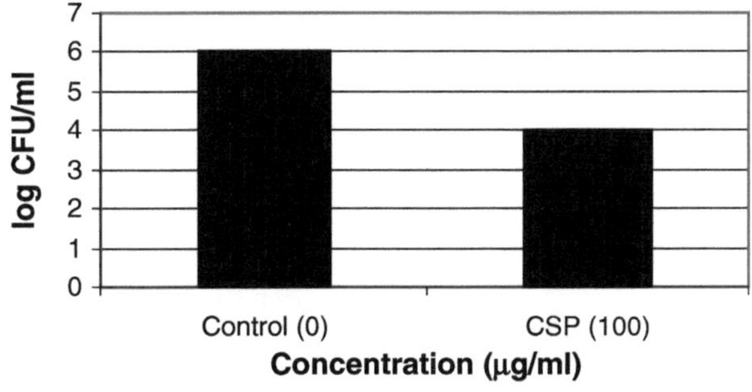

Fig. 24.3. Antibiofilm activity of CSP against biofilm-embedded *S. mutans* on hydroxyapatite disks (as determined by the hydroxyapatite assay. Biofilm-embedded bacteria were measured by CFU counts).

3.5. Confocal Scanning Laser Microscopy (CSLM)

1. The biofilms are grown on HADs as per steps 2–4 from **Section 3.4.**
2. Disks are washed twice in 5 mL PBS (*see* **Note 13**).
3. Disks are removed and transferred to new culture tubes containing 500 μL of the fluorescent stain solution. This lectin conjugate binds to N-acetylglucosamine in biofilms of both Gram-positive and Gram-negative bacteria.
4. The disks are incubated for 2 h (in dark) at 4°C.
5. The disks are washed twice in 5 mL of PBS (*see* **Note 13**).
6. The disks are air dried for 30 min (in dark).
7. Biofilms are observed using an Olympus IX-70 CSLM with an argon laser for excitation at 488 nm (green fluorescence) (6).
8. Images are captured and processed by using Fluoview software and Image Pro-Plus software (**Fig. 24.4**).

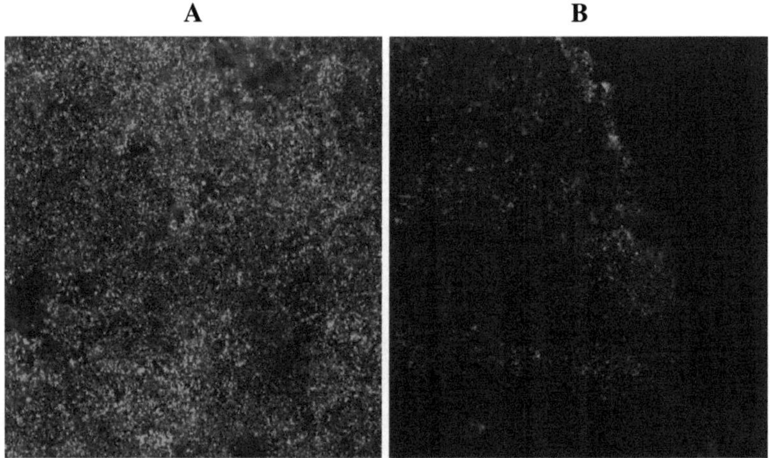

Fig. 24.4. Confocal images of *S. mutans* growing on hydroxyapatite disks in the absence (**a**) and the presence of CSP (**b**) (100 μg/mL).

4. Notes

1. The analogue of CSP has only one amino acid change the arginine (R) to alanine which is reflected by the A̲.
2. To have enough SDM medium with additives for one 96-well plate, you will have to double these amounts.
3. After autoclaving, the SDM medium will have some precipitate, shake well before using to evenly distribute.

4. Filter sterilization is done through a low protein binding hydrophilic polyethersulfone membrane with 0.20 μm pore size into a sterile tube.
5. Before preparing fluorescent stain solution thaw and spin (13,000×g × 30 s) wheat germ agglutinin-Alexa Fluor 488® to precipitate any particulates.
6. Store streptococci cultures no longer than 48 h at 4°C.
7. It is important to optimize the assay with your particular strain of *Streptococcus* spp. The optical density is measured at 600 nm of multiple overnight cultures to obtain optimal cell density. The following are examples of 600 nm readings when 10 μL of overnight culture is added to 990 μL THYE + mucin broth: *S. mutans* = 1.0–1.2 ($\leq 10^7$–10^8 CFU/mL), *S. oralis* = 1.1–1.3 ($\leq 10^7$–10^9 CFU/mL), and *S. sobrinus* = 1.0–1.2 ($\leq 10^7$–10^8 CFU/mL)
8. Multi and single channel pipetting is a crucial aspect to all assays described. For the 96-well assay, when initially removing the planktonic cells it is important to gradually allow the pipette to pull up the planktonic medium as you slowly lower the pipette deeper into the wells, trying to keep the pipette tips in the center of each of the wells. When adding the water or stain solution into the wells for the initial wash as well as subsequent washes and the staining, hold the pipette above the wells and slowly allow the 200 μL of water or stain solution to go into the well a few drops at a time.
9. When air drying overnight, remove the plate lid and wrap the plate with a kimwipe.
10. Pipetting in the 12-well assay is crucial. It is important to remove the planktonic medium as slowly as possible and position the pipette tip in the same position in the well each time. When washing ensure that you dispense the water slowly near the top of the well to not disturb the biofilm. Remove the wash as slowly as possible and position the pipette tip in the same position in the well.
11. To avoid contamination between wells in the 12-well plate, stretch a layer of parafilm over the top of the wells then press lid on tightly.
12. After the sonication, ensure you remove the appropriate quantity from the well/tube by first pipetting a few times to get a good mix and break up of biofilm cells that you will be adding to the dilution tubes. When diluting give tubes firm shaking to create a good dispersal of the biofilm cells as the oral bacteria tend to clump.

13. When rinsing the HAD use sterile tweezers which are angled at the top of the tube to stop the disk from falling out of the tube.

References

1. Håvarstein, L. S. and Morrison, D. A. (1999) Quorum sensing and peptide pheromones in streptococcal competence for genetic transformation. In *Cell–Cell Signaling in Bacteria*. G. M. Dunny and S. C. Winans (Eds.), pp. 9–192. Washington, DC: ASM Press.
2. Li, Y. H., Lau, P. C. Y., Lee, J. H., Ellen, P., and Cvitkovitch, D. G. (2001) Natural genetic transformation of *Streptococcus mutans* growing in biofilms. *J. Bacteriol* **183**, 897–908.
3. Emilson, C. -G. and Krasse, B. (1985) Support for and implications of the specific plaque hypothesis. *Scand. J. Dent. Res.* **93**, 94–104.
4. Cvitkovitch, D. G., Lau, P. C. Y., and Li, Y. H. (2001) Signal peptides, nucleic acid molecules and methods for treatment of caries. Canadian Patent Application 2332733.
5. Sieck, B., Takagi, S., and Chow, L. C. (1990) Assessment of loosely bound fluoride uptake by tooth enamel from topically applied fluoride treatments. *J. Dent. Res.* **69**, 1261–1265.
6. Burton, E., Gawande, P. V., Yakandawala, N., LoVetri, K., Zhanel, G. G., Romeo, T., Friesen, A. D., and Madhyastha, S. (2006) Antibiofilm activity of GlmU enzyme inhibitors against catheter-associated uropathogens. *Antimicrob. Agents. Chemother.* **50**, 1835–1840.

Chapter 25

Characterization of the Leishmanicidal Activity of Antimicrobial Peptides

Juan Román Luque-Ortega and Luis Rivas

Abstract

This chapter describes the basic methodology to assay the activity of antimicrobial peptides (AMPs) on *Leishmania*, a human protozoan parasite. The protocols included can be methodologically divided into two major blocks. The first one addresses the basic technology for growth of the different stages of *Leishmania*, assessment of leishmanicidal activity, and monitoring of plasma membrane permeabilization. The second block encompasses the monitoring of bioenergetic parameters of the parasite, visualization of structural damage by transmission electron microscopy, or those methods more closely related to the involvement of intracellular AMP targets, as subcellular localization of the peptide and induction of parasite apoptosis.

Key words: *Leishmania*, membrane permeabilization, electron microscopy, confocal microscopy, apoptosis, antimicrobial peptide, axenic amastigote.

1. Introduction

Protozoa are one of the most neglected targets for antimicrobial peptides (AMPs). Historically, this is due to several reasons: first, the limited knowledge on the biology of these organisms compared to higher eukaryotes; second, the often difficult conditions for their growth, particularly when scale-up is concerned (1); and third, and most important, the relative lack of interest of big pharma in parasitic diseases, mostly confined to countries with low-income populations that are unable to meet the cost even of classical chemotherapy (2), much less those of a peptide-based treatment. Despite these caveats, protozoa constitute a very

A. Giuliani, A.C. Rinaldi (eds.), *Antimicrobial Peptides*, Methods in Molecular Biology 618,
DOI 10.1007/978-1-60761-594-1_25, © Springer Science+Business Media, LLC 2010

interesting model of pathogen-innate immunity interplay for several reasons: to ensure dissemination they must establish long-term infections (3), where survival to the host immune response – including components of its innate branch (4) – is mandatory (5); their often complex life cycles may involve intracellular stages (6), as well as substantial changes in membrane architecture (7), all of which entail significant changes in their interaction with AMPs. With few exceptions, such as the cyst stage of parasitic protozoa transmitted by environmental contamination, parasitic protozoa are devoid of compactly packed external barriers such as the outer membrane of Gram-negative bacteria, the thick peptidoglycan layer in Gram-positive, or the fungal wall. In the context of AMP action, this situation favors direct interaction with the plasma membrane, without intervening barriers, and thus allows a relatively straightforward interpretation of the results.

The different species of the genus *Leishmania* are responsible for a group of human infections collectively grouped as leishmaniasis. These encompass a wide variety of clinical symptomatologies, from the self-healing forms, associated to some cutaneous leishmaniasis, to the visceral ones, fatal if untreated. The parasite shows a digenic cycle with two major forms: the promastigote, the flagellated form, and the intracellular amastigote form, which lives inside mononuclear phagocytic cells of a wide variety of mammals (8).

Leishmania is in several aspects an atypical target for AMPs: (i) endocytosis and exocytosis processes are restricted to flagellar pocket, a specialized area of the plasma membrane devoid of subpellicular microtubules (9), which somehow limits the repairing capacity of the plasma membrane; (ii) the plasma membrane of the promastigote possesses a thick anionic glycocalix made up mostly of lipophosphoglycan (10) and an abundant external GPI-anchored Zn^{2+} metalloproteinase called Gp63 or leishmanolysin (11), both of which are known to be involved in partial resistance against AMPs (12, 13); (iii) while the amastigote stage would appear more vulnerable than promastigote one due to a much less developed glycocalix, both its location inside the parasitophorous vacuole and likely changes in phospholipid composition of its plasma membrane impose additional hurdles for AMP action.

Although from an anti-*Leishmania* therapeutic perspective, the only data required for a given AMP are its leishmanicidal activity and therapeutic index, our goal in this chapter is to provide the reader with a collection of methods helpful in defining the leishmanicidal mechanisms and establishing a correlation among lethal activity and other effects of the AMPs, such as membrane permeation activity or variation of bioenergetic parameters of the parasite, which can direct further to SAR studies. Furthermore, protocols to assess structural damage to the parasite, based on transmission electron microscopy, or for clues about the involvement of intracellular targets such as subcellular destination of the AMP assessed by confocal microscopy and induction of apoptosis, were also included.

In this chapter, AMP activity will be mostly referred to extracellular parasites, these being not only promastigotes but also pertaining to the axenic *Leishmania pifanoi* amastigote line, whose similarity to the amastigotes isolated from infected macrophages has been exhaustively demonstrated (14–16). We recommend assaying axenic amastigotes, available for a wide variety of *Leishmania* species (17), before performing the time-consuming and labor-consuming AMP tests on *Leishmania*-infected macrophages, despite the much closer similarity of the latter model to the physiology of infection. In the long run, the development of new imaging systems may overcome some of the aforesaid limitations and make high-throughput analysis of infected macrophages a reality (18–23).

2. Materials

2.1. Cell Culture

1. RPMI 1640 medium, supplemented with 10% heat-inactivated fetal calf serum (HIFCS), 25 mM HEPES, 2 mM L-glutamine, 100 units/mL sodium benzyl penicillin, and 48 μg/mL gentamicin at pH 7.2 (RPMI 1640 + HIFCS). Sterilize this medium in a vertical laminar flow hood by filtering through a sterile filtration device fitted with a 0.22 μm nitrocellulose filter. Store at 4°C (*see* **Note 1**).
2. M199 medium, supplemented with 20% HIFCS, 2.5% (w/v) trypticase peptone, 13.8 mM D-glucose, 76 μM hemin, and 48 μg/mL gentamicin sulfate at pH 7.2 (M199 + HIFCS). Sterilize this medium as in Step 1 of **Section 2.1**. Store at 4°C (*see* **Note 2**).
3. Hanks balanced salt solution supplemented with D-glucose (HBSS): 137 mM NaCl, 5.3 mM KCl, 0.4 mM KH_2PO_4, 4.2 mM $NaHCO_3$, 0.4 mM Na_2HPO_4, pH 7.2, 10 mM D-glucose. Adjust to pH 7.2 if necessary. Sterilize as in Step 1 of **Section 2.1** (*see* **Note 3**).

2.2. Cell Proliferation Measurements

1. Sterile Hanks balanced salt solution (HBSS) described in Step 3 of **Section 2.1**.
2. RPMI 1640 medium devoid of phenol red, supplemented with 10% HIFCS, 25 mM HEPES, 2 mM L-glutamine, 100 units/mL sodium benzyl penicillin, 48 μg/mL gentamicin

sulfate, at pH 7.2. Sterilize as in Step 1 of **Section 2.1.** Store at 4°C.

3. M199 + HIFCS medium for amastigotes (described in Step 2 of **Section 2.1**).
4. [3-(4,5-Dimethylthiazol-2-yl)-2,5-diphenyltetrazolium bromide] (MTT) (Sigma Chemical Co., St. Louis, MO, USA) stock solution: 5 mg/mL in sterile HBSS. Store as single-use aliquots at –20°C (this compound is light sensitive). Prepare fresh working solutions at the desired concentration by dilution with HBSS.
5. Sodium dodecyl sulfate (SDS): USP/NF quality, 10% (w/v) aqueous solution.
6. Sterile 96-well polystyrene tissue culture-treated microplates.

2.3. Cytotoxicity Against Intracellular Amastigotes

1. RAW 264.7 murine macrophage cell line (alternatively, mouse peritoneal macrophages can be used) (*see* **Note 4**).
2. Lab-Tek 16-well chamber slide system (Nalge Nunc, Rochester, NY, USA).
3. Sterile RPMI 1640 + HIFCS medium, as described in Step 1 of **Section 2.1.**
4. Sterile RPMI 1640 medium supplemented with 2% HIFCS.
5. Sterile HBSS, as described in Step 3 of **Section 2.1.**
6. Giemsa's azur eosin methylene blue solution (Merck, Darmstadt, Germany).
7. Methanol (Merck).
8. Eukitt® mounting medium (Fluka, Buchs, Switzerland).

2.4. Assessment of Plasma Membrane Permeabilization by AMPs

2.4.1. Entrance of Vital Dyes

1. SYTOX Green (Invitrogen, Carlsbad, CA, USA): 500 μM stock solution in DMSO. Store in small aliquots (20 μL) at –20°C, shielded from light. Prepare fresh working solution (2 μM) in HBSS.
2. HBSS, as described in Step 3 of **Section 2.1.**
3. Triton X-100: 1% (w/v) in H_2O.
4. Black polystyrene 96-well microplates (Nunc, Roskilde, Denmark).
5. Fluorescence microplate reader, equipped with excitation and emission filters of 485 and 520 nm, respectively.

2.4.2. Plasma Membrane Depolarization

1. Bisoxonol [bis-(1,3-diethylthiobarbituric) trimethine oxonol] (Invitrogen): 100 μM stock solution in DMSO. Store at –20°C, shielded from light. Prepare fresh working solution (0.2 μM) by dilution in HBSS.

2. HBSS, as described in Step 3 of **Section 2.1.**
3. Gramicidin D (Sigma): 8 μM stock solution in HBSS (*see* **Note 5**).
4. Black polystyrene 96-well microplates.
5. Fluorescence microplate reader, equipped with excitation and emission filters of 544 and 584 nm, respectively.

2.5. Mitochondrial Membrane Depolarization

1. Rhodamine 123 [2-(6-Amino-3-imino-3*H*-xanthen-9-yl) benzoic acid methyl ester, chloride] (Invitrogen): 0.1 mg/mL stock solution in DMSO. Prepare fresh working solution (0.3 μg/mL) in HBSS.
2. HBSS, as described in Step 3 of **Section 2.1.**
3. KCN: 10 mM stock solution in HBSS, to be made fresh. Beware, this substance is highly toxic. Handling and disposal according to local regulations.

2.6. Oxygen Consumption Rate

1. Respiration buffer: 10 mM Tris–HCl, pH 7.2, 125 mM saccharose, 65 mM KCl, 1 mM $MgCl_2$, 2.5 mM NaH_2PO_4, 0.3 mM EGTA, and 5 mM sodium succinate. Readjust pH to 7.2, if necessary.

2.7. Confocal Microscopy

1. Printed microscope slides (Cel-Line®, Menzel GmbH, Braunschweig, Germany).
2. HBSS, as described in Step 3 of **Section 2.1.**
3. Phosphate buffered saline (PBS) stock solution (2X): 274 mM NaCl, 5.4 mM KCl, 16 mM Na_2HPO_4, 3 mM KH_2PO_4. Adjust to pH 7.2 with HCl.
4. MitoTracker Red CMXRos (Invitrogen): 100 μM stock solution in DMSO. Store at –20°C, shielded from light. Prepare fresh working solution (0.1 μM) in HBSS for each assay.
5. DAPI (4′,6-diamidino-2-phenylindole hydrochloride) (Invitrogen): 10 mg/mL stock solution in Milli-Q ultrapure water. Gentle sonication speeds the solubilization. Prepare fresh working solution (1 μg/mL) in PBS.
6. Paraformaldehyde (PAF): 8% (w/v) stock solution in Milli-Q ultrapure water. To help solubilization heat below 60°C on a stirring hot plate inside a fume hood, as PFA is toxic. Once dissolved, cool to room temperature for use. Prepare working solution (4% PFA) by mixing equal volumes of 8% PFA and of PBS (2X).
7. Mounting medium: add 10 g of Mowiol 4–88 (Calbiochem, La Jolla, CA, USA) to 90 mL of PBS (pH 7.4) and stir for several hours. Then, add 2.5% DABCO

(1,4-diazabicyclo[2.2.2]octane) (Sigma) plus 40 mL glycerol and stir overnight. Centrifuge the solution (40,000×*g* for 30 min), then dispense the clear supernatant into 1–2 mL aliquots and freeze them at –20°C until use. Thaw the aliquot at 37°C before use. Aliquots may be frozen again.

2.8. Transmission Electron Microscopy

1. HBSS, as described in Step 3 of **Section 2.1**.
2. PBS (2X), as described in Step 3 of **Section 2.7**.
3. Glutaraldehyde commercial aqueous solution 25% (v/v): prepare 3% (v/v) solution in PBS fresh for each experiment. Mix equal volumes of the 25% commercial solution and of PBS (2X). Dilute the resulting 12.5% glutaraldehyde solution with PBS (1X) to get a final concentration of 3%.
4. Osmium tetroxide (OsO_4) (Sigma): Due to its toxicity OsO_4 must be manipulated in a fume hood. Prepare an initial 5% (w/v) solution in Milli-Q ultrapure water. Prepare working solution (2.5% w/v) by mixing equal volumes of 5% OsO_4 and of PBS (2X). Prepare the solution the day before use, as OsO_4 crystals are difficult to dissolve. Store at 4°C protected from light.
5. Ethanol solutions: prepare 30, 50, 70, and 90% (v/v) solutions in Milli-Q ultrapure water from ethanol absolute.
6. Propylene oxide (Fluka).
7. 2,4,6-Tri(dimethylamino methyl) phenol (DMP30) (Tousimis, Rockville, MD, USA).
8. Resin A: 6.88 mL of Epon-812 epoxy resin (Tousimis) plus 11.11 mL of dodecenyl succinic anhydride (DDSA) (Tousimis). Prepare in thoroughly cleaned and dried Erlenmeyer flasks. Keep under gentle magnetic stirring until the mixture becomes completely transparent (ca. 30 min).
9. Resin B: 22 mL of Epon-812 plus 19.8 mL methyl-5-norbornene-2,3-dicarboxylic anhydride ultrapure (NMA) (Tousimis). Prepare in thoroughly cleaned and dried Erlenmeyer flasks. Keep under gentle magnetic stirring until the mixture becomes completely transparent (ca. 30 min).
10. Mix resins A and B into a single Erlenmeyer flask and keep under gentle magnetic stirring to get a clear, fully transparent mixture (*see* **Note 6**). The volumes indicated above have been calculated for a set of 24 samples, prepared in 1.5 mL microcentrifuge tubes. If necessary, adjust volumes to your specific requirements. Manipulation must be carried out in a fume hood. Resins are light sensitive.
11. Polyethylene BEEM® pyramidal capsules (Beem Inc., West Chester, PA, USA).

12. Lead citrate (Sigma): prepare a 25 mM solution in Milli-Q ultrapure water. Care should be taken due to lead toxicity.
13. Formvar-carbon coated copper grids (Electron Microscopy Sciences, Hatfield, PA, USA).

2.9. Assessment of Leishmania Apoptosis by Flow Cytometry

1. Sterile HBSS, as described in Step 3 of **Section 2.1**.
2. Sterile RPMI 1640 + HIFCS medium, as described in Step 1 of **Section 2.1**.
3. PBS (1X), obtained from 2X stock solution (*see* Step 3 of **Section 2.7**).
4. Sterile 24-well microplates.
5. *N*-hexadecylphosphocholine (Anatrace Inc, Maumee, OH, USA): 2 mM stock solution in PBS. Store at –20°C. Prepare fresh working solution (30 μM) in PBS for each assay (*see* **Note 8**). This is used as apoptotic stimulus.
6. Staining solution: 3 mg/mL ribonuclease A (Sigma) plus 50 μg/mL propidium iodide (Invitrogen) in PBS (1X). Prepare fresh for each experiment. Consider a 500 μL volume for each sample. Shield from direct light exposure.
7. Ethanol solution: 70% (v/v) in Milli-Q ultrapure water from ethanol absolute. Store at 4°C.

3. Methods

The protocols described below use cell cultures of *Leishmania*, therefore, it is mandatory a strict follow-up of the safety procedures implemented for pathogen and cell culture handling under sterile conditions in your institution. This includes wearing protective clothes, goggles, and gloves; protocols for the correct disposal of biological residues; adequate training of the personnel; and a laboratory with at least a biosafety level 2. For *L. donovani* and *L. braziliensis,* culture and handling regulations may require a level 3 laboratory.

3.1. Cell Culture and Harvesting

3.1.1. Leishmania Culture

1. Cultured *Leishmania* cells are routinely maintained in 25 cm^2 tissue culture flasks. When higher amounts are required, the culture may be scaled up to grow into 1 L Pyrex screw cap bottles on a roller apparatus.
2. Parasite cultures are maintained at 7–15 × 10^6 parasites/mL in the respective growth medium for each stage, i.e., promastigotes in RPMI 1640 + HIFCS, (*see* Step 1 of **Section 2.1**) at 27°C (*see* **Note 9**), and axenic *L. pifanoi* amastigotes

in M199 + HIFCS (*see* Step 2 of **Section 2.1**) at 32°C (*see* **Note 10**).

3.1.2. Leishmania Harvesting

1. Collect parasites at late exponential growth phase (12–15 × 10^6 cells/mL) by centrifugation at 1,000×*g* in a clinical centrifuge, decant supernatant carefully, and wash them twice in HBSS at 4°C. Resuspend the cells in HBSS at a final density of 4 × 10^7 cells/mL, and keep on ice. Unless otherwise stated, these conditions must be maintained for the rest of the experiments.

3.1.3. Macrophages

1. The murine macrophage line RAW 264.7 is routinely cultured as a monolayer in the growth medium described in Step 1 of **Section 2.1**, at 37°C, until cellular confluence is reached.
2. For harvesting and subculturing, the culture medium is removed by careful aspiration, avoiding disturbance of the cell monolayer, replaced by cold HBSS, and the culture flask chilled over ice for 15 min.
3. Detachment of macrophages from plastic culture flasks is carried out with a sterile cell scraper. Alternatively, removal of adherent cells can be obtained by washing the cell monolayer with warm (37°C) HBSS, and further addition of 1 mL (for a 25 cm^2 tissue culture flask) of a sterile solution of 0.25% (w/v) trypsin + 0.2% (w/v) EDTA in RPMI 1640 (without serum). After incubation for 2 min, a massive detachment should be observed in an inverted microscope. Add 10 mL of RPMI + HIFCS (*see* Step 1 of **Section 2.1**) to stop further trypsin activity.
4. Transfer the detached macrophages into a 50 mL sterile tube, fill the rest of the tube with sterile RPMI + HIFCS, and centrifuge at 500×*g* at 4°C. Finally, gently resuspend the pellet in the desired volume of complete growth medium and reseed in new tissue culture flasks.

3.2. Cell Proliferation Measurements Following Exposure to AMPs

Cell counting becomes a tedious practice in those studies with a considerable number of samples. In these cases, the colorimetric measurement of viable cells using the chromophore MTT allows an automated quantification of viable cells, at lower cost than other reagents such as Alamar blue (24) or soluble formazan (XTT) (sodium-2,3-bis-[2-methoxy-4-nitro-5-sulfophenyl]-2*H*-tetrazolium-5-carboxanilide) (25), even when both of them allow direct reading of the soluble final product. The method described exploits the ability of mitochondrial dehydrogenases to reduce the yellow-colored MTT to blue, insoluble formazan (26). The amount of formazan is, hence, directly related to the number of viable cells.

Two complementary method variants following the same procedure may be used, addressing both cell viability (short term) and cell proliferation (long term), which may even use the same sample for the sake of comparison. Whereas in the short-term modality, cell number remains essentially constant and reduction of MTT depends only on viable parasites, in the long-term assay, differential parasite proliferation determines MTT reduction. The protocol is identical for both parasite stages, except for the incubation temperature and growth medium used.

3.2.1. Common Steps

1. Prepare peptide solutions in HBSS at 2X the concentration to be assayed. Sterile conditions must be maintained throughout the whole protocol.
2. In a sterile 96-well microplate, place 60 μL/well of the parasite suspension (4×10^7 cells/mL) in HBSS. Then add 60 μL/well of the corresponding 2X peptide solutions and mix by pipetting. In order to ensure statistical significance, each point must be assayed in triplicate. Include positive control (nontreated parasites) as well as blank wells without parasites to correct for color background.
3. Incubate the samples for 4 h at 27°C or 32°C, for promastigotes and amastigotes, respectively.

3.2.2. Protocol for Promastigote Proliferation (Long-Term AMPs Effect)

1. With a multichannel pipette, transfer 20 μL aliquot from each well of the plate setup in Step 3 of **Section 3.2.1** into another sterile 96-well microplate replica, containing 130 μL/well of RPMI 1640 medium devoid of phenol red, supplemented with 10% HIFCS (*see* Step 2 of **Section 2.2**). Make sure to mix thoroughly each well before transferring the content in order to obtain a homogeneous cellular suspension. Incubate this microplate for 3 days at 27°C, to allow parasite proliferation.
2. Add 50 μL of 2 mg/mL MTT solution to each well and incubate at 27°C. Wait until a well-developed color appears (ca. 90 min) in the positive control wells (untreated parasites).
3. Add 100 μL of 10% (w/v) SDS solution to each well, including those containing the blank, to stop the reaction and dissolve the blue formazan crystals. Full solubilization may take few hours. Incubation of the microplate at 37°C speeds up the process.
4. Read absorbance at 600 nm in a microplate reader fitted with a 595 nm filter. A good linear correlation with cell number under this setting was obtained with a 595 nm absorbance between 0.05–0.90.

3.2.3. Protocol for L. pifanoi Amastigote Proliferation (Long-Term AMPs Effect)

1. Mix thoroughly each well of the plate set up in Step 3 of **Section 3.2.1** by pipetting with a multichannel pipette and transfer 20 μL from each well into another sterile 96-well microplate, containing 130 μL/well of M199 + HIFCS medium (*see* Step 2 of **Section 2.1**). Incubate this microplate for 5 days at 32°C.
2. Transfer the cells from each well into 1.5 mL microcentrifuge tubes. To minimize cell losses, additionally rinse each well with 100 μL HBSS and transfer to the respective tubes (*see* **Note 11**).
3. Centrifuge the tubes in a microcentrifuge (5 min, maximum speed) and remove supernatant.
4. Add 1 mL of HBSS at 4°C to each tube, resuspend the pellet, and centrifuge again as in Step 3 of **Section 3.2.3**.
5. Remove supernatant and resuspend the cells in 100 μL of 0.5 mg/mL MTT solution.
6. Transfer the whole content from each tube to a 96-well microplate. Wait until a well-developed color appears (ca. 2 h) in the positive control wells (untreated amastigotes).
7. Proceed as described in Steps 3 and 4, **Section 3.2.2**.

3.2.4. Protocol for Promastigote/Amastigote Viability (Short-Term AMPs Effect)

1. Add 100 μL per well of 1 mg/mL MTT solution to the plate setup in Step 3 of **Section 3.2.1**.
2. Incubate at 27°C or 32°C (for promastigotes and amastigotes, respectively), shielded from light until color appearance.
3. Proceed as described in Steps 3 and 4, **Section 3.2.2**.

3.3. AMP Cytotoxicity against Intracellular Amastigotes

In the vertebrate host, the amastigote form of *Leishmania* is an obligate intracellular parasite. This precludes a direct interaction with an exogenous antimicrobial peptide. Hence, in order to ensure the feasibility of a given peptide for its development as a new lead, a further assay of its activity on *Leishmania* infected macrophages represents a closer approximation to the physiological condition. Direct examination by microscope of the stained slides and intracellular amastigote counting, constitute the easiest way to assess the efficacy of the peptide on parasite burden reduction. A minimum counting of 200 macrophages from different fields, with their respective intracellular parasites, is recommended to ensure statistical significance.

The protocol described below must be carried out in a vertical laminar flow hood with sterile material.

1. Collect the macrophages, centrifuge at 500×*g* and gently resuspend the cellular pellet in fresh RPMI 1640 + HIFCS (*see* Step 1 of **Section 2.1**) at 10^6 cells/mL.

2. Seed 100 μL/well (10^5 macrophages/well) in a Lab-Tek 16-well chamber slide system. Incubate at 37°C for 2 h to allow attachment of the cells to the glass bottom of the slide.
3. Collect amastigotes from their growth flask by centrifugation at 1,000×*g* and resuspend the pellet in fresh RPMI 1640 + HIFCS medium at a final density of 4×10^6 cells/mL.
4. Remove carefully the medium from the macrophage seeded wells by aspiration, and add immediately 100 μL/well of the 4×10^6 cells/mL amastigote suspension through the wall of the well. Avoid dryness of the well (*see* **Note 12**).
5. Incubate the multiwell slide at 32°C for 4 h to allow phagocytosis of the parasites by macrophages. For amastigotes from species of the ***donovani*** complex, incubation and the rest of this procedure must be carried out at 37°C (*see* **Note 10**).
6. Carefully remove the non-phagocytosed amastigotes by aspiration of the medium from the wells, avoiding perturbation of the macrophage monolayer. Add 100 μL per well of warm RPMI 1640 + HIFCS medium and remove again by aspiration. Repeat the same procedure again. Finally, add 300 μL per well of the RPMI 1640 medium supplemented with 2% HIFCS and leave at 32°C for 4 days to let infection progress.
7. Prepare the peptide solutions at the desired concentration in HBSS (*see* **Note 3**).
8. Carefully remove the medium from the wells by aspiration. Add 300 μL of warm (32°C) HBSS (*see* Step 3 of **Section 2.1**), aspirate, and repeat again.
9. After removal of HBSS, add 100 μL per well of the corresponding peptide solution and incubate for the time set up by the operator at 32°C.
10. After incubation, remove the peptide solution from the wells by aspiration and add 300 μL per well of RPMI medium supplemented with 2% HIFCS (*see* **Note 13**).
11. Incubate the 16-well chamber slide for 48 h at 32°C.
12. Wash the wells twice with 100 μL/well of warm (32°C) HBSS.
13. Air-dry the slides.
14. Add 100 μL of methanol per well, remove it after 1 min, and air-dry the slide.
15. Prepare the Giemsa staining solution by diluting 1 volume of Giemsa stock solution with 20 volumes of distilled water.

16. Add 100 μL of Giemsa staining solution to each well and incubate for 30 min, shielded from light.
17. Rinse the slide with distilled water to remove unspecific cytoplasmic staining of the macrophage. Do not add the water directly to the cell-seeded area as it may detach cells from the preparation.
18. Remove the plastic wells from the slide according to the manufacturer instructions. Make sure to remove any traces of silicone stuck to the slide before adding the mounting medium. Carefully, use a razor blade if necessary.
19. Add a single drop of Eukitt® mounting medium per well and gently place the coverslip over the slide avoiding entrapment of air bubbles. Let dry and observe in an optical light microscope using a 100X oil immersion objective. Slides can be stored at room temperature for months. Calculate the intracellular amastigotes:macrophage ratio. Parasites appear as small violet dots, whereas the macrophage nuclei appear as large, well-stained, irregular, circular-shaped structures (*see* **Note 14**).

3.4. Assessment of Plasma Membrane Permeabilization by AMPs

Perhaps the most documented attribute of AMPs is their interaction with the plasma membrane phospholipids of the target cell, a common step in their mechanism of action, regardless of their wide variety of primary and secondary structures. For many AMPs, this initial interaction is ensued by insertion into the plasma membrane, disrupting its integrity as a permeability barrier and leading to a fast and lethal disruption of internal homeostasis. However, this scenario is increasingly challenged by evidence of other AMPs whose lethal action involves the recognition of intracellular targets. To exert their action, these peptides will translocate across the plasma membrane, through a transient, frequently nonlethal distortion of its structure (27, 28).

In this section, it is described how *Leishmania* plasma membrane integrity can be assessed by two complementary fluorescence techniques: entrance of vital dyes and assessment of membrane depolarization. The entrance of vital dyes is based on the increase of the fluorescence of cationic dyes, such as propidium iodide and SYTOX Green, once they are bound to intracellular nucleic acids. Entrance of these molecules requires severe damage to the plasma membrane due to their relatively high molecular weight (29–31). However, small, transient membrane lesions, which are not big enough to allow the movement of vital dyes such as SYTOX Green (MW = 600) across the membrane, can be monitored using potential-sensitive fluorescent probes such as bisoxonol. These discrete lesions allow permeation of ions, leading to dissipation of ionic gradients across the plasma membrane and hence plasma membrane depolarization (32).

3.4.1. Entrance of the Vital Dye SYTOX Green

In this section plasma membrane integrity is assessed by the entrance of the cationic fluorescent vital dye SYTOX Green, which upon nucleic acid binding enhances its fluorescence more than 500-fold (33, 34).

1. Prepare peptide solutions at 10X the desired final concentration in HBSS (*see* **Note 3**).
2. Mix equal volumes of the parasite suspension (4 × 10^7 cells/mL) and of 2 μM SYTOX Green solution, as described in Step 1 of **Section 2.4.1**; incubate for 5 min at 27°C for promastigotes or 32°C for *L. pifanoi* amastigotes.
3. Transfer 100 μL aliquot per well of the cell suspension obtained in Step 2 into a black polystyrene 96-well microplate.
4. Read fluorescence in a microplate reader, equipped with excitation and emission filters of 485 and 520 nm, respectively. Wait until signal stabilization. Adjust temperature according to the temperature optimal for the specific *Leishmania* stage being assayed (*see* **Note 10**).
5. After fluorescence stabilization, quickly add 10 μL per well of the 10X peptide solutions using a multichannel pipette, and record the increase in fluorescence.
6. Finally, add to each well 10 μL of 1% (w/v) Triton X-100 solution. Wait till fluorescence readout stabilizes and take this value as the maximal permeabilization for each well.

3.4.2. Plasma Membrane Depolarization

This protocol has been designed to monitor plasma membrane depolarization using the potential-sensitive fluorescent probe bisoxonol. Insertion of this anionic fluorescent molecule in the plasma membrane is precluded by the negative membrane potential. Once the plasma membrane potential decreases, the probe inserts into the lipid bilayer and enhances its fluorescence with a red shift. Increased depolarization results in additional dye insertion and enhanced fluorescence and vice versa. The dynamic outcome of depolarization contrasts with the high-affinity binding of SYTOX Green to nucleic acids, which is close to irreversibility; this is an important difference to keep in mind for the interpretation of results.

1. Prepare working peptide solutions in HBSS (*see* **Note 3**) at 10X the desired final concentration.
2. Mix equal volumes of the parasite suspension (4 × 10^7 cells/mL) and of 0.2 μM bisoxonol solution (*see* Step 1 of **Section 2.4.2**). Incubate for 5 min at the growth temperature of the specific *Leishmania* stage assayed (27°C for promastigotes, regardless of the species, 32°C for *L. pifanoi* axenic amastigotes).

3. Transfer 100 μL per well of this parasite suspension into black polystyrene 96-well microplates.
4. Read fluorescence in a microplate reader, equipped with excitation and emission filters of 544 and 584 nm, respectively. Wait until signal stabilization. Set up device at the same temperature as in Step 2 of this protocol.
5. After fluorescence stabilization, quickly add 10 μL per well of the 10X peptide solutions with a multichannel pipette, and record fluorescence. Refer data to maximal depolarization, obtained after addition of 10 μL of 8 μM gramicidin D to each well (*see* **Note 5**).

3.5. Mitochondrial Membrane Depolarization

One of the parameters more closely related to mitochondrial functionality is the mitochondrial electrochemical potential ($\Delta\Psi_m$). Its depolarization is related to several critical processes within the cell, such as dysfunction of oxidative phosphorylation or the development of apoptosis, since in *Leishmania* the extrinsic pathway is absent. Changes in the $\Delta\Psi_m$ are caused by equilibration of H^+/OH^- gradient across the inner mitochondrial membrane. At short times, the permeabilization of the plasma membrane caused by some AMPs may not be immediately reflected in $\Delta\Psi_m$ values, but it does with those AMPs with a mitochondrial target (35).

The variation of $\Delta\Psi_m$ can be easily monitored by using mitochondrial selective fluorescent probes, such as the following protocol, based on the differential accumulation of rhodamine 123 (36). This probe is a cell-permeant, cationic, fluorescent dye that is readily sequestered by metabolically active mitochondria, without significant cytotoxic effects (37).

1. Prepare peptide solutions at 2X the desired final concentration in HBSS (*see* **Note 3**).
2. Mix equal volumes of the parasite suspension (4×10^7 cells/mL) and of 2X peptide solution. Incubate in 1.5 mL microcentrifuge tubes at 27°C in case of promastigotes or 32°C for *L. pifanoi* amastigotes, for 4 h in both cases.
3. Prepare in duplicate a control sample of fully depolarized parasites by mixing equal volumes of the parasite suspension and of 10 mM KCN solution. Incubate as in Step 2.
4. Wash the parasites by adding 1 mL of HBSS warmed at the corresponding temperature defined in Step 2 of this protocol, and centrifuge 5 min at maximum speed in a microcentrifuge. Repeat this procedure once more.
5. Resuspend the cells in 0.3 μg/mL rhodamine 123 solution, at half the volume stated in Step 2 (ca. 4×10^7 cells/mL). Incubate for 10 min in a water bath at 37°C, *regardless of the parasite stage being assayed*, shielded from light.

6. Leave two control samples of parasites without staining to adjust flow-cytometer background.
7. Wash the parasites twice with HBSS (*see* Step 3 of **Section 2.1**) by centrifugation as in Step 4.
8. Resuspend each pellet in 500 μL HBSS and transfer to flow-cytometer tubes.
9. Measure cell fluorescence in a flow cytometer with excitation and emission wavelengths set at 488 and 520 nm, respectively.

3.6. Oxygen Consumption

In *Leishmania*, glycolysis plays a minor role within the energy metabolism, being the oxidative phosphorylation, the main source to fulfill the energetic requirements of the parasite (38). Since the electron transfer throughout the respiratory chain is strictly coupled to mitochondrial ATP-synthesis, inhibition of one process will result in inhibition of the other and vice versa, hence the measurement of the oxygen consumption rate constitutes a parameter tightly related to the oxidative phosphorylation.

The polarographic measurement of dissolved oxygen is the election technique to monitor oxygen consumption in this protocol. The measurement is carried out in a Clark oxygen electrode (39). The device is composed by a platinum electrode connected by a salt bridge to a solid silver electrode, shielded by a layer of insulating material. A saturated KCl solution is held in place over the surfaces of the electrodes by a teflon membrane. The oxygen monitor holds a constant voltage difference across the two electrodes so that the platinum electrode is negatively charged with respect to the silver electrode. The basic principle of this method relies on the fact that the rate at which electrons leak from the platinum electrode is proportional to the concentration of oxygen available in the sample to receive them, as described in the reaction below. The result is an electrical current converted to a voltage by the oxygen monitor:

$$4Ag^+ + 4Ci^- \leftrightarrow 4AgCi + 4e^- \text{(at the silver electrode)}$$

$$4H^+ + 4e^- + O_2 \leftrightarrow 2H_2O \text{ (at the platinum electrode)}$$

In order to calculate the oxygen consumption rate, it is possible to use the values for dissolved oxygen reported by Truesdale and Downing (40). The respiration buffer described in Step 1 of **Section 2.6**, equilibrated with air, in our hands contains 0.237 μmol of molecular oxygen per mL at 25°C (pH 7.4).

1. Set up the Clark electrode following the manufacturer instructions. Note that the electrode needs to be polarized several hours prior to its use.

2. Adjust the temperature of the Clark electrode chamber at 25°C by connecting the jacket into a water circulating bath at the desired temperature.
3. Keep the respiration buffer at the desired temperature (e.g., 25°C for promastigotes) in constant equilibrium with air. For this purpose, maintain the buffer in an open recipient partially immersed in a water bath at the set-up temperature.
4. Collect parasites at late exponential phase by centrifugation as described in **Section 3.1.2**, and wash them twice by centrifugation at 1,000×*g* in respiration buffer at 4°C.
5. Resuspend the pellet at a final density of 10^9 cells/mL in respiration buffer and keep this cellular suspension in ice.
6. Pipette 0.9 mL of respiration buffer into the Clark electrode chamber.
7. Add 100 μL of the parasite suspension prepared in Step 5. Close the chamber immediately to avoid replenishment of consumed oxygen by air.
8. Record the oxygen consumption rate before and after the addition of the peptides under study.

3.7. Confocal Microscopy

The confocal microscopy is the technique of choice to localize the subcellular destination for a given fluorescent-tagged peptide. For AMPs this is especially important to unveil the existence of putative intracellular targets. The key feature of confocal microscopy is its capability to produce in-focus images of relatively thick specimens by selectively collecting light from thin (~1 μm) optical sections, rejecting fluorescent light from other planes within the specimen (41, 42). It is especially useful for co-localization studies by the simultaneous use of several dyes, provided that their respective fluorescence spectra do not overlap. In these experiments, organelle(s) is(are) selectively stained with a specific fluorescent probe and compared with the distribution pattern of a peptide coupled to a different fluorescence tag (35).

The protocol presented herein was designed for a co-localization assay, using three different fluorophores: a fluoresceinated AMP (λ_{ex} 488 nm, λ_{em} 520 nm), MitoTracker Red (λ_{ex} 578 nm, λ_{em} 599 nm) for specific mitochondrial staining, and DAPI (λ_{ex} 358 nm, λ_{em} 461 nm) as a nuclear and kinetoplast dye.

1. Mix equal volumes of the 4 × 10^7 cells/mL parasite suspension with 0.1 μM MitoTracker Red solution as described in Step 4 of **Section 2.7** (*see* **Note 15**). Incubate in darkness for 10 min at 27°C or 32°C in case of promastigotes or amastigotes, respectively.
2. Eliminate the unbound dye from the medium by parasite centrifugation (10 min at 1,000×*g*) and further removal

of the supernatant by aspiration. Wash the resulting pellet as above by adding the same volume of HBSS (described in Step 3 of **Section 2.1**). Resuspend the cell pellet to a density of 4 × 10^7 cells/mL in HBSS.

3. Prepare fluorescein-labelled peptide solutions in HBSS at 2X the desired final concentration (*see* **Note 3**).
4. Mix equal volumes of the parasite suspension (4 × 10^7 cells/mL) and of 2X fluorescein-labelled peptide solution in microcentrifuge tubes. Incubate at 27°C for promastigotes, or at 32°C for amastigotes, for a period of time required by the designed experiment.
5. Stop peptide–cell interaction by adding 1 mL PBS (1X) at 4°C. Centrifuge immediately the parasites for 5 min at maximum speed in a microcentrifuge and remove the supernatant by aspiration. Repeat the whole procedure described in this step once more, but leave 10–15 μL of the last supernatant in the microcentrifuge tube.
6. Resuspend the pellet in the remaining 10–15 μL of PBS, in order to obtain a homogeneous cell suspension, as after fixation cellular aggregation becomes irreversible. It is therefore essential to have a monodisperse cell suspension prior to fixing the cells with paraformaldehyde during the next step.
7. Add to each tube 100 μL of 4% PFA solution in PBS to fix the cells, and keep at room temperature for 20 min.
8. Stop fixation and remove PFA by addition of 1 mL PBS (1X). Spin down parasites by centrifugation for 5 min at maximum speed in a microcentrifuge and remove supernatant. Repeat this step once more.
9. Add to each tube 100 μL of 1 μg/mL DAPI solution in PBS. Incubate at room temperature for 40 min.
10. Wash twice by adding 1 mL of PBS (1X). Collect the parasites as in Step 8 of this protocol. Resuspend the pellet in 30 μL PBS.
11. On a printed microscope slide, pipette 10 μL of each sample per well and air-dry.
12. Add one drop of Mowiol per well, and gently place the coverslip over the slide, avoiding trapping air bubbles. Let it dry and observe under a confocal laser scanning microscope at the corresponding wavelengths, using a 63X oil immersion objective. Dried slides can be stored at 4°C for 30 days.

3.8. Transmission Electron Microscopy

Transmission electron microscopy (TEM) constitutes an excellent tool to observe ultrastructural alterations in *Leishmania* at a thousand times higher magnification than that provided by optical microscopy. Two implicit technical requirements are the obtainment of ultrathin sections, as the specimen must be of such a low density to allow electrons to pass through the tissue, and the use of a heavy metal to contrast the subcellular structures.

The protocol described below requires 5 days to obtain the final epoxy resin blocks ready for processing in the ultramicrotome. Since the TEM works in vacuum, the water contained in the sample must be removed by a gradual dehydration with ethanol. The previous fixation of the cells prevents the disruption of subcellular structures due to dehydration. Sample embedding in the resin must be done gradually, by a progressive increase of the resin percentage into the sample.

1. Prepare peptide solutions in HBSS at 2X the desired final concentration (*see* **Note 3**). Estimate an approximate volume of 500 μL per sample tube.
2. Mix equal volumes (500 μL each) of the parasite suspension (4×10^7 cells/mL) and of the 2X peptide solution. A final volume of 1 mL per sample tube is recommended to ensure an optimal parasite number, as loss of cellular material is frequent through the long procedure. Incubate at 27°C or 32°C for promastigotes or *L. pifanoi* axenic amastigotes, respectively, in microcentrifuge tubes for the desired time (*see* **Note 16**).
3. Spin down the parasites by centrifugation in a microcentrifuge (5 min at full speed).
4. Remove the supernatant by aspiration and resuspend the pellet into 1 mL of PBS (1X). Centrifuge the samples for 5 min at maximum speed in a microcentrifuge. Repeat this step once more (*see* **Note 17**).
5. Resuspend the cellular pellet in the remaining 15 μL of PBS in order to obtain a monodisperse cellular suspension. Parasite clumping will be irreversible after fixation, precluding observation of individual cells at the end of the process.
6. Add to each tube 100 μL of a 3% glutaraldehyde solution in PBS (described in Step 3 of **Section 2.8**) to fix the cells. Incubate samples at room temperature for 90 min.
7. Spin down the cells and wash them as in Step 4 of this protocol; resuspend the pellet in the remaining 15 μL avoiding formation of aggregates.
8. Add to each tube 100 μL of a 2.5% OsO_4 solution in PBS (described in Step 4 of **Section 2.8**); incubate at room temperature for 90 min.
9. Spin down the cells and wash them as in Step 4 of this protocol; resuspend the pellet in the remaining 15 μL avoiding formation of aggregates.

10. Add to each tube 1.5 mL of 30% ethanol. Incubate for 30 min at room temperature.
11. Pellet the cells by 5 min centrifugation at maximum speed in a microcentrifuge. Remove carefully the supernatant by aspiration.
12. Add to each tube 1.5 mL of 50% ethanol. Do not resuspend the pellet. Incubate at room temperature for 30 min and proceed as in Step 11 to pellet the cells.
13. Add to each tube 1.5 mL of 70% ethanol. Do not resuspend the pellet. Incubate for 30 min at room temperature. Once in 70% ethanol, samples can be stored overnight at room temperature.
14. Proceed as in Step 11 to collect parasites. Add to each tube 1.5 mL of 90% ethanol. Do not resuspend the pellet. Incubate at room temperature for 30 min and repeat Step 11.
15. Add to each tube 1.5 mL of absolute ethanol. Do not resuspend the pellet. Keep at room temperature for 30 min and proceed as in Step 11.
16. Add to each tube 0.5 mL of propylene oxide. Do not resuspend the pellet. Keep at room temperature for 1 h. Repeat Step 11. At this point, all traces of water have been removed from the sample, so care should be taken to avoid contact with wet material. This, and all the following steps involving propylene oxide handling, must be carried out inside a fume hood due to the toxicity and volatility of this reagent.
17. Add to each tube 0.5 mL of a solution 3:1 (v/v) propylene oxide:resin (A+B). Do not resuspend the pellet. Incubate at room temperature for 30 min. Prepare the solution 5 min before use. To avoid propylene oxide evaporation, add this reagent over resin (A+B), and keep the solution in a covered flask under soft magnetic stirring until its immediate use. Proceed as in Step 11.
18. Add to each tube 0.5 mL of a freshly prepared solution 1:1 (v/v) propylene oxide:resin (A+B). Do not resuspend pellet. Incubate at room temperature for 30 min. Repeat Step 11.
19. Add to each tube 0.5 mL of a freshly prepared solution 1:3 (v/v) propylene oxide:resin (A+B). Do not resuspend the pellet. Incubate at room temperature for 30 min. Repeat Step 11. If necessary, at this step samples can be stored at room temperature overnight.
20. Pour into each tube 0.5 mL of resin (A+B). Do not resuspend the pellet. Incubate at room temperature for 30 min. Repeat Step 11. Leave 100 μL of the supernatant at the bottom of each tube.

21. Prepare freshly a solution by mixing 0.18 mL DMP30 with 10 mL resin (A+B). Keep under soft magnetic stirring.
22. Carefully transfer the cell pellet from the microcentrifuge tubes into polyethylene BEEM® pyramidal capsules. As this reagent is quite viscous, facilitate transfer by using tips truncated at their end, by cutting with a razor blade. Try to position the pellet as close as possible to the bottom of the capsules.
23. Fill the capsules with the DMP30-resin (A+B) mixture prepared in Step 21, avoiding bubble formation.
24. Close each capsule with its cap and place them carefully with the apex facing down, into 1.5 mL microcentrifuge tubes. Centrifuge at 1,000 × *g* for 20 min in order to pellet the cells at the apex of the pyramidal capsule. To prevent damage of the capsules during centrifugation, a small piece of cotton should be placed at the bottom of the microcentrifuge tubes, so capsules lay over the cotton.
25. Open the capsules, transfer into an appropriate holder and keep them at 70°C for 48 h to allow resin polymerization.
26. Use an ultramicrotome with a gem quality diamond knife to obtain ultrathin sections (<0.1 μm) of each block. Place the sections onto formvar-carbon-coated copper grids. Handle the grids with sharp tweezers, taking care to hold them by their edge.
27. Place one 15-μL drop of 25 mM lead citrate aqueous solution per grid on a Parafilm® strip. Stain the sections by placing the grid over the drop for 5 min, with the sample facing the staining solution. Rinse the grids by immersing several times in Milli-Q ultrapure water (*see* **Note 7**). Finally, air-dry the grids.
28. Observe the samples in an electron microscope, according to manufacturer instructions.

3.9. Assessment of *Leishmania* Apoptosis by Flow Cytometry

The most common outcome for parasites treated with fast-acting membrane active peptides is necrosis, caused by the rapid permeabilization of their plasma membrane (30). Nevertheless, apoptosis as an alternative death mechanism may be more relevant for the aforementioned peptides at sublethal concentrations, for those with slower permeabilization kinetics, or for many others acting on intracellular targets (12). There is an extensive repertoire of techniques monitoring both early and late apoptotic parameters in *Leishmania* (13, 43–45). The analysis of the cell cycle by flow cytometry, based on the individual DNA content, allows a straightforward, fast, and easy to scale-up method to assess *Leishmania* apoptosis. To this end, parasites are stained with propidium iodide (PI), a cationic fluorescent dye, which enhances its fluorescence intensity 20-fold to 30-fold when it binds stoichiometrically to DNA. As double-stranded sections of RNA also bind PI, RNase A is included to eliminate them, increasing specificity of DNA staining. As PI is a non-membrane-permeable dye, ethanol is used to fix and permeabilize cells, making parasite DNA accessible to the dye. This procedure allows discrimination between cells in the G_0/G_1, S, and G_2/M phases of the cell cycle. DNA fragmentation in apoptotic cells impairs PI binding to nucleic acids, leading to a fluorescence intensity lower than that of G_1 cells, with the appearance of the so-called sub-G_1 peak in the flow cytometry histogram (45) (**Fig. 25.1**).

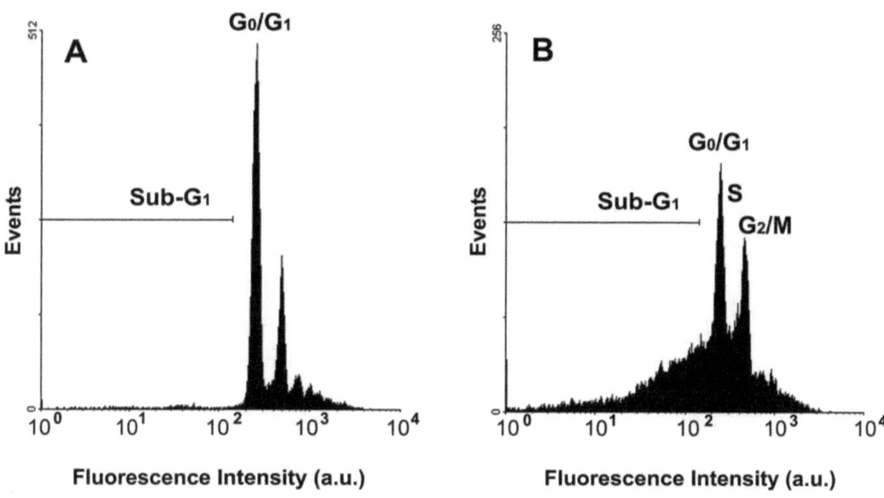

Fig. 25.1. Identification of sub-G_1 peak region by flow cytometry in apoptotic HePC-treated *L. donovani* promastigotes. Parasites were stained with PI in order to study the DNA content along the cell cycle. (**a**) Control promastigotes; (**b**) promastigotes treated with 15 μM HePC overnight. The different phases of the cell cycle are indicated.

1. Until Step 6, the protocol must be carried out in a vertical laminar flow hood to ensure sterile conditions.
2. Prepare working peptide solutions in sterilized HBSS (*see* **Note 3**) at 2X the desired final concentration. A volume of at least 100 μL is recommended to get optimal results.
3. Prepare 30 μM hexadecylphosphocholine in HBSS. This reagent, known commercially as Miltefosine®, is a well-known apoptosis inducer in *Leishmania* (45, 46), thus it is used as positive control.
4. Mix equal volumes (100 μL each) of parasite suspension (4×10^7 cells/mL) and of 2X sterile peptide solution in microcentrifuge tubes. A final volume of 200 μL per sample is recommended to ensure an optimal parasite number. Incubate at the appropriate temperature (27°C or 32°C for promastigotes or *L. pifanoi* axenic amastigotes, respectively) for 4 h, although longer incubation times may be

required for some peptides (*see* **Note 16**). Do the same with the hexadecylphosphocholine control tubes.

5. Transfer the content of each microcentrifuge tube into the corresponding well of a sterile 24-well microplate containing 1 mL of RPMI 1640 + HIFCS medium (described in Step 1 of **Section 2.1**) or M199 + HIFCS medium for amastigote (*see* Step 2 of **Section 2.1**). Incubate overnight at 27°C or 32°C for promastigotes or *L. pifanoi* axenic amastigotes, respectively.
6. Collect the parasites from each well by pipetting their content thoroughly and transfer the resulting parasite suspension into 1.5 mL microcentrifuge tubes.
7. Spin down the parasites by 5 min centrifugation at maximum speed in a microcentrifuge. Remove the supernatant by aspiration and resuspend the pellet in PBS. Centrifuge again and remove the supernatant, leaving at least 15 μL of buffer.
8. Resuspend the pellet in the remaining 15 μL of PBS in order to prevent parasite clumping, as it becomes rather irreversible after ethanol fixation at the next step.
9. Add 200 μL of 70% cold ethanol solution to each tube, mix thoroughly, and keep them at 4°C overnight. If necessary, this cellular suspension can be stored at 4°C for several months.
10. Centrifuge the ethanol-suspended cells 5 min at maximum speed in a microcentrifuge. Remove ethanol by aspiration.
11. Resuspend the pellet in 1 mL PBS, centrifuge again, and remove supernatant as in Step 7. Repeat this procedure once more and finally remove the supernatant. Elimination of any traces of ethanol is essential as it inactivates ribonuclease A activity.
12. Add 500 μL of the staining solution containing ribonuclease A and PI (*see* Step 6 of **Section 2.9**) to each tube, mix thoroughly, and transfer to flow-cytometer tubes. Incubate for 30 min at room temperature, shielded from light.
13. Measure cell fluorescence in a flow cytometer with excitation and emission wavelengths of 488 and 520 nm, respectively.

4. Notes

1. To inactivate fetal calf serum, first ensure that the bottle is completely thawed, and keep it in a water bath at 57°C for 30 min.

2. Hemin must be dissolved in the minimal amount of 4 M KOH. Handle this solution and the solid KOH with care as both are extremely caustic.
3. The electrostatic interaction between the antimicrobial peptides and the phospholipid matrix of the plasma membrane is an essential step in their mechanism of action. It may be severely inhibited by the ionic strength of the medium. Hence, when low ionic strength conditions are required, we recommend the replacement of HBSS by sorbitol buffer (280 mM D-sorbitol, 4.0 mM Na_2HPO_4, 1.0 mM KCl, 4.8 mM $NaHCO_3$,10 mM D-glucose, adjusted to pH 7.2). Sterilize this buffer by filtration through 0.22 μm nitrocellulose filter as described for HBSS.
4. The advantage of tumoral macrophage cell lines relies on the homogeneity and theoretically unlimited number of cells. Nevertheless, be aware that some peptides showed specificity for tumoral over non-transformed cells, and tissue macrophages are terminally differentiated cells with close to nil capacity to proliferate. In this case, the use of peritoneal macrophages or bone-marrow derived murine macrophages is strongly recommended. The method for obtaining these is outside the scope of this chapter, hence, the reader is referred to existing detailed protocols (47, 48).
5. Triton X-100, the detergent used to get the maximal permeabilization for vital dyes, has an extremely low critical micellar concentration. This precludes its use as a positive control for depolarization assays, as bisoxonol binds and inserts into the detergent micelles mimicking the increase in fluorescence obtained with depolarized cells. The use of Gramicidin D, a peptidic ionophore capable to depolarize *Leishmania* at a much lower concentration, eliminates this artifact.
6. The mixture of resin (A+B) must be kept constantly under soft magnetic stirring. Regulate the speed of the magnetic bar so to avoid the formation of bubbles as they interfere with sample embedding. If necessary, the mixture can be stored for 1 day at 4°C, shielded from light.
7. Optionally, a 2% uranyl acetate (Electron Microscopy Sciences, Hatfield, PA) solution in Milli-Q ultrapure water can be used as a contrast enhancer, before lead citrate staining. For this purpose, set up a Petri dish with a Parafilm® strip in it and place some drops (15 μL) of uranyl acetate on it. Place the grids on the drops with the sample facing the solution and incubate for 8 min at room temperature,

shielded from light exposure. Rinse the grids exhaustively by immersion in Milli-Q ultrapure water. Dry completely before staining with lead citrate to avoid formation of precipitates. Be aware that uranyl acetate must be handled as a radioactive material, so take appropriate precautions to avoid contamination. Carry out the experiment and waste disposal in appropriately designated areas, according to the guidelines provided by the local radiation safety office.

8. Parasites treated with a known *Leishmania* apoptotic stimulus are required as positive control. We use hexadecylphosphocholine (Miltefosine®). Nevertheless, it may be substituted with other apoptotic stimuli, including organic Sb(III) compounds, which are active on both stages (49), antimicrobial peptides (12, 13), or other available apoptotic inducers for higher eukaryotic cells, such as campothecin (50).
9. The growth medium may differ in minor requirements for each *Leishmania* species and even strain. Consult the corresponding references to get maximal growth. Nevertheless, for all the promastigotes, usual growth temperatures range from 24 to 27°C.
10. The axenic growth of *Leishmania* amastigotes is much less standardized than that of promastigotes, especially for what concerns growth media composition. Follow the instructions provided by the supplier of a given strain. As a general rule, strains belonging to cutaneous species are grown around 32°C, whereas those from visceral species (*donovani* complex) will require a higher temperature, around 37°C.
11. The presence of hemin in the amastigote growth medium precludes the same strategy used for promastigotes, as described in Step 2 of **Section 3.2.2**, where color developing is carried out in situ.
12. Although in our hands an infection ratio of 1:4 (macrophage:amastigote) resulted optimal, this ratio may be modified as needed or as required by the characteristics of the specific *Leishmania* species or strain under use.
13. If a given peptide causes detachment of the macrophages from the wells, removal of the peptide solution will cause cell losses. As most of the peptides are inactivated by serum components, an alternative method to avoid cell losses is to keep the initial 100 μL of peptide solution and add 100 μL/well of RPMI 1640 culture medium (2X) + 4% HIFCS. The 2X RPMI 1640 medium is not commercially available, but it can be prepared by doubling the amount of powder medium (Gibco, Paisley, UK) commercially

available, plus twice the concentration of supplements described in Step 1 of **Section 2.1** and 48 mM $NaHCO_3$.

14. For massive screening, direct counting of amastigotes in macrophages is an extremely labor-intensive and time-consuming job. Alternatively, incorporation of [^{3}H]methylthymidine or bromodeoxyuridine is recommended. In this case, avoid the use of macrophage cell lines as host cells for *Leishmania* infection, as they proliferate as well and incorporate nucleotides into their DNA. The two aforementioned techniques are based on the fact that tissue macrophages are cells which have undergone terminal differentiation, hence with a proliferation rate close to nil, and minimal DNA synthesis. So the incorporation of nucleotides into DNA measured by radioactivity ([^{3}H]methylthymidine) or color development by ELISA using an anti-bromodeoxyuridine monoclonal antibody can render a detailed estimate of the division rate of intracellular parasites.
15. MitoTracker Red is a red fluorescent dye that selectively stains mitochondria of living cells. Its accumulation depends on mitochondrial membrane potential. Hence, for an accurate staining, MitoTracker Red should be used prior to any other treatment of the cells that may compromise the mitochondrial membrane potential. As the dye binds irreversibly to this organelle, subsequent treatments will not affect mitochondrial labeling by MitoTracker Red.
16. The incubation time depends on the temporal requirements for the peptide to accomplish its lethal mechanisms. For fast-acting peptides the incubation can be carried out in HBSS. However, for slow-acting peptides (>6 h), it is recommended to carry out this step in sterile conditions; if longer, substitute the defined medium for full growth medium, RPMI 1640 + HIFCS for promastigotes (*see* Step 1 of **Section 2.1**), or M199 + HIFCS medium for *L. pifanoi* amastigotes (*see* Step 2 of **Section 2.1**). Be aware of the strong inhibition of the peptide by serum components. It is important to realize that comparison between inhibition of viability/proliferation and subsequent morphological damages and other effects will require identical conditions (time, incubation medium, etc.) in order to draw valid conclusions. An alternative is the use of serum-free culture medium as described for other eukaryotic cells.
17. Due to the high number of steps involving centrifugation and supernatant removal, do not remove the supernatant

exhaustively in order to preserve the pellet. Leave at least 15 μL of supernatant to preclude cell losses throughout the process.

Acknowledgments

This work was supported by research grants from the Consejo Superior de Investigaciones Científicas (CSIC; PIF 80F0171), Fondo de Investigación Sanitaria (FIS PI06115, PS09/01928 and RD 06/0021/0006), Ministry of Science and Innovation PET 2006–0139-01, and European Union (HEALTH-F3–2008-223414). The careful editing of the manuscript by Prof. David Andreu, Pompeu Fabra University, is gratefully acknowledged.

References

1. Udomsangpetch, R., Kaneko, O., Chotivanich, K., and Sattabongkot, J. (2008) Cultivation of *Plasmodium vivax*. *Trends Parasitol.* **24**, 85–88.
2. Hotez, P. J., Molyneux, D. H., Fenwick, A., Kumaresan, J., Sachs, S. E., Sachs, J. D., and Savioli, L. (2007) Control of neglected tropical diseases. *N. Engl. J. Med.* **357**, 1018–1027.
3. Kima, P. E. (2007) The amastigote forms of *Leishmania* are experts at exploiting host cell processes to establish infection and persist. *Int. J. Parasitol.* **37**, 1087–1096.
4. Pays, E., Vanhollebeke, B., Vanhamme, L., Paturiaux-Hanocq, F., Nolan, D. P., and Perez-Morga, D. (2006) The trypanolytic factor of human serum. *Nat. Rev. Microbiol.* **4**, 477–486.
5. Lang, C., Gross, U., and Luder, C. G. (2007) Subversion of innate and adaptive immune responses by *Toxoplasma gondii*. *Parasitol. Res.* **100**, 191–203.
6. Bogdan, C. (2008) Mechanisms and consequences of persistence of intracellular pathogens: leishmaniasis as an example. *Cell Microbiol.* **10**, 1221–1234.
7. McConville, M. J. and Ralton, J. E. (1997) Developmentally regulated changes in the cell surface architecture of *Leishmania* parasites. *Behring Inst. Mitt.* **99**, 34–43.
8. Banuls, A. L., Hide, M., and Prugnolle, F. (2007) *Leishmania* and the leishmaniases: a parasite genetic update and advances in taxonomy, epidemiology and pathogenicity in humans. *Adv. Parasitol.* **64**, 1–109.
9. Morgan, G. W., Hall, B. S., Denny, P. W., Carrington, M., and Field, M. C. (2002) The kinetoplastida endocytic apparatus. Part I: a dynamic system for nutrition and evasion of host defences. *Trends Parasitol.* **18**, 491–496.
10. Naderer, T., Vince, J. E., and McConville, M. J. (2004) Surface determinants of *Leishmania* parasites and their role in infectivity in the mammalian host. *Curr. Mol. Med.* **4**, 649–665.
11. Yao, C., Donelson, J. E., and Wilson, M. E. (2003) The major surface protease (MSP or GP63) of *Leishmania* sp. Biosynthesis, regulation of expression, and function. *Mol. Biochem. Parasitol.* **132**, 1–16.
12. Bera, A., Singh, S., Nagaraj, R., and Vaidya, T. (2003) Induction of autophagic cell death in *Leishmania donovani* by antimicrobial peptides. *Mol. Biochem. Parasitol.* **127**, 23–35.
13. Kulkarni, M. M., McMaster, W. R., Kamysz, E., Kamysz, W., Engman, D. M., and McGwire, B. S. (2006) The major surface-metalloprotease of the parasitic protozoan, *Leishmania*, protects against antimicrobial peptide-induced apoptotic killing. *Mol. Microbiol.* **62**, 1484–1497.
14. Pan, A. A. (1984) *Leishmania mexicana*: serial cultivation of intracellular stages in a cell-free medium. *Exp. Parasitol.* **58**, 72–80.
15. Pan, A. A., Duboise, S. M., Eperon, S., Rivas, L., Hodgkinson, V., Traub-Cseko, Y., and McMahon-Pratt, D. (1993) Developmental life cycle of *Leishmania* – cultivation and characterization of cultured extracellular

amastigotes. *J. Eukaryot. Microbiol.* **40**, 213–223.
16. Pan, A. A. and McMahon-Pratt, D. (1988) Monoclonal antibodies specific for the amastigote stage of *Leishmania pifanoi*. I. Characterization of antigens associated with stage- and species-specific determinants. *J. Immunol.* **140**, 2406–2414.
17. Gupta, N., Goyal, N., and Rastogi, A. K. (2001) In vitro cultivation and characterization of axenic amastigotes of *Leishmania*. *Trends Parasitol.* **17**, 150–153.
18. Lang, T., Goyard, S., Lebastard, M., and Milon, G. (2005) Bioluminescent *Leishmania* expressing luciferase for rapid and high throughput screening of drugs acting on amastigote-harbouring macrophages and for quantitative real-time monitoring of parasitism features in living mice. *Cell Microbiol.* **7**, 383–392.
19. Monte-Alegre, A., Ouaissi, A., and Sereno, D. (2006) *Leishmania* amastigotes as targets for drug screening. *Kinetoplastid Biol. Dis.* **5**, 6.
20. Morales, M. A., Renaud, O., Faigle, W., Shorte, S. L., and Spath, G. F. (2007) Over-expression of *Leishmania major* MAP kinases reveals stage-specific induction of phosphotransferase activity. *Int. J. Parasitol.* **37**, 1187–1199.
21. Paape, D., Lippuner, C., Schmid, M., Ackermann, R., Barrios-Llerena, M. E., Zimny-Arndt, U., Brinkmann, V., Arndt, B., Pleissner, K. P., Jungblut, P. R., and Aebischer, T. (2008) Transgenic, fluorescent *Leishmania mexicana* allow direct analysis of the proteome of intracellular amastigotes. *Mol. Cell Proteomics* 7, 1688–1701.
22. Sereno, D., Roy, G., Lemesre, J. L., Papadopoulou, B., and Ouellette, M. (2001) DNA transformation of *Leishmania infantum* axenic amastigotes and their use in drug screening. *Antimicrob. Agents Chemother.* **45**, 1168–1173.
23. Singh, N. and Dube, A. (2004) Short report: fluorescent *Leishmania*: application to anti-leishmanial drug testing. *Am. J. Trop. Med. Hyg.* **71**, 400–402.
24. Mikus, J. and Steverding, D. (2000) A simple colorimetric method to screen drug cytotoxicity against *Leishmania* using the dye Alamar Blue. *Parasitol. Int.* **48**, 265–269.
25. Williams, C., Espinosa, O. A., Montenegro, H., Cubilla, L., Capson, T. L., Ortega-Barria, E., and Romero, L. I. (2003) Hydrosoluble formazan XTT: its application to natural products drug discovery for *Leishmania*. *J. Microbiol. Methods* **55**, 813–816.
26. Kiderlen, A. F. and Kaye, P. M. (1990) A modified colorimetric assay of macrophage activation for intracellular cytotoxicity against *Leishmania* parasites. *J. Immunol. Methods* **127**, 11–18.
27. Otvos, L., Jr. (2005) Antibacterial peptides and proteins with multiple cellular targets. *J. Pept. Sci.* **11**, 697–706.
28. Shai, Y. (2002) Mode of action of membrane active antimicrobial peptides. *Biopolymers* **66**, 236–248.
29. Chicharro, C., Granata, C., Lozano, R., Andreu, D., and Rivas, L. (2001) N-terminal fatty acid substitution increases the leishmanicidal activity of CA(1–7)M(2–9), a cecropin-melittin hybrid peptide. *Antimicrob. Agents Chemother.* **45**, 2441–2449.
30. Diaz-Achirica, P., Ubach, J., Guinea, A., Andreu, D., and Rivas, L. (1998) The plasma membrane of *Leishmania donovani* promastigotes is the main target for CA(1–8)M(1–18), a synthetic cecropin A-melittin hybrid peptide. *Biochem. J.* **330**, 453–460.
31. Luque-Ortega, J. R., Saugar, J. M., Chiva, C., Andreu, D., and Rivas, L. (2003) Identification of new leishmanicidal peptide lead structures by automated real-time monitoring of changes in intracellular ATP. *Biochem. J.* **375**, 221–230.
32. Vieira, L., Slotki, I., and Cabantchik, Z. I. (1995) Chloride conductive pathways which support electrogenic H^+ pumping by *Leishmania major* promastigotes. *J. Biol. Chem.* **270**, 5299–5304.
33. Lebaron, P., Catala, P., and Parthuisot, N. (1998) Effectiveness of SYTOX Green stain for bacterial viability assessment. *Appl. Environ. Microbiol.* **64**, 2697–2700.
34. Thevissen, K., Terras, F. R., and Broekaert, W. F. (1999) Permeabilization of fungal membranes by plant defensins inhibits fungal growth. *Appl. Environ. Microbiol.* **65**, 5451–5458.
35. Luque-Ortega, J. R., van't Hof, W., Veerman, E. C., Saugar, J. M., and Rivas, L. (2008) Human antimicrobial peptide histatin 5 is a cell-penetrating peptide targeting mitochondrial ATP synthesis in *Leishmania*. *FASEB J.* **22**, 1817–1828.
36. Divo, A. A., Patton, C. L., and Sartorelli, A. C. (1993) Evaluation of rhodamine 123 as a probe for monitoring mitochondrial function in *Trypanosoma brucei* spp. *J. Eukaryot. Microbiol.* **40**, 329–335.
37. Chen, L. B. (1988) Mitochondrial membrane potential in living cells. *Annu. Rev. Cell Biol.* **4**, 155–181.
38. Van Hellemond, J. J. and Tielens, A. G. (1997) Inhibition of the respiratory chain

results in a reversible metabolic arrest in *Leishmania* promastigotes. *Mol. Biochem. Parasitol.* **85**, 135–138.

39. Kielley, W. W. (1963) Preparation and assay of phosphorylating submitochondrial particles: Sonicated mitochondria. *Methods Enzymol.* **6**, 272–277.
40. Truesdale, G. A. and Downing, A. L. (1954) Solubility of oxygen in water. *Nature* **173**, 1236.
41. Shuman, H., Murray, J. M., and DiLullo, C. (1989) Confocal microscopy: an overview. *Biotechniques* **7**, 154–163.
42. Stelzer, E. H., Wacker, I., and De Mey, J. R. (1991) Confocal fluorescence microscopy in modern cell biology. *Semin. Cell Biol.* **2**, 145–152.
43. Sen, N., Das, B. B., Ganguly, A., Mukherjee, T., Tripathi, G., Bandyopadhyay, S., Rakshit, S., Sen, T., and Majumder, H. K. (2004) Camptothecin induced mitochondrial dysfunction leading to programmed cell death in unicellular hemoflagellate *Leishmania donovani. Cell Death Differ.* **11**, 924–936.
44. Verma, N. K., Singh, G., and Dey, C. S. (2007) Miltefosine induces apoptosis in arsenite-resistant *Leishmania donovani* promastigotes through mitochondrial dysfunction. *Exp. Parasitol.* **116**, 1–13.
45. Paris, C., Loiseau, P. M., Bories, C., and Breard, J. (2004) Miltefosine induces apoptosis-like death in *Leishmania donovani* promastigotes. *Antimicrob. Agents Chemother.* **48**, 852–859.
46. Verma, N. K. and Dey, C. S. (2004) Possible mechanism of miltefosine-mediated death of *Leishmania donovani. Antimicrob. Agents Chemother.* **48**, 3010–3015.
47. Chang, K. P., Nacy, C. A., and Pearson, R. D. (1986) Intracellular parasitism of macrophages in leishmaniasis: in vitro systems and their applications. *Methods Enzymol.* **132**, 603–626.
48. Peiser, L., Gordon, S., and Haworth, R. (2002) Isolation of and measuring the function of professional phagocytes: Murine macrophages. *Methods Microbiol.* **32**, 331–358.
49. Sereno, D., Holzmuller, P., Mangot, I., Cuny, G., Ouaissi, A., and Lemesre, J. L. (2001) Antimonial-mediated DNA fragmentation in *Leishmania infantum* amastigotes. *Antimicrob. Agents Chemother.* **45**, 2064–2069.
50. Sen, N., Das, B. B., Ganguly, A., Mukherjee, T., Bandyopadhyay, S., and Majumder, H. K. (2004) Camptothecin-induced imbalance in intracellular cation homeostasis regulates programmed cell death in unicellular hemoflagellate *Leishmania donovani. J. Biol. Chem.* **279**, 52366–52375.

Subject Index

A. Giuliani, A.C. Rinaldi (eds.), *Antimicrobial Peptides*, Methods in Molecular Biology 618, DOI 10.1007/978-1-60761-594-1, © Springer Science+Business Media, LLC 2010

F

G

H

I

K

L

M

N

O

P

Q

R

S

T

V

W

Z

MIX
Papier aus verantwortungsvollen Quellen
Paper from responsible sources
FSC® C105338

If you have any concerns about our products,
you can contact us on
ProductSafety@springernature.com

In case Publisher is established outside the EU,
the EU authorized representative is:
Springer Nature Customer Service Center GmbH
Europaplatz 3, 69115 Heidelberg, Germany

Printed by Libri Plureos GmbH
in Hamburg, Germany